ULTRASOUND AND MICROWAVE FOR FOOD PROCESSING

ULTRASOUND AND MICROWAVE FOR FOOD PROCESSING

Synergism for Preservation and Extraction

Edited by

GULZAR AHMAD NAYIK

Department of Food Science and Technology, Government Degree College, Shopian, Jammu and Kashmir, India

MUHAMMAD MODASSAR AN RANJHA

Institute of Food Science & Nutrition, University of Sargodha, Sargodha, Pakistan

XIN AN ZENG

South China University of Technology, Guangzhou, PR China

SHAFEEQA IRFAN

School of Food and Agricultural Sciences, University of Management and Technology, Lahore, Pakistan

SYEDA MAHVISH ZAHRA

Department of Environmental Design Health and Nutritional Sciences, Allama Iqbal Open University, Islamabad, Pakistan

ELSEVIER

ACADEMIC PRESS

An imprint of Elsevier

elsevier.com/books-and-journals

Academic Press is an imprint of Elsevier

125 London Wall, London EC2Y 5AS, United Kingdom
525 B Street, Suite 1650, San Diego, CA 92101, United States
50 Hampshire Street, 5th Floor, Cambridge, MA 02139, United States
The Boulevard, Langford Lane, Kidlington, Oxford OX5 1GB, United Kingdom

Notices

Knowledge and best practice in this field are constantly changing. As new research and experience broaden our understanding, changes in research methods, professional practices, or medical treatment may become necessary.

Practitioners and researchers must always rely on their own experience and knowledge in evaluating and using any information, methods, compounds, or experiments described herein. In using such information or methods they should be mindful of their own safety and the safety of others, including parties for whom they have a professional responsibility.

To the fullest extent of the law, neither the Publisher nor the authors, contributors, or editors, assume any liability for any injury and/or damage to persons or property as a matter of products liability, negligence or otherwise, or from any use or operation of any methods, products, instructions, or ideas contained in the material herein.

ISBN: 978-0-323-95991-9

For Information on all Academic Press publications visit our website at https://www.elsevier.com/books-and-journals

Publisher: Nikki P. Levy
Acquisitions Editor: Nina, Bandeira
Editorial Project Manager: Kyle Gravel
Production Project Manager: Rashmi Manoharan
Cover Designer: Vicky Pearson Esser

Typeset by Aptara, New Delhi, India
Transferred to Digital Printing 2023

Contents

Contributors

Muhammad Aamir
National Institute of Food science and Technology, University of Agriculture, Faisalabad, Pakistan

Muhammad Afzaal
Department of Food Sciences, Government College University Faisalabad, Faisalabad, Pakistan; Institute of Home and Food Sciences, Government College University, Faisalabad, Pakistan

Muhammad Faizan Afzal
Department of Food Science, Government College University, Faisalabad, Pakistan

Atka Afzal
Institute of Home and Food Sciences, Government College University, Faisalabad, Pakistan

Aftab Ahmed
Department of Food Science, Government College University Faisalabad, Faisalabad, Pakistan; Department of Nutritional Sciences, Government College University, Faisalabad, Punjab, Pakistan

Muhammad Ahtisham Raza
Department of Food Science, Government College University Faisalabad, Faisalabad, Pakistan

Huma Bader Ul Ain
University Institute of Diet & Nutritional Sciences, The University of Lahore, Lahore, Pakistan

Anwar Ali
Department of Epidemiology and Health Statistics, Xiangya School of Public Health, Central South University, China

Rehman Ali
Department of Food Science, Government College University Faisalabad, Faisalabad, Pakistan

Amanjyoti
Department of Food Science and Technology, Chaudhary Devilal University, Sirsa, Haryana, India; Department of Biotechnology, Chaudhary Devilal University, Sirsa, Haryana, India; Department of Biosciences, Jamia Milia University, New Delhi, India; Department of Food Science and Technology, Maharaja Ranjit Singh University, Bathinda, Punjab, India

Mohammad Javed Ansari
Department of Botany, Hindu College Moradabad (Mahatma Jyotiba Phule Rohilkhand University, Bareilly, UP), India

Muhammad Umair Arshad
Department of Food Sciences, Government College University Faisalabad, Faisalabad, Pakistan

Muhammad Sajid Arshad
Department of Food Science, Government College University, Faisalabad, Pakistan

Aasma Asghar
Department of Food Science, Government College University Faisalabad, Faisalabad, Pakistan

Huda Ateeq
Department of Food Science, Government College University Faisalabad, Faisalabad, Pakistan

Afifa Aziz
Department of Food Science, Government College University, Faisalabad, Pakistan

Neha Bakshi
Department of Food and Nutrition and Food Technology, Lady Irwin College, Delhi, India

Shahid Bashir
University Institute of Food Science and Technology, The University of Lahore, Lahore, Pakistan

Maryam Bibi
Department of Food Science, Government College University Faisalabad, Faisalabad, Pakistan

Ahmad Din
National Institute of Food Science & Technology, University of Agriculture Faisalabad, Pakistan

Jogender Duhan
Department of Biotechnology, Chaudhary Devilal University, Sirsa, Haryana, India

Boukaga Farmani
Department of Food Science and Technology, Ahar Faculty of Agriculture and Natural Resources, University of Tabriz, Tabriz, Iran

Umar Farooq
Department of Food Science and Technology, MNS University of Agriculture, Multan, Pakistan

Chasheen Fizza
School of Food Science and Engineering, South China University of Technology, China

Scott W. Fowler
School of Marine and Atmospheric Sciences, Stony Brook University, New York, USA; Institute Bobby, Cap d'Ail, France

Ahmad Gambo
Department of Food Science and Technology, Faculty of Agriculture, Bayero University, Kano, Nigeria

Umar Garba
Department of Food Science and Technology, Faculty of Agriculture, Bayero University, Kano, Nigeria

Antima Gupta
Department of Food Science and Technology, Punjab Agricultural University, Ludhiana, Punjab, India

Farzaneh Hadi Andabjadid
Department of Food Science and Technology, Ahar Faculty of Agriculture and Natural Resources, University of Tabriz, Tabriz, Iran

Hafsa Hanif
Department of Food Science, Government College University Faisalabad, Faisalabad, Pakistan

Muzammal Hussain
Department of Food Sciences, Government College University Faisalabad, Faisalabad, Pakistan; Department of Nutrition and Health Promotion, University of Home Economics, Lahore, Pakistan

Ali Ikram
University Institute of Food Science and Technology, The University of Lahore, Lahore, Pakistan; Department of Food Science, Government College University Faisalabad, Faisalabad, Pakistan

Ali Imran
Department of Food Sciences, Government College University, Faisalabad, Punjab, Pakistan

Balarabe B. Ismail
Department of Food Science and Nutrition, Zhejiang Key Laboratory for Agro-Food Processing, Zhejiang University, Hangzhou, Zhejiang, China; Department of Food Science and Technology, Faculty of Agriculture, Bayero University, Kano, Nigeria

Swati Jain
Department of Food and Nutrition and Food Technology, Lady Irwin College, Delhi, India

G.D.T.M. Jayasinghe
National Aquatic Resources Research & Development Agency, (NARA), Colombo, Sri Lanka; Universite de Pau et des Pays de l'Adour, E2S UPPA,CNRS, IPREM UMR 5254, Helioparc, 64053 Pau, France

B.K.K.K. Jinadasa
National Aquatic Resources Research & Development Agency, (NARA), Colombo, Sri Lanka; Department of Food Science & Technology, Faculty of Livestock, Fisheries & Nutrition, Wayamba University of Sri Lanka, Gonawila (NWP), Sri Lanka

Md. Fahad Jubayer
Department of Food Engineering and Technology, Sylhet Agricultural University, Sylhet, Bangladesh

Md. Mostafa Kamal
Department of Food and Agroprocess Engineering, Bangabandhu Sheikh Mujibur Rahman Science and Technology University, Gopalganj, Bangladesh; Department of Food Processing and Preservation, Hajee Mohammad Danesh Science and Technology University, Dinajpur, Bangladesh

Sudarshanna Kar
Department of Food Process Engineering, National Institute of Technology, Rourkela, India

Naymul Karim
Department of Food Science and Nutrition, Zhejiang University, Hangzhou, China

Satinder Kaur
Department of Food Science and Technology, Punjab Agricultural University, Ludhiana, Punjab, India

Waseem Khalid
Department of Food Science, Government College University, Faisalabad, Pakistan

Muhammad Zubair Khalid
Department of Food Science, Government College University, Faisalabad, Pakistan

Mohsin Khan
Department of Biosciences, Jamia Milia University, New Delhi, India

Safura Kousar
Department of Food Science, Government College University, Faisalabad, Pakistan

Noshiza Majeed
Department of Nutritional Sciences, Government College University, Faisalabad, Punjab, Pakistan

Zahra Maqbool
Department of Food Science, Government College University, Faisalabad, Pakistan

Md. Anisur Rahman Mazumder
Department of Food Technology and Rural Industries, Bangladesh Agricultural University, Mymeningh, Bangladesh

Somaye Mohammadkhani
Department of Food Science and Technology, Ahar Faculty of Agriculture and Natural Resources, University of Tabriz, Tabriz, Iran

Shanza Mukhtar
Department of Nutrition and Dietetics, The University of Faisalabad, Faisalabad, Pakistan

Shamas Murtaza
Department of Food Science and Technology, MNS University of Agriculture, Multan, Pakistan

Anjum Murtaza
Institute of Food Science and Nutrition, University of Sargodha, Sargodha, Pakistan

Hammad Naeem
Department of Food Science and Technology, MNS University of Agriculture, Multan, Pakistan

Siba Prasad Nayak
Department of Food Engineering, National Institute of Food Technology Entrepreneurship and Management, Sonipat, India

Gulzar Ahmad Nayik
Department of Food Science and Technology, Government Degree College, Shopian, Jammu and Kashmir, India

Manju Nehra
Department of Food Science and Technology, Chaudhary Devilal University, Sirsa, Haryana, India

Bushra Niaz
Department of Food Sciences, Government College University Faisalabad, Pakistan

Madiha Khan Niazi
University Institute of Diet & Nutritional Sciences, The University of Lahore, Lahore, Pakistan

Muhammad Nouman
Institute of Home and Food Sciences, Government College University, Faisalabad, Pakistan

Tejasvi Pandey
Department of Forensic science, School of Bioengineering and Biosciences, Lovely Professional University, Phagwara, India

Taru Pant
Department of Food and Nutrition and Food Technology, Lady Irwin College, Delhi, India

Jyotsana Patel
College of Agricultural Engineering, Jawaharlal Nehru Krishi Vishwa Vidyalaya, Jabalpur, India

P.V.V.P. Prudhvi
Department of Food Process Engineering, National Institute of Technology, Rourkela, India

Aiza Qamar
Department of Nutrition and Health Promotion, University of Home Economics, Lahore, Pakistan

Muhammad Abdul Rahim
Department of Food Science, Government College University, Faisalabad, Pakistan

Aishwarya Raman
Department of Food and Nutrition and Food Technology, Lady Irwin College, Delhi, India

Juwel Rana
Department of Nutrition and Food Engineering, Daffodil International University, Dhaka, Bangladesh

Thottiam Vasudevan Ranganathan
Department of Food Processing Technology, School of Agriculture and Bioscience, Karunya Institute of Technology and Sciences, Coimbatore, Tamil Nadu, India

Husnain Raza
Shenzhen Key Laboratory of Marine Microbiome Engineering, Institute for Advanced Study, Shenzhen University, Shenzhen, China

Farhan Saeed
Department of Food Sciences, Government College University Faisalabad, Faisalabad, Pakistan; Institute of Home and Food Sciences, Government College University, Faisalabad, Pakistan

Md. Saifullah
School of Environmental and Life Sciences, College of Engineering, Science and Environment, University of Newcastle, Ourimbah, NSW, Australia

Muhammad Awais Saleem
Department of Nutritional Sciences, Government College University, Faisalabad, Punjab, Pakistan

Ayesha Sameen
National Institute of Food Science and Technology, University of Agriculture, Faisalabad, Pakistan

Kawaljit Sandhu
Department of Food Science and Technology, Maharaja Ranjit Singh University, Bathinda, Punjab, India

Anuradha Sandhu
Department of Forensic science, School of Bioengineering and Biosciences, Lovely Professional University, Phagwara, India

Faiz ul Hassan Shah
Department of Food Science and Technology, The Islamia University of Bahawalpur, Bahawalpur, Pakistan

Yasir Abbas Shah
Department of Food Science, Government College University Faisalabad, Faisalabad, Pakistan; Institute of Home and Food Sciences, Government College University, Faisalabad, Pakistan

Muhammad Shahbaz
Department of Food Science and Technology, MNS University of Agriculture, Multan, Pakistan

Piyush Sharma
Department of Food Process Engineering, National Institute of Technology, Rourkela, India

Aditi Sharma
Department of Forensic science, School of Bioengineering and Biosciences, Lovely Professional University, Phagwara, India

Umber Shehzadi
Department of Food Science, Government College University Faisalabad, Faisalabad, Pakistan

Mohammad Rezaul Islam Shishir
Shenzhen Key Laboratory of Marine Microbiome Engineering, Institute for Advanced Study, Shenzhen University, Shenzhen, China

Arashdeep Singh
Department of Food Science and Technology, Punjab Agricultural University, Ludhiana, Punjab, India

Ifra Tariq
Institute of Home and Food Sciences, Government College University, Faisalabad, Pakistan

Tabussam Tufail
University Institute of Diet & Nutritional Sciences, The University of Lahore, Lahore, Pakistan

Tanazzam Tufail
Faisalabad Medical University, Faisalabad, Pakistan

Muhammad Umair
Department of Food Science and Engineering, College of Chemistry and Engineering, Shenzhen University, Shenzhen, China

Ifrah Usman
Institute of Home and Food Sciences, Government College University, Faisalabad, Pakistan

Numra Waris
Department of Food Science, Government College University Faisalabad, Faisalabad, Pakistan

Janifer Raj Xavier
Defence Food Research Laboratory, Defence Research and Development Organization, Mysore, Karnataka, India

CHAPTER ONE

Sonication: An overview

Huma Bader Ul Ain[a], Tabussam Tufail[a], Farhan Saeed[b], Muhammad Umair Arshad[b], Muhammad Afzaal[b], Tanazzam Tufail[c], Ahmad Din[d], Madiha Khan Niazi[a] and Muzzamal Hussain[b]

[a]University Institute of Diet & Nutritional Sciences, The University of Lahore, Lahore, Pakistan
[b]Department of Food Sciences, Government College University Faisalabad, Faisalabad, Pakistan
[c]Faisalabad Medical University, Faisalabad, Pakistan
[d]National Institute of Food Science & Technology, University of Agriculture Faisalabad, Faisalabad, Pakistan

1.1 Introduction

Sonication is the technique of using sound waves to disrupt or homogenize anything, generally a chemical solution or biological medium. Sonication employs pulsed, high-frequency sound waves to lyse cells. Direct and indirect sonications are the two most commonly utilized sonication techniques. A titanium probe is introduced into the sample vessel during direct sonication. The probe transmits ultrasonic energy directly into the sample, allowing for quick processing. A sonicator is the name for this device. An ultrasonic bath or a cup horn sonicator can be used for indirect sonication. This approach eliminates the need for a probe to come in close touch with the sample, reducing the risk of cross contamination. The ultrasonic energy is transferred into the sample containers through a bath of water. Through cavitation, both direct and indirect sonication can lyse cells. When a fast pressure shift happens in a liquid, it creates "liquid-free zones" or cavities. Shock waves are created throughout the liquid when these holes are collapse, causing the cells to burst.

Ultrasonication is a one-of-a-kind homogenization process that may be used in a huge range of applications. In the base fluid, it is a process that breaks big particles into smaller fragments or more uniformly sized particles (Liu, 2019; Karami, 2019). Sound energy is used to stir the nanoparticles in the suspension, resulting in sonication of the nanofluid (Mahbubul, 2016; Noroozi et al., 2014; Leena and Srinivasan, 2015). Ultrasonication is a homogenization procedure that uses ultrasonic rates/frequencies greater than 20 kHz. Bath-type and probe-type sonicators are two types of sonicators that are regularly used. High-intensity probe sonicators have been found to be more effective than bath-type sonicators (Green and Hersam, 2010;

Ultrasound and Microwave for Food Processing: Synergism for Preservation and Extraction.
DOI: https://doi.org/10.1016/B978-0-323-95991-9.00016-3

Son et al., 2020). You may learn more about low-power ultrasonography in food applications at a couple of other websites (Awad et al., 2012; Mulet et al., 2002; Welti-Chanes et al., 2017; Watson et al., 2017). High-power ultrasound at a lower frequency has been used to change the physicochemical characteristics and structures of food (about 40 kHz) (Paniwnyk, 2014), resulting in beneficial effects on food preservation, quality-maintenance, and food safety (Alarcon-Rojo et al., 2019; Ercan and Soysal, 2013; Nicolau-Lapeña et al., 2019). Furthermore, rice-starch and rice grain were shown to have functionalizing properties.

Ultrasonication is a nonthermal approach and food science and technology are two fields where this term is utilized. Food fermentation (Ojha et al., 2017), antimicrobial efficacy (Sango et al., 2014), bioactive chemical extract method (Chemat et al., 2017; Hasan et al., 2017), food deterioration has all been well investigated (Kentish and Feng, 2014; Knorr et al., 2004; Masjid et al., 2015; Pingret et al., 2013). Similarly, a number of researches on the ultrasonic therapy's effect on starches have been carried out using data from a variety of sources. (Bemiller, 2018; Bemiller and Huber, 2015; Zhu, 2015), there is no comprehensive meta-analysis for ultrasonic-treated (UT) rice.

One of the most current technologies for decreasing processing, enhancing quality, and maintaining food safety is the ultrasonic approach. Ultrasound consists of mechanical waves and is divided into two frequency ranges. In the food sector, it is a crucial field of research and development. Low-voltage electricity is a type of electricity that is used in a variety of applications. High-energy ultrasound is defined as ultrasound with frequencies more than 100 kHz but less than 1 Wcm^2. Ultrasound has a frequency of greater than 100 kHz but less than 1 Wcm^2 (high power, high intensity). The frequency range of ultrasound is 20 to 500 kHz, with a power density of more than 1 Wcm^2.

Ultrasonic technology has a frequency. Acidity, hardness, sugar content, maturity, and other physicochemical qualities of food are all determined using as an analytical tool, high-frequency ultrasound is used. Low-frequency ultrasound producing alterations in the media in which microorganisms flourish, such as pressure, shear, and temperature. It also has ability to create cavitations, which may be utilized to kill bacteria in food. The frequency range for ultrasonic frequencies is 20 kHz to 500 MHz, which is the standard range.

Crystallization, bleaching, freezing, extraction, degassing, drying, filtering, emulsification, sterilizing, cutting, and other food processing procedures all employ ultrasound. Ultrasound is used to preserve a wide variety of goods in the food business, including cereals, gels, proteins, honey, vegetables and fruits, microbial inactivation, enzymes, cereal technology, water treatment, and dairy technology. Fresh vegetables and fruits, processing of cheese, food gels, aerated, and frozen meals, as well as bulk and emulsified fat-based foods, are examples of industrial items like cooking oils, bread, and cereals items all employ ultrasonication for quality control before and after harvest. Two further uses include detecting honey adulteration and analyzing the condition, size, and kind of protein aggregation. Traditional food processing technologies no longer deliver optimal results, as a result of increased consumer demand and tougher food and environmental constraints, necessitating the development of new and powerful technologies. For a few years, ultrasonic has been employed in the food sector as a nondestructive, environmentally friendly, rapid, adaptable, and promising technology.

Ultrasound also facilitates in the interfacial transfer of molecules, making bioactive extraction more efficient from plants and animals. The extraction method enhances the amount of data extracted material's physical and chemical qualities while also boosting the extraction yield. According to Sun et al. (2020), protein extracted by ultrasonication has better particle size, emulsification power, and structure. After being sonicated for 30 minutes at 20 kHz, particles removed by ultrasonication with a small particle size and a larger helix form suggested better emulsifying potential. Ultrasonic technology may be used to freeze, dry, and thaw food (Chow et al., 2005; Miles et al., 1999; Cheng et al., 2014). Ultrasonication may also be used to degas carbonated beverages and is a great alternative to pasteurization and sterilization for lowering microbial burden in food and food products (Bhargava et al., 2021). Ultrasound has demonstrated its effectiveness in the food sector in a variety of ways, including extraction, enhanced synthesis, food preservation, physical, and chemical quality enhancement. However, before being used on a large scale, the therapy must be tested on bulk food to ensure its efficacy.

1.2 Ultrasonication in food processing

1.2.1 Ultrasonication in meat technology

Ultrasonic treatment has been used to reduce meat roughness, identify the composition of meat products and promote cattle genetic development programmes. Therapy of ultrasound shown to aid in the digestion of a wide range of foods. It was first used in the meat business in the 1950s. Ultrasound (low intensity) is frequently utilized to improve flavor, softness, and quality, all of which are critical for consumer acceptance. Recent investigations have demonstrated that high-intensity ultrasound may tenderize (Peña-González et al., 2017), brine (Carcel et al., 2007), cook (Chemat and Khan, 2011), thaw (Miles et al., 1999), freeze (Zheng and Sun, 2006), and suppress microbes on fresh meat (Alarcon-Rojo et al., 2019; Caraveo et al., 2015). The Semitendinosus and Longissimus muscles of cattle are now being studied to see how high-intensity ultrasound affects meat products (Caraveo et al., 2015; Wang et al., 2018; Barekat and Soltanizadeh, 2018; Amiri et al., 2018). The utilization of high-intensity ultrasound, as well as microscopic alterations, changes in salt dynamics, water, and meat tissue texture, are all addressed in these studies. Ultrasound has an impact on oxidative stability, sensory qualities, and color of meat. Tissue tenderness and water dynamics are both affected by ultrasound (Ojha et al., 2020).

The impact of ultrasonography on pH levels was studied, and discovered that after ultrasonic treatment, Meat's pH level can rise from its prerigour mortis value. Changes in protein structure that causes ionic function configurations to alter, could be the cause. This might be due to cells releasing ions into the cytoplasm, or it could be due to changes in protein structure, causing ionic function configurations to alter, leading to a change in muscle pH is already over (Stadik and Dolatowski, 2011). While ultrasonic enhances meat suppleness and lowers ageing time, it has little effect on other quality indices (Peña-González et al., 2017; Chang et al., 2015). The rupture of the protein's myofibrillar structure, collagen macromolecule fragmentation, protein migration, and other variables all accelerate proteolysis, when meat is ultrasonically treated. Cavitation causes muscle injury by passing stress waves across it.

Ultrasonic intensity is related to the distance between fibers, and excessive salt levels are connected to myofibril rupture (Kang et al., 2016). Wet and dry salts are used in meat marinades. Ultrasonography has been found to be efficient in the wet marination of pigs. This explains why ultrasound-treated beef has a higher salt content (Li et al., 2020). The water dynamics of ultrasound-treated beef are a source of debate, with some scientists reporting a rise (Wang et al., 2018) and others alleging a loss (Wang et al., 2018; Gomez-Salazar et al., 2018). The difference might be related to the ultrasonic treatment approach, duration, or intensity levels used (Alarcon-Rojo et al., 2019). The color of the meat has an impact on consumer acceptability. When it comes to fresh meat, a cherry red color is ideal. The color of meat is unaffected by ultrasonography according to Peña-González et al. (2017), because the heat generated is inadequate for protein or pigment denaturation. The color of the ultrasound-treated beef was less bright than the control (Stadnik and Dolatowski, 2011). As a result, ultrasonography enhances overall color variations while reducing oxymyoglobin and metmyoglobin growth.

Ultrasound has an antibacterial impact and extends the shelf life of meat products (Kang et al., 2016). The antibacterial effect is influenced by the length of microbes (Mason, 1998). Ultrasound may potentially produce nucleation during the freezing process at temperatures greater than those used to freeze meat. This might be used to control ice crystal dispersion, as well as the size and time needs of frozen meals (Miles et al., 1999). Ultrasonication may also be used to break down ice crystals into nuclei, increasing the number of nuclei while reducing the crystal size. In sanitary applications, ultrasound is widely utilized to increase the efficacy of sanitary chemicals (Pohlman et al., 1997; Pagan et al., 1999). Tenderization occurs when ultrasonic therapy causes myofibrillar proteins to be released. Water binding capacity, softness, and cohesiveness are all improved with ultrasonic tenderization (McClements, 1995).

Fragrance, flavor, look, tenderness, and juiciness of meat indicate its quality. Mechanical pounding is a time-honored technique for tenderizing meat and enhancing the flavor of lower-quality steaks. This approach has also been demonstrated to be successful with high-power ultrasound. Ultrasound has two effects: it can harm muscle cells or it can increase enzyme activity, causing a biochemical response (Boistier-Marquis et al., 1999). Tenderize chicken, veal, or cattle with ultrasonic technology (Pohlman et al., 1997). Tenderization was discovered as a side effect of employing heat and ultrasound to sterilize meat during the research (Pagan et al., 1999). Using high-energy ultrasound, it was feasible to cut the heat treatment time in half. These studies have led to the conclusion that researchers discovered that combining ultrasound with mild heating (thermosonication or pressure mano-sonication), also known as pressure mano-sonication or thermosonication, might increase the fatal impact on bacteria in a variety of food testing systems. Processed meats are made using ultrasound technology. Myofibrillar proteins produced during processing generate a protein gel that keeps muscle pieces in recombined meat products like beef rolls together (McClements, 1995). Swirling meat chunks and sprinkling them with salt does this. When the pieces are crushed, a sticky exudate forms, which binds them together (Vimini et al., 1983). In samples that were treated with salt and then sonicated, all of the features improved. As a result, ultrasonography can aid in improving the physical qualities of beef products, such as water binding and suppleness.

1.2.2 Ultrasonication in fruit and vegetable processing

The ultrasonication technique is employed in the food industry to preserve quality characteristics in fruits and vegetables before and after harvest. It may be used to wash fruits and vegetables instead of hand-washing them. Ultrasonication has been shown to keep freshness in orange juice, Guava juice, and tomato juice (Cheng et al., 2007). Ultrasound therapy can also be used to offset the nutritional losses caused by blanching, resulting in a gain from both treatments. Employed ultrasonication cleaners (20–400 kHz) to efficiently produce contamination-free fruits and vegetables, and at 40 kHz, it was used to reduce decay and infection while retaining quality in strawberry fruits.

Fresh fruits and vegetables decay fast, making it challenging to maintain their freshness. Ultrasound has been found as a viable method for reducing losses both before and after harvest. Because of the pores and gaps in fruits and vegetables, ultrasound is scattered and so attenuated, making it difficult to interpret ultrasound data. According to studies, ultrasound can change the hardness of fresh fruits and vegetables, which changes depending on the intensity. In other cases, color changes were seen, perhaps as a result of the inactivation of the enzyme's phenolperoxidase and polyphenol oxidase. These enzymes cause browning and colour changes in frozen and fresh vegetables. Sonication may have antibacterial characteristics (Nicolau-Lapeña et al., 2019; Mizrach et al., 1991; Kuo et al., 2008). Oil purity, quality, and composition may all be evaluated using ultrasound velocity measurements. This method of washing fresh fruits and vegetables is ideal since it saves time, preserves colour and nutrients, and is easy to use. Using ultrasonic on fresh food can reduce or prevent the growth of biofilms or germs, according to Chen et al. (2020). Blunt enzymes generated by ultrasound prevent heat-sensitive molecules from losing weight. It may also be used to tenderize and cure fresh vegetables.

Also investigated were the effects in orange juice, the effects of sonication on pectin methyl esterase activity and cloud stability. With highest energy density sonication and a long treatment period, the highest percentage of pectin methyl esterase activity inactivation was 62%. Inactivation of pectin methyl esterase activity and particle size reduction are required for sonicated orange juice cloud stability. To obtain the desired cloud stability were used for sonication (Tiwari et al., 2009). Tiwari et al. investigated the impact of amplitude level and sonication period on juice quality parameters (2008). The pH, Brix, and titratable acidity all changed only slightly. The browning index, cloud value, and color on the other hand, all degraded as a result. On Alicyclo bacillusacidoterrestris in apple juice, ultrasonic treatments had higher power levels and longer processing times. After 30 minutes and 60 minutes of treatment with 300-W ultrasound when 600-W was employed, the lowest D value of 36.18 minutes was reached. After ultrasonic treatments, changes in sugar content, acidity, haze, and juice browning were seen, but the juice quality did not deteriorate (Yuan et al., 2009).

It has been demonstrated that nonthermal power ultrasound may be used to inactivate germs in fruit juices. Sonication alone has been shown to be a successful way for inactivating yeast in tomato juice to the necessary level of inactivation (Adekunte et al., 2010). Using a combination of ultrasound and heat, the inactivation rates of both PME and PG were enhanced (thermosonication, Terefe et al., 2009). The quality of strawberry juice was studied as a function of amplitude and sonication duration (Tiwari et al., 2008). The effects of carbonation and sonication on guava juice quality were studied. Ascorbic acid levels were found to be considerably greater in carbonated and sonicated samples than in control samples. As the juice was carbonated, the number of nuclei available for cavitation increased, allowing dissolved oxygen to be removed (Cheng et al., 2007).

Cheila et al. (2020) revealed that employing ultrasonic technology to extract bioactives from velame leaves is more ecologically friendly. In 39.5 minutes, indirect ultrasonication increased the yield of extracted bioactives by 94%. Ultrasound has been demonstrated to be more effective than conventional techniques for extracting oil from olive fruit, soybeans, and flaxseed (Cavallo et al., 2020; Juliano et al., 2017). It can also extract bioactive compounds from plant, fruit, and vegetable (Iftikhar et al., 2020; Qin et al., 2021; Wu et al., 2020; Martinez-Ramos et al., 2020). Mothibe et al. (2014) employed ultrasonication as a pretreatment step before dehydrating apples and that the dried apple had an excellent texture. As the treatment period proceeded, the amount of soluble solids in the apple decreased. The process of dry food acquiring moisture is known as rehydration (Tian et al., 2016).

According to Tao et al. (2019), the ultrasound-rehydrated white cabbage rehydrated faster than the untreated sample. Carrots and green peppers have been rehydrated in similar methods (Ricce et al., 2016; Szadzinska et al., 2017). It is also utilized as a freeze-drying and convection-drying preparation (Rojas et al., 2020; Prosapio and Norton, 2018). Ultrasonication was shown to prevent retrogradation and despite the fact that molecular degradation was not observed using Fourier transform infrared spectroscopy (Nie et al., 2019).

1.2.3 Ultrasonication in cereal products

Cereal-based foods include biscuits, morning cereals, and bars, as well as bakery items, the most common of which being bread. There is a link between rheology and ultrasonic attenuation and velocity, according to Ross et al. (2004). This was done by comparing the mixing extent of three different wheat doughs using traditional rheology and ultrasonic techniques, proving the value of ultrasound in dough quality control. A different research project investigated the use of ultrasonography to characterize the fermentation stage of the bread-making process (Elmehdi et al., 2003). To combat the continuing during fermentation, dough expands both chemically and physically, a low-frequency ultrasonic method with two large sensors was proposed (Skaf et al., 2009). This enabled the examination of technical components that impact the dough development process, as well as the assessment of dough physical attributes and the calculation of critical periods. Porosity is a significant physical–mechanical characteristic of bakery products that is linked to their quality. To improve the rheology and texture of bread, air must be given throughout the mixing and holding of the dough until formation. These air bubbles have a significant impact on sound attenuation and velocity, both of which are controlled by wave frequency (Elmehdi et al., 2003; Petrauskas, 2007). Devised a quick and cost-effective transmission ultrasonic velocity approach for determining how protein and starch gelatinization affects the quality of wheat dough. Low-frequency, nondestructive, and quick-direct ultrasonic testing can be used to assess the textural qualities of bread products.

Ultrasound can increase the stiffness of the dough, the texture of the bread, and the color component. It also leads to higher-quality products with improved visual and sensory qualities (Jalali et al., 2020). The use of ultrasound to treat wheat flour dough reduced surface hydrophobicity while also aligning the secondary structures of gliadin, gluten, and glutenin are all terms for the same thing. Gluten's structural qualities were also improved as a result of this (Zhang et al., 2020). According to another study (Zhu and Li, 2019), the functional qualities of quinoa flour include crispiness, which is a hallmark of biscuits and other cereal-based meals. Based on the displacement/force and acoustic attenuation behavior of food products, Leeds University developed an acoustic envelop detector to evaluate crispiness

(Chen et al., 2019). In waffles, donuts, pancakes, and cupcakes, ultrasound mayto test the qualities of the batter including viscosity, rheology, and density, as well as cake aspects like symmetry, height, density, and volume index (Salazar et al., 2004).

Fox et al. (2004) created a low-cost ultrasonic device for monitoring batters while they are being mixed for their unique purpose, emphasizing the use of ultrasonic devices to monitor batters while they are being mixed. By examining compressibility variations, the batter's acoustic impedance was determined. Additional ultrasonic testing found a link between batter consistency and acoustic impedance. Bonto et al. (2021) studied the impact of the effects of ultrasonication on rice starch and grain textural, morphological, and physicochemical properties. Ultrasonication created fissures and fractures in rice starch in a biphasic system, allowing for increased water absorption and changing the rice starch's textural properties. Ultrasonic treatment increased rice palatability by increasing the generation of health-related metabolites, boosting the absorption of micronutrient fortifiers, and softening the texture of wholegrain rice.

1.2.4 Ultrasonication in dairy technology

Ultrasound is a useful tool for retaining nutrients, which leads to a longer life span and product of high grade. It is less expensive, faster, and simpler to use. It has been shown to be effective in the dairy industry for microbial inactivation. Microorganisms' cellular membranes are damaged by intracellular cavitation. As a result, free radicals are formed, and DNA is damaged (Guimaraes et al., 2019). It also reduces fermentation time by speeding up lactose breakdown by releasing the enzyme lactase and boosting acid generation. Bioactive peptides and oligosaccharides are improved while lactose content is reduced. Ultrasound therapy improves the organoleptic and nutritional aspects of fermented milk products (Huang et al., 2019).

Ultrasound also speeds the breakdown of protein structure, reducing the time it takes for cheese to mature and boosting the textural, organoleptic, and nutritional qualities of cheese (Gomez et al., 2008). Ultrasound with a high-intensity has been shown to shrink fat globules, enhance viscosity, minimize syneresis, boost gel strength, and speed up fermentation when used in the production of yoghurt (Akdeniz et al., 2019). The texture and organoleptic characteristics of ice cream are improved when it is treated with ultrasound during the manufacturing process. In cavitations, the bubbles burst, generating a localized increase in pressure that favors ice nucleation and the disintegration of big crystals (Minj et al., 2020). In the dairy and beverage industries, ultrasonic-assisted filtration is also very efficient and crucial. In cheese manufacturing, the membrane filtering method is utilized to entirely separate milk protein from other milk solids (Saxena et al., 2009).

1.2.5 Ultrasonication in oil technology

Ultrasonication is used to speed up the transesterification process; ultrasonic irradiation is also used. Bioactives extraction, emulsification, heating, debittering, and improved synthesis and other processes all employ frequencies in the 20 kHz–100 kHz range in food processing. Jadhav et al. (2021) reported on the use of sonication in the synthesis of designer lipids as an excellent approach for higher yield. In 6 hours of reaction time, the authors were able to obtain a maximum output of 92%. The creation of high-energy spots hastens mass transfer and, as a result of this, increased energy, the reaction requires less time to complete. Classical synthesis takes longer than ultrasonication-assisted synthesis (Jadhav and Annapure, 2021).

Oil dynamic rheology and composition, emulsion oil content and droplet size, and partially crystalline emulsion solid fat content can all be determined using ultrasound. It can take precise measurements quickly and easily, is nondestructive and noninvasive, works both online and offline, and is fairly priced. As a result, ultrasonic methods for evaluating the composition of fats and oils will be a beneficial supplement to existing analytical techniques (McClements and Povey, 1992). Some off-flavors may be noticed during the ultrasonic processing of lipid-containing meals, resulting in customer displeasure. Using two distinct ultrasound horns, Pingret et al. (2012) demonstrated the consequences of ultrasound therapy on sunflower oil (titanium and pyrex). Traditional physicochemical criteria and volatile profile were used to identify and quantify the generated radicals. The findings show a rise in generated radicals as well as changes in the physicochemical qualities of sonicated oils, indicating that the oils have been oxidized.

1.2.6 Advantages of ultrasonication

Ultrasound is a green, eco-friendly technology that has helped the food industry enhance a range of procedures. It's also an excellent alternative to a variety of old heat-based technologies that decrease product quality. Ultrasonication produces a higher-quality product at lower temperatures while also improving heat and mass transfer rates. Ultrasound accelerates freezing, resulting in smaller crystal sizes, faster drying, and thawing procedures. It also accelerates the filtering process, extending the filter's life. Ultrasound is a low-cost manufacturing method that allows for rapid processing. It increases product output and enhances process efficiency by removing the requirement for process stages. Improved organoleptic qualities, hardness, and texture improve the end product's quality and purity. It also helps to preserve the nutritious properties of the product while also extending its shelf life.

When ultrasound is employed at high intensities, heat is produced as a result of the increased temperature, which has a negative effect on the organoleptic and nutritional qualities of the food. Furthermore, there is no evidence that ultrasound can inhibit microbial and enzymatic inactivation. Inactivation may come from the synergistic effect. Furthermore, high-powered ultrasound has the potential to physically and chemically destroy food. Cavitation generates free radicals, which result in lipid oxidation, bad tastes and odors, protein denaturation, and ascorbic acid degradation, resulting in a drop in total phenolic content. When ultrasound is combined with warmth and pressure, free radicals are produced, which speed up processes that may alter protein structure. As a result, the strength of ultrasound must be modified prior to use. Advantages of Ultrasound applications are listed below:

- Ultrasound waves are nontoxic, harmless, and favorable to the environment.
- Microbial inactivation by combining ultrasonication with other non-thermal methods has been found to be effective.
- Ultrasonication has a cheap operating cost, is simple to use, and produces a high power output.
- Ultrasonication does not need high-tech equipment or a diverse set of technologies.
- When compared to other extraction procedures, ultrasound gives a higher yield and rate of extraction.
- Ultrasonication results in less taste loss, greater consistency (viscosity, homogeneity), and considerable energy savings.
- Processing, extraction, emulsification, preservation, and homogenization are just a handful of the food industry's ultrasonic uses.

1.3 Limitations of sonication

Ultrasonication, despite its numerous advantages, has a number of disadvantages, including:

- Ultrasound induces inactivation of released products due to shear stress induced by shock wave spirals (mechanical effects).

- Ultrasound necessitates significant energy input, causing industrialists to hesitate using this method on a large scale.
- Ultrasound has physiochemical effects that can create off-flavors, physical changes, and component degradation, all of which can compromise food quality.

  Ultrasonication produces radicals as a result of critical temperature and pressure conditions, which are responsible for dietary component alterations. The radicals created in the medium settle on the cavitation bubble's surface, activating radical chain reactions that yield degradation products, resulting in significant product quality flaws.
- The frequency of ultrasound waves can influence mass transfer resistance.
- According to the properties of the medium, ultrasonic power is responsible for material change. As a result, in order to get optimum outcomes, this power must be limited in the food sector.
- One of the disadvantages of sonic cell lysis is the generation of heat. To avoid overheating, ultrasonic pulses are delivered in short bursts to a sample that is immersed in an ice bath. Sonication is a time-consuming method that is only suitable for small batches.
- Another drawback of sonication is that it does not lyse all cells equally. Plant cells are more difficult to break than animal cells because their cell walls are substantially more resistant to cell lysis sonication.

1.4 Conclusion

Ultrasound is a green technology since it is nontoxic and environmentally friendly. It saves a significant amount of energy while significantly boosting output. In science and food technology, ultrasound is utilized to detect foreign extraneous components, as well as to assess the composition of foods (fruits, vegetables, and dairy products), in canned and dairy foods. Despite the fact that considerable research has been done on ultrasound technologies in food technology, there is still much more work to be done in order to develop industrial-automated ultrasound systems that save time, money, and energy while ensuring the highest output of high-value and safe food items.

References

Adekunte, A., Tiwari, B.K., Scannell, A., Cullen, P.J., O'Donnell, C., 2010. Modelling of yeast inactivation in sonicated tomato juice. Int. J. Food Microbiol. 137, 116–120.

Akdeniz, V., Akalın, A.S., 2019. New approach for yoghurt and ice cream production: high-intensity ultrasound. Trends Food Sci. Technol. 86, 392–398.

Alarcon-Rojo, A.D., Carrillo-Lopez, L.M., Reyes-Villagrana, R., Huerta-Jiménez, M., Garcia-Galicia, I.A., 2019. Ultrasound and meat quality: a review. Ultrason. Sonochem. 55, 369–382.

Amiri, A., Sharifian, P., Soltanizadeh, N., 2018. Application of ultrasound treatment for improving the physicochemical, functional and rheological properties of myofibrillar proteins. Int. J. Biol. Macromol. 111, 139–147.

Aouzale, N., Chitnalah, A., Jakjoud, H., Kourtiche, D., Nadi, M., 2010. Comparison Between VHDL-AMS and PSPICE Modeling of Ultrasound Measurement System For Biological Medium. Springer, Berlin, Heidelberg, pp. 529–532.

Ashokkumar, M., 2015. Applications of ultrasound in food and bioprocessing. Ultrason. Sonochem. 25, 17–23.

Awad, T.S., Moharram, H.A., Shaltout, O.E., Asker, D., Youssef, M.M., 2012. Applications of ultrasound in analysis, processing and quality control of food: a review. Food Res. Int. 48, 410–427.

Azam, S.R., Ma, H., Xu, B., Devi, S., Siddique, M.A.B., Stanley, S.L., 2020. Efficacy of ultrasound treatment in the removal of pesticide residues from fresh vegetables: a review. Trends Food Sci. Technol. 97, 417–432.

Barekat, S., Soltanizadeh, N., 2018. Effects of ultrasound on microstructure and enzyme penetration in beef longissimus lumborum muscle. Food Bioprocess Technol. 11 (3), 680–693.

Bemiller, J.N., 2018. Physical Modification of Starch. Elsevier Ltd., USA.

Bemiller, J.N., Huber, K.C., 2015. Physical modification of food starch functionalities. Annu. Rev. Food Sci. Technol. 6, 19–69.

Bhargava, N., Mor, R.S., Kumar, K., Sharanagat, V.S., 2021. Advances in application of ultrasound in food processing: a review. Ultrason. Sonochem. 70 (105293).

Boistier-Marquis, E., Lagsir-Oulahal, N., Callard, M., 1999. Applications des ultrasons de puissances en industries alimentaires. Industries Alimentaires et Agricoles 116, 23–31.

Bonto, A.P., Tiozon, R.N., Sreenivasulu, N., Camacho, D.H., 2021. Impact of ultrasonic treatment on rice starch and grain functional properties: a review. Ultrason. Sonochem. 71 (105383).

Caraveo, O., Alarcon-Rojo, A.D., Renteria, A., Santellano, E., Paniwnyk, L., 2015. Physico-chemical and microbiological characteristics of beef treated with high intensity ultra-sound and stored at 4 °C. J. Sci. Food Agric. 95 (12), 2487–2493.

Cárcel, J.A., Benedito, J., Bon, J., Mulet, A., 2007. High intensity ultrasound effects on meat brining. Meat Sci. 76 (4), 611–619.

Carovac, A., Smajlovic, F., Junuzovic, D., 2011. Application of ultrasound in medicine. Acta Informatica Medica 19, 168.

Cavallo, C., Carlucci, D., Carfora, V., Caso, D., Cicia, G., Clodoveo, M.L., 2020. Innovation in traditional foods: a laboratory experiment on consumers' acceptance of extra-virgin olive oil extracted through ultrasounds. NJAS - Wageningen J. Life Sci. 92, 100336.

Chang, H.J., Wang, Q., Tang, C.H., Zhou, G.H., 2015. Effects of ultrasound treatment on connective tissue collagen and meat quality of beef semitendinosus muscle. J. Food Qual. 38 (4), 256–267.

Cheila, C.B., dos Anjos, G.L., Nóbrega, R.S.A., da S. Magaton, A., de Miranda, F.M., Dias, F., 2020. Greener ultrasound-assisted extraction of bioactive phenolic compounds in Croton heliotropiifolius Kunth leaves. Microchem. J. 159 (105525).

Chemat, F., Khan, M.K., 2011. Applications of ultrasound in food technology: processing, preservation and extraction. Ultrason. Sonochem. 18 (4), 813–835.

Chemat, F., Rombaut, N., Sicaire, A.G., Meullemiestre, A., Fabiano-Tixier, A.S., Abert-Vian, M., 2017. Ultrasound assisted extraction of food and natural products, mechanisms, techniques, combinations, protocols and applications, a review. Ultrason. Sonochem. 34, 540–560.

Chemat, F., Zill-E-Huma, Khan, M.K., 2011. Applications of ultrasound in food technology: processing, preservation and extraction. Ultrason. Sonochem. 18, 813–835.

Chen, F., Zhang, M., Yang, C.H., 2019. Application of ultrasound technology in processing of ready-to-eat fresh food: a review. Ultrason. Sonochem. 104953.

Chen, F., Zhang, M., Yang, C.H., 2020. Application of ultrasound technology in processing of ready-to-eat fresh food: a review. Ultrason. Sonochem. 63 (104953).

Cheng, L.H., Soh, C.Y., Liew, S.C., Teh, F.F., 2007. Effects of sonication and carbonation on guava juice quality. Food Chem. 104, 1396–1401.

Cheng, X.F., Zhang, M., Adhikari, B., 2014. Effect of ultrasonically induced nucleation on the drying kinetics and physical properties of freeze-dried strawberry. Drying Technol. 32, 1857–1864.

Chow, R., Blindt, R., Chivers, R., Povey, M., 2005. A study on the primary and secondary nucleation of ice by power ultrasound. Ultrasonics 43, 227–230.

Elmehdi, H.M., Page, J.H., Scanlon, M.G., 2003. Using ultrasound to investigate the cellular structure of bread crumb. J. Cereal Sci. 38 (1), 33–42.

Ercan, S., Soysal, C., 2013. Use of ultrasound in food preservation. Nat. Sci. 5 (5), 13.

Fox, P., Smith, P.P., Sahi, S., 2004. Ultrasound measurements to monitor the specific gravity of food batters. J. Food Eng. 65 (3), 317–324.

Gallo, M., Ferrara, L., Naviglio, D., 2018. Application of ultrasound in food science and technology: a perspective. Foods 7, 164.

Gómez, M., Oliete, B., García-Álvarez, J., Ronda, F., Salazar, J., 2008. Characterization of cake batters by ultrasound measurements. J. Food Eng. 89 (4), 408–413.

Gomez-Lopez, V.M., Orsolani, L., Martinez-Yepez, A., Tapia, M.S., 2010. Microbiological and sensory quality of sonicated calcium-added orange juice. LWT—Food Sci. Technol. 43, 808–813.

Gómez-Salazar, J.A., Ochoa-Montes, D.A., Cerón-García, A., Ozuna, C., Sosa-Morales, M.E., 2018. Effect of acid marination assisted by power ultrasound on the quality of rabbit meat. J. Food Qual. 1–6.

Green, A.A., Hersam, M.C., 2010. Emerging methods for producing monodisperse graphene dispersions. J. Phys. Chem. Lett. 1 (2), 544–549.

Guimarães, J.T., Balthazar, C.F., Scudino, H., Pimentel, T.C., Esmerino, E.A., Ashokkumar, M., 2019. High-intensity ultrasound: a novel technology for the development of probiotic and prebiotic dairy products. Ultrason. Sonochem. 57, 12–21.

Hasan, M.M., Bashir, T., Bae, H., 2017. Use of ultrasonication technology for the increased production of plant secondary metabolites. Molecules 22, 1046.

Huang, D., Men, K., Li, D., Wen, T., Gong, Z., Sunden, B., Wu, Z., 2019. Application of ultrasound technology in the drying of food products. Ultrason. Sonochem. May 1; 63 (104950).

Iftikhar, M., Zhang, H., Iftikhar, A., Raza, A., Begum, N., Tahamina, A., 2020. Study on optimization of ultrasonic assisted extraction of phenolic compounds from rye bran. LWT 134 (110243).

Jadhav, H.B., Annapure, U., 2021. Process intensification for synthesis of triglycerides of capric acid using green approaches. J. Ind. Chem. Soc. 98 (100030).

Jadhav, H.B., Gogate, P.R., Waghmare, J.T., Annapure, U.S., 2021. Intensified synthesis of palm olein designer lipids using sonication. Ultrason. Sonochem. 73 (105478).

Jalali, M., Sheikholeslami, Z., Elhamirad, A.H., Khodaparast, M.H.H., Karimi, M., 2020. The effect of the ultrasound process and pre-gelatinization of the corn flour on the textural, visual, and sensory properties in gluten-free pan bread. J. Food Sci. Technol. 57 (3), 993–1002.

Juliano, P., Bainczyk, F., Swiergon, P., Supriyatna, M.I.M., Guillaume, C., Ravetti, L., 2017. Extraction of olive oil assisted by high-frequency ultrasound standing waves. Ultrason. Sonochem. 38, 104–114.

Kang, D.C., Wang, A.R., Zhou, G.H., Zhang, W.G., Xu, S.M., Guo, G.P., 2016. Power ultrasonic on mass transport of beef: effects of ultrasound intensity and NaCl concentration. Innov. Food Sci. Emerg. Technol. 35, 36–44.

Karami, H., 2019. The thermophysical properties and the stability of nanofluids containing carboxyl-functionalized graphene nano-platelets and multi-walled carbon nanotubes. Int. Commun. Heat Mass Transfer 108.

Kentish, S., Feng, H., 2014. Applications of power ultrasound in food processing. Annu. Rev. Food Sci. Technol. 5, 263–284.

Knorr, D., Zenker, M., Heinz, V., Lee, D.U., 2004. Applications and potential of ultrasonics in food processing. Trends Food Sci. Technol. 15, 261–266.

Kuo, F.J., Sheng, C.T., Ting, C.H., 2008. Evaluation of ultrasonic propagation to measure sugar content and viscosity of reconstituted orange juice. J. Food Eng. 86 (1), 84–90.

Leena, M., Srinivasan, S., 2015. Synthesis and ultrasonic investigations of titanium oxide nanofluids. J. Mol. Liq. 206, 103–109.

Li, D., Zhao, H., Muhammad, A.I., Song, L., Guo, M. & Liu, D. 2020, 'The comparison of ultrasound-assisted thawing, air thawing and water immersion thawing on the quality of slow/fast freezing bighead carp (Aristichthys nobilis) fillets', Food Chem., no. 126614.

Liu, W., 2019. A novel comprehensive experimental study concerned graphene oxide nanoparticles dispersed in water: synthesise, characterisation, thermal conductivity measurement and present a new approach of RLSF neural network. Int. Commun. Heat Mass Transfer 109.

Mahbubul, I., 2016. Influence of ultrasonication duration on rheological properties of nanofluid: an experimental study with alumina–water nanofluid'. Int. Commun. Heat Mass Transfer 76, 33–40.

Majid, I., Nayik, G.A., Nanda, V., 2015. Ultrasonication and food technology: a review. J. Sci. Food Agric. 1 (1071022).

Martínez-Ramos, T., Benedito-Fort, J., Watson, N.J., Ruiz-López, I.I., Che-Galicia, G., Corona-Jiménez, E., 2020. Effect of solvent composition and its interaction with ultrasonic energy on the ultrasound-assisted extraction of phenolic compounds from mango peels (Mangifera indica L.). In: Food Bioprod Process, 122, pp. 41–54.

Mason, T.J., 1998. Power ultrasound in food processing-the way forward. In: Ultrasound Food Processing, pp. 105–126.

McClements, D.J., 1995. Advances in the application of ultrasound in food analysis and processing. Trend Food Sci. Technol. 6, 293–299.

McClements, D.J., Povey, M.J., 1992. Ultrasonic analysis of edible fats and oils. Ultrasonics 30 (6), 383–388.

Miles, C.A., Morley, M.J., Rendell, M., 1999. High power ultrasonic thawing of frozen foods. J. Food Eng. 39, 151–159.

Minj, J., Sudhakaran, A., Kumari, A., 2020. Basic facts about dairy processing and technologies. In: Minj, J., Sudhakaran, V.A., Kumari, A. (Eds.), Dairy Processing: Advanced Research to Applications. Springer, Singapore, pp. 1–24.

Mizrach, A., 2008. Ultrasonic technology for quality evaluation of fresh fruit and vegetables in pre-and postharvest processes. Postharvest Biol. Technol. 48 (3), 315–330.

Mizrach, A., Galili, N., Rosenhouse, G., Teitel, D.C., 1991. Acoustical, mechanical, and quality parameters of winter-grown melon tissue. Trans. ASAE 34 (5), 2135–2138.

Mothibe, K.J., Zhang, M., Mujumdar, A.S., Wang, Y.C., Cheng, X., 2014. Effects of ultrasound and microwave pretreatments of apple before spouted bed drying on rate of dehydration and physical properties. Drying Technol. 32, 1848–1856.

Mulet, A., Benedito, J., Golás, Y., Carcel, J.A., 2002. Noninvasive ultrasonic measurements in the food industry. Food Rev. Int. 18, 123–133.

Nicolau-Lapeña, I., Lafarga, T., Viñas, I., Abadias, M., Bobo, G., Aguiló-Aguayo, I., 2019. Ultrasound processing alone or in combination with other chemical or physical treatments as a safety and quality preservation strategy of fresh and processed fruits and vegetables: a review. Food Bioprocess Technol. 1–20.

Nicolau-Lapeña, I., Lafarga, T., Viñas, I., Abadias, M., Bobo, G., Aguiló-Aguayo, I., 2019. Ultrasound processing alone or in combination with other chemical or physical treatments as a safety and quality preservation strategy of fresh and processed fruits and vegetables: a review. Food Bioprocess Technol. 12, 1452–1471.

Nie, H., Li, C., Liu, P.H., Lei, C.Y., Li, J.B., 2019. Retrogradation, gel texture properties, intrinsic viscosity and degradation mechanism of potato starch paste under ultrasonic irradiation. Food Hydrocolloids 95, 590–600.

Noroozi, M., Radiman, S., Zakaria, A., 2014. Influence of sonication on the stability and thermal properties of Al2O3 nanofluids. J. Nanomater. 2014, 1–10.

Ojha, K.S., Aznar, R., O'Donnell, C., Tiwari, B.K., 2020. Ultrasound technology for the extraction of biologically active molecules from plant, animal and marine sources. Trends Anal. Chem. 122, 115663.

Ojha, K.S., Mason, T.J., O'Donnell, C.P., Kerry, J.P., Tiwari, B.K., 2017. Ultrasound technology for food fermentation applications. Ultrason. Sonochem. 34, 410–417.

Pagan, R., Manas, P., Alvarez, I., Condon, S., 1999. Resistance of Listeria monocytogenes to ultrasonic waves under pressure at sublethal (manosonication) and lethal (manothermosonication) temperatures. Food Microbiol. 16, 139–148.

Paniwnyk, L., 2014. Application of ultrasound. In: Sun, D.-W. (Ed.), Emerging Technologies in Food Processing. Elsevier, London, U.K., pp. 271–291.

Patist, A., Bates, D., 2008. Ultrasonic innovations in the food industry: from the laboratory to commercial production. Innov. Food Sci. Emerg. Technol. 9, 147–154.

Peña-González, E.M, Alarcón-Rojo, A.D, Rentería, A., García, I., Santellano, E., Quintero, A., Luna, L., 2017. Quality and sensory profile of ultrasound-treated beef. Ital. J. Food Sci. 29 (3).

Petrauskas, A., 2007. Evaluation of porous food products by using ultrasonic methods. Ultragarsas "Ultrasound" 62 (3), 20–25.

Pingret, D., Durand, G., Fabiano-Tixier, A.S., Rockenbauer, A., Ginies, C., Chemat, F., 2012. Degradation of edible oil during food processing by ultrasound: electron paramagnetic resonance, physicochemical, and sensory appreciation'. J. Agric. Food Chem. 60 (31), 7761–7768.

Pingret, D., Fabiano-Tixier, A.S., Chemat, F., 2013. Degradation during application of ultrasound in food processing: a review. Food Control 31, 593–606.

Pohlman, F.W., Dikeman, M.E., Zayas, J.F., 1997. The effect of low intensity ultrasound treatment on shear properties, color stability and shelf-life of vacuum packaged beef semitendinosus and biceps femoris muscles. Meat Sci. 45, 329–337.

Prosapio, V., Norton, I., 2018. Simultaneous application of ultrasounds and firming agents to improve the quality properties of osmotic + freeze-dried foods. LWT 96, 402–410.

Qin, L., Yu, J., Zhu, J., Kong, B., Chen, Q., 2021. Ultrasonic-assisted extraction of polyphenol from the seeds of Allium senescens L. and its antioxidative role in Harbin dry sausage. Meat Sci. 172, 108351.

Ricce, C., Rojas, M.L., Miano, A.C., Siche, R., Augusto, P.E.D., 2016. Ultrasound pre-treatment enhances the carrot drying and rehydration. Food Res. Int. 89, 701–708.

Roberts, R.T., 1991. Sound for processing food. Nutr. Food Sci. 91, 18–19.

Rojas, M.L., Augusto, P.E.D., Cárcel, J.A., 2020. Ethanol pre-treatment to ultrasound-assisted convective drying of apple. Food Sci. Emerg. Technol. 61 (102328).

Ross, K.A., Pyrak-Nolte, L.J., Campanella, O.H., 2004. The use of ultrasound and shear oscillatory tests to characterize the effect of mixing time on the rheological properties of dough. Food Res. Int. 37 (6), 567–577.

Salazar, J., Turó, A., Chávez, J.A., Garcıa, M.J., 2004. Ultrasonic inspection of batters for ´on-line process monitoring. Ultrasonics 42 (1–9), 155–159.

Sango, D.M., Abela, D., Mcelhatton, A., Valdramidis, V.P., 2014. Assisted ultrasound applications for the production of safe foods. J. Appl. Microbiol. 116, 1067–1083.

Saxena, A., Tripathi, B.P., Kumar, M., Shahi, V.K., 2009. Membrane-based techniques for the separation and purification of proteins: an overview. Adv. Colloid Interface Sci. 145, 1–22.

Skaf, A., Nassar, G., Lefebvre, F., Nongaillard, B., 2009. A new acoustic technique to monitor bread dough during the fermentation phase. J. Food Eng. 93, 365–378.

Smith, N.B., Cannon, J.E., Novakofski, J.E., McKeith, F.K., O'Brien, W.D., 1991. Tenderization of semitendinosus muscle using high intensity ultrasound. Ultrason. Sonochem. 1371–1374.

Soltani Firouz, M., Farahmandi, A., Hosseinpour, S., 2019. Recent advances in ultrasound application as a novel technique in analysis, processing and quality control of fruits, juices and dairy products industries: a review. Ultrason. Sonochem. 57, 73–88.

Son, Y., No, Y., Kim, J., 2020. Geometric and operational optimization of 20-kHz probe-type sonoreactor for enhancing sonochemical activity. Ultrason. Sonochem. 65.

Stadnik, J., Dolatowski, Z.J., 2011. Influence of sonication on Warner-Bratzler shear force, colour and myoglobin of beef (m. semimembranosus). Eur. Food Res. Technol. 233 (4), 553.

Sun, X., Zhang, W., Zhang, L., Tian, S. & Chen, F. 2020, 'Molecular and emulsifying properties of arachin and conarachin of peanut protein isolate from ultrasound-assisted extraction', LWT, Vol. 132, no. 109790.

Szadzińska, J., Łechtańska, J., Kowalski, S.J., Stasiak, M., 2017. The effect of high power airborne ultrasound and microwaves on convective drying effectiveness and quality of green pepper. Ultrason. Sonochem. 34, 531–539.

Tao, Y., Han, M., Gao, X., Han, Y., Show, P.L., Liu, C., 2019. Applications of water blanching, surface contacting ultrasound-assisted air drying, and their combination for dehydration of white cabbage: drying mechanism, bioactive profile, color and rehydration property. Ultrason. Sonochem. 53, 192–201.

Terefe, N.S., Gamage, M., Vilkhu, K., Simons, L., Mawson, R., Versteeg, C., 2009. The kinetics of inactivation of pectin methylesterase and polygalacturonase in tomato juice by thermosonication. Food Chem. 117, 20–27.

Tian, Y., Zhao, Y., Huang, J., Zeng, H., Zheng, B., 2016. Effects of different drying methods on the product quality and volatile compounds of whole shiitake mushrooms. Food Chem. 197, 714–722.

Tiwari, B.K., O'Donnell, C.P., Muthukumarappan, K., Cullen, P.J., 2008. Effect of ultrasound processing on the quality and nutritional properties of fruit juices. Stewart Postharvest Rev. 4 (5), 1–6.

Tiwari, B.K., O'Donnell, C.P., Muthukumarappan, K., Cullen, P.J., 2009. Inactivation kinetics of pectin methylesterase and cloud retention in sonicated orange juice. Innov. Food Sci. Emerg. Technol. 10, 166–171.

Tiwari, B.K., O'Donnell, C.P., Patras, A., Cullen, P.J., 2008. Anthocyanin and ascorbic acid degradation in sonicated strawberry juice. J. Agric. Food Chem. 56, 10071–10077.

Vimini, R.J., Kemp, J.D., Fox, J.D., 1983. Effects of low frequency ultrasound on properties of restructured beef rolls. J. Food Sci. 48, 1572–1573.

Wang, A., Kang, D., Zhang, W., Zhang, C., Zou, Y., Zhou, G., 2018. Changes in calpain activity, protein degradation and microstructure of beef M. semitendinosus by the application of ultrasound. Food Chem. 245, 724–730.

Watson, N.J., Povey, M.J.W., Parker, N.G., 2017. Acoustic microscopy. In: Villamiel, M, GarcíaPérez, JV, Montilla, A, Carcel, JA, Benedito, J. (Eds.), Ultrasound in Food Processing: Recent Advances. John Wiley & Sons Ltd, USA, pp. 230–251.

Welti-Chanes, J., Morales-de la Peña, M., Jacobo-Velázquez, D.A., Martín-Belloso, O., 2017. Opportunities and challenges of ultrasound for food processing: an industry point of view. In: Bermudez-Aguirre, D. (Ed.), Ultrasound: Advances for Food Processing and Preservation, pp. 457–497.

Wu, L., Li, L., Chen, S., Wang, L., Lin, X., 2020. Deep eutectic solvent-based ultrasonic-assisted extraction of phenolic compounds from Moringa oleifera L. leaves: optimization, comparison and antioxidant activity. Sep. Purif. Technol. 247 (117014).

Yuan, Y., Hu, Y., Yue, T., Chen, T., Lo, Y.M., 2009. Effect of ultrasonic treatments on thermoacidophilic alicyclobacillus acidoterrestris in apple juice. J. Food Process. Preserv. 33, 370–383.

Zhang, Y., Li, Y., Wang, H., Oladejo, A.O., Zhang, H., Liu, X., 2020. Effects of ultrasound assisted freezing on the water migration of dough and the structural characteristics of gluten components. J. Cereal Sci. 94 (102893).

Zheng, L., Sun, D.W., 2006. Innovative applications of power ultrasound during food freezing processes—a review'. Trends Food Sci. Technol. 17 (1), 16–23.

Zhu, F., 2015. Impact of ultrasound on structure, physicochemical properties, modifications, and applications of starch. Trends Food Sci. Technol. 43, 1–17.

Zhu, F., Li, H., 2019. Modification of quinoa flour functionality using ultrasound. Ultrason. Sonochem. 52, 305–310.

Zhu, F., Li, H., 2019. Modification of quinoa flour functionality using ultrasound. Ultrason. Sonochem. 52, 305–310.

CHAPTER TWO

Microwave: An overview

Neha Bakshi, Swati Jain, Aishwarya Raman and Taru Pant
Department of Food and Nutrition and Food Technology, Lady Irwin College, Delhi, India

2.1 Introduction

In the last few years, the processing of food materials using microwave technology has gained remarkable attention, and is one of the leading alternative technologies to conventional food processing. Microwave ovens are an essential part of the household in the 21st century, and the crucial aspect supporting this increasing demand of microwave heat processing is its rapidness which matches well with the fast paced lifestyle of people nowadays. Further factors are low maintenance, faster heating time, minimal nutrient losses and ease of operation (Chandrasekaran et al., 2013; Datta and Rakesh, 2013). Due to these advantages, the food processing industry has seen a wide use of microwave technology.

Microwaves are defined as electromagnetic waves with bandwidth that ranges from 300 MHz to 300 GHz (Chandrasekaran et al., 2013). They have been successfully utilized for various industrial applications and, further recently, in many food processing industries, like cooking, blanching, dehydration, thawing, pasteurization, or sanitization of a variety of food materials (Contreras et al., 2017). Moreover, the food processing industry not only utilizes microwave technology for preservation and processing but also for extraction of food materials. Microwave-assisted extraction (MAE) is one such method that has been used extensively, as a probable alternative for the purpose of extracting bioactive compounds from food residues. Additionally, MAE technique offers various advantages in the extraction of materials, as compared to other methods (Sanchez-Aldana, 2013), like faster processing time, higher rate of extraction along with reduced cost and lower energy consumption (Yan et al., 2010; Hao et al., 2002).

This chapter covers the fundamentals of microwave technology and the corresponding electromagnetic theory, along with the history and evolution of this technology. It will also cover the general configuration of the microwave system, the mechanism of generation of heat as well as the dielectric properties of foods. It also provides a review of the principles and

Ultrasound and Microwave for Food Processing: Synergism for Preservation and Extraction.
DOI: https://doi.org/10.1016/B978-0-323-95991-9.00014-X

the industrial applications of microwaves for food processing, preservation and extraction. Further, the chapter focuses on the research and development in future applications of microwave systems.

2.1.1 Definition

Microwaves are electromagnetic wave radiation, synchronized perpendicularly on magnetic and electric fields which propagate at the speed of light (Purcell, 2013). The frequency ranges from 300 MHz to 300 GHz, and the wavelength of 1–1 mm (Hitchcock, 2004; National Association of Broadcasters et al., 2007; Pozar, 1990; Sorrentino and Bianchi 2010). Microwaves fall under the category of nonionizing radiations in the electronic spectrum.

2.1.2 History and applications of microwaves

The development of dielectric heat technology in the food industry started in the 1940s, when British physicist John T. Randall and Henry A. Boot developed a high-power vacuum tube known as the cavity magnetron, which led to the discovery of microwaves (Osepchuk, 1984). Microwave technology was initially developed for radar during World War II (Cooper, 2009).

During 1945-1946, Percy Spencer of the Raytheon Manufacturing Corporation was credited for the development of the first microwave oven (Osepchuk, 1984); while the first patent of microwave heating was issued much later, in 1952. Raytheon Manufacturing Corporation was a pioneer in the development and manufacturing of first microwave ovens. In the 1950s, General Electric, Litton Corporation and Philips had also taken a keen interest in this appliance, and were soon selling commercial microwave ovens (Cooper, 2009). In 1955, Tappan Manufacturing Company, under the license from Raytheon, was the first to market microwave ovens for households (Reynolds, 1989). Later on, Amana acquired Raytheon Manufacturing Company, and introduced the first Amana Radarange in 1967, which caused dramatic growth for consumer microwave ovens in the market (Osepchuk, 1984). Eventually, through the 1980 and 1990s, the prices of microwave ovens continued to fall (Cooper, 2009), while the sales rapidly grew (Fig. 2.1).

2.2 General configuration of microwave system

2.2.1 Electromagnetic theory

A nineteenth-century physicist, James Maxwell devised a theory that analyzed the relationship between electricity and magnetism, and also described that visible light consists of electromagnetic waves.

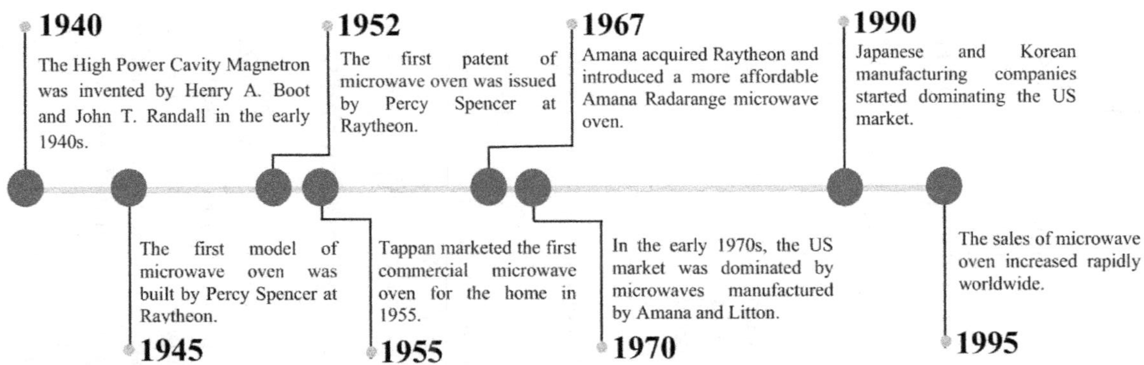

Figure 2.1 Timeline and history of the microwave oven.

The equations given by Maxwell laid down the foundation for the classical electromagnetic episode. It was illustrated how the electric fields induced magnetic fields and vis-a-vis (Du, 2014). These fields travel in the form of waves, and are known as electromagnetic waves or electromagnetic radiations. They are at right angles to each other and are perpendicular to the direction of the wave.

The theoretical analysis of the microwave components is determined by the Maxwell equations:

$$\nabla \times E = \frac{\delta B}{\delta t}, \nabla.B = 0 \text{ (Siores and Do Rego, 1995)}$$

$$\nabla \times H = \frac{\delta D}{\delta t} + I, V.D = \rho$$

where, **E**, the electric field vector,

D, the electric flux density vector,

I, the current density vector,

H, the magnetic field vector,

B, the magnetic flux density vector, and

$\boldsymbol{\rho}$, the charge density.

Microwave oven consists of three primary components (Thostenson and Chou, 1999):

1. The microwave source that generates the electromagnetic energy,
2. The transmission lines are used to transfer the electromagnetic radiation from the source to the applicator, and
3. The applicator, where the material either absorbs or reflects the radiation.

2.2.2 Mechanism of heat generation and the dielectric properties of foods

An electron tube known as magnetron produces microwave radiation inside the oven furnace. These radiations are reflected within the oven by the metal interior and are absorbed by food material (James and James, 2014). Therefore, heat is generated volumetrically after the microwaves interact with food particles. Dielectric heating encompasses microwave and radio frequency heating (Chatterjee and Abraham, 2018). When the waves come in contact with a dielectric material such as food, they give up their energy, which finally causes the rise in temperature within the food (James and James, 2014).

When the food is exposed to microwaves, it results in increased agitation of molecules, which leads to conversion of electromagnetic energy to thermal energy, thus producing heat (Tang and Resurreccion, 2009). Further, the rate of generation of heat and the degree of absorption are determined by the frequency and intensity of the field along with dielectric properties of the material (Venkatesh and Raghavan, 2004).

It is crucial to understand the dielectric properties of food materials and mechanism of microwave heating, as they describe the interaction of food materials with electromagnetic waves (Ryynänen, 1995), and the subsequent absorption of energy. In addition, they are crucial for designing and stimulating microwave processing equipment (Nelson, 2015).

The major dielectric parameters are the dielectric loss factor (ε'') and constant (ε'), which are the imaginary and real components of the relative complex permittivity (ε_r), respectively (Guo et al., 2010).

The dielectric constant can be defined as the capacity of a material to accumulate microwave energy, while the dielectric loss factor is the capability of a material to give off microwave energy as heat (Sumnu and Sahin, 2005), and the association between them is given by the following equation:

$$\varepsilon_r = \varepsilon' - j\varepsilon''$$

These dielectric properties are depended on factors, like the physical structure, chemical composition, and the nature of the material (Calay et al., 2007; Sosa-Morales et al., 2010); for example, dissociated salts and free water have a high dielectric activity, while associated salts and bound water have a low dielectric activity (Lewis, 1996). It also depends upon electromagnetic heating conditions like temperature and frequency (Nelson, 2015).

2.2.3 Microwave interaction with dielectric materials

Microwaves consist of perpendicularly oscillating electric and magnetic fields. Microwave heating occurs when dielectric materials interact with electromagnetic waves, as mentioned previously in the chapter, and the heat is generated volumetrically. Thus, the dielectric characteristics of food components are significant since they impact the efficiency of the heating process (Curet et al., 2014).

There are two mechanisms through which heating takes place through microwave systems: dipolar rotation and ionic conduction (Franco et al., 2015; Fig. 2.2). Dipolar interactions occur due to the molecular motion and intermolecular friction of food particles when they are penetrated by microwaves (Salazar González et al., 2012). Water is an example of dipolar rotation because it contains permanent dipoles that have a tendency to fix themselves when exposed to a changing electric field, resulting in

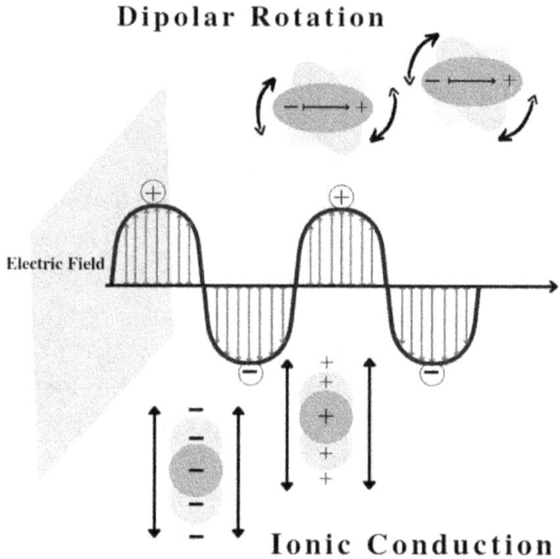

Figure 2.2 Microwave interactions with dielectric materials – dipolar rotation and ionic conduction *(Source: Adapted from Martins et al. (2019)).*

orientation polarization (Metaxas and Meredith, 1983). As the water molecule rotates back and forth it generates heat, while constantly trying to align its dipole with the alternating field (Chemat et al., 2015).

If there are free ions in the reaction mixture, then microwave energy can be transferred through ionic conduction. In this, the charged particles in food will experience a force that will accelerate the particles back and forth. Heat is generated when the accelerated particles collide with the surrounding units, causing them to agitate (Buffler, 1993).

2.3 Microwave for food processing, preservation, and extraction

2.3.1 Microwave processing of different food matrices

2.3.1.1 Changes during microwave processing

2.3.1.1.1 Cereals grains and pulses

Impact on sensory properties: On heat processing by microwave ovens, a study reported that cereal grains and pulses showed changes in their physical structure and physicochemical qualities (Lewandowicz et al., 2000). It also states that microwave irradiation reduces the solubility, crystallinity, and swelling characteristics of corn and wheat starches. However, drastic uncontrolled microwave treatment compromised the sensorial (textural), nutritional (loss of nutrients and minerals), and germination features of wheat (Barba et al., 2020).

Microwave heating influences the temperature and water content of the starches, and a strong association was found between the water content and the rise in temperature (Lewandowicz et al., 1997); and the most pronounced change was seen in the case of potato starch as its crystal structure changed.

Impact on nutritional and functional value: It has been observed that the functional properties of pulses and cereals improved after microwave processing. It was reported that after microwave processing of sorghum, the amount of soluble, insoluble and total dietary fiber were significantly increased (Li et al., 2021).

Processing through the microwave also results in retaining various vitamins and minerals in cereals and pulses. In case of chickpeas, the losses in minerals and B-vitamins were lesser when cooked by microwaving in comparison to traditional methods (El-Adawy, 2002). It has also been noted by Kjeldahl's method, that controlled microwave drying does not alter the protein content (Velu et al., 2006).

However, microwave processing exhibits some unfavorable changes as well. Kamble et al. (2020) reported that, in case of semolina, an increase in the microwave heating caused a significant decrease in the in-vitro protein digestibility. The total phenolic content also decreased, although the antioxidant activity improved as microwave heating progressed. Heat treatment of cereal grains and pulses produced contrasting effect on zinc and iron bioaccessibility; wherein the bioaccessibility of zinc decreased, and on the other hand, iron bioaccessibility was significantly enhanced upon heat treatment (Hemalatha et al., 2007).

2.3.1.1.2 Fruits and vegetables

Impact on sensory properties: Sensory properties are crucial in the evaluation of food quality. Certain distinctive characteristics of fruits and vegetables affect their quality after processing. Many properties, like rehydration ability, density, color, appearance and shrinkage are important in the case of fruits and vegetables.

It has been discussed previously in this chapter, a crucial attribute of microwave technology is that they reduce the processing time. This seems to improve the nutritional, organoleptic, and functional properties, with a positive effect on the color and textural characteristics (Igual et al., 2010), causing minimal deterioration in the final product (Sattar et al., 2019). Fresh juice was compared with a sample pasteurized by microwave technology followed by conventional methods. From a sensory point of view, no difference was observed in both the samples. However, a small change in color was detected in samples pasteurized by microwave (Igual et al., 2014).

Contreras et al. (2005), stated that while microwave processing improves the sensory properties of fruits and vegetables, however, the increased temperatures during processing causes pigment degradation. For example, loss of anthocyanin and an enhanced luminosity were seen when higher microwave processing time was applied in case of strawberry (Contreras et al., 2005).

Effect on functional and nutritional value: Plants contain natural antioxidants or bioactive compounds which are known for its health benefits. Thermal processing techniques like microwaves help to minimize nutrient losses and achieve commercial sterility. Nevertheless, thermal processing can still alter the overall quality of food materials.

In the last few years, various studies have investigated the nutritional properties of fruits and vegetables processed by microwave heating, like broccoli (Zhang and Hamauzu, 2004), apples (Vadivambal and Jayas 2007), potato, carrot, and cabbage (Faller and Fialho, 2009), peas and turnips (Sultana et al., 2008). However, both positive and negative effects have been outlined based on the differences in processing environment and characteristics of fruits and vegetables.

A previous study by Gerard and Roberts (2004) reported that microwave thermal processing has been successful in increasing the concentration of phenolic compounds in fruit juices. Similarly, another study stated that, in apples, microwave drying was effective in increasing the phenolic content and decreasing the browning activity (Vadivambal and Jayas, 2007). It was also stated that microwave processing leads to significant reductions in the color, appearance, antioxidant capacity and polyphenols of vegetables. The results further confirmed major losses in the antioxidant and total phenolic content (Jaiswal and Abu-Ghannam 2013).

2.3.1.1.3 Meat and meat products

Impact on sensory properties: The spoilage of meat occurs due to the growth of microorganisms, which makes the meat unfit for human consumption. Microbial spoilage may result in changes in the odor, quality, color and pH (Mills et al., 2014). It has been outlined that the sensory characteristics of meat can be greatly improved by cooking, thus making it more palatable and also improving its appearance (de Andrade et al., 2016). Further, it was reported that microwave heating enabled better sensory and physicochemical properties, like desirable flavor and juiciness of meat stored in for a month (Augustyńska-Prejsnar et al., 2018).

Impact on nutritional and functional value: Meat contains high biological value protein (Battaglia Richi et al., 2015) with essential amino acids and good amino acid digestibility (de Gavelle et al., 2018); and for this reason researchers are analyzing the consequences of microwave cooking on its nutritive and functional value.

During higher heating temperatures, microwave technology has been reported to cause less denaturation of meat proteins, as compared to traditional methods (Hassoun et al., 2020; Roberts and Lawrie, 2007). A study by Domínguez et al. (2015) stated that microwaved samples of cooked meat showed higher protein index than raw meat. Also, microwave heating preserves lysine more than conventional cooking (Jouquand et al., 2015). In addition, microwave heating produced less compositional changes than those produced by other methods.

Furthermore, Dong et al. (2021) reported the effects of microwave processing at high temperature on in-vitro protein digestibility. A significant decrease in the in-vitro protein digestibility and total soluble protein content was observed. It seems that the effect of microwave thermal processing protein hydrolysis depends on the origin of proteins as well as the processing conditions.

2.3.1.1.4 Milk and milk products

Impact on sensory properties: Milk and milk products are a good source of essential nutrients, like protein, calcium, and vitamin D (Borad et al., 2017). Before consumption, milk is invariably processed; and it has been reported that microwave heating does not affect the taste and smell of milk (Valero et al., 2000).

Impact on nutritional and functional value: Many studies have reported that microwave heating did not induce protein degradation at low temperatures; however, at higher temperatures, significant changes were observed in the secondary and tertiary structures (Gomaa et al., 2016; Izquierdo et al., 2005; Munir et al., 2019). Similarly, a previous study by Munir et al. (2019) also reported that high power microwave treatment denatured whey proteins rapidly by deactivating enzymes. However, low-power controlled irradiation has been reported to improve enzyme activity.

2.3.1.2 Advantages and disadvantages of microwave processing

Processing of foods using microwave technology has several advantages compared to conventional heating techniques as reported by some researchers (Salazar González et al., 2012; Chandrasekaran et al., 2013; Vadivambal and Jayas, 2007; Kalla and R., 2017).

Advantages:

1. Volumetric heating: In a microwave, volumetric heating takes place, which allows homogeneous thermal heating of the food material. Moreover, very high temperatures can be attained with simple and relatively low power microwave generators.
2. Time saving: It significantly results in lesser thermal processing times, resulting in more fresh-like taste and improved texture in food products.
3. Better quality of Food: Microwave heating is more efficient in retaining the nutritional components of the foods. And as the target temperature is attained fast, thus reducing the harmful effects of microwave heating on food.
4. Energy efficient: It has been reported that energy efficiency is higher with microwaves than with conventional heating. Microwave radiation helps in higher energy savings as microwaves heat up the sample and not the equipment, and thus there is less energy consumption.
5. Clean energy: It does not produce harmful gasses or any pollutants and waste products.
6. It is suitable for sensitive, high viscosity, and multiphase fluids.
7. Small space requirement, low energy consumption, and maintenance cost.

Disadvantages:

1. The microwave oven uses electricity, which is high in cost.
2. Microwave heating is not suitable for bread baking as staling of bread may occur quickly as compared to conventional techniques.
3. The erratic temperature dispensation can result in cold and hot spots mainly in solids and semi-solid products.
4. Microwave heating may cause uneven heating on the edges which may result in off-flavors.

2.3.2 Microwave in food preservation: Techniques and their principles

Preservation of food is the technique of handling and treating food materials, with the aim of preserving its qualities, and reducing the spoilage to prevent food borne illnesses. It is accomplished by controlling or inhibiting the external contaminants and/or internal enzymatic reactions that could alter the nutritional and organoleptic quality of food material (Chemat et al., 2017). New and innovative technologies are continuously researched so as to maintain the nutritional properties of the food materials. Microwave heating is an emerging alternative to direct heating conventional methods (Laguerre and Mazen Hamoud-Agha, 2020). Microwaves can preserve food by drying, pasteurization, and sterilization of foods using electromagnetic radiation.

2.3.2.1 Microwave-assisted drying

Drying is the oldest method used to preserve food materials and prevent microbial degradation. It is used to provide shelf stability to food products by reducing the water activity, which also makes the food material microbiologically and enzymatically stable. Food dehydration or drying is widely used for economic purposes by converting the food into light weight products in order to minimize the transport costs (Woodroof and Luh, 1986).

In drying of food materials, microwave technology is quite different from that of conventional methods. In the latter, the heat is transferred to the inward from outside, thus moisture initially gets removed from the surface. On the contrary, during microwave drying, heat is generated in the interior of the material, thereby resulting in a faster rise in temperature. This method utilizes hot air with microwave heating to increase the drying rate.

In the food industry, microwave-assisted drying (MWD) is used as an efficient and innovative technology for in-depth volumetric heating and shorter processing time, which results in less consumption of energy (Holtz et al., 2009; Tulasidas, 1995). Further advantages include less material used up in production, faster and accurate electronic control, less space requirement, as well as improved quality of the product in terms of color, texture, nutritional value, and better rehydration properties of the dried products (Ji et al., 2012; Arikan et al., 2012; Mujumdar, 2015).

Microwaves are often combined with other methods too, like freeze and vacuum drying as they are relatively time-consuming. Microwave drying may cause physical damage to the food products like off color, scorching, and uneven heat distribution. The use of vacuum in microwave drying is considered to be an alternative for reducing the above mentioned shortcomings (Gunasekaran, 1990). Because of the rapid drying rate, the aroma is retained and the shrinkage is less (Erle, 2005). Furthermore, due to the air being excluded during microwave vacuum drying, oxidation reactions are also minimized (Gunasekaran, 1990), thereby maintaining the color, texture and flavor of the dried product (Chandrasekaran et al., 2013).

Further, commercial applications of microwave-vacuum dehydration have been studied with an array of products like apples and strawberries (Erle and Schubert, 2001), banana (Mousa and Farid, 2002), dried garlic slice (Cui et al., 2003), and mushrooms (Argyropoulos et al., 2011) among others.

Freeze drying is a low-temperature dehydration technique, and is used to preserve heat-sensitive foods. In microwave freeze drying, heat can be applied simultaneously throughout the sublimation process under vacuum conditions (Laguerre and Mazen Hamoud-Agha, 2020). This process preserves the physical structure of the product which minimizes the losses of valuable compounds, and also maintains the porosity of the matrix, allowing the food product to rehydrate quickly (Beaudry et al., 2004). However, this technique is also quite expensive, as the cost of microwave freeze drying has been estimated to be 5–10 times more than hot air drying, thereby limiting its use to high value products (Duan et al., 2007).

2.3.2.2 Microwave pasteurization and sterilization

Pasteurization is a method which utilizes mild heat to destroy pathogenic microorganisms, and to deactivate enzymes to make food safe for human consumption. In pasteurization processes, temperatures, and treatment time vary from a few seconds to an hour.

During conventional pasteurization, continuous heat-transfer mechanisms are employed; whereas, in microwave pasteurization, electromagnetic waves with frequency between 2450 MHz and 915 MHz (Swart et al., 2003) are used to heat the food directly. The energy required to heat a particular food is determined by the amount of energy lost in the process, owing to the material packaging; which reduces the microwave's penetration capabilities (Swart et al., 2003). Another important principle is that microwave energy produces heat through its effect on dipolar molecules such as water molecules. It causes the molecules to oscillate, thereby inducing intermolecular friction which produces heat. Therefore, in irradiated foods, heat is produced from the inside.

Microwave pasteurization has been applied mainly to milk and fresh fruit juices to eliminate pathogenic microorganisms. However, the main drawback of this treatment is that it does not eliminate bacterial spores, thus making the product unstable at room temperature. Although, the shelf life can be extended by 2–6 weeks in refrigerated storage settings.

Sterilization process is more severe than pasteurization, as it provides commercial sterility to the food products, thus imparting long-term shelf stability (Ahmed and Ramaswamy, 2007). In conventional heating of foods, heat is transferred mostly through conduction from the surface to the center, which is frequently referred to as the cold point. This may lead to overcooking of the food's surface as more severe conditions are required to reach the goal temperature at the cold point; however, microwave sterilization, due to its direct and volumetric heating, overcomes the limitations of traditional heating (Laguerre and Mazen Hamoud-Agha, 2020).

Microwave sterilization or pasteurization can be described by different techniques, like electroporation, selective heating, cell membrane rupture, and magnetic field coupling (Kozempel et al., 1998). For example, during electroporation method, the electric potential causes the cell membrane to leak due to development of pores. Further, in the selective heating mechanism, only the microbes are destroyed as they reach a higher temperature which kills them.

Numerous studies have shown the benefits of employing microwave technology to pasteurize and sterilize food (Guo et al., 2017; Salazar-González et al., 2012; Laguerre and Mazen Hamoud-Agha, 2020; Zhou et al., 2010). However, there are certain limitations, like, uneven distribution of temperature, which results in hot and cold spots in microwave-heated products (Vadivambal and Jayas, 2010). Another drawback of microwave sterilization is the nonavailability of actual temperature profiles of the material (Ahmed and Ramaswamy, 2007). During microwave heating, the real temperature distribution cannot be estimated on the product, as the heating may vary and might also change during the treatment. Thus, the researchers have observed inconsistent results in this field.

2.3.2.3 Superiority of microwaves over conventional heating techniques to preserve food quality

Over several decades, microwave heating technology has provided us with vast applications in the field of food processing. The various applications include drying, sterilization, pasteurization, thawing, baking, tempering and so on (Gupta and Wai Leong Eugene, 2007; Metaxas and Meredith, 1983). As mentioned before in this chapter, microwave technology has gained significant recognition in the processing of food materials, chiefly due to its ability to achieve higher temperatures, lesser thermal processing times, uniform volumetric heating, and low maintenance (Salazar-González et al., 2012).

Furthermore, studies have shown that microwave cooking retains the food quality to a larger extent than traditional heating procedures

(Vadivambal and Jayas, 2007; Salazar-González et al., 2012); and it also enhances the shelf life of the products (Salazar-González et al., 2012).

The volumetric heating feature of microwave oven offers many advantages over traditional heating; these include uniform heating rate, reduced cooking time, selective heating, high efficiency, among others (Vadivambal and Jayas, 2007; Salazar-González et al., 2012; Chandrasekaran et al., 2013). In addition, faster heating and the quality of the food are the most acknowledged features of microwaves (Gowen et al., 2006; Huang et al., 2007; Queiroz et al., 2008).

2.3.3 Microwave-assisted extraction

Microwave-assisted extraction (MAE) is a technique that uses microwave radiation to heat solvents in contact with a sample for the extraction of analytes (Sparr Eskilsson and Björklund, 2000; Jocelyn Paré et al., 1994). It is considered a relatively newer technique for the extraction of soluble substances using microwave radiation (Jocelyn Paré et al., 1994).

The use of microwave energy in chemical analysis emerged in 1975 (Abu-Samra et al., 1975). It was utilized to digest biological samples followed by elemental analysis. However, the development of MAE started much later, and was initially used for the environmental analysis by extraction of organic compounds like organometallic pollutants (Ganzler et al., 1986). Over time, the applications of this technology increased and expanded to other fields such as food, clinical and pharmaceutical products.

2.3.3.1 Principle of microwave-assisted extraction

MAE is a process that uses nonionizing electromagnetic radiation to disrupt the structure of cells in the sample matrix (Leonelli et al., 2012), thus promoting the extraction of the compounds of interest from the sample.

Contrary to the traditional heating methods, microwaves heat the solution directly. Thus, thermal gradients are minimized and the heating rate is faster (Veggi et al., 2012). It also offers many other advantages like reduced equipment size and increase in extracted yield (Esquivel-Hernández et al., 2017; Chemat et al., 2015). The time for extraction of bioactive compounds in case of MAE is also less than conventional methods (Esquivel-Hernández et al., 2017; Michalak and Chojnacka, 2014).

Extractions are performed in either closed or open vessels, where the sample and solvent are mixed and then subjected to the radiation. In MAE, the time taken is less because the sample is directly heated, while with traditional techniques, the extraction takes time as it is required to heat up the vessel before it can be transferred to the sample (Veggi et al., 2012).

2.3.3.2 Applications of microwave-assisted extraction

Microwave-assisted extraction technology has been mostly utilized for the extraction of complex samples, but its applicability is continuously growing to other areas of research as well, such as extraction of bioactive compounds and analysis of food (Llompart et al., 2018).

Many studies have reported on the microwave extractions of organic pollutants from food (Chung and Chen, 2011; Wilkowska and Biziuk, 2010; Guo et al., 2019). Further, MAE has also been used in analysis of food substances, such as green extraction of phytochemicals and natural compounds (phenolic compounds, carotenoids, ascorbic acid and aroma compounds) obtained from plant sources (Chan and Ismail, 2009; Maran et al., 2015).

The current scenario of this method gives a promising sign for the utilization of green extraction in the food industry, and more research is expected to enhance the outline and scale up of the novel extraction frameworks.

2.4 Applications of microwave technology for food processing and preservation

Microwave technology is employed quite extensively in the food industry. Here, the relative benefits of microwave energy make it a suitable and preferred alternative to other traditional methods of thermal treatment. Some of these advantages have been discussed previously, namely, reduced processing cost and time, high degree retention of the sensorial attributes, and resulting in very few alterations in the food nutritional value. Hence, a number of applications have been developed for microwave technology on an industrial scale. These are briefly discussed below.

2.4.1 Baking

A combination of convection and radiation from the oven walls, transmit heat out of the heating media to the surface of the product in traditional baking, which is subsequently routed to the center. But, in microwave baking, the food is surrounded by ambient air that is not heated by microwaves. In fact, when microwaves interact with charged particles as well as with polar molecules, heat is generated inside the product and transmitted through the food substance.

As discussed previously, heating within a microwave is induced by the interaction of dielectric material with alternating electromagnetic fields. The potential of microwave energy to provide more consistent and speedier volumetric heating than conventional heating assists in microwave baking. Further, the direct coupling of microwave energy without having to heat the surrounding air shortens the process time. These, combined with dielectric properties and penetration depth, all contribute to the utilization of microwaves for baking.

The major benefit gained from microwave baking is the speed. The rapid heat generation in a microwave oven lends itself to a faster heating time. A second benefit is that the end product offered by microwave baking has a superior nutritional content than achieved through traditional process. The disadvantages are precise control of cooking time, and browning and crispness that does not occur in microwave processed food. Other quality-related issues noted in such products are rapid staling, a stiff, tough and firm texture, and a resultant dry product (Sumnu, 2001). Flavors generated due to browning are also not observed in products obtained from microwave baking. These shortcomings can be explained by the fact that, browning processes and starch gelatinization, for example, are kinetic reactions that are affected not only by temperature but also by baking time, and in the rapid heating process of microwave baking, there may not be enough time for these kinetic reactions to be fully realized. Hence, the crucial process reactions in baking, namely enzyme conversion of starch, starch gelatinization, sufficient expansion of dough/batter, browning, creation of crust, and the ultimate formation of the dough/batter into a rigid crumb arrangement might not achieve completion in the course of microwave baking (Yolacaner et al., 2017).

These challenges are overcome by typically combining microwave baking with conventional baking. In such combined approaches, low levels of microwave heating is used and the process parameters of the accompanying heating technology are adjusted to achieve the requisite qualities of crispness, color and softness, among others. Heating techniques including hot air and infrared are suited for such combined approaches, as their functionality is achieved at a time and spatial location different from that of microwave heating, thereby allowing for targeted heating. This synergy imparts a superior spatial uniformity in addition to a surge in the general heating rate (Datta et al., 2005).

Further research is needed to fully comprehend the interaction of different ingredients with microwave energy and the impact of integrated microwave technology on quality characteristics of baked products such as cookies, cakes, muffins, etc.

2.4.2 Blanching

The process of blanching is a preparatory one, carried out ahead of the procedures such as drying, frying, canning, and freezing being undertaken (Xiao et al., 2017). It entails the food material being submerged in hot water or steam. It is the process of inactivation of heat-resistant enzymes, which also provides the benefits of color preservation, product cleansing, reduction in the microbial load, exhaustion of gas from plant tissue, product preheating (prior to processing) (Shaheen et al., 2012), delivering carotenoids, and ensuring they become more bioavailable and extractable (Arroqui et al., 2002).

Traditional blanching techniques have certain disadvantages like leaching of soluble nutrients and vitamins, softening of tissue, and waste water generation. This has led to the emergence of microwave blanching which is acknowledged as a technique that brings down the time and energy required to inactivate enzymes due to the volumetric heating-based treatment of solid foods for a short time and at high temperatures, which is critical in the retention and preservation of heat-sensitive nutrients and other bioactive food components (Dorantes-Alvarez et al., 2017).

Several studies have been undertaken to analyze the results of microwave blanching of foods, like mushroom (Devece et al., 1999), carrots (Kidmose and Martens, 1999; Lemmens et al., 2009), peas (Lin and Brewer, 2005), peanuts (Schirack et al., 2006), pepper (Dorantes-Alvarez et al., 2011), sweet potato (Liu et al., 2015), red bell pepper (Wang et al., 2017), red beet (Latorre et al., 2013), shiitake mushrooms (Lentinula edodes) (Chen et al., 2018), mango (Isabel et al., 2018), and green beans (Ruiz-Ojeda and Penas, 2013). The results have led to the understanding that microwave blanching of foods offers several advantages over conventional blanching, such as reduced leaching of vitamins, minerals and sugars, lesser requirement of energy and water and the shorter time required for processing.

2.4.3 Drying

Drying is an energy-intensive process that extracts moisture from a product to a predetermined level. The chief objective of drying food is to preserve it. Microwave drying entails different heating mechanisms from that of conventional drying. During microwave heating, the generation of internal heat dispenses vapor pressure in the interior of the substance, which pushes any moisture present, to the surface, as a result of which case hardening is avoided (Turner and Jolly, 1991). Volumetric heating and internal generation of vapor is also a benefit of microwave heating as it leads to a decreased drying time which consequently enhances the product quality. Thus, microwave drying offers significantly enhanced drying rates and improved product quality, without an accompanying rise in the surface temperature. For products such as cabbages, mushrooms and carrots, all of which have high moisture content, this technique is particularly suitable as water's high dielectric property is capable of swiftly absorbing the microwave radiation (Prakash et al., 2003).

There are several heating mechanisms for drying food materials, namely hot air-drying, vacuum-drying, freeze-drying, solar drying, and osmotic drying. Microwave-induced heating may be successfully combined with all these techniques to achieve better results of drying such as increase in the quality of the product by preservation of the sensorial attributes and nutritional values.

In drying processes, microwaves can be operated as either the sole or secondary source of heat generation. During this process, they can be utilized in several ways, including continuous, intermittent/pulsed, or temperature-controlled applications. Air-drying, vacuum-drying, and freeze-drying are the three most notable and significant uses of microwaves as the source of heat in such drying procedures (Gaukel, 2017).

2.4.3.1 Microwave-assisted hot air-drying

During the combined process of microwave and hot air drying, airflow created while using microwaves as the heat source facilitates in elimination of residual moisture from the surface of the food substance, whereas the volumetric heat provided by the microwaves results in the moisture moving quickly from the interior to the exterior of the food substance (Gaukel, 2017). This combination decreases drying time, as demonstrated in several studies: a 25%–90% reduction in carrot slices (Prabhanjan et al., 1995), 64.3% in bananas (Maskan, 2000), 40%–89% in kiwifruit, 80% in blueberries (Zielinska et al., 2015), olives (Gogus and Maskan, 2001), asparagus (Nindo et al., 2003), 60%–82.5% reduction in cranberries (Beaudry et al., 2003), upto 80% in garlic (Sharma and Prasad, 2006), and 75% in oyster mushrooms (Bhattacharya et al., 2015), pumpkin (Alibas, 2007), and upto 70% in collard greens (Alibas, 2009).

2.4.3.2 *Microwave vacuum-drying*

Microwave vacuum-drying can be understood as a very effective process that allows for decreased processing times producing a product with enhanced quality. The moisture-elimination occurs at temperatures below the product's boiling point in ambient conditions by creating, in the absence of oxygen, a vacuum at a relatively reduced drying time, and resulting in the prevention of product oxidation (Orsat et al., 2017). This results in maintaining sensorial attributes of the substance all through the process of drying. The reduction of the water's boiling point causes the water molecules to diffuse and evaporate due to the reduced pressure, and water vapor collects at the product's surface. Some studies on microwave-assisted vacuum drying are: apple and kiwi (Kiranoudis et al., 1997), bananas (Drouzas and Schubert, 1996), beetroot (Figiel, 2010), cranberries (Yongsawatdigal and Gunasekaran, 1996), grapes (Clary et al., 2005), and mango (Pu and Sun, 2015, 2016, 2017).

2.4.3.3 *Microwave freeze-drying*

In freeze-drying (FD), the substance is first frozen, then dried in a vacuum, by use of sublimation. Microwave freeze-drying (MFD) is an efficient drying method which primarily operates below water's freezing point. Freeze-drying of foods and other products is an expensive technique. With progress in research, this method can be replaced by MFD technology, which requires less time for processing and consumes lower amounts of energy (Orsat et al., 2017).

2.4.4 Frying

Deep-fat frying is a popular, quick, and easy method of cooking. Under this technique, the food substance is placed in hot oil (170°C–190°C) resulting in a thorough removal of moisture from within the food by vaporization and its transmission through the surface of the substance (Ni and Datta, 1999). The technique has many appeals such as efficiency of heat transfer and specific organoleptic characteristics that are preferred by consumers, namely color, flavor, palatability and texture (a crisp but porous structure). These are an outcome of several chemical reactions, including gelatinization, browning and denaturation, among others (Tangduangdee et al., 2003). But the technique has some undesirable effects such as over-absorption of oil, the presence of acrylamide (especially with regard to starch-heavy food items like potatoes) in the resulting fried item, and the breakdown of lipids that takes place during the process of frying.

The use of microwave energy allows for raising the temperature inside the material, and when combined with the other source of heat, that is, hot oil, may create products of equal or superior quality, at costs that are similar to, or less than the conventional procedures. Microwave-induced heating has already been studied as a viable substitute for the frying process, with research focusing on the benefits of reduced frying time (Sahin et al., 2007), and deriving fried foods with decreased levels of oil absorption (Oztop et al., 2007), reduced lipid degradation (Gharachorloo et al., 2010), and reduced content of acrylamide (Sahin et al., 2007; Barutcu et al., 2009). These studies need to be built upon to eventually adopt the microwave assisted frying technique on a larger scale, especially considering the public health concerns surrounding conventional deep-fat frying.

2.4.5 Tempering and thawing

Before being sent for further processing, frozen food products are tempered or thawed. The temperature of the product is the main distinction between the two techniques. When a frozen product is thawed, the temperature is kept close to 0°C, while when tempering, the temperature is kept lower than the freezing point (between −5°C and −2°C). Furthermore, thawing is the technique whereby frozen materials from ice are released by allowing them to remain at room temperature (above 0°C), in order for mechanical processes including slicing and cutting, as well as molding of the food material can occur. Conventional thawing methods have some inherent disadvantages, namely the requirement of a large space and a greater duration of time for thawing. These have the possibility of leading to chemical and biological degradation of the product. Microwave thawing works by directing electromagnetic waves at the food material through waveguides, rather than using conductors or electrodes. It takes less time, but it has the disadvantages of possibly uneven or runaway heating. Upon comparison with conventional thawing (with water bath), microwave assisted thawing has been observed to be a superior and more effective technique which produces a final product of better quality and with a more homogeneous structure.

Tempering is the process of raising the temperature of a substance, followed by completely thawing it (Shaheen et al., 2012). The process also helps avoid concerns like drip losses and bacterial development that arise with thawing (James and James, 2002). Microwave-based tempering is among the most productive and viable implementations of microwave heating. It is utilized to develop a consistency in the distribution of temperature without any accompanying heat-related abuse of the substance and with lower drip losses (Orsat et al., 2017). It is possible to achieve successful tempering within minutes by using a microwave system, as opposed to the longer duration of 1–14 days that are typically required by conventional industrial air tempering systems (James et al., 2017).

2.5 Research and development toward novel and future applications

Microwave technology is recognized as one of the emerging novel technologies for food processing. In surveys conducted by a team of researchers on food experts' views on commercial applications of emerging and novel technologies within the food processing industry and their development potential, high-pressure processing and microwaves emerged as the most relevant and vital technologies of recent times (Jermann et al., 2015). In the preceding sections, we looked at how microwaves are operated for a number of varied purposes in the food industry—-baking, cooking, blanching, drying, pasteurization, extraction, tempering, and thawing. Beyond the food industry, electromagnetic waves in the microwave frequency band also have applications in a wide array of fields such as communications- cellular phones, TV broadcasts, navigation of ships and airplanes, radar for weather observation, and aerospace communications; pharmaceuticals, biotechnology, metallurgy, ceramics, plastics, forestry, petroleum, and chemicals.

This section discusses the industrially available microwave applicators and recent developments in applications of microwave technology across the food industry.

2.5.1 Industrial microwave applicators

Any industrial process technology has to optimize process-specific requirements to be able to achieve the maximum rate of energy and production efficiency. A key feature of that is the availability of the right equipment. A microwave system generally comprises three fundamental components: the microwave source, the applicator, and the waveguide. Beyond these components, other elements are the transformer, oven cavity, and the material. Step-up transformers are used to power the microwave source. The microwave cavity contains the food material which requires heating and consists of

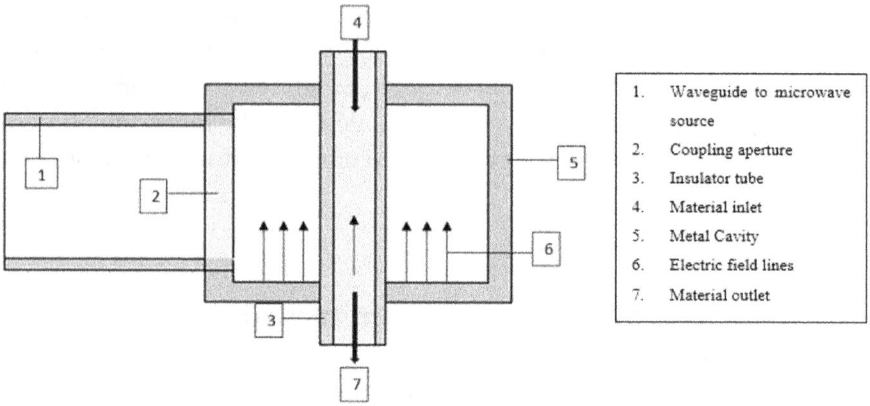

Figure 2.3 Schematic diagram of a single-mode applicator *(Source: Adapted from Regier et al. (2017)).*

reflective walls to raise and optimize the potency of the heating process (Muthukumarappan and Swamy, 2019).

Industrial microwave systems have commonly used magnetron tubes as the microwave source. Notably, the use of solid-state microwave generators is fast emerging as an alternate microwave source for industrial systems. Waveguides are the hollow conductors used for guiding electromagnetic waves of higher frequencies (such as microwaves). Applicators can be understood as devices that are used to heat food samples by subjecting them to a microwave field within a modulated environment. These applicators can be conveyor operated, batch, or a combination of both. The most popularly used applicators within microwave systems are the single-mode and multimode applicators, and traveling wave devices.

2.5.1.1 Single mode applicator

A feeding waveguide with a tuning aperture combined with a moderately compact microwave resonator of proportions in the range of the wavelength, make up a single mode applicator (Orsat et al., 2017; Fig. 2.3). The resonator generates heat by producing a specific electric field pattern. The material is held in a single-mode cavity (it may be rectangular or cylindrical in shape) in specific points where the electric field is acute enough to achieve maximum heating. As it is a single-mode cavity, within a predetermined space of a particular volume only the singular mode of propagation is permitted, which allows microwave energy to be distributed efficiently. Single mode applicators have some advantages. These are listed below:

- It is possible to achieve high fields

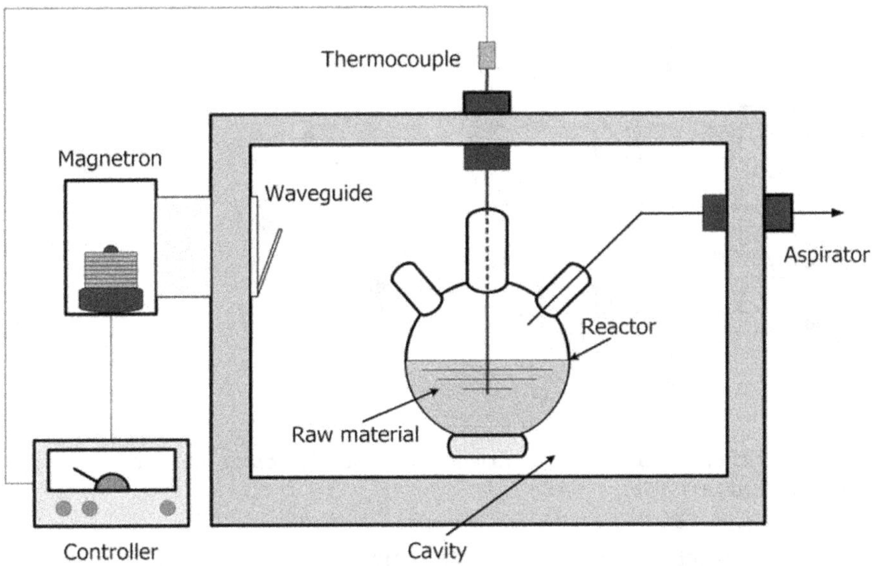

Figure 2.4 Schematic diagram of a multimode applicator *(Source: Adapted from Karisma et al. (2020)).*

- Applicators can function in standing or traveling wave modes
- Fields are well defined
- Fields can be tailored to match the product geometry
- Both low and high-loss materials can be heated by this applicator
- Continuous product flow is supported by the applicators
- The process is highly efficient

However, there are also certain challenges associated with single mode applicators. The small size of the applicator means that the breadth of the food substance needs to be less than half of the arriving microwave energy's wavelength. The most simple and widely used single-mode cavity functions at the TM_{010} mode (Regier et al., 2017). It is also physically the most compact.

2.5.1.2 Multimode applicator

Multimode applicators are used most commonly in industrial microwave systems. They are used for processing bulk materials/samples, whose general proportions are too big (greater than the working frequency's wavelength) to allow for use in a single-mode oven. The applicator generally looks like a rectangular metal box within which the food is kept (Fig. 2.4). With the help of a waveguide, the microwave energy penetrates the applicator and travels across several reflections off the walls of the cavity. These reflected waves of

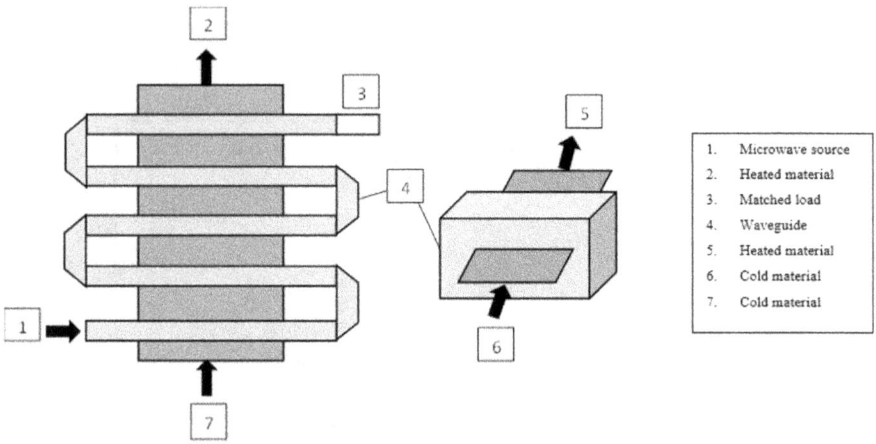

Figure 2.5 Schematic diagram of a traveling wave device *(Source: Adapted from Roussy and Pearce (1995)).*

energy then reach the interface of the sample and establish an electric field distribution both inside the cavity as well as the sample. This supports stable modes of conduction of the electric field at different locations and is hence referred to as a multimode applicator.

Multimode applicators are observed to be particularly well-suited for substances which have a high factor for dielectric loss and which engage greater volumes of the applicator (Orsat et al., 2017). They are adaptable to batch or continuous product flow, but continuous mode is more appropriate for carrying out industrial applications as high yields are required.

One challenge of multimode applicators is the difficulty in achieving uniform heating and to overcome this, a number of techniques can be employed. The employment of metallic mode stirrers, for example, ensures that all potential modes are stimulated. Others include product motion, scanning of surface to aim the energy at specific points of interest, and, in some situations hybrid heating through conventional heating to reduce surface losses.

2.5.1.3 Traveling wave device

When the food sample to be treated by heat is presented through wall slots and the waveguide is stopped by a matched load, it can be operated as an applicator for microwave-based heating (Fig. 2.5; Regier et al., 2017). A waveguide is set up with the traveling wave applicator in such a manner that the material to undergo processing travels through one end while the source (microwave) is located at the other. As it passes along the waveguide, the sample absorbs microwave energy. The conveyor belt speed is controlled such that the material goes through the waveguide at a steady speed. As a result, every point faces an equal amount of energy of the microwave field, and a singular pass of the substance through the waveguide can provide a consistent and constant transmission of microwave energy to the substance.

2.5.2 Novel and future applications

The following sections highlight some of the latest developments in applications of microwaves—-particularly in the food industry.

2.5.2.1 Combined and hybrid microwave-assisted drying

Microwave drying has piqued the interest of researchers for several years. This approach has become popular due to enhanced drying rates and product quality when compared to the use of hot air. However, due to limitations such as high investment costs and difficulties regulating the temperature of the substance at the last step of the drying procedure, other approaches—-combined and hybrid MWD—-have emerged (Zhang et al., 2006; Zielinska et al., 2019).

Hybrid MWD can enhance the quality attributes of food materials. Some examples of hybrid drying technologies are: microwave-assisted inert medium fluidized bed drying, microwave-assisted vacuum drying, microwave-assisted atmospheric freeze drying, and microwave-assisted vacuum freeze drying.

A combination of microwaves with spouted bed drying has been known to refine the otherwise uneven heating of dried particles. Use of microwave-assisted inert medium fluidized bed drying can achieve a decreased drying time and uniformity in distribution of the temperature (Zielinska et al., 2019).

In addition to these hybrid technologies, to amplify the benefits of distinct drying techniques various combined MWD methods have been developed. Some of these include microwave vacuum drying in combination with: hot air convective drying, osmotic dehydration, and a combination of all three together.

MWD systems, both hybrid and combination, are yet to complete the initial phase of development and are mostly employed on a pilot-scale or in a research capacity. Research in key areas such as advancing combined and hybrid MWD processes, evolution of material properties during hybrid and combined drying, validation of the best combinations of heat sources, types of dryers, and order of combination, as well as the development of appropriate equipment, will determine the future scope.

2.5.2.2 Microwave-assisted thermal sterilization

Microwave-assisted thermal sterilization (MATS) is an emerging technique that integrates the energy from long wavelength microwaves (915 MHz) with hot water immersion to sterilize the food material inside polymeric packages (Soni et al., 2020). It effectively combines thermal and microwave energies to offer microbial inactivation at par with that offered by thermal sterilization. It also reduces the damaging effects of heat on the elements of the food.

Sterilization by MATS has marked benefits over conventional sterilization. The use of hot water as an interactive medium to heat improves the nonuniform heating that may be caused by conventional microwaving. In comparison to traditional sterilization, combined microwaving with hot water shortens the amount of time the food is exposed to fatal temperatures. The polymer packaging designed for conventional retorting has been reported to work better in the use of MATS (Tang et al., 2014). Flexible packaging would also enable both liquid and solid food to be handled as full meals sealed within multicompartment trays, allowing multiple types of food to be sterilized in the same packet (Tang, J. 2015).

Research was undertaken to study the longevity of conventional-retorted chicken breast against chicken breast treated by MATS before and after storage at 37.8°C over a period of 6 months. The findings of the study revealed a generally higher quality and flavor score for the MATS treated product (Zhang et al., 2016).Ultimately, the MATS technology lends itself to better nutrient retention, reduced sensory effects, and increased storage life.

The capacity to manage heat-sensitive foods, clean label credentials, obtaining fresh-like quality, application of flexible packaging, as well as the possibility for mathematical models to resolve validation challenges instead of cumbersome laboratory assays, are some of the upcoming and possible fields of research with MATS approach (Soni et al., 2020). Research should also focus on addressing the challenges of packaging material recycling and manufacturing costs, as well as the potential for food safety concerns due to the risk of packaging material contaminants migrating into the food held inside.

2.5.2.3 Microwave-assisted pasteurization

Similar to the MATS technology, a microwave-assisted pasteurization system (MAPS) comprises four segments: preheating, microwave heating, holding, and cooling, each integrated with a temperature control-equipped water circulation system (Bornhorst et al., 2017). The main distinction is that MAPS uses flowing water at 70°C–90°C in the preheating portion, whereas MATS uses overpressure to heat the water over the boiling point of 100°C. Trays or pouches containing the foods travel across the four sections of the system in stainless steel carriers. As the use of MAPS does not require overpressure, it is more compatible and much easier to utilize in the development of ongoing commercial applications, as well as being more cost effective for usage in relatively smaller businesses.

The biggest challenge in developing MAPS was in determining the possible location of a cold spot in a specific product, and significant progress has been made in the creation of new tools to help with this. Various model food systems for determining the pattern of heating in microwave pasteurization systems have recently been created (for processing temperatures of 70°C–100°C). This can then be studied in combination with a computerized vision system to enable for the quick detection of the hot and cold regions. Mashed potato, egg white, and gellan gel with M-2 chemical marker precursors are among the model foods that have been developed thus far (Tang et al., 2018).

Microwave-assisted pasteurization is particularly viable as a processing option for production of chilled or frozen ready-to-eat meals. Further research is needed to make the application commercially viable on a large-scale.

2.5.2.4 Microwave-assisted extraction (MAE) and integrated technologies

Among the new and upcoming methods for sorting, segregating and extracting various components of foods, microwave-assisted chemical extraction is the most diffused method. This is because the method is highly efficient, eco-friendly, economical, with a notable reduction in recovery time and a relative ease in the assembly of the technique at both small and large scales (Ekezie et al., 2017). Microwave-assisted extraction has traditionally been applied for extricating complex samples, but with continued research, its relevance, applicability and scope has forayed to include analysis of the environment, food assessment, and for drawing out bioactive substances from natural sources.

Some of the integrated technology enhancements have been briefly discussed below.

2.5.2.4.1 Microwaves and negative pressure cavitation (MW-NPC)

This type of extraction (NPC) is typically based on the formation, proliferation, and disintegration of an infinite amount of minuscule vapor bubbles caused by negative pressure at either the liquid or the liquid-solid interface, resulting in a high-energy output. A combined microwaves and negative pressure cavitation system, simultaneously employing the two techniques, intensified the output for bioactive compounds 20-O-galloyl, hyperin and chimaphilin from pyrola (Yao et al., 2015) and phenolic compounds from pyrola (Zhang et al., 2013). It was found that extraction time can be shortened by as much as 40%. Thus, the combined MW-NPC system has proven to be highly efficient, while also being environmentally friendly and economic. It is applicable for various extractions, particularly for the recovery of thermolabile compounds.

2.5.2.4.2 Ultrasonic microwave-assisted extraction (UMAE)

Ultrasonication of the microwave-assisted extraction has been exhibited to contribute mechanical and thermal effects, which lead to lower energy consumption, reduced time for extraction, lower levels of solvent consumption, and a more enhanced extraction output as compared to those obtained from conventional extraction techniques or even these two techniques used individually. Thus, this combined technique cumulatively boosts extraction efficiency.

The studied compounds include pectin from pomelo peels (Liu et al., 2017); polyphenols, flavonoids, triterpenoids, and vitamin C from *Clinacanthus nutans* (Qun et al., 2017); juglone from walnut green husk (Xu et al., 2016); polydatin, resveratrol, emodin-8-O-β-D glucoside and emodin from *Polygonum cuspidatum* (Gao et al., 2016); and glucosinolates, vitamin C, phenolics and sulforaphane from cabbage outer leaves (Pongmai et al., 2015), among others.

2.5.2.4.3 Microwave-assisted subcritical and supercritical fluid extraction

The characteristic quality of a substance that is gained when it is exposed to temperatures and pressure levels over its critical point is known as supercritical state (SC). Their solvation capacity, viscosity, and surface tension make SC fluids ideal for extracting oil and fat components from food. Subcritical water extraction uses water as an extractant that is superheated to values from 100 to 374°C at extreme levels of pressure to ensure that liquids remain in the same state (Matusiewicz and Ślachciński, 2014). It is particularly suited for extracting both polar and nonpolar chemical substances.

Both these technologies, integrated with microwave assisted extraction have been shown to increase yield and save energy thereby increasing efficiency in several studies. The studies are listed in Table 2.1 for further

Table 2.1 List of research conducted on microwave-assisted subcritical and supercritical fluid extraction.

Compound	Matrix	Technology	Author(s)
Polyphenols	Roselle seeds	MW-SWE	Yusoff & Leo 2017
Polyphenols	Defatted rice bran	MW-SWE	Wataniyakul et al., 2012
inorganic-metallic constituents	Several	MW-SWE	Matusiewicz & Ślachciński, 2014
Alkaloids	Gynura segetum	MW-SCCO$_2$	Lu et al., 2014
Fatty acids	Chlorella vulgaris	MW-SCCO$_2$	Dejoye et al., 2011

Source: Ekezie et al., 2017.

reference:

However, despite these forays, these particular combinations of extraction technologies are required to be studied and researched further.

2.5.2.4.4 Microwave-assisted enzymatic extraction (MAEE)

Good selectivity, sharp specificity, and the potential to activate specific reactions are all intrinsic features of enzymes. These characteristics are the reason why enzyme-assisted extraction (EAE) is conducive to effective extraction.

Application of microwave irradiation during EAE supplies energy, thereby allowing the rotation of dipoles and inducing breakdown of the cell matrix (Adetunji et al., 2017). MAEE results in rapid and successful removal of the target compound from the specified matrix, according to studies. Table 2.2 is a collection of recent studies on this subject for further reference.

2.5.2.4.5 Microwave hydrodiffusion and gravity extraction (MHG)

For extraction, an MHG device operates at atmospheric pressure and combines heat from microwaves with the earth's gravitational attraction. When solvent is not present, plant materials are set down inside the microwave reactor, wherein the heat generated by internal moisture within the matrix breaks down the cell wall and oleiferous receptacles, expelling essential oil from the plant material, which then falls from the microwave reactor due to gravity (Perino et al., 2016). Many studies have explored the advantages of MHG as opposed to simple extraction but have so far yielded uneven results.

Thus, further research has to be carried out for validation and sustainability of the integrated extraction systems. Efforts should be focused toward

Table 2.2 List of recent studies on microwave-assisted enzymatic extraction.

Compound	Matrix	Enzyme	Author(s)
Oil	*Lavandula angustifolia*	Cellulase and hemicellulase	Rashed et al., 2017
Fish Protein hydrolysates	Rainbow trout	Alcalase	Nguyen et al., 2017
Protein	Australian rock lobster shells	Alcalase	Nguyen et al., 2016
Antioxidants	*Nitraria tangutorum* juice by-products	Cellulase	Wu et al., 2015
Polysaccharides	*Pyropia yezoensis*	Carborhydrases, iscozyme, celluclast, termamyl, AMG and ultraflo	Lee et al., 2016

Source: Ekezie et al., 2017.

process optimization, resolving any associated theoretical and technical issues, and achieving commercial adaptability of such improved MAE systems for large scale application.

2.5.2.5 IR-microwave combination

Although not a new concept, the combination of infrared technology with microwaves has certain applications which are yet to be realized to their potential. Integrating microwaves with other heating technologies, such as infrared, can improve microwave heating uniformity and manage moisture mobility (avoiding the surface accumulation of moisture found in microwave heating), all while boosting heating speed. The use of infrared microwaves improves the quality and prolongs the shelf life of microwave baked goods. This combination also has potential for use in roasting of foods and achieving acceptable quality.

Despite these advancements in the application of microwaves, the technology holds more potential for research and development and experts agree on the possibility of identifying further innovative areas of application in the food industry (Orsat et al., 2017; Das and Banik, 2021). For future research, a special emphasis may be laid on the development of additional mathematical models and optimization studies which may lead to an improved understanding and utilization of the technology (Guzik et al., 2021).

2.6 Conclusion

This chapter throws light on the basics and advanced concepts of microwave technology. Microwave technology brings along its own set of the advantages and disadvantages over conventional methods. It is used in an array of food processing and preservation processes—-blanching, drying, baking, pasteurization, sterilization, thawing/tempering, frying, and microwave-assisted chemical extraction. Research indicates that microwave-based processing is superior to traditional processes in both key aspects—-quality of the food (improved texture, stronger structure, improved color parameters, and reduced weight losses) and its nutritional value (higher retention values of bioactive compounds and vitamins). With growing modernization and increased demand for quicker food processing methods, this technology has gained popularity and has shown great potential for future applications.

References

Abu-Samra, A., Morris, J.S., Koirtyohann, S.R., 1975. Wet ashing of some biological samples in a microwave oven. Anal. Chem. 47 (8), 1475–1477.

Adetunji, L., Adekunle, A., Orsat, V., Raghavan, V., 2017. Advances in the pectin production process using novel extraction techniques: a review. Food Hydrocoll. 62, 239–250.

Ahmed, J., Ramaswamy, H., 2007. Microwave pasteurization and sterilization of foods. In: Shafiur Rahman, M. (Ed.), Handbook of Food Preservation, Second Edition, Food Science and Technology. CRC Press, Florida, pp. 691–711.

Alibas, I., 2007. Microwave, air and combined microwave–air-drying parameters of pumpkin slices. LWT - Food Sci. Technol. 40 (8), 1445–1451.

Alibas, I., 2009. Microwave, vacuum, and air drying characteristics of collard leaves. Drying Technol. 27 (11), 1266–1273.

Amiri Chayjan, R., Kaveh, M., Khayati, S., 2015. Modeling drying characteristics of hawthorn fruit under microwave-convective conditions: hawthorn fruit drying under microwave convection. J. Food Process. Preserv. 39 (3), 239–253.

Argyropoulos, D., Heindl, A., Müller, J., 2011. Assessment of convection, hot-air combined with microwave-vacuum and freeze-drying methods for mushrooms with regard to product quality: characterization of dried mushrooms produced by different techniques. Int. J. Food Sci. Technol. 46 (2), 333–342.

Arikan, M.F., Ayhan, Z., Soysal, Y., Esturk, O., 2012. Drying characteristics and quality parameters of microwave-dried grated carrots. Food Bioprocess Technol. 5 (8), 3217–3229.

Arroqui, C., Rumsey, T., Lopez, A., Virseda, P., 2002. Losses by diffusion of ascorbic acid during recycled water blanching of potato tissue. J. Food Eng. 52 (1), 25–30.

Augustyńska-Prejsnar, A., Ormian, M., Sokołowicz, Z., 2018. Physicochemical and sensory properties of broiler chicken breast meat stored frozen and thawed using various methods. J. Food Qual. 2018, 1–9.

Barba, A.A., Naddeo, C., Caputo, S., Lamberti, G., d'Amore, M., Dalmoro, A., 2020. Microwave treatments of cereals: effects on thermophysical and parenchymal-related properties. Foods 9 (6), 711.

Barutcu, I., Sahin, S., Sumnu, G., 2009. Acrylamide formation in different batter formulations during microwave frying. LWT - Food Sci. Technol. 42 (1), 17–22.

Battaglia Richi, E., Baumer, B., Conrad, B., Darioli, R., Schmid, A., Keller, U., 2015. Health risks associated with meat consumption: a review of epidemiological studies. Int. J. Vitam. Nutr. Res. 85 (1–2), 70–78.

Beaudry, C., Raghavan, G., Rennie, T., 2003. Microwave finish drying of osmotically dehydrated cranberries. Drying Technol. 21 (9), 1797–1810.

Beaudry, C., Raghavan, G.S.V., Ratti, C., Rennie, T.J., 2004. Effect of four drying methods on the quality of osmotically dehydrated cranberries. Drying Technol. 22 (3), 521–539.

Bhattacharya, M., Srivastav, P., Mishra, H., 2015. Thin-layer modeling of convective and microwave-convective drying of oyster mushroom (Pleurotus ostreatus). J. Food Sci. Technol. 52 (4), 2013–2022.

Borad, S.G., Kumar, A., Singh, A.K., 2017. Effect of processing on nutritive values of milk protein. Crit. Rev. Food Sci. Nutr. 57 (17), 3690–3702.

Bornhorst, E., Tang, J., Sablani, S., Barbosa-Cánovas, G., 2017. Thermal pasteurization process evaluation using mashed potato model food with Maillard reaction products. LWT - Food Sci. Technol. 82, 454–463.

Buffler, C.R., 1993. Microwave Cooking and Processing: Engineering Fundamentals for the Food Scientist. AVI Publication, New York.

Calay, R.K., Newborough, M., Probert, D., Calay, P.S., 2007. Predictive equations for the dielectric properties of foods. Int. J. Food Sci. Technol. 29 (6), 699–713.

Chan, K.W., Ismail, M., 2009. Supercritical carbon dioxide fluid extraction of Hibiscus cannabinus L. seed oil: a potential solvent-free and high antioxidative edible oil. Food Chem. 114 (3), 970–975.

Chandrasekaran, S., Ramanathan, S., Basak, T., 2013. Microwave food processing - a review. Food Res. Int. 52 (1), 243–261.

Chatterjee, A., Abraham, J., 2018. Microbial contamination, prevention, and early detection in food industry. In: Microbial Contamination and Food Degradation. Elsevier, Academic Press, Massachusetts, pp. 21–47.

Chemat, F., Fabiano-Tixier, A.S., Vian, M., Allaf, T., Vorobiev, E., 2015. Solvent-free extraction of food and natural products. Trends Anal. Chem. 71, 157–168.

Chemat, F., Rombaut, N., Meullemiestre, A., Turk, M., Périno-Issartier, S., Fabiano-Tixier, A.S., et al., 2017. Review of green food processing techniques. Preservation, transformation, and extraction. Innov. Food Sci. Emerg. Technol 41, 357–377.

Chen, S., Xu, J., Chen, G., Hu, Q., Zhao, L., 2018. Influence of microwave blanching on arsenic speciation and bioaccessibility in lentinus edodes. Anal. Lett. 51 (13), 2141–2155.

Chung, S.W.C., Chen, B.L.S., 2011. Determination of organochlorine pesticide residues in fatty foods: a critical review on the analytical methods and their testing capabilities. J. Chromatogr. 1218 (33), 5555–5567.

Clary, C., Gamache, A., Cliff, M., Fellman, J., Edwards, C., 2005. Flavor and aroma attributes of Riesling wines produced by freeze concentration and microwave vacuum dehydration. J. Food Process. Preserv. 30 (4), 393–406.

Contreras, C., Benlloch-Tinoco, M., Rodrigo, D., Martínez-Navarrete, N., 2017. Impact of microwave processing on nutritional, sensory, and other quality attributes. In: The Microwave Processing of Foods. Elsevier, pp. 65–99.

Contreras, C., Martín, M.E., Martínez-Navarrete, N., Chiralt, A., 2005. Effect of vacuum impregnation and microwave application on structural changes which occurred during air-drying of apple. Food Sci. Technol. 38 (5), 471–477.

Cooper, N., 2009. Microwave Ovens in Development of Packaging and Products for Use in Microwave Ovens. Elsevier, Woodhead Publishing, Cambridge, pp. 105–128.

Cui, Z.W., Xu, S.Y., Sun, D.W., 2003. Dehydration of garlic slices by combined microwave-vacuum and air drying. Drying Technol. 21 (7), 1173–1184.

Curet, S., Rouaud, O., Boillereaux, L., 2014. Estimation of dielectric properties of food materials during microwave tempering and heating. Food and Bioprocess Technol. 7 (2), 371–384.

Das, A., Banik, B., 2021. Future trends in microwave chemistry and biology. In: Das, A., Banik, B. (Eds.), Microwaves in Chemistry Applications Fundamentals, Methods, and Future Trends, First Edition. Elsevier, Amsterdam, pp. 375–384.

Datta, A., Geedipalli, S. and Almeida, M., 2005. Microwave combination heating. Food Technology Magazine, [online] (59). Available at: https://www.ift.org/news-and-publications/food-technology-magazine/issues/2005/january/features/microwave-combination-heating. [Accessed October 20, 2021].

Datta, A.K., Rakesh, V., 2013. Principles of microwave combination heating: principles of microwave combination heating. Compr. Rev. Food Sci. Food Saf. 12 (1), 24–39.

de Andrade, J.C., de Aguiar-Sobral, L., Ares, G., Deliza, R., 2016. Understanding consumers perception of lamb meat using free word association. Meat Sci. 117, 68–74.

de Gavelle, E., Huneau, J.F., Mariotti, F., 2018. Patterns of protein food intake are associated with nutrient adequacy in the general French adult population. Nutrients 10 (2), 226.

Dejoye, C., Vian, M., Lumia, G., Bouscarle, C., Charton, F., Chemat, F., 2011. Combined extraction processes of lipid from chlorella vulgaris microalgae: microwave prior to supercritical carbon dioxide extraction. Int. J. Mol. Sci. 12 (12), 9332–9341.

Devece, C., Rodríguez-López, J., Fenoll, L., Tudela, J., Catalá, J., de los Reyes, E., García-Cánovas, F., 1999. Enzyme inactivation analysis for industrial blanching applications: comparison of microwave, conventional, and combination heat treatments on mushroom polyphenoloxidase activity. J. Agric. Food Chem. 47 (11), 4506–4511.

Domínguez, R., Borrajo, P., Lorenzo, J.M., 2015. The effect of cooking methods on nutritional value of foal meat. J. Food Compos. Anal. 43, 61–67.

Dong, X., Wang, J., Raghavan, V., 2021. Impact of microwave processing on the secondary structure, in-vitro protein digestibility and allergenicity of shrimp (Litopenaeus vannamei) proteins. Food Chem. 337, 127–811.

Dorantes-Alvarez, L., Jaramillo-Flores, E., González, K., Martinez, R., Parada, L., 2011. Blanching peppers using microwaves. Procedia Food Sci. 1, 178–183.

Dorantes-Alvarez, L., Ortiz-Moreno, A., Guzmán-Gerónimo, R., Parada-Dorantes, L., 2017. Microwave-assisted blanching. In: Regier, M., Knoerzer, K., Schubert, H. (Eds.), The Microwave Processing of Foods, Second Edition. Woodhead Publishing Series in Food Science, Technology and Nutrition, Woodhead Publishing, Duxford, pp. 179–199.

Drouzas, A., Schubert, H., 1996. Microwave application in vacuum drying of fruits. J. Food Eng. 28 (2), 203–209.

Du, Y., 2014. Electromagnetic theory and wave propagation. In: Njoku, EG (Ed.), Encyclopedia of Remote Sensing, Encyclopedia of Earth Sciences Series. Springer, New York, NY, pp. 150–158.

Duan, X., Zhang, M., Mujumdar, A.S., 2007. Studies on the microwave freeze drying technique and sterilization characteristics of cabbage. Drying Technol. 25 (10), 1725–1731.

Ekezie, F., Sun, D., Cheng, J., 2017. Acceleration of microwave-assisted extraction processes of food components by integrating technologies and applying emerging solvents: a review of latest developments. Trends Food Sci. Technol. 67, 160–172.

El-Adawy, T.A., 2002. Nutritional composition and antinutritional factors of chickpeas (Cicerarietinum L.) undergoing different cooking methods and grmination. In: Plant Foods Hum. Nutr. 57 (1), 83–97.

Erle, U., 2005. Drying using microwave processing. In: The Microwave Processing of Foods. Elsevier, Woodhead Publishing Ltd, Cambridge, pp. 142–152.

Erle, U., Schubert, H., 2001. Combined osmotic and microwave-vacuum dehydration of apples and strawberries. J. Food Eng. 49 (2–3), 193–199.

Esquivel-Hernández, D.A., Ibarra-Garza, I.P., Rodríguez-Rodríguez, J., Cuéllar-Bermúdez, S.P., Rostro-Alanis, M.D.J., Alemán-Nava, G.S., et al., 2017. Green extraction technologies for high-value metabolites from algae: a review. Biofuels Bioprod. Biorefin. 11 (1), 215–231.
Faller, A.L.K., Fialho, E., 2009. The antioxidant capacity and polyphenol content of organic and conventional retail vegetables after domestic cooking. Food Res. Int. 42 (1), 210–215.
Figiel, A., 2010. Drying kinetics and quality of beetroots dehydrated by combination of convective and vacuum-microwave methods. J. Food Eng. 98 (4), 461–470.
Franco, A.P., Yamamoto, L.Y., Tadini, C.C., Gut, J.A.W., 2015. Dielectric properties of green coconut water relevant to microwave processing: effect of temperature and field frequency. J. Food Eng. 155, 69–78.
Ganzler, K., Salgó, A., Valkó, K., 1986. Microwave extraction. J. Chromatogr. 371, 299–306.
Gao, F., Zhou, T., Hu, Y., Lan, L., Heyden, Y., Crommen, J., Lu, G., Fan, G., 2016. Cyclodextrin-based ultrasonic-assisted microwave extraction and HPLC-PDA-ESI-ITMSn separation and identification of hydrophilic and hydrophobic components of Polygonum cuspidatum: a green, rapid and effective process. Ind. Crops Prod. 80, 59–69.
Gaukel, V., Siebert, T., Erle, U., 2017. Microwave-assisted drying. In: Regier, M., Knoerzer, K., Schubert, H. (Eds.), The Microwave Processing of Foods, Second Edition. Woodhead Publishing Series in Food Science, Technology and Nutrition, Woodhead Publishing, Duxford, pp. 152–178.
Gerard, K.A., Roberts, J.S., 2004. Microwave heating of apple mash to improve juice yield and quality. Food Sci. Technol. 37 (5), 551–557.
Gharachorloo, M., Ghavami, M., Mahdiani, M., Azizinezhad, R. , 2010. The effects of microwave frying on physicochemical properties of frying and sunflower oils. J. Am. Oil Chem. Soc. 87 (4), 355–360.
Gögüs, F., Maskan, M., 2001. Drying of olive pomace by a combined microwave-fan assisted convection oven. Nahrung/Food 45 (2), 129–132.
Gomaa, A., Nsonzi, F., Sedman, J., Ismail, A., 2016. Enhanced unfolding of bovine β-Lactoglobulin structure using microwave treatment: a multi-spectroscopic study. Food Biophys. 11 (4), 370–379.
Gowen, A., Abu-Ghannam, N., Frias, J., Oliveira, J., 2006. Optimisation of dehydration and rehydration properties of cooked chickpeas (Cicer arietinum L.) undergoing microwave–hot air combination drying. Trends Food Sci. Technol. 17 (4), 177–183.
Gunasekaran, S., 1990. Grain drying using continuous and pulsed microwave energy. Drying Technol. 8 (5), 1039–1047.
Guo, Q., Sun, D.W., Cheng, J.H., Han, Z., 2017. Microwave processing techniques and their recent applications in the food industry. Trends Food Sci. Technol. 67, 236–247.
Guo, W., Pan, B., Sakkiah, S., Yavas, G., Ge, W., Zou, W., et al., 2019. Persistent organic pollutants in food: contamination sources, health effects and detection methods. Int. J. Environ. Res. Public Health 16 (22), 4361.
Guo, W., Wang, S., Tiwari, G., Johnson, J.A., Tang, J., 2010. Temperature and moisture dependent dielectric properties of legume flour associated with dielectric heating. Food Sci. Technol. 43 (2), 193–201.
Gupta, M., Wong, W.L.E., 2007. Microwaves and Metals. John Wiley and Sons (Asia) Ltd, Singapore.
Guzik, P., Kulawik, P., Zając, M., Migdał, W., 2021. Microwave applications in the food industry: an overview of recent developments. Crit. Rev. Food Sci. Nutr. 1–20.
Hao, J., Han, W., Huang, S., Xue, B., Deng, X., 2002. Microwave-assisted extraction of artemisinin from Artemisia annua L. Sep. Purif. Technol. 28 (3), 191–196.
Hassoun, A., Ojha, S., Tiwari, B., Rustad, T., Nilsen, H., Heia, K., et al., 2020. Monitoring thermal and non-thermal treatments during processing of muscle foods: a comprehensive review of recent technological advances. Appl. Sci. 10 (19), 6802.

Hemalatha, S., Platel, K., Srinivasan, K., 2007. Influence of heat processing on the bioaccessibility of zinc and iron from cereals and pulses consumed in India. J. Trace Elem. Med. Biol. 21 (1), 1–7.
Hemis, M., Raghavan, G.S.V., 2014. Effect of convective air attributes with microwave drying of soybean: model prediction and experimental validation. Drying Technol. 32 (5), 543–549.
Hitchcock, R.T., 2004. Radio-Frequency and Microwave Radiation, Third Edition. American Industrial Hygiene Association, Fairfax, VA.
Holtz, E., Ahrné, L., Karlsson, T.H., Rittenauer, M., Rasmuson, A., 2009. The role of processing parameters on energy efficiency during microwave convective drying of porous materials. Drying Technol. 27 (2), 173–185.
http://www.crcnetbase.com/doi/abs/10.1201/9781420017373.ch28. (Accessed October 19, 2021.).
Huang, Y., Sheng, J., Yang, F., Hu, Q.H., 2007. Effect of enzyme inactivation by microwave and oven heating on preservation quality of green tea. J. Food Eng. 78 (2), 687–692.
Igual, M., Contreras, C., Camacho, M.M., Martínez-Navarrete, N., 2014. Effect of thermal treatment and storage conditions on the physical and sensory properties of grapefruit juice. Food Bioprocess Technol. 7 (1), 191–203.
Igual, M., García-Martínez, E., Camacho, M.M., Martínez-Navarrete, N., 2010. Effect of thermal treatment and storage on the stability of organic acids and the functional value of grapefruit juice. Food Chem. 118 (2), 291–299.
Isabel, I., Lilia, A., Ulf, S., 2018. Enhancing the retention of -carotene and vitamin c in dried mango using alternative blanching processes. Afr. J. Food Sci. 12 (7), 165–174.
Izquierdo, F., Ali, I., Gomez, R.R., Ramaswamy, H.S., Yaylayan, V., 2005. Effects of high pressure and microwave on pronase and α-chymotrypsin hydrolysis of β-lactoglobulin. Food Chem. 92 (4), 713–719.
Jaiswal, A.K., Abu-Ghannam, N., 2013. Degradation kinetic modelling of color, texture, polyphenols and antioxidant capacity of York cabbage after microwave processing. Food Res. Int. 53 (1), 125–133.
James, S., James, C., 2002. Thawing and tempering. In: Meat Refrigeration, First Edition. Woodhead Publishing Series in Food Science, Technology and Nutrition, pp. 159–189.
James, S., James, C., Purnell, G., 2017. Microwave-assisted thawing and tempering. In: Regier, M., Knoerzer, K., Schubert, H. (Eds.), The Microwave Processing of Foods, Second Edition. Woodhead Publishing Series in Food Science, Technology and Nutrition, pp. 252–272.
James, S.J., James, C., 2014. Cooking of meat: heat processing methods. In: Encyclopedia of Meat Sciences. Elsevier, Cambridge, pp. 385–390.
Jermann, C., Koutchma, T., Margas, E., Leadley, C., Ros-Polski, V., 2015. Mapping trends in novel and emerging food processing technologies around the world. Innov. Food Sci. Emerg. Technol. 31, 14–27.
Ji, H., Du, A., Zhang, L., Li, S., et al., 2012. Effects of drying methods on antioxidant properties and phenolic content in white button mushroom. Int. J. Food Eng. 8 (3).
Jocelyn Paré, J.R., Bélanger, J.M.R., Stafford, S.S. , 1994. Microwave-assisted process (MAPTM): a new tool for the analytical laboratory. Trends Anal. Chem. 13 (4), 176–184.
Jouquand, C., Tessier, F.J., Bernard, J., Marier, D., Woodward, K., Jacolot, P., et al., 2015. Optimization of microwave cooking of beef burgundy in terms of nutritional and organoleptic properties. Food Sci. Technol. 60 (1), 271–276.
Kalla, A.M., R, D., 2017. Microwave energy and its application in food industry: a review. Asian J. Dairy Food Res. 36, 37–44.
Kamble, D.B., Singh, R., Kaur, B.P., Rani, S., Upadhyay, A., 2020. Effect of microwave processing on physicothermal properties, antioxidant potential, in vitro protein digestibility and microstructure of durum wheat semolina. J. Food Meas. Charact. 14 (2), 761–769.

Karisma, A., Shinokawa, Y., Fukasawa, T., Ishigami, T., Fukui, K., 2020. Synthesis of nicuzn ferrite nanoparticles from metallic nitrate solutions using the microwave direct denitration method and evaluation of its properties. Part. Sci. Technol. 39 (4), 427–435.

Kidmose, U., Martens, H., 1999. Changes in texture, microstructure and nutritional quality of carrot slices during blanching and freezing. J. Sci. Food Agric. 79 (12), 1747–1753.

Kiranoudis, C., Tsami, E., Maroulis, Z., 1997. Microwave vacuum drying kinetics of some fruits. Drying Technol. 15 (10), 2421–2440.

Kozempel, M.F., Annous, B.A., Cook, R.D., Scullen, O.J., Whiting, R.C., 1998. Inactivation of microorganisms with microwaves at reduced temperatures. J. Food Prot. 61 (5), 582–585.

Laguerre, J.C., Mazen Hamoud-Agha, M., 2020. Microwave heating for food preservation. In: Socaci, S.A., et al. (Eds.), Food Preservation and Waste Exploitation. IntechOpen.

Latorre, M., de Escalada Plá, M., Rojas, A., Gerschenson, L., 2013. Blanching of red beet (*Beta vulgaris* L. var. conditiva) root. Effect of hot water or microwave radiation on cell wall characteristics. LWT - Food Sci. Technol. 50 (1), 193–203.

Lee, J., Kim, H., Ko, J., Jang, J., Kim, G., Lee, J., Nah, J., Jeon, Y., 2016. Rapid preparation of functional polysaccharides from pyropia yezoensis by microwave-assistant rapid enzyme digest system. Carbohydr. Polym. 153, 512–517.

Lemmens, L., Tibäck, E., Svelander, C., Smout, C., Ahrné, L., Langton, M., Alminger, M., Van Loey, A., Hendrickx, M., 2009. Thermal pretreatments of carrot pieces using different heating techniques: effect on quality related aspects. Innovative Food Sci. Emerg. Technol. 10 (4), 522–529.

Leonelli, C., Veronesi, P., Cravotto, G., 2012. Microwave-Assisted extraction: an introduction to dielectric heating. In: Chemat, F., Cravotto, G. (Eds.), Microwave-Assisted Extraction for Bioactive Compounds. Food Engineering Series. Springer US, Boston, MA, pp. 1–14.

Lewandowicz, G., Fornal, J., Walkowski, A., 1997. Effect of microwave radiation on physico-chemical properties and structure of potato and tapioca starches. Carbohydr. Polym. 34 (4), 213–220.

Lewandowicz, G., Jankowski, T., Fornal, J., 2000. Effect of microwave radiation on physico-chemical properties and structure of cereal starches. Carbohydr. Polym. 42 (2), 193–199.

Lewis, M.J., 1996. Electrical properties. In: Physical Properties of Foods and Food Processing Systems. Elsevier, Woodhead Publishing, Cambridge, pp. 366–412.

Li, R., Dai, L., Peng, H., Jiang, P., Liu, N., Zhang, D., et al., 2021. Effects of microwave treatment on sorghum grains: effects on the physicochemical properties and in vitro digestibility of starch. J. Food Process Eng. 44 (10), 13804.

Lin, S., Brewer, M., 2005. Effects of blanching method on the quality characteristics of frozen peas. J. Food Qual. 28 (4), 350–360.

Liu, P., Mujumdar, A., Zhang, M., Jiang, H., 2015. Comparison of three blanching treatments on the color and anthocyanin level of the microwave-assisted spouted bed drying of purple flesh sweet potato. Drying Technol. 33 (1), 66–71.

Liu, Z., Qiao, L., Yang, F., Gu, H., Yang, L., 2017. Brönsted acidic ionic liquid based ultrasound-microwave synergistic extraction of pectin from pomelo peels. Int. J. Biol. Macromol. 94, 309–318.

Llompart, M., Garcia-Jares, C., Celeiro, M., Dagnac, T., 2018. Microwave-Assisted extraction. Reference Module in Chemistry, Molecular Sciences and Chemical Engineering. Elsevier, Cambridge, B9780124095472144000, https://linkinghub.elsevier.com/retrieve/pii/B9780124095472144427 (Accessed on 03.10.21).

Lu, Q.F., Pan, L.F., Chen, M., Qiu, Y., Xie, B.H., 2014. Research on the microwave-assisted supercritical CO_2 extraction of alkaloids from gynura segetum (Lour.) merr. Adv. Mater. Res. 988, 390–396.

Maran, J.P., Swathi, K., Jeevitha, P., Jayalakshmi, J., Ashvini, G., 2015. Microwave-assisted extraction of pectic polysaccharide from waste mango peel. Carbohydr. Polym. 123, 67–71.

Martins, C.P.C., Cavalcanti, R.N., Couto, S.M., Moraes, J., Esmerino, E.A., et al., 2019. Microwave processing: current background and effects on the physicochemical and microbiological aspects of dairy products: potential of microwave heating technology of dairy product. Compr. Rev. Food Sci. Food Saf. 18 (1), 67–83.

Marzec, A., Kowalska, H., Zadrożna, M., 2010. Analysis of instrumental and sensory texture attributes of microwave-convective dried apples: microwave-convective dried apples. J. Texture Stud. 41 (4), 417–439.

Maskan, M., 2000. Microwave/air and microwave finish drying of banana. J. Food Eng. 44 (2), 71–78.

Maskan, M., 2001. Drying, shrinkage and rehydration characteristics of kiwifruits during hot air and microwave drying. J. Food Eng. 48 (2), 177–182.

Matusiewicz, H., Ślachciński, M., 2014. Development of a one-step microwave-assisted subcritical water extraction for simultaneous determination of inorganic elements (Ba, Ca, Cu, Fe, Mg, Mn, Na, Pb, Sr, Zn) in reference materials by microwave induced plasma spectrometry. Microchem. J. 115, 6–10.

Metaxas, A.C., Meredith, R.J., 1983. Industrial Microwave Heating. Peregrinus. P., on behalf of the Institution of Electrical Engineers, London, UK.

Michalak, I., Chojnacka, K., 2014. Algal extracts: technology and advances. Eng. Life Sci. 14 (6), 581–591.

Mills, J., Donnison, A., Brightwell, G., 2014. Factors affecting microbial spoilage and shelf-life of chilled vacuum-packed lamb transported to distant markets: a review. Meat Sci. 98 (1), 71–80.

Monteiro, R.L., Carciofi, B.A.M., Laurindo, J.B., 2016. A microwave multi-flash drying process for producing crispy bananas. J. Food Eng. 178, 1–11.

Mousa, N., Farid, M., 2002. Microwave vacuum drying of banana slices. Drying Technol. 20 (10), 2055–2066.

Mujumdar, A.S., (Ed.) 2015. Handbook of Industrial Drying, 4th ed, CRC Press, Taylor and Francis Group, Boca Raton.

Munir, M., Nadeem, M., Qureshi, T.M., Leong, T.S.H., Gamlath, C., et al., 2019. Effects of high pressure, microwave and ultrasound processing on proteins and enzyme activity in dairy systems - a review. Innov. Food Sci. Emerg. Technol. 57 (4), 102192.

Muthukumarappan, K., Swamy, G., 2019. Microwave processing of foods. In: Kutz, M. (Ed.), Handbook of Farm, Dairy and Food Machinery Engineering, Third Edition. Academic Press, London, pp. 417–438.

Nelson, S.O., 2015. Dielectric properties of selected food materials. In: Dielectric Properties of Agricultural Materials and Their Applications. Elsevier, Academic Press, Cambridge, pp. 147–165. https://linkinghub.elsevier.com/retrieve/pii/B9780128023051000117 (Accessed on 02.10.21)

Nguyen, E., Jones, O., Kim, Y., San Martin-Gonzalez, F., Liceaga, A., 2017. Impact of microwave-assisted enzymatic hydrolysis on functional and antioxidant properties of rainbow trout Oncorhynchus mykiss by-products. Fish. Sci. 83 (2), 317–331.

Nguyen, T., Zhang, W., Barber, A., Su, P., He, S., 2016. Microwave-Intensified enzymatic deproteinization of Australian rock lobster shells (Jasus edwardsii) for the efficient recovery of protein hydrolysate as food functional nutrients. Food Bioprocess Technol. 9 (4), 628–636.

Ni, H., Datta, A., 1999. Moisture, oil and energy transport during deep-fat frying of food materials. Food Bioprod. Process. 77 (3), 194–204.

Nindo, C., Sun, T., Wang, S., Tang, J., Powers, J., 2003. Evaluation of drying technologies for retention of physical quality and antioxidants in asparagus (Asparagus officinalis, L.). LWT - Food Sci. Technol. 36 (5), 507–516.

Orsat, V., Raghavan, G., Krishnaswamy, K., 2017. Microwave technology for food processing: an overview of current and future applications. In: Regier, M., Knoerzer, K., Schubert, H.

(Eds.), The Microwave Processing of Foods, Second Edition. Woodhead Publishing Series in Food Science, Technology and Nutrition, Woodhead Publishing, Duxford, pp. 100–116.

Osepchuk, J.M., 1984. A history of microwave heating applications. IEEE Trans. Microwave Theory Tech. 32 (9), 1200–1224.

Oztop, M., Sahin, S., Sumnu, G., 2007. Optimization of microwave frying of potato slices by using Taguchi technique. J. Food Eng. 79 (1), 83–91.

Périno, S., Pierson, J., Ruiz, K., Cravotto, G., Chemat, F., 2016. Laboratory to pilot scale: microwave extraction for polyphenols lettuce. Food Chem. 204, 108–114.

Pongmalai, P., Devahastin, S., Chiewchan, N., Soponronnarit, S., 2015. Enhancement of microwave-assisted extraction of bioactive compounds from cabbage outer leaves via the application of ultrasonic pretreatment. Sep. Purif. Technol. 144, 37–45.

Pozar, D.M., 1990. Microwave Engineering, Fourth Edition, Addison-Wesley, Reading, Mass.

Prabhanjan, D., Ramaswamy, H., Raghavan, G., 1995. Microwave-assisted convective air drying of thin layer carrots. J. Food Eng. 25 (2), 283–293.

Prakash, S., Jha, S., Datta, N., 2003. Performance evaluation of blanched carrots dried by three different driers. J. Food Eng. 62 (3), 305–313.

Pu, Y., Sun, D., 2015. Vis–NIR hyperspectral imaging in visualizing moisture distribution of mango slices during microwave-vacuum drying. Food Chem. 188, 271–278.

Pu, Y., Sun, D., 2016. Prediction of moisture content uniformity of microwave-vacuum dried mangoes as affected by different shapes using NIR hyperspectral imaging. Innovative Food Sci. Emerg. Technol. 33, 348–356.

Pu, Y., Sun, D., 2017. Combined hot-air and microwave-vacuum drying for improving drying uniformity of mango slices based on hyperspectral imaging visualisation of moisture content distribution. Biosyst. Eng. 156, 108–119.

Purcell, E.M., 2013. Electricity and Magnetism, Third Edition. Cambridge University Press, Cambridge.

Queiroz, C., Lopes, M.L.M., Fialho, E., Valente-Mesquita, V.L., 2008. Polyphenol oxidase: characteristics and mechanisms of browning control. Food Rev. Int. 24 (4), 361–375.

Qun, Y., Chuan, L., Zhenhua, D., Bing, L., Weiwen, D., Feifei, S., 2017. Ultrasonic microwave-assisted extraction of polyphenols, flavonoids, triterpenoids, and vitamin c from Clinacanthus nutans. Czech J. Food Sci. 35 (No. 1), 89–94.

Rashed, M., Tong, Q., Nagi, A., Li, J., Khan, N., Chen, L., Rotail, A., Bakry, A., 2017. Isolation of essential oil from Lavandula angustifolia by using ultrasonic-microwave assisted method preceded by enzymolysis treatment, and assessment of its biological activities. Ind. Crops Prod. 100, 236–245.

Regier, M., Knoerzer, K., Schubert, H., 2017. Introducing microwave-assisted processing of food: fundamentals of the technology. In: Regier, M., Knoerzer, K., Schubert, H. (Eds.), The Microwave Processing of Foods, Second Edition. Woodhead Publishing Series in Food Science, Technology and Nutrition, pp. 1–22.

Reynolds, L., 1989. The History of the Microwave Oven. Microwave World. 10 (5), 11–15.

Roberts, P.C.B., Lawrie, R.A., 2007. Effects on bovine dorsi muscle of conventional and microwave heating. Int. J. Food Sci. Technol. 9 (3), 345–356.

Roussy, G., Pearce, J., 1995. Foundations and Industrial Applications of Microwave and Radio Frequency Fields. Wiley, Chichester.

Ruiz-Ojeda, L., Peñas, F., 2013. Comparison study of conventional hot-water and microwave blanching on quality of green beans. Innovative Food Sci. Emerg. Technol. 20, 191–197.

Ryynänen, S., 1995. The electromagnetic properties of food materials: a review of the basic principles. J. Food Eng. 26 (4), 409–429.

Sahin, S., Sumnu, G., Oztop, M., 2007. Effect of osmotic pretreatment and microwave frying on acrylamide formation in potato strips. J. Sci. Food Agric. 87 (15), 2830–2836.

Salazar-González, C., Martín-González, M.F.S., López-Malo, A., Sosa-Morales, M.E., 2012.

Recent studies related to microwave processing of fluid foods. Food Bioprocess Technol. 5 (1), 31–46.

Sanchez-Aldana, D., Aguilar, C.N., Nevarez-Moorillon, G.V., Esquivel, J.C.C., 2013. Comparative extraction of pectic and polyphenols from mexican lime pomace and bagasse. Am. J. Agric. Biol. Sci. 8 (4), 309–322.

Sattar, S., Ahmad, T., Nisa, M., Imran, M., Holmes, M., Maycock, J., et al., 2019. Microwave processing impact on physicochemical and bioactive attributes of optimized peach functional beverage. J. Food Process. Preserv. 43 (7), e13952.

Schirack, A., Drake, M., Sanders, T., Sandeep, K., 2006. Impact of microwave blanching on the flavor of roasted peanuts. J. Sens. Stud. 21 (4), 428–440.

Shaheen, M., El Massry, K., El-Ghorab, A., Anjum, F., 2012. Microwave applications in thermal food processing. In: The Development and Application of Microwave Heating. InTech, pp. 3–16.

Sharma, G., Prasad, S., 2006. Optimization of process parameters for microwave drying of garlic cloves. J. Food Eng. 75 (4), 441–446.

Siores, E., Do-Rego, D., 1995. Microwave applications in materials joining. J. Mater. Process. Technol. 48 (1–4), 619–625.

Śledź, M., Nowacka, M., Wiktor, A., Witrowa-Rajchert, D., 2013. Selected chemical and physico-chemical properties of microwave-convective dried herbs. Food Bioprod. Process. 91 (4), 421–428.

Soni, A., Smith, J., Thompson, A., Brightwell, G., 2020. Microwave-induced thermal sterilization- a review on history, technical progress, advantages and challenges as compared to the conventional methods. Trends Food Sci. Technol. 97, 433–442.

Sorrentino, R., Bianchi, G., 2010. Microwave and RF Engineering. Wiley, Chichester, West Sussex, U.K.

Sosa-Morales, M.E., Valerio-Junco, L., López-Malo, A., García, H.S., 2010. Dielectric properties of foods: reported data in the 21st century and their potential applications. Food Sci. Technol. 43 (8), 1169–1179.

Sparr Eskilsson, C., Björklund, E., 2000. Analytical-scale microwave-assisted extraction. J. Chromatogr. 902 (1), 227–250.

Sultana, B., Anwar, F., Iqbal, S., 2008. Effect of different cooking methods on the antioxidant activity of some vegetables from Pakistan. Int. J. Food Sci. Technol. 43 (3), 560–567.

Sumnu, G., 2001. A review on microwave baking of foods. Int. J. Food Sci. Technol. 36 (2), 117–127.

Sumnu, G., Sahin, S., 2005. Recent developments in microwave heating. In: Emerging Technologies for Food Processing. Elsevier, pp. 419–444. https://linkinghub.elsevier.com/retrieve/pii/B9780126767575500189 (Accessed on 02.10.21).

Swart, G.J., Blignaut, C.M., Jooste, P.J., 2003. Pasteurization: other pasteurization processes. In: Encyclopedia of Food Sciences and Nutrition. Elsevier, pp. 4401–4406. https://linkinghub.elsevier.com/retrieve/pii/B012227055X008920 (Accessed on 19.10.21).

Tang, F., Xia, W., Xu, Y., Jiang, Q., Zhang, W., Zhang, L., 2014. Effect of thermal sterilization on the selected quality attributes of sweet and sour carp. Int. J. Food Prop. 17 (8), 1828–1840.

Tang, J., 2015. Unlocking potentials of microwaves for food safety and quality. J. Food Sci. 80 (8), Chicago, pp. E1776–E1793.

Tang, J., Hong, Y., Inanoglu, S., Liu, F., 2018. Microwave pasteurization for ready-to-eat meals. Curr. Opin. Food Sci. 23, 133–141.

Tang, J., Resurreccion, F.P., 2009. Electromagnetic basis of microwave heating. In: Development of Packaging and Products for Use in Microwave Ovens. Elsevier, Woodhead Publishing, Cambridge, pp. 3–38e. https://linkinghub.elsevier.com/retrieve/pii/B9781845694203500019 (Accessed on 02.10.21)

Tangduangdee, C., Bhumiratana, S., Tia, S., 2003. Heat and mass transfer during deep-fat frying of frozen composite foods with thermal protein denaturation as quality index. Sci. Asia 29 (4), 355–364.

Thostenson, E.T., Chou, T.W., 1999. Microwave processing: fundamentals and applications. Composites, Part A 30 (9), 1055–1071.

Tulasidas, T., 1995. Combined convective and microwave drying of grapes. Drying Technol. 13 (4), 1029–1031.

Turner, I., Jolly, P., 1991. Combined microwave and convective drying of a porous material. Drying Technol. 9 (5), 1209–1269.

Vadivambal, R., Jayas, D.S., 2007. Changes in quality of microwave-treated agricultural products—a review. Biosyst. Eng. 98 (1), 1–16.

Vadivambal, R., Jayas, D.S., 2010. Non-uniform temperature distribution during microwave heating of food materials—a review. Food Bioprocess Technol. 3 (2), 161–171.

Valero, E., Villamiel, M., Sanz, J., Martnez-Castro, I., 2000. Chemical and sensorial changes in milk pasteurised by microwave and conventional systems during cold storage. Food Chem. 70 (1), 77–81.

Veggi, P.C., Martinez, J., Meireles, M.A.A., 2012. Fundamentals of microwave extraction. In: Chemat, F., Cravotto, G. (Eds.), Microwave-Assisted Extraction for Bioactive Compounds. Food Engineering Series. Springer US, Boston, MA, pp. 15–52.

Velu, V., Nagender, A., Prabhakara Rao, P.G., Rao, D.G., 2006. Dry milling characteristics of microwave dried maize grains (Zea mays L.). J. Food Eng. 74 (1), 30–36.

Venkatesh, M.S., Raghavan, G.S.V., 2004. An overview of microwave processing and dielectric properties of agri-food materials. Biosyst. Eng. 88 (1), 1–18.

Wang, H., Zhang, M., Mujumdar, A.S., 2014. Comparison of three new drying methods for drying characteristics and quality of shiitake mushroom (Lentinus edodes). Drying Technol. 32 (15), 1791–1802.

Wang, J., Yang, X., Mujumdar, A., Wang, D., Zhao, J., Fang, X., Zhang, Q., Xie, L., Gao, Z., Xiao, H., 2017. Effects of various blanching methods on weight loss, enzymes inactivation, phytochemical contents, antioxidant capacity, ultrastructure and drying kinetics of red bell pepper (Capsicum annuum L.). LWT 77, 337–347.

Wataniyakul, P., Pavasant, P., Goto, M., Shotipruk, A., 2012. Microwave pretreatment of defatted rice bran for enhanced recovery of total phenolic compounds extracted by subcritical water. Bioresour. Technol. 124, 18–22.

Wilkowska, A.M., Biziuk, M., 2010. Rapid method for the determination of organochlorine pesticides and PCBs in fish muscle samples by microwave-assisted extraction and analysis of extracts by gc-ecd. J. AOAC Int. 93 (6), 1987–1994.

Williams, E.A., Jones, G.A. National Association of Broadcasters, 2007. Engineering Handbook. Focal Press, Amsterdam.

Woodroof, J.G., Luh, B.S., (Eds.), 1986. Commercial Fruit Processing. Second Edition, AVI Publication, Co, Westport, Conn.

Wu, D., Gao, T., Yang, H., Du, Y., Li, C., Wei, L., Zhou, T., Lu, J., Bi, H., 2015. Simultaneous microwave/ultrasonic-assisted enzymatic extraction of antioxidant ingredients from Nitraria tangutorun Bobr. juice by-products. Ind. Crops Prod. 66, 229–238.

Xiao, H., Pan, Z., Deng, L., El-Mashad, H., Yang, X., Mujumdar, A., Gao, Z., Zhang, Q., 2017. Recent developments and trends in thermal blanching – a comprehensive review. Inf. Process. Agric. 4 (2), 101–127.

Xu, M., Yang, X., Fu, M., 2016. Combined ultrasonic and microwave method for juglone extraction from walnut green husk (Juglans nigra). Waste Biomass Valorization 7 (5), 1159–1166.

Yan, M.M., Liu, W., Fu, Y., Zu, Y.G., Chen, C.Y., Luo, M., 2010. Optimisation of the microwave-assisted extraction process for four main astragalosides in radix astragali. Food Chem. 119 (4), 1663–1670.

Yao, X., Zhang, D., Luo, M., Jin, S., Zu, Y., Efferth, T., Fu, Y., 2015. Negative pressure cavitation-microwave assisted preparation of extract of pyrola incarnata fisch. rich in hyperin, 2′-O-galloylhyperin and chimaphilin and evaluation of its antioxidant activity. Food Chem. 169, 270–276.

Yolacaner, E., Sumnu, G., Sahin, S., 2017. Microwave-assisted baking. In: Regier, M., Knoerzer, K., Schubert, H. (Eds.), The Microwave Processing of Foods. Second Edition. Woodhead Publishing, Duxford, pp. 117–141.

Yongsawatdigul, J., Gunasekaran, S., 1996. Microwave-Vacuum drying of cranberries: part I. energy use and efficiency. J. Food Process. Preserv. 20 (2), 121–143.

Yusoff, N., Leo, C., 2017. Microwave assisted extraction of defatted roselle (Hibiscus sabdariffa L.) seed at subcritical conditions with statistical analysis. J. Food Qual. 2017, 1–10.

Zhang, D., Hamauzu, Y., 2004. Phenolics, ascorbic acid, carotenoids and antioxidant activity of broccoli and their changes during conventional and microwave cooking. Food Chem. 88 (4), 503–509.

Zhang, G., Hu, M., He, L., Fu, P., Wang, L., Zhou, J., 2013. Optimization of microwave-assisted enzymatic extraction of polyphenols from waste peanut shells and evaluation of its antioxidant and antibacterial activities in vitro. Food Bioprod. Process. 91 (2), 158–168.

Zhang, H., Tang, Z., Rasco, B., Tang, J., Sablani, S., 2016. Shelf-life modeling of microwave-assisted thermal sterilized mashed potato in polymeric pouches of different gas barrier properties. J. Food Eng. 183, 65–73.

Zhang, M., Tang, J., Mujumdar, A., Wang, S., 2006. Trends in microwave-related drying of fruits and vegetables. Trends Food Sci. Technol. 17 (10), 524–534.

Zhou, B.W., Shin, S.G., Hwang, K.H., Ahn, J.H., Hwang, S., 2010. Effect of microwave irradiation on cellular disintegration of gram positive and negative cells. Appl. Microbiol. Biotechnol. 87 (2), 765–770.

Zielinska, M., Ropelewska, E., Xiao, H., Mujumdar, A., Law, C., 2019. Review of recent applications and research progress in hybrid and combined microwave-assisted drying of food products: quality properties. Crit. Rev. Food Sci. Nutr. 60 (13), 2212–2264.

Zielinska, M., Sadowski, P., Błaszczak, W., 2015. Freezing/thawing and microwave-assisted drying of blueberries (Vaccinium corymbosum L.). LWT - Food Sci. Technol. 62 (1), 555–563.

CHAPTER THREE

Synergism of ultrasound and microwave for food processing, preservation, and extraction

Aftab Ahmed[a], Farhan Saeed[b], Muhammad Afzaal[b], Ali Imran[b], Muhammad Awais Saleem[a], Noshiza Majeed[a] and Mohammad Javed Ansari[c]

[a]Department of Nutritional Sciences, Government College University, Faisalabad, Punjab, Pakistan
[b]Department of Food Sciences, Government College University, Faisalabad, Punjab, Pakistan
[c]Department of Botany, Hindu College Moradabad (Mahatma Jyotiba Phule Rohilkhand University, Bareilly, UP), India

3.1 Introduction

Chemical compound and organic or inorganic substance preparation are usually complemented by the idea of conserving resources by improving reactivity circumstances and/or introducing innovative processes and techniques that are more ecologically acceptable. Ionic liquids and a solvent-free method are two of such approaches; however, microwave and ultrasonic waves have recently shown to be viable solutions for reducing energy consumption and improving process regulation. This overview begins with a discussion of the fundamentals of microwave and ultrasonic interactions with material before moving on to various laboratories and commercial uses of each kind of stimulation. It's also feasible to boost the impacts of microwaves' high-frequency electromagnetic fields or sonochemistry's cavitational energy by mixing them with other severe circumstances like plasma screens, air pressure, and ultraviolet. Eventually, the utilization of microwaves and ultrasound together in an unitary plant is presented. This unique technology combination was used to combine the impacts of increased energy with better material mobility (Chemat et al., 2011).

Microwave is being used for processing of food specially tempering and defrosting purposes since long by generation of high temperatures. Employ of ultrasounds in food manufacturing is new and is used in combination with microwave technology to achieve better results and develop foods in which their natural taste and flavor is intact without the addition of

Ultrasound and Microwave for Food Processing: Synergism for Preservation and Extraction.
DOI: https://doi.org/10.1016/B978-0-323-95991-9.00017-5

preservatives and chemical additives. These are preferred over the conventional food processing techniques as they have minimal adverse effects on the environment. Microwave and ultrasound significantly differ in terms of nature but are used simultaneously to form simultaneous microwave and ultrasound irradiation (SMUI) technology that significantly optimizes reaction parameters. Combination of microwave and ultrasound provide a likelihood to build up food processing procedures which are effective, cost efficient and environment pleasant (Misra and Schlüter, 2016).

3.2 Ultrasound in food processing

Nowadays ultrasound is also widely used in food processing, its preservation and even extraction commercially to lower the cost and time of processing and augment the excellence and shelf life of the foodstuffs. Ultrasound waves when enter any liquid medium they lead to the formation of small bubbles that rapidly grow and collapse. This process is repeated with each sound wave and is called cavitations. Cavitations gives ultrasound waves their antimicrobial activity (Rawson, 1998).

3.2.1 Cooking

The characteristics of ultrasound to transfer heat uniformly throughout any food material is preferred over the conventional cooking methods by which the food maybe over cooked externally and under cooked internally and high temperatures reduce the appearance, flavor and nutritional value of the food. Ultrasound cooking has a high cooking speed retrain moisture, flavor and color of food in lesser time (Chemat et al., 2011).

3.2.2 Freezing/crystallization

Process of crystallization and freezing are intertwined as freezing is followed by crystallization, use of ultrasonic cavitation accelerate the cooling process by formation of finer ice crystals that consequence in the diminution of freezing moment. The food stored by ultrasonic freezing provides the advantage of maintaining the dietary worth, shade and flavor of food (Zheng et al., 2006).

3.2.3 Drying

Ultrasound is used to stimulate the drying process of fish, fruit, meat and vegetables and keeping intact the characteristics like taste, shade and dietary worth of the food items (Huang et al., 2020).

3.2.4 Pickling/marinating

Traditional pickling process is achieved by addition of high quantity of salt that reduces the crunch of vegetable, removes water from the pickling vegetables, tarnish its color and is a very time consuming procedure (Cárcel et al., 2007). All these drawbacks can be overcome by using ultrasound for the process of pickling and marinating. Low salt, crunchy and fresh looking pickles can be made by use ultrasound in lesser time (Hatloe, 1995).

3.2.5 Degassing/defoaming

Ultrasounds are used for degassing and defoaming of liquids as the little bubbles hovering in liquid move towards exterior and discharge gas abandoned in liquid resulting in nonfoamed carbonated drinks. This process is mainly done where the foam interacts and lowers the quality of the products (Matsuura et al., 1994).

3.2.6 Filtration

Filtration is the process by which solids are separated out from their mother liquor. Ultrasound is effectively used for the purpose as it reduces the polarization in concentrations, thus enhancing the process of filtration. The velocity of ultrasound is optimized as it is of utmost importance in order to avoid the damages to the food product and keep its quality high (Kyllönen et al., 2005).

3.2.7 Demolding and extrusion

When any food material is cooked in a vessel some of the food remains attached to the container even after the food is removed from the container. This remaining food adds to the food wastage and enhances the cleaning cost and time of the cooking containers. Ultrasound is now used to replace the demolding process by coupling the mold and the ultrasound source. High frequency relative movement between mold and food material ensures no residual material on the mold so reducing the food wastage, cleaning time, and cost.

3.2.8 Defrosting/thawing

Thawing is a very time-consuming procedure if allow to complete on its own that can be reduce by use of microwave and ultrasound as heat energy is produced inside the food material lowering the thawing time many folds (Kyllönen et al., 2005).

3.2.9 Meat tenderization

Ultrasound acts as a meat tenderizer by the breakdown of muscular integrity or by enhancing the enzymatic reaction.

3.2.10 Sterilization/pasteurization

The ultrasound is use as a strong antimicrobial agent due to its cavitation power. Small bubbles are consistently grown and broken inside the microbial cell leading to the disruption of microbe's cell membrane thus killing the microbial cell (Baboli, 2019).

3.2.11 Emulsification/homogenization

Traditionally surfactants are used to retard the collection of dispersed matter in liquid ensuing in configuration of emulsions. These surfactants are chemical in nature thus manipulate the worth, flavor and the shelf life of foodstuffs (Guimarães et al., 2019). The same can be achieved by use of high intensity ultrasound to produce cavitation resulting in production of stabilize emulsions. This influence is widely exploited in industries for the production of ketchup, fruit juices, mayonnaise, and homogenized milk (Akdeniz et al., 2019).

3.2.12 Oxidation

Ultrasound is also used in oxidation of food in a variety of food industries. The key function of ultrasound oxidation is found in brewing industry for the purpose of aging the alcoholic beverages that normally is a very time consuming procedure.

3.2.13 Depolymerization

Ultrasound is use for polymer degradation, this can be achieved either by chemical or mechanical means (cavitation phenomena). Generally the depolymerization of starch is achieved by ultrasound in order to bring permanent changes in the food processing (Liu et al., 2020).

3.2.14 Advantages of ultrasound in food processing

There is a spacious assortment of advantages of use of ultrasound in food sector as extraction of phytocomplexes without altering their properties, reduction of extraction time, cost reduction. Ultrasounds are used as bactericidal without the use of any chemical additives. Ultrasound are used in heating of food items, drying processes, activation and deactivation of drying processes (Bhargava et al., 2021).

3.2.15 Disadvantages of ultrasound in food processing

In spite of range of advantages of ultrasound usage it also has disadvantages like the formation of free radicals that may pose health risks to the consumers. Excessive use of ultrasound may result in discoloration and deflavoring of the food items. Initially it can be very costly as specialized equipment's are needed (Bhargava et al., 2021).

3.3 Microwave in food processing

In the 1930s, dielectric heating applications in the food sector were first developed in the radio frequency range (Püschner, 1966). The expected increase in energy transfer rate resulted in an increase in frequency: microwaves. Although the initial copyright for a manufacturing conveyor belt microwave method has awarded in 1952 (Spencer, 1952), the initial purpose was not filed till 10 years afterward. It was prompted by the necessity to produce high-power microwave sources.

Aeration potato chips, precooking poultry and bacon, tempering frozen meals, as well as drying pasta were the initial important uses (Decareau, 1985). Although the early uses were only temporary successes, because the quality increase produced by the microwave method could be swiftly achieved by an additional cost-effective modification of the traditional approach, the other techniques survived moreover are still utilized in manufacturing.

3.3.1 Microwave heating in food processing

More the route of abundant decades, microwave heating had a number of usages in the world of food grounding. It is employed in food processing for a diversity of functions, counting aeration, pasteurization, and sterilization, thawing, tempering, and baking of food ingredients (Metaxas et al., 1983; Gupta et al., 2007). Microwave cookery has developed in fame in food processing manufacturing owing to its capacity to produce elevated heating rates, appreciably diminish cookery time, and supply extra standardized heating, protected management, convenience of process, along with low-priced preservation.

Furthermore, compared to traditional heating throughout food preparation or reheating, microwave heating might impact the taste and dietary characteristics of foodstuff to a smaller stage (Vadivambal et al., 2010). Microwaves are electromagnetic waves by means of frequencies ranging from 300 MHz to 300 GHz. Industrial microwave machines run at frequencies of 915 MHz and 2.45 GHz, while household microwave maneuver at 2.45 GHz (Datta at al., 2000).

3.3.1.1 *The heating principle of the microwave*

The capacity of resources to soak up microwave radiation along with translates it to warm up triggers microwave heating. Dipolar and ionic operations are chiefly accountable for microwave heating of food substance. Owing to the dipolar nature of water, the occurrence of dampness or water induces dielectric heating. When water molecules are revealed to an oscillating electric field, the enduringly polarized dipolar molecules endeavor to realign themselves in way of electric ground. This rearrangement occurs at a million times per second since of an elevated frequency electric ground, triggering interior resistance of molecules and volumetric heating of the stuff. Microwave heating might potentially be caused by the oscillatory movement of ions in food that produces temperature in incidence of a high-frequency oscillating electric field (Datta et al., 2000).

Microwave heating and heat scattering are prejudiced by a number of elements, mainly noteworthy of which is dielectric uniqueness along with infiltration deepness. Expressive a material's dielectric uniqueness can assist you figure out how fine it transforms microwaves to heat.

3.3.2 Usage for microwaves in food industry

Because there are so many microwave ovens in homes, the food business not just employs microwaves in favor of processing; however furthermore produces items along with qualities specifically for microwave heating. Manufactured engineering, often known as formulation, is a method of improving a product.

3.3.2.1 *Baking and cooking*

On an industrial scale, comprehensive references to the baking of bread, cakes, and pastries using microwaves may be discovered. An increase in throughput is accomplished by speeding up the baking process, with no new room requirements for microwave power generators. Microwaves are employed in baking in conjunction with traditional or infrared surface baking, which eliminates problem of crust development.

Unlike traditional baking, microwave heating inactivates the enzyme quickly enough to stop severe starch crash while also producing enough CO_2 along with steam to generate a very absorbent result (Decareau and Peterson, 1986). The principal employ of microwaves in the baking business nowadays is microwave concluding, where the limited heat conductivity of the conventional procedure results in much longer baking periods. (Pre-)cooking is a distinct procedure that may be hastened by using microwave heating. It has been verified to be effectual for (pre)cooking chicken, beef patties, and bacon (Helmar, 2007). Microwaves are the primary source of energy used to provide fat and coagulate proteins by raising the temperature. A constant air flood removes the exterior water at the same time.

3.3.2.2 Tempering and thawing

In comparison to most other food processing activities, thawing and tempering have received far fewer concentration in the writing. Here are just a handful controlled thawing systems in commercial use. The amount and form of frozen meat provided to the industry vary from full hindquarters of cattle to tiny lamb breasts and chicken pieces, while the bulk of stuff is "boned-out" and packed in 15 cm thick boxes weighing between 20 and 40 kg. Fruit and vegetables are usually packaged in boxes, sacks, or tubs, and juice is stored in big barrels. Because little processes could grip frozen stuff, it is normally thawed prior to being processed further.

Whilst the entire of the substance reached 0°C in addition to there is no loose ice, thawing is considered complete. It is the lowest temperature at which boned beef or other goods may be sliced. Lesser temperatures (−5 to −2°C) are permissible for manufactured goods that will be mechanically chopped, although such material will be tempered rather than thawed. Tempering is merely the first part of a complete thawing process, thus the two procedures should not be confused. The setback of the freezing procedure is generally referred to as thawing.

Nevertheless, there is a concern with thawing that does not happen instantly in the cold product and therefore the bulk of germs that initiates food illness or spoilage originate on food surfaces. Surface temperatures drop fast during the freezing process, limiting bacterial proliferation and causing bacteria to go entirely dormant below −10°C. These identical surface regions are the first to warm up during the thawing process, allowing bacterial proliferation to resume. Surface spoiling can occur on big items subjected to protracted uncontrolled thawing cycles previous to the center areas have entirely thawed.

Traditional thawing and tempering techniques warm up the exterior before relying upon transmission to carry the heat to the product's core. Electromagnetic radiations like microwaves are used to generate heat inside the foodstuff. When choosing a thawing or tempering method for industrialized

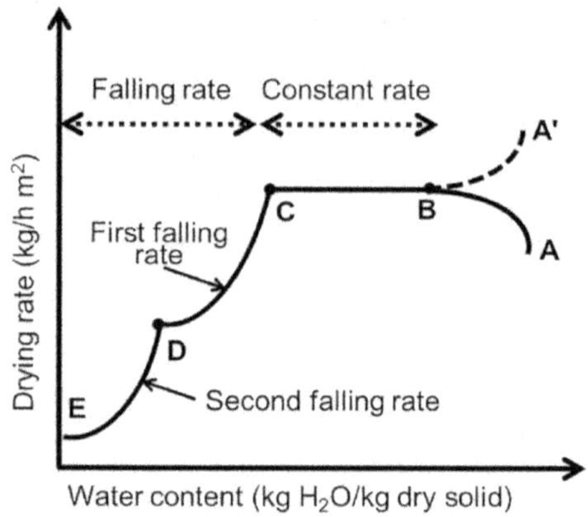

Figure 3.1 Drying rate curve as a function of moisture content. A–B: initial adjustment period; B–C: constant rate; C–D: first falling rate period; D–E: second falling rate period.

relevance, equilibrium should be made among thawing duration, product look along with bacteriological situation, processing issues like effluent discarding, and the respective systems' capital and operational expenses. Thawing time is the most important parameter for system selection among these factors.

According to a study, research has demonstrated that thawing rate has little influence on the volume of drip in beef. Numerous scholars came to the conclusion that rapid thawing rates would consequence in greater drip, whereas others came to the conflicting conclusion. Increased drip loss was observed when thawing periods from −8 to 0°C were less than 1 minute or higher than 2000 minutes (James et al., 2002). As a result, the findings are contradictory, and no meaningful propose statistics intended for improving a thawing method is available. Ice formation throughout freezing damages cell structure in fish, fruit, and vegetables, and fluids are decreased when thawing. Heating homogeneity and control of the final temperature are critical in microwave tempering operations, since localized melting can lead to a thermal runaway phenomenon.

3.3.2.3 Drying

Initially, let's know the benefits of microwave drying; next, let us just glance at another more traditional technique of air drying. As illustrated in Fig. 3.1, a standard food drying curve could be divided into three phases. A steady drying time per unit cross-sectional area characterizes a first phase.

Throughout this epoch, exterior is reserved moist by continuous capillary-driven stream of water from inside molecule. Warmth, humidity levels, and air movement all seem to be factors that manipulate and limit the drying rate throughout the "period under analysis" (Erle, 2000).

Microwaves are used in drying to reduce processing time that would have otherwise been impeded via weak heat conductivity, particularly within substances having low humidity. Longer extraction times or excessive heat flow could impair the senses and produce nutritional problems, both of which can be prevented. Additional advantage has been the possibility of preventing surface stiffening due to further uniform drying without significant humidity fluctuations. Microwave drying could be accomplished under one of 2 directions: at air pressure or is under vacuum.

Combination microwave-air-dryers are often extra popular with in food business, and they could be classified as just a parallel or in series fusion of two procedures. Sequential heated air and microwave drying are used in the drying of spaghetti as well as the production of dehydrated onion, although ultimate dryness of French fries was still only sometimes successful in the late 1960s and early 1970s. Combining microwave and vacuum drying seems to have some potential as well. Microwave-assisted freeze drying has also been carefully investigated, and yet no feasible commercial application has been established owing to elevated expenses as well as a minute marketplace for freeze-dried foodstuffs (Knutson et al., 1987).

Microwave vacuum dehydration under pressure greater than the three-fold point of water has greater market viability than microwave vacuum dryness with pressure less than the tripled point of water. Microwave radiation eliminates excessive heat transmission and conducting reluctance, allowing for speedier dryer periods. Higher drying speed results in decreased contraction as well as the preservation of water insoluble materials, as seen in Fig 3.1. As parsley, for instance, a bulk of essential oils typically located inside a distinct phase with such elevated boiling temperature. During speedy drying conditions, only a small quantity of absorbed volatile essential oils being eliminated, but there isn't enough time to fix the remaining oil inside the isolated stage (Erle, 2000).

The preservation of water soluble scents, such as those found in apples, is less favorable because microwave energy creates many vapor bubbles, giving volatile fragrances a vast surface area to evaporate. Low pressures, on the other hand, limit product temperatures to lesser levels as extended because a certain quantity of liberated water are available, which aids in the preservation of temperature-sensitive compounds such as vitamins and colors. As a result, in some situations, the excellent quality of the goods may make this somewhat costly technique cost-effective.

Microwave vacuum drying concentrates or indeed powders fruit juices, and cereals have been dehydrated quickly without sprouting. Another novel and effective technique would be the coupling of preair dehydration, discontinuous microwave vacuum drying, and postair drying. It's mostly utilized to create rehydration-friendly dried fruits and vegetables. Because of the rapid loss of moisture, the microwave vacuum approach exposes the cellular components (puffing) thus creates an open porous structure just after shape is formed by case hardening due to standard air-drying. After that, postdrying is used to decrease the relative humidity to the required amount.

3.3.2.4 Microwave-dried food product reliability

Reliability is often intermediate amid air dry and freeze dry items. Drying periods may be cut in half, which is good for the color and aroma. Microwave-dried apples and carrots were tested for color and porosity by Krokida and Maroulis (1999). When potatoes that had been air-dried and those that had been microwave-dried were compared; they discovered that the latter had less shrinkage and better rehydration (Khraisheh et al., 2004).

For example, microwave-vacuum-dried cranberries were reported to have higher texture and color over air-dried counterparts (Yongsawatdigul and Gunasekaran, 1996). When it comes to the retention of scent, there are two main scenarios that must be distinguished. Aroma molecules are found in modest numbers in most meals, therefore they are expected to dissolve in the water phase. The volatile nature of the fragrance molecule in water is critical in this case.

Considering as you smell through your noses instead of your tongues, fragrance particles must be rather volatile; alternatively, these would remain inside the food throughout digestion thus contributing very little to the odor. To put it another way, if a water phase (such as a meal) or a gas phase coincide, then scent particles will prefer the gas phase. In air drying, the smell particles might permeate from the particles' external surface. Similarly, water molecules are expelled here. As a consequence, all fragrance here on foodstuff particle's surface would be diminished; however the wastage would not be larger than any of those generated by the internal capillary water movement. As a consequence, the wastes of water as well as fragrance are related.

3.3.3 Microwave heating applications: Usage, benefits, and drawbacks

Pasteurization and sterilization, as well as combination procedures such microwave vacuum drying, are all used nowadays. Two causes for such comparably delayed acceptance of food industry microwave technologies are the food industry's conservatism (Decareau, 1985) as well as its relatively small research budget. As a consequence, fixing problems using microwave heating applications is difficult. Another of the primary challenges seems to be that they need a high level of technical understanding to produce successful results.

Unlike traditional heating systems, where adequate results may frequently be attained simply via intuition, effective microwave application outcomes generally need a great deal of knowledge or expertise to understand and mitigate problems such as uneven heating and thermal runaway. Microwave heating has another drawback over traditional heating in that it requires electrical energy that is the mainly costly source of energy. Microwave heating, on the other hand, provides a variety of quantitative and qualitative benefits over traditional heating techniques, making it a real consideration. One significant benefit is the source of heat, which is the product itself. As a result, minor heat conductivities or heat transfer coefficients have less of an impact. As a result, larger portions may be cooked in less time and at a more consistent temperature. These benefits frequently result in higher output.

3.3.4 Microwave radiations for sewage disposal

Several industrialized operations include the development along with succeeding discarding of garbage that comes at a significant financial and environmental cost. Furthermore, some waste items, such as asbestos-containing materials or nuclear plant wastes, might be harmful. In this scenario, regulatory requirements are very stringent in order to ensure the operators' safety, and choosing an inertization technique becomes conciliation among security concerns, energy assessments, and cost considerations. As a result, the waste treatment must be assessed in terms of the ultimate product's nation.

The disposal of waste materials has been a major issue in recent years, as the rapid growth in their production has not been matched by an increase in the number of permitted landfills. Furthermore, current regulations do not always allow for the recycling of all types of garbage, particularly where toxic or perilous elements are concerned (Oda et al., 1992). However, given the current year's output of wastes such as ashes, as well as the widespread occurrence on the land of asbestos-containing products, it appears hard to solve this ecological problem just by dump removal. To address this issue, it is required to research and develop alternate methods for treating and reusing waste material components, such as transforming them into secondary raw materials and, if feasible, restore them to their original use.

Even though it arrives out of the same production process and therefore relates to a certain classification, effluent could be assumed of as a multi-component substance with such a broad variety of compounds (like ashes, nuclear waste, asbestos-containing materials). Just the existence of a few of these products are frequently packaged qualifies the combination as a waste product. As a result, a technology that allows for quantitative remediation of the waste's "undesired" portion could save a lot of time and money, particularly for substances having limited heat transmission (Marucci et al., 2000). Microwaves could be a viable option towards meeting this need, particularly whereas if solid wastes matrices have dielectric properties that differ significantly from that of unwanted components.

3.3.5 Microwave-processed foodstuff is secure for consumers

Since the microwave power is converted to heat as soon as it is engrossed by meal, it could not turn the food radioactive or polluted. There is no lingering radiation in food after the microwave energy has been switched off and item has been taken from the oven. A microwave oven is similar to an electric light that turns off when it is turned off in this sense.

3.4 Food preservation with ultrasound

Customers in the food sector developed into additional worried about the sanitary elements of food production. As a result, the phrases "safety" and "transparency" in manufacturing procedures have become crucial within the sector. A number of people believe with the intention of achieving the finest in food preservation is an art. Heat treatment, such as pasteurization and sterilization, was once the preferred approach for destroying both microorganisms along with enzymes, the second is the cause of food degradation.

Its efficacy, nevertheless, is reliant on the handling heat and duration, which results in nutritional loss, the creation of unpleasant smells and colors, and the degradation of food's organoleptic qualities. As a consequence, nonthermal technologies are gaining popularity. The current issue is to mix moderate thermal preservation techniques with emerging bacterial destructive techniques. The technological advances under development make sure food preservation devoid of using preservatives whereas maintain dietary value and organoleptic uniqueness (consistency, shade, flavor) while consuming less energy, being more cost effective, being environmentally friendly, and having a high level of protection.

A few of these innovative technologies include ultrasound processing. While its use in food processing remains new, it has been demonstrated that high-intensity ultrasonic waves may burst cells and disintegrate enzymes, and even that low-intensity ultrasound might alter cellular metabolism. Ultrasonication, when used in conjunction with heat, may speed up the sterilization of meals, reducing the time and severity of heat treatments, as well as the harm that produced. Ultrasound sterilization has several benefits beyond thermal sterilization, including less taste loss, increased uniformity, and considerable energy conservation.

3.4.1 Food preservation and ultrasonic phenomenon

The principal causes of food degradation include microbes as well as enzymes. Meal is an excellent substrate for organisms such as bacteria and spores because they require surroundings highly nutritious plus moisture to develop and reproduce. But on other side, enzymes found naturally in foodstuff smash the nutrients, such as hydrolytic enzymes break down lipids and proteolytic enzymes breaking down peptides. Thermal processing denatures enzymes and destroys vegetative bacteria and certain spores. Nevertheless, the quantity of nutritional degradation, the creation of unpleasant tastes, and the deterioration of functional qualities of food items generally related to the duration and heat of the procedure.

A few of the novel preservation approaches that potentially decrease bacterial growth involves ultrasound. High-intensity ultrasound has shown to damage biological cells on its own. This could speed up the sterilization of foodstuffs when used in conjunction using thermal treatment. As a consequence, the time and severity of the heat processing, as well as the resulting impairments, are reduced. Ultrasound has been shown to burst cells at reasonably high acoustic power inputs (Chisti, 2003). The cell could be down regulated at quite a lower level than of required to damage them. The weakening of cellular membrane, localized heat, and generation of free radicals are the major mechanisms of bacterial death (Butz and Tauscher, 2002).

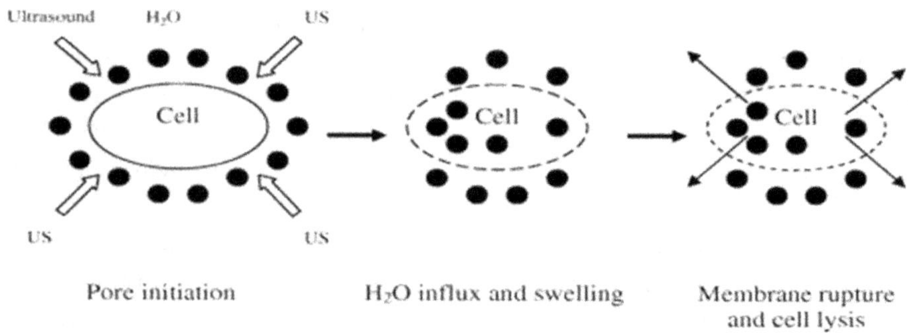

Figure 3.2 Applications of ultrasound in food technology: Processing, preservation and extraction.

There are two forms of acoustic cavitations: transient and steady (Mason et al., 2003). The latter happened whenever gas or vapor-filled cavitations bubbles oscillate irregularly and eventually collapse. Relative high pressures and temperatures result, causing biological cells to dissolve and enzymes to deform. High shear stresses and fluid jets inside the solvent employed are also produced by the exploding bubble, which could have enough energy to physiologically destroy the cell wall and cell membrane. Bubbles which oscillate in a predictable pattern over several acoustic cycles are referred to be stable cavitations. Microstreaming in the surrounding liquid is caused by the bubbles, which might stress microbiological organisms.

The formation of intracellular cavitations has just been linked to the inactivation impact of ultrasound, and all these mechanical shocks could alter cellular structural and functional elements up to the point of cell lysis (Fig. 3.2).

Spores seem to become more robust than vegetative stages, and ultrasound has been observed to depolymerize enzymes, rendering them inactive. The structure of ultrasonic waves, exposure duration with the microorganisms, the kind of microbes, amount of food processed, content of a meal, as well as the heat are all important processing elements. Nevertheless, while employing ultrasound alone, the impacts are not strong enough to kill microbes effectively (Butz and Tauscher, 2002). Ultrasound could be used with temperature, pressure, or both to produce this. The following are examples of applications that use a combination of preservation methods:

- ➢ Manosonication is a practice that combines ultrasound with pressure (MS).
- ➢ Thermosonication is a practice that combines ultrasound and heat (TS).
- ➢ Manothermosonication is a practice that combines ultrasound, pressure, and heat (MTS).

Table 3.1 Ultrasound's impacts when combined by heat and pressure.

Deactivation by	Vegetative cells	Spores	Enzymes
Ultrasound only (US)	+	−	−
Ultrasound and heat (MT)	+	+	−
Ultrasound and heat and pressure (MTS)	+	+	+

When contrasted to each therapy utilized alone, the amalgamation of heat and ultrasound is far additional efficient in terms of treatment duration and energy usage (Table 3.1).

3.4.2 Application inextricably linked to food

Particular appliance of ultrasound in food preservation has following classifications.

3.4.2.1 Down regulation of microbes

Microbes might not all respond to ultrasonic therapy in much the same manner, according to research. The efficiency of inactivation of micxx1robes is influenced by many variables (Piyasena et al., 2003).

- ➢ Ultrasound wave intensity
- ➢ Contact or retention time
- ➢ Quantity of foodstuff processed
- ➢ Content of the food
- ➢ Heat at which the food was treated.

The efficiency of such parameters is often influenced either by microbes' kind, size, and dimension (Table 3.2) (Heinz et al., 2003). Cells that are greater are much more susceptible than those that are smaller. This is most likely owing to the fact that they have a bigger surface area. Gram-positive bacteria are identified to be more challenging as compare to gram-negative bacteria, presumably owing to their stiffer cell walls that shield organisms from the impacts of ultrasound (Drakopoulou et al., 2009). Variations in cellular susceptibility might possibly be related to the more densely adhering coating of peptidoglycans in gram-positive cells, as per the research. As of the link amid cell surface and volume, cocci are more resilient than bacilli in terms of morphology. Lastly, spores are far extra complicated to kill than vegetative cells in the development phase.

Ultrasound has been used to immobilize a range of bacteria. Some of these have been investigated in culture medium, while others have been investigated in food, utilizing ultrasonography in combination or alone. *Saccaromyces cerevisiae* and *Escherichia coli* are the most commonly investigated

Table 3.2 Microorganisms' susceptibility is affected by a variety of factors.

Type of microorganism	Shape	Cell size (μm)	Structure of cell wall
Listeria monocytogenes	Short rods 	0.4–0.5 × 0.5–2	Gram-positive
Saccharomyces cerevisiae	Ellipsoidal shape 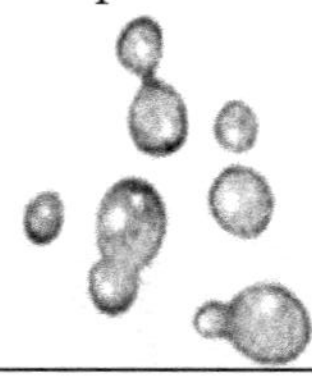	3–15 × 2–8	Gram-positive

(*continued on next page*)

Staphylococcus aureus	Small, irregular coccoid 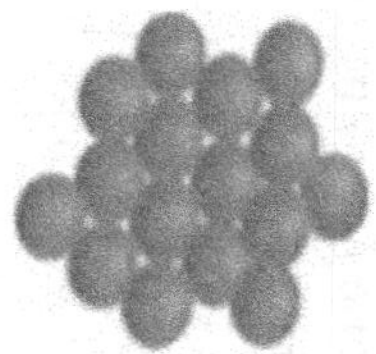	0.8–1	Gram-positive
Salmonella typhimurium	Straight rods 	0.7–1.5 × 2–5	Gram-negative
Escherichia coli	Straight rods 	1–1.5 × 2–6	Gram-negative

microorganisms, not just in the area of power ultrasonography, as well as in other food preservation technologies. Because of its bigger size, the latter has indeed been proven to be less susceptible to ultrasonography than some other vegetative cells. This microorganism's deactivation has indeed been demonstrated in dietary models like freshwater, phosphate buffers, and Sabouraud broth (S. Guerrero et al., 2001). *Staphylococcus aureus, Pseudomonas fluorescens,* and *E. coli* have been shown to be inactivated in water and phosphate buffers (Cameron et al., 2009).

The deactivation of *Listeria innocua* and *mesophilic* bacteria in uncooked entire milk was studied using ultrasonication combined with temperature (Bermdez-Aguirre et al., 2009). When ultrasound was used in conjunction with heat, the death rates were higher as when the heat was used alone, indicating a synergistic rather than additive impact. With Bacillus species, ultrasound exhibited a high degree of deactivation under various treatment settings and medium. *Listeria monocytogenes* were shown to be inactivated by high-power ultrasonic waves at ambient temperature and pressure, having decimal reduction values of 4.3 minutes. Nevertheless, this might be enhanced by raising either the pressure (manosonication) or the sonication power. Manothermal sonication (MTS) was shown to be more successful than other methods of deactivation (Pagan et al., 1999).

The decimal reduction number declined rapidly over 1.5 minutes to 1.0 minute by merging sub-lethal warmth and greater pressures of 200 kPa. The use of manosonication rather than heat alone just to neutralize bacteria has the benefit of a much reduced recovery rate and did not show up to be reliant on the treatment media used. The decimal reduction times of *Streptococcus faecium, Listeria monocytogenes, Salmonella enteritides,* and *Aeromonas hydrophila* were reduced by manosonication at 40°C and 200 kPa (Pagan et al., 1999).

Such rates were further decreased by increasing the ultrasonic amplitude. The decimal reduction rates were similarly reduced when the pressure was increased from 200 to 400 kPa. Heat resistance was twice as high in cells grown at 37°C as it was in cells produced at 4°C. Cavitations are much more difficult to create at greater viscosities; however, the eventual collapse of the bubble is much more violent. Researchers identified a linear link between power entering the media and decimal reduction times using a calorimetric approach of measuring the power used straight to the material (Manas et al., 2000). This implies that increased ultrasonic strength might be used to produce higher death rates, which would be directly proportional to the frequency and concentration of the ultrasonic power going through the system. The integration of TS, MS, or MTS approaches with certain innovative technologies has recently been demonstrated to be successful in the inactivation of microorganisms (Walkling-Ribeiro et al., 2009).

Furthermore, the TS/PEF mixture has demonstrated that orange juice may be stored for a long time without losing its sensory characteristics. In compared to hot water, steam vacuum, and lactic acid treatments, the combination of steam plus ultrasound was determined to become the most expense way for inactivating Salmonella (Lawson et al., 2009). However, the synergistic benefit of mixing ultrasonic and nonthermal plasma treatments with aeration was not detected experimentally, indicating that the combination impact may not always be lucrative (Chen et al., 2009). Ultrasound therapy not only helps to immobilize germs, but this also helps to enhance the eminence of the finished product (Riener et al., 2010).

3.4.2.2 Inhibition of spores

High heat and osmotic pressures, as well as elevated/low pHs as well as psychological stresses, are not a problem for microbiological spores. Because of decomposition and toxicity, bacterial spores which stay alive heat handling may drastically limit life span of thermally treated foods. Bacillus and Clostridium endospores are extremely resistant to severe temperatures; for instance, *Bacillus thermophilus* spores can be killed by heating at 100°C for 4 hours.

Denaturement of *Bacillus subtilis* spores is incredibly difficult. Like a consequence, spores can frequently use to evaluate the efficacy of various treatment/sterilization techniques, as their great resistance allows the effectiveness of the procedure to be measured against the bacillus spore's death rate. When the influence of a 20 kHz ultrasonic probe was combined with the impact of static pressure on specimens, it was discovered that boost in static pressure lead to an augment in spore deactivation. Over 99% of the spores were inhibited after a 12-minute manosonication treatment at 500 kPa. The amount of inhibition was enhanced by increasing the amplitude of ultrasonic vibration of the transducer, that is, the acoustic force going through the system; for instance, a 20 kHz probe at 300 kPa, 12 minutes sonication at 90 m amplitude inhibited 75% of the spores. Extend the frequency to 150 m results in spore inactivation of 99.5%. Lastly, comparing to thermal treatment solely, raising the treatment's thermal temperature resulted in higher rates of deactivation at 300 kPa. *Bacillus cereus* and *Bacillus licheniformis* spores have just been discovered to be difficult to treat with ultrasonic waves singly (Palacios et al., 1991).

Cryptosporidium parvum, a protozoan, has been identified as a cause of the water-borne illness Cryptosporidiosis in humans, with several instances reported in the media over the previous decade. These epidemics have happened in both Europe and the United States, and they are not limited to underdeveloped nations. The major worry of water providers is that traditional treatment procedures are insufficient in avoiding cryptosporidiosis spread through water. The most regularly used disinfection treatments in water treatment, such as chlorination and ultraviolet radiation, are unsuccessful, however ozone, in conjunction with ultrasound, appears to be able to eliminate a few of these oocysts (Biwater Treatment Ltd, 1993).

3.4.2.3 Down regulation of enzymes

The enzyme must maintain its original shape to avoid denaturation. The three-dimensional molecular structure of globular proteins is stabilized by hydrophobic interactions, hydrogen bonding, van der Waals contacts, ion paring, electrostatic forces, and steric restrictions. Enzymes should be inactivated or their activity lowered to stabilize some food ingredients. In reality, proteolysis produced by enzymes such as proteases may induce taste and dark pigmentation deficiencies. Thermal treatment is a simple way to inactivate enzymes. Unfortunately, in some circumstances, enzymes' great heat tolerance might be an issue since heat can alter meal qualities such as taste, coloring, and nutritional content. It is one of the reasons why high-power ultrasound, or sonic waves exceeding 20 kHz, is becoming more popular as an alternate way of enzyme inactivation.

Ultrasonic waves have a wide range of impacts upon proteins. Whenever oxygen is available, polymeric globular proteins are broken down in components, as well as the quaternary structure is not reversible. Lipoproteins could be partially dilapidated, while proteins could be hydrolyzed and polypeptide chains disrupted if an ultrasonic irradiation goes prolonged sufficient. This production of free radicals via cavitations has now been linked towards the effect of a gas just on degree of enzyme inactivation. The susceptibility towards ultrasounds is determined by the treatment parameters and the enzyme's composition. By practice, ultrasonication combined with the other therapies is much more successful in inactivating dietary enzymes. By reality, when matched to ultrasound alone, MTS therapy is more efficient. Numerous enzymes are inactivated by MTS therapies at room temperature and/or in less time as thermal processing around the same temperatures.

Heating was commonly used to preserve orange juice, although many thermoresistant enzymes are challenging to deform with only warmth.

Pectin methylesterase (PME) in citrus is well guarded versus thermoinactivation by its own substrates pectin. Conversely, the existence of such a compound has no influence upon MTS therapy, so enzyme inhibition remains successful (Vercet et al., 1999). Manothermosonication with pressures of 200 kPa over citrate buffer and orange juice was used to cure PME found in oranges. Its deactivation of an enzyme by MTS at 35.5°C was calculated to be seven times more than by heat treatment only.

The level of polymerization of pectic compounds determines overall viscosity of cooked tomatoes, which is one of the most significant quality parameters (Vercet et al., 2001). The viscosity of tomatoes decreases throughout conversion due to the synergistic activity of endopolygalacturonases (PGs) and pectin methylesterase (PME). Such enzymes depolymerize pectin chains in pulp or serum, resulting in a reduction in resultant viscosity. Throughout processing, certain enzymes are freed and must be inactivated as soon as feasible. To inactivate pectic enzymes, tomatoes are often quickly cooked at temperatures ranging from 82 to 104°C. Slicing or smashing occurs shortly after this phase. When MTS was used to inactivate PGs and PME, it was more effective than just heating them. When ultrasound and enhanced pressure were combined at 37°C, larger levels of deactivation were attained than when heat therapy was used at 62.5°C alone (Lopez et al., 1998).

By use of thermosonication were investigated in order to achieve fewer harsh thermal procedures than traditional blanching and, as just a result, in create an novel and healthful freezing foodstuff, watercress (Nasturtium officinale), with few color or taste alterations throughout its life span (Cruz et al., 2006). The main enzymes are involved with in browning of fresh fruit and veg are peroxidase and polyphenol oxidase. Browning seems to be a significant limiting element in the acceptability of customers. The researchers discovered that using ultrasound and ascorbic acid together had a synergistic suppressive impact on multiple enzymes involved in enzymatic browning (Jang and Moon, 2011).

3.4.3 Ultrasound application

Ultrasound seems to be a popular topic with in food sector these times. Ultrasonic machinery may be used in companies, so it is recognized like a sustainable new technology because of its significance towards environmental sustainability. There are three types of ultrasonic approaches: (1) Direct application towards the material, (2) Apparatus connection, and (3) Immersion inside an ultrasound cleaner (Chemat et al., 2011). By addition, depending to the power produced mostly by acoustic waves, ultrasonic operations within food sector generally separated into two distinct groups.

The sound power (W), sound energy density (Ws/m^3), and sound intensity (W/m^2) of low and high energy ultrasounds are used to classify them.

Ultrasound operations with low power are done using frequencies greater than 100 kHz and intensities less than 1 W/cm^2. Low intensity ultrasound uses a low power level to ensure that it must be noninvasive and thus does not alter any physiochemical characteristics of food. Within food sector, low-intensity ultrasound being commonly utilized as diagnostic purposes to obtain knowledge on the physicochemical qualities of products, which including content, morphology, and physiological state (Jayasooriya et al., 2004).

Ultrasonic treatments with high power are often done using frequencies around 18 and 100 kHz with intensities more than 1 W/cm^2 (normally in the range of 10–1000 W/cm^2). Ultrasonic waves can cause devastation at such a power level owing to their mechanical, structural, or biochemical impacts. Since many decades, high-intensity ultrasound has now been utilized to make emulsions, disturb cells, as well as disperse aggregated substances. It is now employed for quite a variety of reasons, including crystallization process optimization and management, liquid food degassing, enzyme inactivation, improved dryness as well as filtering, and the stimulation of oxidation process (Zheng et al., 2006).

3.5 Microwave in food preservation

3.5.1 Brief introduction to microwave heating

Microwaves (MWs) comprise of electromagnetic (EM) waves that travel on the velocity of light in open room and therefore be synchronized perpendicular oscillations of electromagnetic fields. The frequency (around 300MHz and 300GHz) and wavelength of MWs are defined (ranging from 1 m to 1 mm). Five wavelengths (433, 896, 915, 2375, and 2450 MHz) are approved for MW thermal activities, depending on the nations and areas. Household appliances use the 2450 MHZ wavelength exclusively.

3.5.1.1 Microwave heating methods

The contact of a wave with stuff is determined by the wave's own properties (frequency, wavelength) as well as the material's nature, notably its absolute permittivity, a complicated quantity that controls how the substance accumulates and dissipates the electrical energy of the EF. For more comprehensive information, users might check additional specialist publications (Datta et al., 2001). We can quickly describe the actual absorptive, or dielectric invariable, of a substance, that signifies the material's capability to hoard electrical power, as well as the efficient dissipation aspect, that conveys the resources capability to take up wave energy as well as disperse it as heat through dielectric recreation along with ionic transmission.

Whenever a stuff containing gratis charges (ions) and polar molecules is exposed to an EF, the ions will travel at an accelerated pace proportional to their charge, causing collisions and, as a consequence, a transformation of kinetic energy into heat (ohmic heating). Similarly, the polar molecules of this substance, which were previously randomly orientated, will now be oriented according to the field's polarity. Such compounds will spin to stay aligned on the EF if it is an alternative. This dipolar rotation will cause friction among the molecules, resulting in internal heat production (dielectric heating).

3.5.1.2 Wave's transmission and absorption in a substance

As an EM wave is aimed towards a substance, a portion of it is reflected on the surface, as the rest enters the medium and is absorbed. The inner EF's amplitude and hence its power are reduced as a result of the wave's absorption during its crossing. The use of the Maxwell equations to calculate the internal electric field is advised for tiny or thin materials. For bigger entities, Lambert's law (exponential EF decaying) can be employed.

The penetration deepness is measured while the distance through the material for which the incident wave's 64 percent incident power had been engrossed. This deepness is determined by the material's dielectric characteristics as well as the wavelength. At the identical circumstances, the deepness of penetration at 915 MHz is greater than the penetration depth at 2450 MHz. Once the loss factor is large, greater power is consumed (Datta et al., 2001).

3.5.2 Drying in the microwave

3.5.2.1 Introduction

Among the most common techniques for preserving food and avoiding microbial damage is drying. Food drying can be employed for economic reasons, such as downsizing a food to save transportation expenses (e.g., milk powder), or for consumer reasons, such as producing novel textures and/or goods (e.g., prunes). By eliminating part of the water from the food product, the water activity (Aw) is reduced. The sorption isotherm curve, which depicts the connection among a product's moisture content and its water activity, may be used to calculate how much water has to be eliminated to ensure microbial viability.

In general, the water activity threshold beyond which pathogenic bacteria cannot proliferate more is 0.85–0.86, for yeasts is 0.7 and 0.6 for molds. It is worth noting that Aw not only regulates the rate of chemical and metabolic processes inside the food, but it also effects the proliferation of microorganisms. The nonenzymatic browning (Maillard reaction) is most active around Aw of 0.6–0.7, whereas lipid oxidation occurs more often at extremely low and very high Aw levels. As a result, these factors should be used to alter the ultimate moisture of the product. The array of Aw between 0.3 and 0.4 is commonly desired for optimal dried product stability. Drying has essentially little deadly impact on the microbes near on the item owing to inadequate product temperatures achieved while drying, which is an important point to underline. To minimize moisture restoration and maintain the dried product's quality, appropriate packing is required.

Depending on the type of the commodity to be dried (fluid, doughy, solid, or particle) as well as the intended final attributes, industrial drying procedures vary greatly. In the food industry, hot air (HA) direct drying is the mainly widespread dehydration process. Unfortunately, this approach has a number of drawbacks, including poor terminal quality and poor operational performance. A heat relocate from a warm as well as dried up environment medium to the product and subsequently a mass move from the product to the nearby environment are the two processes in a traditional drying technique.

In common, exterior temperature and mass relocate may be readily regulated by careful monitoring of drying air properties; hence, interior temperature and mass transport is the drying operation's preventive step and efficient driving power. With time, the pace of drying slows down, making removal of water more difficult. As a result, traditional HA drying necessitates the use of harsh situation, especially at the conclusion of the process, resulting in over-heating and over-drying of the product surface. Engineers had been tremendously driven to design as well as suggest innovative drying procedures such as MW drying as a result of these factors. Dry potatoes chips, pasta, and snacks, for example, have all been effectively dried using MW drying (Schiffmann et al., 2001). The literature resurrected several MW drying investigations (Chandrasekaran et al., 2013).

3.5.2.2 Microwave drying methods in combined effect

In spite of its numerous rewards greater than traditional techniques, MW drying had a number of critical flaws, as discussed above. In comparison to MW drying only, MW drying coupled through other traditional heating

Figure 3.3 Stainless Steel Microwave Vacuum Dryer/Microwave Vacuum Drying Machine.

methods improves drying competence and dried quality of products. MW-assisted hot air (HA) drying, MW vacuum drying, and MW freeze drying are some of the most common uses of integrated MW drying.

3.5.2.2.1 MW-assisted hot air drying

Utilization of MW power (interior heating) in combination through warm air circulation is a smart way to get around some of the issues that occur with using these two approaches independently. MW heating could be used at beginning of the drying procedure to quickly heat the product's inside layers. Whenever the temperature profile is determined in the second phase of the drying process, MW heating may be used to drive vapor out of the product, resulting in a spongy makeup. MW heating might also be used at end of the drying procedure to limit mass relocate and increase drying rate by eliminating bound water (Zhang et al., 2006). For instance when MW drying was coupled with HA drying at end, the drying occasion was lowered by 63% when compared to convective air drying (Maskan, 2000). Other fruits that have been reported to benefit from this procedure include blueberries (Zielinska et al., 2016), macadamia nuts (Silva et al., 2006), and green peas (Chahbani et al., 2018).

3.5.2.2.2 MW vacuum drying

To retain the purity (color and taste) of dehydrated items, MW heating could be used with a vacuum to lower boiling point of water and avoid oxidation reactions, Fig. 3.3 shown a vacuum microwave dryer. Apple slices (Schulze et al., 2014), pumpkin (Lemos et al., 2018), and cranberries were

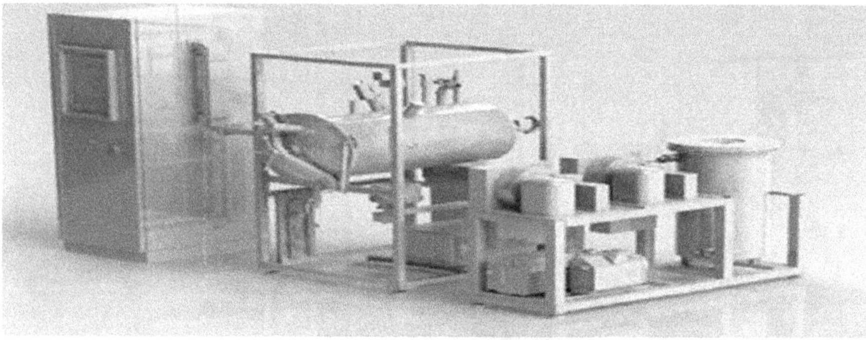

Figure 3.4 Microwave freeze dryer of fruits and vegetables.

effectively dried using this approach. In conditions of energy usage, drying time, and product eminence, this approach outperforms MW air drying (Zielinska et al., 2017).

3.5.2.2.3 MW freeze drying

Lyophilization, often referred as freeze drying, seems to be a low-temperature drying method that includes initially freezing the item and then sublimating the ice under low-pressure circumstances to remove the ice. It is utilized in the food and pharmaceutical sectors to dehydrate especially heat-sensitive products. This approach keeps the structure intact while reducing the amount of important chemicals lost. Fig. 3.4 shows microwave freeze dryer of fruits and vegetables.

However, the expensive pricing of this technique limits its application. In the case of MW freeze drying, MW heating could be used simultaneously with sublimation to provide heat under vacuum circumstances. After a typical lyophilization procedure, MW heating can be used individually. Owing to its low processing temperature and absence of oxygen in the processing atmosphere, MW freeze drying has a number of benefits (Li et al., 2018; Kröncke et al., 2018). Thermal runaway may occur because the MW energy does not interrelate with frozen water, resulting in poor product quality (Chandrasekaran et al., 2013).

3.5.3 Foods pasteurization and sterilization using microwaves

Several commodities are pasteurized or sterilized in order to increase their storage life. Pasteurization's major aims are to kill vegetative harmful microorganisms and disable certain enzymes in foodstuffs. Pasteurization temperatures and action times are resolute mostly by the product's type, pH, and target microbe. The material is heated to 60–85°C for a duration ranging as of a small number of seconds to an hour in most pasteurization techniques. Pasteurization necessitates refrigeration storage (3–4°C) for a 2–6 week shelf life.

Bacterial spores are destroyed during sterilization, which is similar to pasteurization. Conduction from the exterior to the core, frequently referred to as the "cold" point, is the primary mode of heat transmission in solid or semisolid objects. These results to the application of more harsh conditions in order to attain the required temperature at the freezing point, resulting in surface overcooking and a decrease of product quality. Thermal treatment optimization (i.e., optimizing bacteria deactivation while reducing nutritional degradation) is so critical. This is not, obviously, a simple process, but it is always a technological and scientific difficulty. The immediate and substantial contact among MWs and foodstuff allows MW heating to overcome the limitations established by traditional heating's sluggish thermal diffusion (Ahmed et al., 2007).

3.5.3.1 Microwave disinfection of microorganisms

Several researches have shown that employing microwave heating for pasteurization along with sterilization of food is successful (Salazar 2012 and Shaheen et al., 2012). Moreover, *Bacillus cereus, Campylobacter jejuni, Clostridium perfringens, Escherichia coli, Enterococcus faecalis, Listeria monocytogenes, Staphylococcus aureus, and Salmonella* have all been inhibited by MW heating (Gedikli et al., 2008; Zhou et al., 2010).

3.5.3.2 Microwave heating's thermal and athermal impacts on microbes

The investigation of microbial annihilation processes throughout MW heating had piqued curiosity. The presence of nonthermal (athermal) impacts of microwaves, in particularly, is a point of contention. Numerous ideas had offered to elucidate how electromagnetic fields might render bacteria inactive at temperatures below deadly. The interaction of microwaves with specific cellular elements might cause this effect. Many researches, on the other hand, have disputed the nonthermal deadly impact of MWs. Most studies use experimental evaluations of conventional and MW inactivation under identical heating settings to differentiate among thermal and nonthermal effects (Heddleson and Doores 1994; Valsechi et al., 2004).

Fujikawa et al. (1992) found no difference among deactivation procedures within a standard water bath and in a MW oven for *E. coli*. (Welt et al., 1994) created a machine to assess the potential nonthermal impacts of microwaves. They examined the inactivation of *Clostridium sporogenes* spores in a model medium in favor of usual and MW action under the same time and temperature conditions. The data show that there is no nonthermal impact. Microwaves, on the other hand, have been shown in multiple investigations to have a greater bactericidal impact (Barnabas et al., 2010). The inactivation of *E. coli* K12 subjected to MW radiation was greater than that achieved in a water bath at the same temperature (45 and 50°C), according to Sato et al. (1996).

Microwaves have a better bactericidal impact, according to certain writers. Kozempel et al. (2000) built an experimental system to detect a nonthermal action of microwaves on microorganisms at low temperatures. Microwaves provide quick energy to the food system, which is combined with fast thermal energy removal. A double-tube heat exchanger is employed, which is positioned within a constant MW channel. The outside pipe may be microwaved, whereas the internal pipe is made of stainless steel and was utilized to quickly cool the system to stay it at 45°C.

In diverse solutions, no nonthermal impact of MW radiation was seen on yeast, Pediococcus sp., *Escherichia coli, Listeria innocua*, or *Enterobacteraerogenes*. In deadly conditions, however, the author claims that MWs can get better or exacerbate the thermal impact. In the same vein, Ramaswamy et al. (2002) discovered that at sublethal temperatures (<40°C), inactivation of *S. cerevisiae* inoculated in apple juice treated by means of steam, burning water, or MW was not statistically different. MW radiations, on the other hand, increased inactivation for the same fatal temperature circumstances (55–65°C).

Several researches have been conducted on the interactions between MWs and various biological elements like as DNA, the cell membrane, enzymes, and proteins. Kakita et al. (1995) investigated the survival of the *Lactobacillus casei*-specific bacteriophage PL-1 under MW irradiation. When MW heating was used instead of conventional heating, more viral DNA fragmentation was seen.

3.5.3.3 Microwave microbial decontamination's benefits and drawbacks

Inside reality, due of the intrinsic differences in heating principles among both the two techniques, comparing the efficiency of MW to conservative heating is challenging (Shamis et al., 2012).

A gradual heat swap happens amid the heating standard and the manufactured goods in traditional heating. The physical features of the product influence heat diffusion inside it (specific heat, thermal conductivity, porosity, etc.). Foodstuffs, on the other hand, are known to be poor thermal conductors, and heat dispersion in the direction of the freezing region is typically sluggish. MW heating, on the other hand, is normally quicker than traditional heating. The heat is created within the meal itself. This straight volumetric heating shortens the processing period, allowing more minerals, sensitive vitamins, and aromatic compounds to be retained. Microwave-processed meals may have a superior texture, flavor, and appearance than conventionally processed foods. For instance, Lau and Tang (2002) presented twice quicker MW pasteurization treatment of pickled asparagus at 915 MHz.

When compared to a traditional water bath treatment, this benefit considerably reduced asparagus thermal deterioration. Likewise, MW pasteurization of foiegras was shown to be satisfactory, with a 50% time savings and improved organoleptic quality as compared to a conventional approach (Massoubre, 2003). Furthermore, diverse packaging equipment may be pasteurized using microwaves (Tang et al., 2008). This technique may also be used in conjunction with added techniques, like wise infrared heating for surface cooking (Datta and Rakesh, 2013).

Despite the many benefits of MW decontamination technique, heating heterogeneity is a severe issue that results in insufficient microorganism inactivation. Pathogens such as Salmonella spp. (Heddleson et al., 1996) and *Listeria monocytogenes* have been shown to survive in meals cooked in microwave ovens (Farber et al., 1998).

The impact of temporary MW exposures on *Escherichia coli* K12 along with *Campylobacter jejuni* inoculation on chicken flesh was examined by Goksoy et al. (2000). The temperature difference among different portions of the sample was 20°C. The specimens exposed to MWs exhibited symptoms of incomplete cooking zones, according to the authors. There was no indication that MWs had any bactericidal impact on bacteria after a brief exposure (up to 30 seconds). These findings were validated by Apostolou et al. (2005). *E. coli* O157: H7 was not eliminated by MW heating of chicken parts. There was also a large temperature range (from 67 to 92°C).

Heterogeneous heating, on the other hand, results in a significant loss of food quality. Localized overheating frequently causes irreversible color changes where the temperature is higher (Tang 2015). Due to wave reflection, these phenomena are most noticeable at the product's corners and edges. The problem of cold spot localization is another issue with MW decontamination. The cold point of a typical thermal process is well characterized and generally positioned in the product's core. One point of temperature surveillance within the product throughout MW pasteurization is insufficient to assure food safety (Hamoud-Agha et al., 2014).

Schnepf and Barbeau (1989), for example, looked at the deactivation of injected Salmonella in chicken. Their findings revealed that monitoring interior temperature throughout MW action does not represent surface inactivation, which occurs at a lower temperature. The chemical indicator approach established by Kim and Taub (1993) was effectively employed to find the cold spot (Auksornsri et al., 2018). Numerical simulations, when combined with experimental studies, are highly suggested for locating the cold spot and obtaining an accurate research in order to create a realistic MW decontamination procedure (Hamoud-Agha et al., 2014).

3.5.4 Application of microwave in food preservation

In-pack protection of packed meals which be heated constantly on a conveyor belt, batch-wise sterilization, and tube in-flow sterilization of semi-fluid pump able foods are all applications for microwave pasteurization and sterilization (Raaholt, 2020).

Microwave drying takes twenty times less time as for traditional drying methods. Further benefits include large savings in terms of built space, power and steam use, and the usage of washing chemicals and maintenance. The product quality has increased, the look has enhanced, and harmful bacteria (staphylococci and salmon) have been eradicated additional carefully. Such facilities are utilized to dry pasta and produce "chips." Microwaves might as well be utilized to dry vegetable products, with a dehydration time of around 20 minutes and a humidity reduction of 60%–20% (Project ADER 7.5.1, 2020).

Drying is a common activity in industry operations that involves using heat to detach a liquid from a solid, resulting in the liquid phase evaporating. Drying is an energy-intensive process, and dryers are costly pieces of apparatus, thus it must be done as efficiently as possible. Due to the selective nature of microwave contact with the object to be dried, the drying process using microwaves, which employs a fundamentally different mechanism of energy relocate, could be more favorable than traditional drying (Radoiu, 2020).

Soups, tomato extracts, vegetable proteins, meat and liver extracts, instant drinks, coffee extracts, malt and hop extracts, chocolate mass, vitamin extracts, enzymes, and antibiotics, and microorganism cultures can all be vacuum dehydrated using microwaves.

Microwaves have been proven to inactivate anti-nutritional elements de several agri-food items, particularly those of a protein type, because the food is heated through its whole bulk. Following 240 seconds of microwave treatment, thermo-stable aflatoxins in peanut mass can be removed. Microwaves can also be used to lyse microorganisms in conjunction with other methods such enzymes, chemicals, ultrasound, psychological stresses, pressure, and thermal shocks.

Microwaves are used for 150 seconds to thermolyze a variety of microbial cells, including *Trichoderma album, Geotrihumcandium, and Saccharomyces cerevisiae*, allowing the release of cellular components, which is critical for plasmolysates, autolyses, and protein hydrolysates. Microwaves can be used for analytical purposes such as determining the dry matter composition of milk products, molasses, and preserving cans, as well as determining the moisture content of granular and powdered items (Project ADER 7.5.1, 2020).

3.6 Ultrasound-assisted extraction

Due to the drawbacks of existing extraction techniques, the food and chemical industries have been forced to develop innovative "green" extraction methods that customarily use fewer solvent and energy, like as microwave extraction. Partition under severe or nonclassical settings is a hot subject in practical study and manufacturing at the present.

3.6.1 Techniques of extraction

Ultrasounds are mechanical waves which must propagate via an elastic material. The frequency of the wave is the distinction among sound and ultrasounds; sound waves are at human earring frequencies (from 16 Hz to 16–20 kHz), but ultrasounds have frequencies that are higher than human earring but lower than microwave frequencies (from 20 kHz to 10 MHz). When a sound wave travels from an elastic material, it causes particles to move longitudinally. In essence, the sound wave's source acts as a piston on the medium's surface (Mason, 1990).

This is resulting in a sequence of compression and rarefaction phases into the media. Whenever the piston is in its opened position it produces a compression into the medium and while the piston is in its contracted position it causes a rarefaction phase. Each medium has a crucial molecular spacing: underneath this number, the liquid stays unchanged; beyond this amount, the liquid breaks down, allowing voids to form in the liquid. If the rarefaction cycle is solid sufficient inside the situation of ultrasounds, the distance (d) among adjacent molecules can reach or even surpass the liquid's critical molecular distance. The cavitations bubbles that are sensitive to the ultrasonic impact are the voids generated in the medium. In reality, these cavitations bubbles can expand during rarefaction periods and contract during compression phases.

Whenever the volume of these bubbles reaches a crucial threshold, they burst and unleash a tremendous quantity of energy during a compression cycle. In an ultrasonic bath at ambient temperature, the temperature and pressure at the moment of collapse had been calculated to be up to 5000 K and 2000 atmospheres. This forms hotspots, which can drastically increase the chemical reactivity in the medium. When these bubbles collide with a solid material's surface, the tremendous pressure and temperature produced create micro jets that are aimed at the solid surface. These micro jets are accountable for ultrasounds' degreasing impact on metallic surfaces that is commonly employed for cleaning materials. The separation of vegetal chemicals is another use of micro jets in the food business. A cavitations bubble as shown in Fig. 3.5 can be formed near to the plant material surface (a), therefore this bubble collapses (b) during a compression cycle, resulting in a micro jet directed at the plant matrix (b and c). The high pressure and temperature used in this procedure will dissolve the plant matrix's cell walls, allowing its contents to escape into the medium (d). This is an intriguing technology for extracting ingredients from natural items.

3.7 Synergism of ultrasound and microwave in food processing

Ultrasound, particularly high-intensity (frequencies among 18 and 100 kHz) ultrasound, is extensively used in food processing, including food fermentation (Ojha et al., 2017), cutting (Liu et al., 2015), crystallization (Povey, 2017), food preservation (Demirdöven, 2008), and drying (Musielak et al., 2016). The goal of using high-intensity ultrasounds is to modify the medium in which the waves travel and enhance the competence

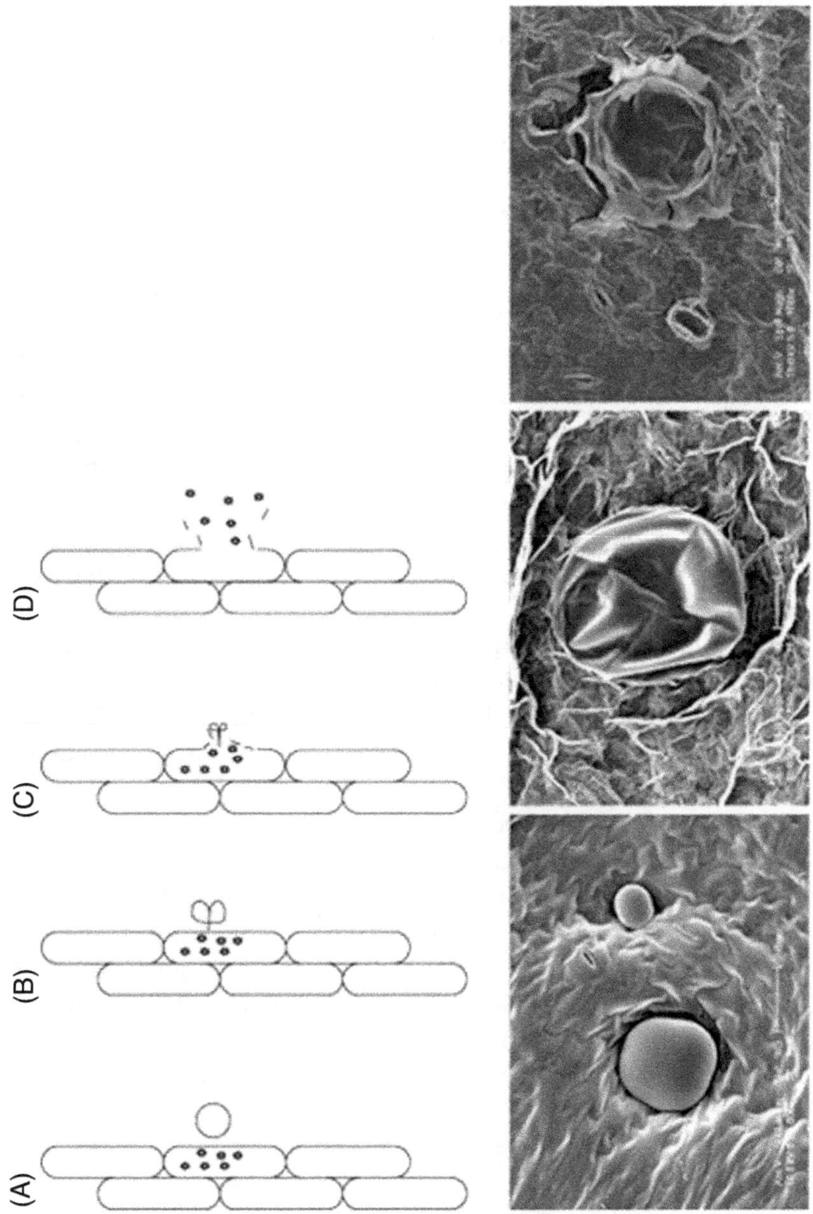

Figure 3.5 Cavitation bubbles burst and plant stuff spills out. An example of basil essential oil extraction.

of the food processing (Musielak et al., 2016). Ultrasound-enhanced heat and mass transmission when it passed through a medium.

The main reason for the improvement was that acoustic cavitations caused by ultrasonic waves made water evacuation easier (Chemat and Khan, 2011; Fan et al., 2017). The heating result, which is produced by the waste, of mechanical energy, and the "sponge effect," that is created by the alternate compression and expansion cycles formed, both has an impact. The "sponge" effect, in particular, could effect in the creation of tiny channels and, as a result, easier substance exchange due to the difference in force among the material's inner and outer surfaces (Yao, 2016).

High-intensity ultrasounds were used in two ways during the drying and dehydration process: ultrasound aided drying and as a pretreatment. Ricce et al. (2016) discovered that pretreating carrot slices with ultrasound improved their drying and rehydration abilities. According to Deng (2008), drying apples using ultrasonic pretreatment can reduce water activity while also saving energy. Chen et al. (2016) found that the drying time was shortened and the quality of the dried carrot slices was improved. The use of ultrasound in conjunction with a drying process was also investigated. Bantle and Eikevik (2011) used a mix of ultrasonic and freeze drying to dry the peas. Ultrasound had a favorable influence on the drying rate, according to the findings. Rodrguez (2014) found that using ultrasound at the right temperature and acoustic density throughout convective drying of apple slices can reduce the loss of whole polyphenol and flavonoid. As a result, ultrasonography helped to improve the quality of drying items.

Vacuum frying has long been used as a method of drying vegetables and fruits for a variety of delectable treats. The color of vegetable chips is an essential food quality trait that influences customer decisions (Nowacka et al., 2017). Low-oil goods are also a highlight for customers when it comes to their health. In compared to vacuum frying, the quality criteria of frying chips, such as color, moisture content, oil content, and structure, are considerably improved when employing microwave-assisted vacuum frying technology (Quan et al., 2014; Su et al., 2015; Pedrós-Garrido et al., 2017; Su et al., 2017).

The microwave is utilized as a heating foundation during this procedure to reduce the frying temperature and hence enhance the quality (Su et al., 2015). In addition, in the convective drying process, a mixture of ultrasound and microwave has indeed been effectively employed. Quality of the product and drying effectiveness may both be considerably improved. The finest color parameter of green pepper was attained while the ultrasound and microwave were employed concurrently throughout convective drying, according to Szadzinska et al. (2017). Microwave and ultrasound-aided convective drying of strawberries considerably enhanced heat and mass transfer competence, according to Szadziska (2016). Moreover, no one has yet documented using ultrasound and microwave at the same time in the vacuum frying procedure.

3.8 Synergism of ultrasound and microwave in food extraction

One of the most successful hybridized techniques (UMAE) that provides rapid and efficient extraction is combining UAE with microwave-aided extraction (MAE) (Sillero et al., 2020). Numerous authors have explored the use of combined UMAE irradiation in natural extraction (Cravotto and Binello, 2010), but the hybrid procedure's thoughtful potential has yet to be fully exploited. Because of its great efficiency and much quicker extraction time, we consider UMAE has enormous promise in academic and industry research. Using recent process intensification technologies, this is a cost-effective way for quick sample processing. Furthermore, when nonmetallic horns are just utilized at modest strength, double synchronous irradiation has additional and synergistic impacts on the vegetal matrix extraction phenomena.

According to Cravotto and Cintas, horns constructed of Pyrex, quartz, or Peek can be safely used up to 90 W, after which the material's internal structure is irrevocably ruined. However, it has a minor disadvantage in that UMAE needs low power levels from only two single energy sources. Via the cavitations phenomenon, ultrasound considerably improves the target extraction component. The action of mechanical ultrasonication enhances the removal of soluble chemicals by the plant body due to cell wall breakage, boosting mass movement, and allowing solvent access to cell material. In the intervening time, the whole sample is rapidly heated by microwave, causing dissociated molecules to migrate. Irradiation also promotes solvent penetration into matrices, enhancing analyte solvation while also improving the solubility of target components (Cravotto and Cintas, 2007).

Garcia-Vaquero et al. (2020) discovered that when UMAE was compared to UAE and MAE independently, UMAE extracted a greater quantity of phenolic compound and antioxidant from brown macroalgae. In addition,

Wang et al. (2020) investigated UMAE for flavonoid compound extraction from Eucommia ulmoides leaves and found that UMAE extracted 2.454% higher yield than ultrasonic and microwave extraction. Cravotto et al. (2008) effectively applied UMAE as a complementary approach for extracting oils from vegetable sources, such as soybean germ and DHA-rich farmed seaweed.

3.8.1 UAE in compared to MAE

Microwaves allow extraction and distillation to be finalized in minutes rather than hours, with elevated production, dropping solvent intake, rationalizing manipulation and work-up, increasing final product purification, eradicating waste water post-treatment, by using a fraction of the energy required for traditional separation processes like as distillation or solvent extraction. Essential oils, fragrances, pigments, antioxidants, and other chemical substances have all been successfully derived from several of matrices. More efficient heating, rapid energy transfer, lowered thermal gradients, specific heating, reduced equipment size, rapid response to process heating control, quicker start-up, increased production, and eradication of process steps are some of the benefits of using microwave energy, which is a noncontact heat source (Chemat and Lucchesi, 2006).

MAE is gaining popularity because it enables for the effective extraction of precious chemicals from rock-hard samples using microwave radiation Pan et al. (2003) discovered that MAE extraction increases the competence of polyphenol and caffeine extraction from green tea leaves. The moisture content of specimens, on the other hand, is a determining criterion for yield restoration. The recovery yield declines considerably when utilizing dry materials. In addition, Molins et al. (1997) found that using hexane as the only solvent in the presence of a totally dry sample was ineffective. Whenever the solvent is lacking a substantial dipole moment for microwave energy absorption, MAE efficiency is often poor. As a result, the procedures are constrained by solvents and the composition of the solid substance.

Ultrasound-assisted extraction is a promising new technique that can speed up heat and mass transfer and has already been applied in the extraction area. Ultrasound waves influence the physical and chemical characteristics of the exposed plant material, and their cavitational impact stimulates the discharge of extractable chemicals and increases mass transfer by destroying plant cell walls. UAE is a safe procedure that eliminates the need for a big amount of solvent while also reducing the amount of time spent working. Ultrasounds have been used in the field of plant extraction for a long time.

Table 3.3 Advantages and drawbacks of UAE and MAE extraction techniques.

	UAE	MAE
Brief description	Utilizing a US probe or a US bath, the sample is submerged in solvent and subjected to ultrasound.	Microwave energy is applied to a sample that has been submerged in a solvent.
Extraction time	10–60 min	3–30 min
Sample size	1–30 g	1–10 g
Solvent use	50–200 mL	10–40 mL
Investment	Low	Moderate
Advantages	Easy to use	Quick It's simple to use Use of solvents is moderate
Drawbacks	A significant amount of solvent is used. Filtration is necessary	Microwave energy must be absorbed by the extraction solvent. Filtration is necessary.

Ultrasound has long been known to have a considerable impact on the pace of several chemical and culinary processes.

The exploit of ultrasound for the extraction of natural compounds which would normally take hours or days to complete utilizing customary procedures has received a lot of interest. Full extractions can now be finished in minutes by means of high reproducibility using ultrasound, lowering solvent expenditure, simplifying manipulation and work-up, improving ending result pureness, eradicating waste water post-treatment. Numerous classes of food components, including fragrances, pigments, antioxidants, and other organic compounds, had successfully extracted and studied from a range of matrices (Luque-Garci'a et al., 2003).

In compared to these innovative extraction technologies, Table 3.3 clearly shows the advantages of UAE. UAE might potentially be simpler and quicker than MAE (Luque-Garci'a et al., 2003).

3.9 HAACP for ultrasound and microwave food processing operation

Hazard analysis critical control point is a system for the identification and assessment of risks, hazards in the field of food manufacturing, storage and distribution, HACCP also aims to reduce eliminate a control these hazards. In terms of ultrasound usage the quality of equipment used physical properties of food, time of ultrasound exposure, temperature, and nature of probe need to be monitored to not reach or exceed the critical control point limit. These parameters differ for different food material and need to be adjusted accordingly in order to avoid the risks of hazards (Bhargava et al., 2021).

3.10 Conclusion

The microwaves have been used extensively in food processing from the past few decades and have proved to be time saving, economical and safe during the course of their use. There are possibilities of some side effects that are linked with misuse of this technique but are of less importance when compared with the benefits of this technique. The use of ultrasonic has recently been introduced for different food application and is gaining interest for its added benefits when used in combination with microwave technique. The synergistic effect of ultrasounds and microwaves has successfully been employed in food processing, extraction of phytochemicals and microbial distraction. But this area has a vast potential to explore by utilizing this technique for designing and formulating innovative food processing strategies.

References

Ahmed, J., Ramaswamy, H.S., Raghavan, V.G., 2007. Dielectric properties of butter in the MW frequency range as affected by salt and temperature. J. Food Eng. 82 (3), 351–358.

Akdeniz, V., Akalın, A.S., 2019. New approach for yoghurt and ice cream production: high-intensity ultrasound. Trends Food Sci. Technol. 86, 392–398.

Apostolou, I., Papadopoulou, C., Levidiotou, S., Ioannides, K., 2005. The effect of short-time microwave exposures on Escherichia coli O157: H7 inoculated onto chicken meat portions and whole chickens. Int. J. Food Microbiol. 101 (1), 105–110.

Auksornsri, T., Bornhorst, E.R., Tang, J., Tang, Z., Songsermpong, S., 2018. Developing model food systems with rice based products for microwave assisted thermal sterilization. LWT 96, 551–559.

Baboli, Z.M., 2019. Design and Experimental Validation of an Ultrasonic Reactor For Rapid Pasteurization of Fruit Juices (Doctoral Dissertation. North Carolina Agricultural and Technical State University.

Bantle, M., Eikevik, T.M., 2011. Parametric study of high-intensity ultrasound in the atmospheric freeze drying of peas. Drying Technol. 29 (10), 1230–1239.

Barnabas, J., Siores, E., Lamb, A., 2010. Non-thermal microwave reduction of pathogenic cellular population. Int. J. Food Eng. 6 (5), 1556–1878.

Bermúdez-Aguirre, D., Corradini, M.G., Mawson, R., Barbosa-Cánovas, G.V., 2009. Modeling the inactivation of Listeria innocua in raw whole milk treated under thermo-sonication. Innovative Food Sci. Emerg. Technol. 10 (2), 172–178.

Bhargava, N., Mor, R.S., Kumar, K., Sharanagat, V.S., 2021. Advances in application of ultrasound in food processing: a review. Ultrasonicssonochemistry 70, 105293.

Biwater Treatment Limited, 1993. Registered in England No. 929686. Biwater House, Station Approach, Dorking, Surrey, RH4 1TZ.

Butz, P., Tauscher, B., 2002. Emerging technologies: chemical aspects. Food Res. Int. 35 (2-3), 279–284.
Cameron, M., McMaster, L.D., Britz, T.J., 2009. Impact of ultrasound on dairy spoilage microbes and milk components. Dairy Sci. Technol. 89 (1), 83–98.
Cárcel, J.A., Benedito, J., Bon, J., Mulet, A., 2007. High intensity ultrasound effects on meat brining. Meat Sci. 76 (4), 611–619.
Chahbani, A., Fakhfakh, N., Balti, M.A., Mabrouk, M., El-Hatmi, H., Zouari, N., Kechaou, N., 2018. Microwave drying effects on drying kinetics, bioactive compounds and antioxidant activity of green peas (Pisumsativum L.). Food Biosci. 25, 32–38.
Chandrasekaran, S., Ramanathan, S., Basak, T., 2013. Microwave food processing—a review. Food Res. Int. 52 (1), 243–261.
Chemat, F., Khan, M.K., 2011. Applications of ultrasound in food technology: processing, preservation and extraction. Ultrasonicssonochemistry 18 (4), 813–835.
Chemat, F., Lucchesi, M., 2006. Microwave-assisted extraction of essential oils. In: Loupy, A. (Ed.), Microwaves in Organic Synthesis. WILEY-VCH GmbH & Co. KGaA, Weinheim, pp. 959–983.
Chen, C.W., Lee, H.M., Chen, S.H., Chen, H.L., Chang, M.B., 2009. Ultrasound-assisted plasma: a novel technique for inactivation of aquatic microorganisms. Environ. Sci. Technol. 43 (12), 4493–4497.
Chen, Z.G., Guo, X.Y., Wu, T., 2016. A novel dehydration technique for carrot slices implementing ultrasound and vacuum drying methods. Ultrasonicssonochemistry 30, 28–34.
Chisti, Y., 2003. Sonobioreactors: using ultrasound for enhanced microbial productivity. Trends Biotechnol. 21 (2), 89–93.
Cravotto, G. and Binello, A., 2010. Innovative techniques and equipments for flavours extraction-Application and effectiveness of ultrasound and microwaves.
Cravotto, G., Boffa, L., Mantegna, S., Perego, P., Avogadro, M., Cintas, P., 2008. Improved extraction of vegetable oils under high-intensity ultrasound and/or microwaves. Ultrasonicssonochemistry 15 (5), 898–902.
Cravotto, G., Cintas, P., 2007. The combined use of microwaves and ultrasound: improved tools in process chemistry and organic synthesis. Chemistry Eur. J. 13 (7), 1902–1909.
Cruz, R.M., Vieira, M.C., Silva, C.L., 2006. Effect of heat and thermosonication treatments on peroxidase inactivation kinetics in watercress (Nasturtium officinale). J. Food Eng. 72 (1), 8–15.
Datta, A.K., Anantheswaran, R.C., 2001. The Handbook of Microwave Technology for Food Applications, 1st ed. Marcel Dekker, NewYork Epub ahead of print 2001.
Datta, A.K., Davidson, P.M., 2000. Microwave and radio frequency processing. J. Food Sci. 65, 32–41.
Datta, A.K. and Nelson, S.O., 2000. Fundamental physical aspects of microwave absorption and heating in handbook of microwave technology for food applications. CHIPS.
Datta, A.K., Rakesh, V., 2013. Principles of microwave combination heating. Compreh. Rev. Food Sci. Food Saf. 12 (1), 24–39.
Decareau, R.V., 1985. Microwaves in the Food Processing Industry. Academic Press Inc, Orlando.
Decareau, R.V., Peterson, R.A., 1986. Microwave processing and engineering. Ellis Horwood Ser. Food Sci. Technol. 40 (6), 99.
Demirdöven, A., Baysal, T., 2008. The use of ultrasound and combined technologies in food preservation. Food Rev. Int. 25 (1), 1–11.
Deng, Y., Zhao, Y., 2008. Effect of pulsed vacuum and ultrasound osmopretreatments on glass transition temperature, texture, microstructure and calcium penetration of dried apples (Fuji). LWT 41 (9), 1575–1585.

Drakopoulou, S., Terzakis, S., Fountoulakis, M.S., Mantzavinos, D., Manios, T., 2009. Ultrasound-induced inactivation of gram-negative and gram-positive bacteria in secondary treated municipal wastewater. Ultrasonicssonochemistry 16 (5), 629–634.

Erle, U., 2000. Untersuchungen zur Mikrowellen-Vakuumtrocknung von Lebensmitteln. Universität KarlsruheShaker Verlag, Aachen Ph.D.Thesis.

Fan, K., Zhang, M., Mujumdar, A.S., 2017. Application of airborne ultrasound in the convective drying of fruits and vegetables: a review. UltrasonicsSonochemistry 39, 47–57.

Farber, J.M., D'AOUST, J.Y., Diotte, M., Sewell, A., Daley, E., 1998. Survival of Listeria spp. on raw whole chickens cooked in microwave ovens. J. Food Prot. 61 (11), 1465–1469.

Fujikawa, H.I.R.O.S.H.I., Ushioda, H., Kudo, Y., 1992. Kinetics of Escherichia coli destruction by microwave irradiation. Appl. Environ. Microbiol. 58 (3), 920–924.

Garcia-Vaquero, M., Ummat, V., Tiwari, B., Rajauria, G., 2020. Exploring ultrasound, microwave and ultrasound–microwave assisted extraction technologies to increase the extraction of bioactive compounds and antioxidants from brown macroalgae. Marine drugs 18 (3), 172.

Garcia-Vaquero, M., Ummat, V., Tiwari, B., Rajauria, G., 2020. Exploring ultrasound, microwave and ultrasound–microwave assisted extraction technologies to increase the extraction of bioactive compounds and antioxidants from brown macroalgae. Marine drugs 18 (3), 172.

Gedikli, S, Tabak, Ö, Tomsuk, Ö, et al., 2008. Effect of microwaves on some gram negative and gram positive bacteria. Journal of Applied Biological Science 2, 67–71.

Göksoy, E.O., James, C., Corry, J.E.L., 2000. The effect of short-time microwave exposures on inoculated pathogens on chicken and the shelf-life of uninoculated chicken meat. J. Food Eng. 45 (3), 153–160.

Guerrero, S., López-Malo, A., Alzamora, S.M., 2001. Effect of ultrasound on the survival of Saccharomyces cerevisiae: influence of temperature, pH and amplitude. Innovative Food Sci. Emerg. Technol. 2 (1), 31–39.

Guimarães, J.T., Balthazar, C.F., Scudino, H., Pimentel, T.C., Esmerino, E.A., Ashokkumar, M., Freitas, M.Q., Cruz, A.G., 2019. High-intensity ultrasound: a novel technology for the development of probiotic and prebiotic dairy products. Ultrasonicssonochemistry 57, 12–21.

Gupta, M., Paul, S., Gupta, R., 2007. General Characteristics and Applications of Microwaves in Organic Synthesis. J. Heterocycl. Chem. 44, 1023–1025.

Hamoud-Agha, M.M., Curet, S., Simonin, H., Boillereaux, L., 2014. Holding time effect on microwave inactivation of Escherichia coli K12: experimental and numerical investigations. J. Food Eng. 143, 102–113.

Hatloe, J., 1995. Methods for pickling and/or marinating non-vegetable foodstuff raw material. Int. Pat. WO 95, 18537.

Heddleson, R.A., Doores, S., 1994. Factors affecting microwave heating of foods and microwave induced destruction of foodborne pathogens–a review. J. Food Prot. 57 (11), 1025–1037.

Heddleson, R.A., Doores, S., Anantheswaran, R.C., Kuhn, G.D., 1996. Viability loss of Salmonella species, Staphylococcus aureus, and Listeria monocytogenes in complex foods heated by microwave energy. J. Food Prot. 59 (8), 813–818.

Helmar, S.M., 2007. The microwave processing of foods. Woodhead Publishing Limited and CRC Press 20312.

Huang, D., Men, K., Li, D., Wen, T., Gong, Z., Sunden, B., Wu, Z., 2020. Application of ultrasound technology in the drying of food products. Ultrasonicssonochemistry 63, 104950.

James, S.J., James, C., 2002. Thawing and tempering. In: Meat Refrigeration. Woodhead Publishing, US, pp. 159–190.

Jang, J.H., Moon, K.D., 2011. Inhibition of polyphenol oxidase and peroxidase activities on fresh-cut apple by simultaneous treatment of ultrasound and ascorbic acid. Food Chem. 124 (2), 444–449.

Jayasooriya, S.D., Bhandari, B.R., Torley, P., D'arcy, B.R., 2004. Effect of high power ultrasound waves on properties of meat: a review. Int. J. Food Prop. 7 (2), 301–319.

Kakita, Y., Kashige, N., Murata, K., Kuroiwa, A., Funatsu, M., Watanabe, K., 1995. Inactivation of Lactobacillus bacteriophage PL-1 by microwave irradiation. Microbiol. Immunol. 39 (8), 571–576.

KIM, H.J., Taub, I.A., 1993. Intrinsic chemical markers for aseptic processing of particulate foods. Food Technol. (Chicago) 47 (1), 91–99.

Knutson, K.M., Marth, E.H., Wagner, M.K., 1987. Microwave heating of food. Lebensmittel-Wissenschaft+ Technologie 20 (3), 101–110.

Kozempel, M., Cook, R.D., Scullen, O.J. and Annous, B.A., 2000.Development of a process for detecting nonthermal effects of microwave energy on microorganisms at low temperature.

Krokida, M.K., Maroulis, Z.B., 1999. Effect of microwave drying on some quality properties of dehydrated products. Drying Technol. 17 (3), 449–466.

Kröncke, N., Böschen, V., Woyzichovski, J., Demtröder, S., Benning, R., 2018. Comparison of suitable drying processes for mealworms (Tenebriomolitor). Innovative Food Sci. Emerg. Technol. 50, 20–25.

Kyllönen, H.M., Pirkonen, P., Nyström, M., 2005. Membrane filtration enhanced by ultrasound: a review. Desalination 181 (1-3), 319–335.

Lau, M.H., Tang, J., 2002. Pasteurization of pickled asparagus using 915 MHz microwaves. J. Food Eng. 51 (4), 283–290.

Lawson, L.G., Jensen, J.D., Christiansen, P., Lund, M., 2009. Cost-effectiveness of Salmonella reduction in Danish abattoirs. Int. J. Food Microbiol. 134 (1-2), 126–132.

Lemos, R, Varaschim, J, Tribuzi, G, et al., 2018. Effect of multi-flash drying and microwave vacuum drying on the microstructure and texture of pumpkin slices. LWT—Food Science and Technology. 96, 612–619.

Li, C.U.I., Niu, L.Y., Li, D.J., Liu, C.Q., Liu, Y.P., Liu, C.J., Song, J.F., 2018. Effects of different drying methods on quality, bacterial viability and storage stability of probiotic enriched apple snacks. J. Integr. Agric. 17 (1), 247–255.

Liu, B., Du, B., Sun, Y., Zhu, M., Yang, Y., Wang, X., Zhou, J., 2020. Ultrasound acoustic cavitation enhances depolymerization of organosolv lignin to phenolic monomers and low molecular weight lignin bio-oils. Fuel Process. Technol. 203, 106387.

Liu, L., Jia, W., Xu, D., Li, R., 2015. Applications of ultrasonic cutting in food processing. J. Food Process. Preserv. 39 (6), 1762–1769.

Lopez, P., Vercet, A., Sanchez, A.C., Burgos, J., 1998. Inactivation of tomato pectic enzymes by manothermosonication. ZeitschriftfürLebensmitteluntersuchung und-Forschung A 207 (3), 249–252.

Luque-Garcıa, J.L., De Castro, M.L., 2003. Ultrasound: a powerful tool for leaching. TrAC Trends Anal. Chem. 22 (1), 41–47.

Manas, P., Pagan, R., Raso, J., Sala, F.J., Condon, S., 2000. Inactivation of salmonella enteritidis, salmonella typhimurium, and salmonella senftenberg by ultrasonic waves under pressure. J. Food Prot. 63 (4), 451–456.

Helmer, S., Marc, R., 2007. The Microwave Processing of Foods. Woodhead Publishing Limited and CRC Press, p. 20312.

Marucci, G., Annibali, M., Carboni, G., Gherardi, G., Ragazzo, G., Siligardi, C., Veronesi, P., Lusvarghi, L. and Rivasi, M.R., 2000. Characterization of microwave inertized asbestos containing materials.

Maskan, M., 2000. Microwave/air and microwave finish drying of banana. J. Food Eng. 44 (2), 71–78.

Mason, T.J., 1990. Chemistry With Ultrasound. *Elsevier Applied Science*, New York.

Mason, T.J., Paniwnyk, L., Chemat, F., 2003. Ultrasound as a preservation technology. In: Food Preservation Techniques. Elsevier, pp. 303–337.

Massoubre, E., 2003. Micro onde pour une bonne qualitésensorielle. Viandesetproduitscarnés (Aubière) 23 (2), 49–52.

Matsuura, K., Hirotsune, M., Nunokawa, Y., Satoh, M., Honda, K., 1994. Acceleration of cell growth and ester formation by ultrasonic wave irradiation. J. Ferment. Bioeng. 77 (1), 36–40.

Metaxas, A.A. and Meredith, R.J., 1983. Industrial microwave heating (No. 4). IET.

Misra, N.N., Schlüter, 2016. In: Cullen, P.J. (Ed.), Cold Plasma in Food and Agriculture: Fundamentals and Applications, Academic Press, London.

Molins, C., Hogendoorn, E.A., Heusinkveld, H.A.G., Zoonen, P.V., Baumann, R.A., 1997. Microwave assisted solvent extraction (MASE) of organochlorine pesticides from soil samples. Int. J. Environ. Anal. Chem. 68 (2), 155–169.

Musielak, G., Mierzwa, D., Kroehnke, J., 2016. Food drying enhancement by ultrasound–a review. Trends Food Sci. Technol. 56, 126–141.

Nowacka, M., Tylewicz, U., Romani, S., Dalla Rosa, M., Witrowa-Rajchert, D., 2017. Influence of ultrasound-assisted osmotic dehydration on the main quality parameters of kiwifruit. Innovative Food Sci. Emerg. Technol. 41, 71–78.

Oda, S.J., 1992. Dielectric processing of hazardous materials-present and future opportunities. In: MRS Online Proceedings Library (OPL), p. 269.

Ojha, K.S., Mason, T.J., O'Donnell, C.P., Kerry, J.P., Tiwari, B.K., 2017. Ultrasound technology for food fermentation applications. Ultrasonicssonochemistry 34, 410–417.

Pagan, R., Manas, P., Alvarez, I., Condon, S., 1999. Resistance ofListeriamonocytogenesto ultrasonic waves under pressure at sublethal (manosonication) and lethal (manothermosonication) temperatures. Food Microbiol. 16 (2), 139–148.

Palacios, P., Burgos, J., Hoz, L., Sanz, B., Ordonez, J.A., 1991. Study of substances released by ultrasonic treatment from Bacillus stearothermophilus spores. J. Appl. Bacteriol. 71 (5), 445–451.

Pan, X., Niu, G., Liu, H., 2003. Microwave-assisted extraction of tea polyphenols and tea caffeine from green tea leaves. Chem. Eng. Process. 42 (2), 129–133.

Pedrós-Garrido, S., Condón-Abanto, S., Beltrán, J.A., Lyng, J.G., Brunton, N.P., Bolton, D., Whyte, P., 2017. Assessment of high intensity ultrasound for surface decontamination of salmon (S. salar), mackerel (S. scombrus), cod (G. morhua) and hake (M. merluccius) fillets, and its impact on fish quality. Innovative Food Sci. Emerg. Technol. 41, 64–70.

Piyasena, P., Mohareb, E., McKellar, R.C., 2003. Inactivation of microbes using ultrasound: a review. Int. J. Food Microbiol. 87 (3), 207–216.

Povey, M.J., 2017. Applications of ultrasonics in food science-novel control of fat crystallization and structuring. Curr. Opin. Colloid Interface Sci. 281–286.

Project ADER 7.5.1, 2020. Research on the application of emerging technologies based on minimal processing methods in the industrialization of horticultural products. Sectoral Plan of the Romanian Ministry of Agriculture and Rural Development [Google Scholar].

Püschner, H.A., 1966. Heating With Microwaves. Philips Technical Library, Berlin.

Quan, X., Zhang, M., Zhang, W., Adhikari, B., 2014. Effect of microwave-assisted vacuum frying on the quality of potato chips. Drying Technol. 32 (15), 1812–1819.

Raaholt, B.W., 2020. Influence of food geometry and dielectric properties on heating performance. Development of Packaging and Products for Use in Microwave Ovens (73-93). Woodhead Publishing.

Radoiu, M., 2020. Microwave drying process scale-up. Chem. Eng. Process. 155, 108088.

Ramaswamy, H.S., Koutchma, T., Tajchakavit, S., 2002. Enhanced thermal effects under microwave heating conditions. In: Engineering and Food for the 21st Century. CRC Press, pp. 769–792.

Rawson, F.F., 1998. 14 An introduction to ultrasonic food cutting. Ultrasound Food Process 2, 254.

Ricce, C., Rojas, M.L., Miano, A.C., Siche, R., Augusto, P.E.D., 2016. Ultrasound pre-treatment enhances the carrot drying and rehydration. Food Res. Int. 89, 701–708.

Riener, J., Noci, F., Cronin, D.A., Morgan, D.J., Lyng, J.G., 2010. A comparison of selected quality characteristics of yoghurts prepared from thermosonicated and conventionally heated milks. Food Chem. 119 (3), 1108–1113.

Rodríguez, Ó., Santacatalina, J.V., Simal, S., Garcia-Perez, J.V., Femenia, A., Rosselló, C., 2014. Influence of power ultrasound application on drying kinetics of apple and its antioxidant and microstructural properties. J. Food Eng. 129, 21–29.

Salazar-González, C., San Martín-González, M.F., López-Malo, A., Sosa-Morales, M.E., 2012. Recent studies related to microwave processing of fluid foods. Food Bioprocess Technol. 5 (1), 31–46.

Salazar-González, C., San Martín-González, M.F., López-Malo, A., Sosa-Morales, M.E., 2012. Recent studies related to microwave processing of fluid foods. Food Bioprocess Technol. 5 (1), 31–46.

Sato, S., Shibata, C., Yazu, M., 1996. Nonthermal killing effect of microwave irradiation. Biotechnol. Tech. 10 (3), 145–150.

Schiffmann, R.F., 2001. Microwave processes for the food industry. In: Handbook of Microwave Technology for Food Application. CRC Press, pp. 331–370.

Schnepf, M., Barbeau, W.E., 1989. Survival of Salmonella typhimurium in roasting chickens cooked in a microwave, convection microwave, and a conventional electric oven. J. Food Saf. 9 (4), 245–252.

Schulze, B., Hubbermann, E.M., Schwarz, K., 2014. Stability of quercetin derivatives in vacuum impregnated apple slices after drying (microwave vacuum drying, air drying, freeze drying) and storage. LWT 57 (1), 426–433.

Shaheen, M.S., El-Massry, K.F., El-Ghorab, A.H., Anjum, F.M., 2012. Microwave applications in thermal food processing. The Development and Application of Microwave Heating 3–16.

Shamis, Y., Croft, R., Taube, A., Crawford, R.J., Ivanova, E.P., 2012. Review of the specific effects of microwave radiation on bacterial cells. Appl. Microbiol. Biotechnol. 96 (2), 319–325.

Sillero, L., Prado, R., Labidi, J., 2020. Simultaneous microwave-ultrasound assisted extraction of bioactive compounds from bark. Chem. Eng. Process. Process Intens. 156, 108100.

Silva, F.A., MarsaioliJr, A., Maximo, G.J., Silva, M.A.A.P., Goncalves, L.A.G., 2006. Microwave assisted drying of macadamia nuts. J. Food Eng. 77 (3), 550–558.

Spencer, P.L., Raytheon manufacturing co, 1952. Means for treating food stuffs. U.S. Patent 2,605,383.

Su, Y., Zhang, M., Zhang, W., 2015. Effect of low temperature on the microwave-assisted vacuum frying of potato chips. Drying Technology 34 (2), 227–234.

Su, Y., Zhang, M., Fang, Z., Zhang, W., 2017. Analysis of dehydration kinetics, status of water and oil distribution of microwave-assisted vacuum frying potato chips combined with NMR and confocal laser scanning microscopy. Food Res. Int. 101, 188–197.

Szadzińska, J., Kowalski, S.J., Stasiak, M., 2016. Microwave and ultrasound enhancement of convective drying of strawberries: experimental and modeling efficiency. Int. J. Heat Mass Transfer 103, 1065–1074.

Szadzińska, J., Łechtańska, J., Kowalski, S.J., Stasiak, M., 2017. The effect of high power airborne ultrasound and microwaves on convective drying effectiveness and quality of green pepper. UltrasonicsSonochemistry 34, 531–539.

Tang, J., 2015. Unlocking potentials of microwaves for food safety and quality. J. Food Sci. 80 (8), 1776–1793.

Tang, Z., Mikhaylenko, G., Liu, F., Mah, J.H., Pandit, R., Younce, F., Tang, J., 2008. Microwave sterilization of sliced beef in gravy in 7-oz trays. J. Food Eng. 89 (4), 375–383.

Vadivambal, R., Jayas, D.S., 2010. Non-uniform temperature distribution during microwave heating of food materials—a review. Food Bioprocess Technol. 3 (2), 161–171.

Valsechi, OAA, Horii, J, De Angelis, DF, et al., 2004. The effect of microwaves on microorganisms. Arquivos do Instituto Biológico 71, 399–404.

Vercet, A., Lopez, P., Burgos, J., 1999. Inactivation of heat-resistant pectinmethylesterase from orange by manothermosonication. J. Agric. Food Chem. 47 (2), 432–437.

Walkling-Ribeiro, M., Noci, F., Riener, J., Cronin, D.A., Lyng, J.G., Morgan, D.J., 2009. The impact of thermosonication and pulsed electric fields on Staphylococcus aureus inactivation and selected quality parameters in orange juice. Food Bioprocess Technol. 2 (4), 422–430.

Wang, X., Peng, M.J., Wang, Z.H., Yang, Q.L., Peng, S., 2020. Ultrasound-microwave assisted extraction of flavonoid compounds from Eucommiaulmoides leaves and an evaluation of their antioxidant and antibacterial activities. Arch. Biol. Sci. 72 (2), 211–221.

Welt, B.A., Tong, C.H., Rossen, J.L., Lund, D.B., 1994. Effect of microwave radiation on inactivation of Clostridium sporogenes (PA 3679) spores. Appl. Environ. Microbiol. 60 (2), 482–488.

Yao, Y., 2016. Enhancement of mass transfer by ultrasound: application to adsorbent regeneration and food drying/dehydration. Ultrasonicssonochemistry 31, 512–531.

Yongsawatdigul, J., Gunasekaran, S., 1996. Microwave-vacuum drying of cranberries: part I. Energy use and efficiency. J. Food Process. Preserv. 20 (2), 121–143.

Zhang, M., Tang, J., Mujumdar, A.S., Wang, S., 2006. Trends in microwave-related drying of fruits and vegetables. Trends Food Sci. Technol. 17 (10), 524–534.

Zheng, L., Sun, D.W., 2006. Innovative applications of power ultrasound during food freezing processes—a review. Trends Food Sci. Technol. 17 (1), 16–23.

Zhou, B.W., Shin, S.G., Hwang, K., Ahn, J.H., Hwang, S., 2010. Effect of microwave irradiation on cellular disintegration of gram positive and negative cells. Appl. Microbiol. Biotechnol. 87 (2), 765–770.

Zielinska, M., Michalska, A., 2016. Microwave-assisted drying of blueberry (Vacciniumcorymbosum L.) fruits: drying kinetics, polyphenols, anthocyanins, antioxidant capacity, colour and texture. Food Chem. 212, 671–680.

Zielinska, M., Ropelewska, E., Markowski, M., 2017. Thermophysical properties of raw, hot-air and microwave-vacuum dried cranberry fruits (Vacciniummacrocarpon). LWT 85, 204–211.

CHAPTER FOUR

Synergistic effects of microwaves and sonication in dairy industry

Shamas Murtaza[a], Muhammad Shahbaz[a], Anjum Murtaza[b], Ayesha Sameen[c], Umar Farooq[a], Faiz ul Hassan Shah[d], Hammad Naeem[a] and Mohammad Javed Ansari[e]
[a]Department of Food Science and Technology, MNS University of Agriculture, Multan, Pakistan
[b]Institute of Food Science and Nutrition, University of Sargodha, Sargodha, Pakistan
[c]National Institute of Food Science and Technology, University of Agriculture, Faisalabad, Pakistan
[d]Department of Food Science and Technology, The Islamia University of Bahawalpur, Bahawalpur, Pakistan
[e]Department of Botany, Hindu College Moradabad (Mahatma Jyotiba Phule Rohilkhand University Bareilly), India

4.1 Introduction

Mankind has been processing food since ancient times. Some the most ancient techniques involve drying, cooking by heat, salting, etc. Increase in population, the need to increase the shelf life of foods also increased drastically to reduce postharvest losses. The concern of food safety has also been rising with every passing day. In the 19th and 20th century pasteurization, canning, freezing, etc. were the techniques introduced to enhance the shelf life of food products. Traditionally, heat treatment is used to kill harmful bacteria and ensure that milk is safe for human consumption, but heat treatment also causes loss of nutrients and sometimes heat treatment alone is not sufficient to ensure milk safety for consumption. Prolonged heating causes loss of polypeptides and vitamins, as well as causes unwanted changes in flavor and color. In order to overcome these challenges, new process and techniques are required. With the increase in demand of milk and milk products, safer and more efficient techniques are also required.

Among the novel techniques being explored, microwave technique has shown desirable results. With the advancement in research, it also well known that microwave technique is a safe technique having no known adverse effects on food products. Microwave technique does not have the drawbacks of conventional heat processing techniques. This makes microwave technique more suitable to use in milk industry. Talking about novel techniques, sonication (or ultrasonication), that is, the use of ultrasound waves, is a great nonthermal technique. The fascinating thing about this

Ultrasound and Microwave for Food Processing: Synergism for Preservation and Extraction.
DOI: https://doi.org/10.1016/B978-0-323-95991-9.00018-7

technique is the provision of same beneficial results as of thermal processing with none of the draw backs associated with thermal processing. Moreover, sonication has number of other benefits in gelation processes, fermentation etc. which will be discussed further in the chapter. All the associated benefits of microwave technique and sonication technique are beneficial to be used in dairy industry to overcome existing challenges.

Microwaves and sonication techniques have shown to impart synergistic effects in milk processing industries. Both of these techniques have been shown to improve process efficiencies as well as provide better quality of end products. With the increasing demands and rising economical stresses it is imperative that novel techniques must explore and utilized for the betterment of mankind. This chapter will be divided into two parts for better understanding of beneficial roles of both microwaves and sonication in dairy industry.

4.2 Effects of microwaves in dairy industry

Proteins, carbohydrates, fats/lipids, and minerals, as well as a range of other ingredients that may or may not be included in small amounts, are all important components of various foods. All of these food components contribute to the functional features and overall quality of food products. On the other hand these components are vulnerable to deterioration and microbial attack. Hence, there arose the need to find techniques that can alter the composition of food with minimal loss of nutritional and quality attributes. Milk or dairy is an excellent example of complex multinutrient food.

Because of milk's extraordinary nutritional levels and moisture percentage, milk has a very favorable environment for the growth of pathogenic microbes, and then milk is rapidly subjected to spoilage. The chief nutritional constituents of milk are carbohydrate (e.g., lactose), fat, protein (casein and whey protein), minerals, and vitamins. Some parameters, as in types and intensities of various heat treatments, can affect its composition. The impacts of various food processing/preservation methods should be carefully assessed to ensure that a food product's quality is preserved (Martins et al., 2019).

Heat treatment is traditionally used to kill harmful bacteria and ensure that milk is safe for human consumption. Milk is usually contaminated with microbes during process of milking (milking utensil/machine, water, animal dung, and environmental dust particles). Heating of raw milk is the important technological part in the agro-food industry, and specifically at milk processing sectors. The major disadvantage of commercial heat treatment is that it degrades the quality and nutritional content of milk due to prolonged heating and subsequent loss of polypeptides and vitamins, as well as causing unwanted variations in other attributes such as flavor and color. Due to the widespread use of traditional thermal processing techniques within milk and dairy industries, a certain fact has come to light that the high temperatures used to inactivate bacterial cells also cause breakdown of volatile and heat sensitive components that results in great loss of desirable sensory and nutritional aspects of milk-based/dairy products.

The search for processing alternatives that meet customer needs, meet the safety parameters, and healthy products are always in demand. Food experts are investigating preservation factors as part of their research for potential pasteurization and sterilization alternatives. In recent years, many milk pasteurization alternatives have been studied more thoroughly than others, with significant and encouraging findings (Barbosa-Canovas and Bermudez-Aguirre, 2010). Food preservation techniques such as ultrasound, infrared, and microwave have been widely employed to preserve and retain nutritional components of food products having optimum efficiency. Furthermore, such techniques are indicated for food item changes and interaction (Li et al., 2021). Microwave heating is a cutting-edge alternative to traditional thermal processing. There has been a lot of interest in adopting this technology to improve the microbiological safety, shelf life, and functional properties of food products in recent years (Gholamhosseinpour et al., 2020).

Microwave technology is a volumetric heating method that evenly distributes heat throughout the food product's volume. This is due to the interaction of microwaves with polar water molecules and charged ions in food. Because microwave radiation is selectively absorbed by locations with higher moisture content, temperature, and moisture profiles become more uniform. Microwave cooking has emerged as one of the most promising food processing methods (Kalla and Devaraju, 2017). Microwaves (MW) and ultrasound (US) are attaining admiration in sector of milk, dairy or dairy-based industries owing to their effects of pasteurization with the least amount of heat input and the least amount of quality degradation. These methods change the physicochemical characteristics of milk proteins (casein and whey) and enzymes (Munir et al., 2019).

Microwave heating systems are well-known for their reliability and ability to retain heat-unstable nutrients such as ascorbic acid and vitamins B, as well as dietary antioxidants, carotenoids and flavonoids (Kalla and Devaraju, 2017). Microwaves provide a number of advantages. The most important feature is decreasing the heating process time (by up to 50%) and lowering the cost of the procedure. Furthermore, it enables the processes to be more efficient while retaining the high quality (Guzik et al., 2021). The microwave heating technology has been successfully employed in a variety of industries to prepare a range of food. When compared to traditional procedures, microwave heat treatments of foods promise to successfully eradicate harmful bacteria while also reducing processing time without compromising product quality. The advanced level of living and increased income of customers have led to the appeal for new food processing techniques. Microwave heating treatments have become more important as a result of technical advancements and dietary guidelines. Microwaves and heat caused folding-unfolding of caseins (proteins) in the existence of different saccharides. It was observed that MW treatment at the same bulk temperature as conventional heating generated casein unfolding. Furthermore, acidic gels generated with microwaves treated caseins solutions were a lot harder as well as had higher microstructure rigidity than heated caseins, owing to increased contacts and crosslinking between thiol-groups and residues (hydrophobic) in unfolded casein-proteins. When exposed to MW, whey proteins similarly unfold and destroyed (denature). When utilizing microwave technique or direct heating at temperatures above 70°C, whey proteins denature. Whey proteins denature more quickly in microwave-cooked milk (over 30 minutes) than in milk prepared at the same temperature with traditional heating. With extremely small process times, degree of denaturation of whey protein produced by microwave heating and UHT treatments remained the same (at equivalent thermal inputs) (20 seconds) (Clare et al., 2005).

Microwaves are created in ovens when discontinuous electric field is applied. The waves produced have wavelengths less than 1 cm and frequencies around of 2.45 Giga-Hz. This is categorized as nonionizing radiations that is unable to break down covalent bonding. MW has the capacity to modify milk components both thermally and nonthermally (Munir et al., 2019). The dissemination of microwaves and their interfaces with food products are usually governed by dielectric characteristics, while heat transfer is managed by fundamental heat and mass transport processes. MW may cause both thermal and nonthermal changes in milk. When microwaves collide with water dipoles, the molecules try to restructure themselves to match those of the microwaves, generating friction. Nonthermal effects relate to changes in milk caused only by molecular rearrangement, whereas thermal effects refer to changes caused by molecular friction-induced localized heat. The relative influence on protein and enzyme function is currently under debate since chemical rearrangement and heat production occur at the same time. The intermolecular spacing between proteins alters as a result of molecular rearrangement during MW treatment, as well as their quaternary and tertiary structures. Length, strength, shape, and frequency of the transmitted waves governs how microwaves will denature/destroy sensitive durable casein micelles and whey protein components (El Mecherfi et al., 2011).

In comparison to typical thermal procedures, dielectric heating by microwave oven increases volumetric heating, resulting in superior quality products with higher thermal efficiency. The microwave (MW) heating system has been found to produce high-quality milk-based products with longer shelf lives, demonstrating that it is a viable alternative to traditional heat treatment. The fundamental disadvantage of microwaves heating technique is that it causes nonuniform temperature dispersion in cold and hot locations, especially in semi-solids and solids; nevertheless, MW technique of heating for a fact, is appropriate for several liquid meals, exclusively in a continuous fluid-systems (Martins et al., 2019). MWs, unlike traditional convection and conduction methods, may penetrate the food, absorb it, and then convert it to heat quickly, resulting in volumetric heat generation and fast heating rates. The following are some of the major advantages of using a microwave in the dairy industry: Faster heating: high heating performance (at least 80%); volumetric heat generation; Process time is reduced by a fourth compared to typical heating methods; Better quality preservation in several processed food products: the rapid heating rate quickly achieves the desired temperature, reducing heat adverse impact on the food and minimizing negative sensory and nutritional effects. Suitable application for thermal sensitive, high viscosity, and multiphase fluids; there is also a prospect of safe handling of pasteurized product after packaging (Martins et al., 2019).

i. Microwave enters the food product and, therefore, cooking occurs throughout the entire volume of food internally, uniformly, and quickly, which significantly decreases the processing time and energy.
ii. Suitable geometry for clean-in-place (CIP) system.
iii. Inexpensive system maintenance.

4.2.1 Electrothermal technique

Microwaves can be categorized as an electrothermal technique for the pasteurization of foods. Microwaves are effective for the inactivating of pathogenic microbes such as *Salmonella spp.*, *E. coli*, and *L. monocytogenes*. Microwave pasteurization of milk for 2.5 minutes resulted in a 97% reduction in bacterial load, according to research. Microwave heating of milk for

3 minutes C. jejuni, 8 minutes completely inactivated *Y. enterocolitica,* and 10 minutes completely inactivated *L. monocytogenes* (Kalla and Devaraju, 2017).

As described in a study, while assessing diverse fat contents in various samples of milk with 2.5%, 3%, and 3.5% fat and acid contents of 1.60, 1.70, and 1.80 g/L, *Listeria monocytogenes* (4A) was inactivated in less than 35 seconds (i.e., seven log-reductions) while using microwaves at 2450.0 MHz. Higher the fat contents and nonfat solids, higher the inactivation of *L. monocytogenes* (Barbosa-Canovas and Bermudez-Aguirre, 2010). This implies that microwaves have greater efficiency owing to their electrothermal properties.

4.2.2 Ionic interaction and dipolar rotation

Ionic interaction and dipolar rotation are the two main methods by which microwave radiation is absorbed by food. Ionic polarization occurs when an electrical field is applied to food solutions containing ions, causing the ions to move faster owing to their intrinsic charge and collide with other ions. As a result of the collisions, the kinetic energy of ionic mobility is transformed into thermal energy. Water is the main component in food that promotes dipolar rotation. Because of their dipole nature, water molecules try to follow the electric field associated with electromagnetic radiation as it oscillates at microwave frequency. The absorbed energy from microwaves may raise the temperature of the product to a level that kills microbes. Despite the fact that several studies have demonstrated that the thermal effect is the most essential component in microbial mortality, some researchers have attempted to explain the nonthermal effects of microwaves on microbial inactivation (Bozkurt-Cekmer and Davidson, 2017).

4.2.3 Time efficient

Due to faster cellular protein breakdown, microwave heating takes less treatment durations to achieve the same microbial inactivation as traditional thermal pasteurization. As a result, MW-assisted pasteurization can help to reduce the intensity of certain harmful metabolic processes in milk that occur during thermal pasteurization. When compared to UHT treated milk, milk pasteurized using microwaves offers improved sensory attributes in terms of odor, caramelization, astringency, and fatty tastes due to a lower amount of lipid breakdown (Clare et al., 2005).

4.2.4 Effect of microwave on milk proteins

The most significant structural as well as chemical modifications that follow microwave treatment include protein aggregation and denaturation, Maillard reaction and isomerization of lactose. These are typically related to heat treatment. Conventional heating methods show greater changes in aforementioned properties as compared to microwave heating (Martins et al., 2019).

Even though energy supplied by microwave radiations can dislocate weak hydrophobic bonds, they are not sufficient to cause breakdown of covalent-bonds within or between amino acids. Studies hence report that the proteins denaturation as well as aggregation caused by microwaves, reduces the amount of soluble amino-acids in milk while causing no alterations in the total amino acids pool. The soluble or free amino acids content was significantly reduced under microwave treatment (at 2.45 GHz, at temp of 68°C and for 40 seconds), however total amino-acid pool remained unaffected. Microwave pasteurization takes less time than traditional heat/thermal pasteurization procedures to achieve the same level of microbial inactivation due to faster cellular protein breakdown. As a result, numerous adverse metabolic activities that occur within milk during heat/thermal pasteurization procedures may be reduced by microwave pasteurization. Acceptable sensory attributes of milk were obtained when pasteurized using microwave technique, owing to a lower levels of lipids degradation. Better sensory features were also stated, with respect to smell/odor, extent of caramelization, fatty flavors and astringency as compared to ultra-high temperature treated milks. The activity of numerous added (proteolytic) enzymes (trypsin, chymotrypsin, pepsin, and other gastro-intestinal enzymes) became more noticeable throughout the incubation when milk was preheated in the microwave (to reach a temperature of 60°C in under 5 minutes) (Munir et al., 2019). Microwaves can also be utilized to polymerize casein proteins using microbe based MT-Gase (transglutaminase), it was found that microwaves can greatly hasten microbial transglutaminase-induced polymerization process in milk. It was exhibited the reaction time was reduced from 3 hours to 1 hours by using a 30 watt microwaves treatment (Chen and Hsieh, 2016).

In an interesting research study, Cheese was made from different milk samples that were subjected to diverse pretreatments (C: untreated control, U.S-1: ultrasonication, specific-energy 23 J/g, frequency 20 kHz; U.S-2: ultrasonication, specific-energy: 41 J/g, Frequency 20 kHz; HPP: high-pressure process 400 MPa for 15 minutes, at 40°C; MW: microwave less than 40°C, specific-energy: 86.5 J/g) and were analyzed after process of ripening for about 0–9 months. The results exhibited that proteolysis rate during cheese making and succeeding ripening was expressively affected by various pretreatments. ACE inhibitory and antioxidant potential of cheeses made from these pretreated milk samples significantly amplified in order as follows: US-2 > HPP > US-1 > MW > C (Munir et al., 2020).

In a recent study, it was discovered that the activity of protease in microwaved milk was 1.3–1.8 times greater than the values obtained from the other samples after 24 hours of fermentation. The microwave treatment generated less whey denaturation, namely -β-lactoglobulin, as compared to traditional ways of heating, preserving whey protein's sensitivity to proteolytic enzymes. The peptides formed when these proteins are proteolyzed can boost the development of lactic-acid bacteria. During fermentation, the amount of peptides in each sample rose. The ultrasound plus microwave treated milk samples always constituted higher quantities of desirable peptides. During the fermentation process, the autoclaved milk samples had much lower peptide levels than the other samples. As a result, the peptide content of the subject samples was enhanced when microwave and ultrasound treatments were combined or microwave (alone) treatments were used. Furthermore, microwave treatments resulted in enhanced protein hydrolysis due to increased protease activity and the formation of unfolded proteins/peptides. This in turn leads to produce more peptides and thus better proliferation of lactic-acid bacteria. (Gholamhosseinpour et al., 2020).

4.2.5 Rapid microwave moisture determination

Whole milk-powder is a convenient product which is widely utilized by the consumers as a substitute of fresh milk. During production of Whole milk-powder, milk is carried through a group of procedures, that is, forcing, agitation, accumulation, heating, homogenization, spray drying, etc. The water percentage in milk powder, which is a primary parameter to determine the notable quality of milk or milk products and also determines the cost of production, must be accurately controlled during the step of spray-drying. Quick and accurate determination of water percentage of milk powder can be found utilizing the resonant microwave sensor (Yang et al., 2016).

4.2.6 Continuous flow microwave heating

"Continuous flow microwave heating (CFMH)" may be a viable alternative to overcome the limits of traditional pasteurization processes, since it reduces heat-sensitive component degradation and preserves the sensory qualities of milk-based products. It was proved in a study that microwave pasteurization at 120°C for 10 seconds) produced lower deleterious effects on product quality (Li et al., 2021). While discussing CFMH, a European project called

Micro-Milk has developed a new continuous microwave system. Micro-Milk project goals are: (1) Develop an electrothermal prototype of microwave cavity to facilitate uniform and controlled heating, (2) decrease cleaning times and minimize the sources of contamination of milk occurring through equipment, (3) reduce the cost of milk treatment, (4) improve the nutritional, organoleptic and microbiological quality as well as the storage life of milk and milk products (Martins et al., 2019).

4.2.7 Checking adulteration using microwaves

Researchers have studied the impact of water percentage (or concentrations) in relation to milk freshness using the dielectric attributes of the milk product to check adulterations. When compared to a sample of diluted milk, it was clear that undiluted or unadulterated milk had lower relative electric permittivity and the highest loss aspect, and the depth of diffusion rose with cumulative water percentage. It was determined that fluctuations in dielectric attributes were adequate to detect adulteration that can lead to possible economic loss, decline in quality of the final milk product and a serious risk for consumers (Murthy et al., 2017).

4.2.8 Limitations

One disadvantage of using microwaves to pasteurize the food is the nonuniform heating or uneven temperature distribution throughout this process. Heat is the most important weapon for inactivating cells using microwaves. As a result, the effect on bacteria is similar to the effect seen when cells are inactivated using a traditional thermal/heat technique, which results in denaturation of proteins and cell walls (Barbosa-Canovas and Bermudez-Aguirre, 2010).

4.3 Effects of sonication in dairy industry

Consumer desire for tasty, healthier, and less processed meals develops; the development of unique dairy products becomes increasingly important. Sonication, often known as ultrasonic processing, is an alternate food processing procedure that can help milk and dairy products enhance their technical and functional features. Ultrasound has been used in the dairy sector to improve milk quality and processes such as homogenization, emulsification, the creation of yoghurt and fermented beverages, and food safety. Aside from slight color differences and some protein oxidation, there are few drawbacks to using high-intensity ultrasonication (HIU). Ultrasonically treated milk might result in an undesirable end product with unpleasant tastes. To emphasize the sensory characters that have the most influence on quality, more studies on proteolysis, lipolysis, and fatty acid oxidation are needed. Ultrasound is ineffective for decontaminating milk on its own, but when paired with heat, pathogens can be effectively eradicated (Carrillo-Lopez et al., 2021). Wood and Loomis discovered that power ultrasonography was beneficial in the mixing of oil-in-water emulsions in a 1927 research of the physicochemical effects of power ultrasonography in an oil bath. Ultrasound has been used in commercial food processing for a variety of applications, including energy-efficient food material cutting, homogenization of food products like milk and mayonnaise, flavonol extraction from plant parts, degassing, cleaning, microbial deactivation, and enzyme inactivation, among others. Ultrasonication has been shown to be as effective as or more effective than typical food processing techniques at all scales, including nanotechnological, microtechnological, and macrotechnological (Kehinde et al., 2021).

Food processing has an essential role to overcome the challenges of food safety and security by reduction of waste and prevention of spoilage in foods. Ultrasound technology has modernized food processing industries having wide applications in several processes/techniques while serving as a good low cost and sustainable alternative technique. Sonication of ultrasound technique is a nondestructive technique that offers several benefits as in rapid process, improved efficiency of process, elimination of certain steps all the while giving better quality of products and retention of product's features (nutrition value, texture, organoleptic characteristics, etc.) with enhanced shelf life (Bhargava et al., 2021). The usage of nonthermal treatments and methods has been increasing because of ever growing demand for supreme quality foods comprising of natural tastefulness, flavors, and basically being free from chemicals or additives as well as preservative compounds. Advanced food processing techniques, for example, ultrasound, microwave and more, are extensively applied for the valorization and preservation of food nutrients during food processing with very high efficiency. Nonetheless, these treatments cause food elements to change and interact in different ways. Because of lower processing temperatures, less energy use, and less loss of important nutrients and flavor, power ultrasound may be utilized as a processing method for microbial inactivation in the milk industry, offering a superior quality than typical heat treatment. Scholars have recently indicated an interest in using ultrasonography to help the dairy industry reach its full potential.

Ultrasound has been used to modify milk components, increase milk and milk product processing efficiency, improve food safety and microbial destruction, perform nondestructive analysis, clean equipment surfaces, and manage waste, according to numerous researches. Some of the technologies, such as ultrasound-assisted UHT package sealing and an HC-based yoghurt enhancement method, have been successfully adopted in the dairy industry. However, further research is needed to determine the industrial application of these technologies (Sutariya et al., 2018). Amid the several nonthermal methods, the sonication or ultrasound processing has upheld its name to being very valuable. Ultrasound process, used alone or combined with further processing techniques, yields substantial positive outcomes for the quality of foods; therefore, it is now considered an efficient method. Food processing performed using the mechanisms of ultrasound/sonication are said to be affected partly by cavitation (phenomenon) and mass-transfer enhancements. It is now considered to be one of the promising and emerging technologies and being applied proficiently in the food industries for numerous processes as in freezing, drying, separation, filtration, extraction, sterilization, and emulsion. Several researches have noted that ultrasound points to enhancements in the existing efficiencies of available processes and increases the quality aspects of the food products. This chapter will discuss the synergistic effects produced by the utilization of high-intensity ultrasonics/sonication (Singla and Sit, 2021).

Combined with other technologies, ultrasound processing offers a number of benefits, including faster energy, food quality retention, mass transfer, reduced concentration and thermal gradients, increased production, effective extraction, and a cost-effective alternative to conventional thermal techniques. In the food industry, ultrasound has demonstrated its utility in terms of preservation, extraction, and processing. Ultrasound is quickly becoming a very promising improvement tool in the agricultural/food business, where it is being utilized to optimize a variety of operations (Singla and Sit, 2021).

Both the food industry and customers, ultrasound provides a number of benefits over conventional pasteurization. While shorter processing times can assist to inactivate hazardous pathogens, it's also worth noting that processing has improved the quality of milk significantly. For example, sonication affects the size and microstructure of fat globules, resulting in greater homogenization and a better (i.e., whiter) milk color (Barbosa-Canovas and Bermudez-Aguirre, 2010). This technology might be used to speed up operations, decrease energy needs, enhance productivity, and produce higher-quality food materials in the realms of fruits, juices, and dairy products, according to recent research and advancements in the field of ultrasonic applications. Because of its versatility in terms of frequency and power, as well as its simplicity of use and human safety, ultrasound is a useful device for analyzing, regulating, and processing fruits, juices, and dairy products (Firouz et al., 2019). Many industrial food operations now employ ultrasound in the high-frequency bands. Ultrasound technology is being used by food processors to replace some antiquated techniques for inspecting food quality in production chains, utilizing ultrasound's nondestructive nature at high frequencies. Many aspects of food may be determined before, during, and after processing using fundamental scientific ideas that characterize wave behavior. High-intensity ultrasound at low frequencies, on the other hand, has lately become one of the most intriguing disciplines for food scientists. When combined with temperature, pressure, and other preservation parameters, sound waves have been shown to aid microbial and enzymatic inactivation. Ultrasound-processed items benefit from improved texture, yield, and color. As a result, ultrasonography is increasingly being used to pasteurize products including milk, fruit juices, and other liquid food items (Bermudez-Aguirre et al., 2011). Different studies present different processing parameters for the ultrasonication of foods. There are considerable variations among pulsations, time, intensity, and probes. However, all these studies show favorable results in terms of physicochemical properties of food in the dairy industry. Quality acceptance as well as consumer preference is highly dependent on the physicochemical parameters in food such as texture and color. Moreover, these parameters are the determinants of nutrients retention, prolonged shelf life of product, and further processing conditions. Important physicochemical parameters include oxidative stability, characteristics of fat globule and pH among others. Optimization of these parameters has always been the focus of researchers. One potential and tested technique providing favorable results in the physicochemical parameters is the HIU. However, no sufficient studies are available on the comparative analysis of HIU with other techniques such as pasteurization and micro-fluidization. HIU when applied on milk and milk-based beverages has resulted in improved quality of the physicochemical parameters. After the promising results of HIU, evaluation has recently been done on the bovine milk, sheep milk, buffalo milk, and supplemented milk. Overall, the promising results of HIU have been confirmed related to the milk particles such as fat crystals and fat droplets and changes in the curd matrix. Numerous physical benefits are there when there are changes in the size of milk components, for example, increase in specific surface area, accelerated gel hardness, and formation, smaller particle size distribution of fat, and reduction in curd firmness (Carrillo-Lopez et al., 2021).

The ultrasonic settings utilized in the study looked to be promising for inactivating bacteria in semi-skimmed sheep milk, resulting in minimal contamination that is safe to drink. This finding is interesting since, when compared to traditional treatment; equivalent bacterial inactivation was obtained with the added benefit of low temperature processing. In comparison to HTST pasteurization, US-treated sheep milk may be favorable since ultrasound processing can maintain certain lactic acid bacteria while inactivating a sufficient proportion of contaminating bacteria in frozen stored milk. Furthermore, no significant differences in protein or free amino acid profiles of pasteurized and US-treated semi-skimmed sheep milk were found, suggesting that this method might be used to create sheep milk products such cheeses while retaining product quality. Overall, the sheep milk sector should adopt ultrasound processing (Balthazar et al., 2019). Although the influence of sonication on the structural and functional qualities of cream has been extensively researched, little is known about the effect of sonication on milk and cream homogenization. The rennet and acid gelation features of raw milk, ultrafiltration retentate, and cream samples were studied after being sonicated at 20 kHz. For comparison, a high-pressure homogenization at 80 bar was also done. The size of fat globules was reduced after sonication of raw milk and retentate samples. Fat globules in cream samples sonicated at 10°C, on the other hand, flocculated to create grapelike structures, but cream samples sonicated at 50°C did not. After high pressure homogenization at 50°C, similar flocculated structures were observed, but not at lower temperatures. This demonstrates how sonication technology may help save energy by allowing for cream homogenization at low temperatures. Instead of fat globules being absorbed into the gel matrix, a gel made with cheese-milk and sonicated cream resulted in the separation of a fat layer. The rennet gelation characteristics of the sonicated and homogenized samples were much superior than the native control sample, with shorter coagulation durations and reduced syneresis (Chandrapala et al., 2016).

Modifications in milk proteins caused by sonication can increase yield, physicochemical properties, and sensory aspects of dairy products. The gel structure became more linked and regular when sonicated milk was used to manufacture yoghurt than when untreated milk was utilized. Firm rennet gels and faster gelation rates have also been seen in US-treated milk and protein suspensions. Gels with greater elastic moduli result from such dense microstructures (Yanjun et al., 2014). Whey protein casein crosslinking in

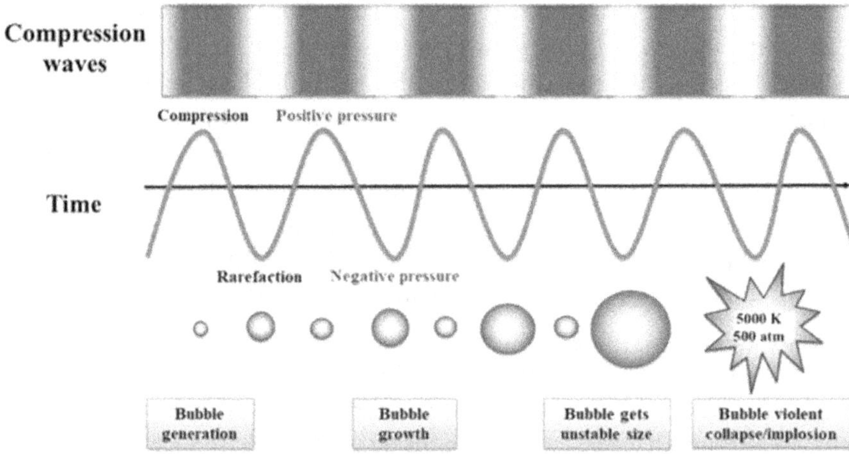

Figure 4.1 Ultrasonic cavitation (Soria and Villamiel, 2010).

milk induced by thermosonication (24 kHz, 100–150W, 70°C) was employed to enhance production and decrease serum separation in yogurt drinks. Similar thermosonication processes can help soft cheese and cream cheese yields. Ultrasound, microwave, and high-pressure treatments can affect the physiochemical and functional characteristics of food components, with varied consequences depending on the processing parameters and food matrix. The influence of cutting-edge technology on food components might be favorable or detrimental. Food component interaction may be increased without destroying the compounds' important structures by modifying processing parameters, monitoring, and adjusting.

Ultrasound is defined as sound waves with a frequency greater than 18 kHz (US). When ultrasound waves move through a material, they cause mechanical vibrations, acoustic streaming, and acoustic cavitation. Mechanical vibrations can impact the shape and size of solid particles, whereas acoustic streaming can facilitate or accelerate mass transfer through a medium (Tho et al., 2007).

Cavitation effect and instrument settings, low frequency/high energy ultrasound from 20 to 100 kHz, particularly 20, 25, and 40 kHz, is favored for food component alteration (Fig. 4.1). Ultrasound at 20–25 kHz has also been discovered to change the size and structural conformation of protein and carbohydrate particles, improving their functional characteristics in food systems. Under the right ultrasonic conditions, food component interaction can be improved without changing the compounds' basic structures.

(Li et al., 2021). Ultrasound has two extra actions that mostly modify native casein and whey protein confirmation as compared to HPP and MW: (1) Chemical destruction of proteins caused by low-power high-frequency ultrasound generated radicals, and (2) disintegration of previously denatured protein clumps in suspensions caused by high-power low-frequency ultrasonic produced shearing effect. Shock waves and microjets, for example, enhance the force with which aggregates in suspension collide. When protein clumps with weaker hydrophobic interactions or Van der Waal connections come into contact, they can be disrupted, allowing US to homogenize protein solutions. Hydrophobic interactions, for example, maintain protein clumps (0.2–10 m) together in whey protein concentrates (WPC) produced from cheese whey. Researchers employed a sonicator to sonicate reconstituted 5% (w/w) WPC solution for 60 minutes at 20 kHz and 31 watts, reducing avg. particle sizes from approx. 200 to 125 nanometer. Correspondingly, the aggregate size of reconstituted whey protein isolate suspensions may be decreased by 50% by employing a 20 kHz 34 Wcm^2 US for 2 minutes (O'sullivan et al., 2016). Sonicated, milk protein concentrate solutions (20 kHz, 300 W electric power), sodium caseinate solutions (34 W cm^2, 20 kHz), and denatured casein-whey protein combinations all reduced aggregates to similar levels. (Leong et al., 2018). The effects of probe (20 kHz, 43–48 watts per cm^2) and the bath (40 kHz, 1 watt per cm^2) ultrasonication on aggregate reduction in WPC were investigated, with probe based sonication having a larger but intense region of cavitation. As a result, while considering aggregate size reduction, the amount of energy wasted by cavitation per unit area is crucial (Jambrak et al., 2014). US may impact protein confirmation at greater applied radiation intensities. Cavitation-induced shear disturbed casein micelles, resulting in micelle size decrease, Researchers worked with reconstituted micellar casein powder solutions ranging pH of six and 12 (35 kHz, 200 watt output power, 6 hours). It was noticed that a reduction in the particle size of skim milk in the early stages of sonication (22.5 kHz, 50W for up to 10 minutes) and it was concluded that ultrasonication could dissociate the casein micelles (Madadlou et al., 2009). It has been discovered that when milk was sonicated at high pH for 15 minutes (20 kHz 286 kJ kg1 power) and then neutralized, casein micelle integrity was certainly impaired. Because casein micelles expand due to electrostatic repulsion at higher pH, they are more sensitive to the mechanical forces of ultrasonic cavitation (Liu et al., 2014). Despite the fact that the impact of US on the structure of individual caseins has yet to be thoroughly studied, structural alterations in whey proteins treated with US were observed. At very short treatment times (less than 5 minutes), it was also demonstrated that sonic shear (from US treatment at 20 kHz, 31 W) may unwind whey proteins and expose hidden hydrophobic residues to the environment. After a longer treatment (>10 minutes), hydrophobic interactions triggered whey protein aggregation (Chandrapala et al., 2011). According to a recent investigation, the effects of ultrasound (20 kHz, 75.6 J/mL) on the secondary structure of whey proteins in whey protein-rich settings. When whey was sonicated, they found that the -LG dimers separated. In the monomers, the number of random coil configurations grew as well. Sonication reduced the number of free sulfhydryl groups in BSA isolate solutions. An increase was observed in thiol-content which was the result of proteins unfolding process while working with whey protein concentrates (Silva et al., 2018). The functional alterations in dairy streams are caused by substance homogenization and conformational changes protein of proteins generated as a function of the sonication treatment. For example, reconstituted-proteins suspensions become solubilized completely and the turbidity and viscosity diminishes whenever the proteins aggregate sizes are lowered (Shen et al., 2017). Protein denaturation and subsequent aggregation have also been linked to major changes in viscosity and solubility in the ultrasonication. For example, it was found that ultrasonication method (30 kHz, 73–78 watts per cm^2, 15 minutes) decreased the solubility character as well as the flowability of subjected whey-protein isolates solution (Jambrak et al., 2011).

Depending on the application, ultrasonography in food preservation may be split into two categories: directly linked to food and indirectly related to food. Microbial inactivation, spore inactivation, and enzyme inactivation are all directly linked to food. Indirectly related to the food, one of the most well-known industrial uses of power ultrasound is surface cleaning, which has shown to be highly successful (Chemat and Khan, 2011). According to Shanmugam et al. (2012), ultrasonic processing is a potential approach that might be utilized in current food processing/preservation industries. Shear force of sonic cavitation impart modest modifications to the milk, indicating the possibility of improving this technology for industrial uses. The optimization procedure required is determined by the functionality and end products, such as bioactive food emulsions. The power density, temperature, processing time, and other parameters may all be fine-tuned to produce the best results. The impact of ultrasound on the physical and functional qualities of skim milk was analyzed. Pasteurized homogenized skim milk was treated with 20 kHz ultrasound at 20 and 41 W for varied time intervals up to 60 minutes under controlled temperature settings. The turbidity was reduced, but the viscosity of the milk remained unaffected. After 60 minutes of sonication, the casein micelles, fat globules, and soluble particles were 174 nm, 157 nm, and 89 nm at 20 W and 172 nm, 175 nm, and 84 nm at 41 W, respectively, compared to 178 nm, 170 nm, and 136 nm for unsonicated samples. During the first 30 minutes of sonication, the whey proteins in the milk were denatured and produced soluble whey–whey/whey–casein aggregates, which then interacted with casein micelles to generate micellar aggregates. Some whey proteins from these aggregates were partially disrupted by prolonged sonication. (Shanmugam et al., 2012). Ultrasound technology allows for the manufacture of a microbiologically safe and stable product that can be kept for more than 16 days at 4°C. Despite the apparent modest drop in protein level, other criteria including butter fat availability, color, appearance, and homogeneity have improved. By refining and producing new commodities, minor modifications in other physical–chemical features of thermo-sonicated milk can be produced, allowing the product to be sold as a beverage as well as a potential dairy component.

Bermdez et al. (2009) discovered that ultrasonic treated milk had higher butter fat content and lower nonfat solids. Pasteurization of whole milk was done using (heat) pasteurized milk and second technique was thermo-sonication. The batch system pasteurization was utilized for heat treated sample and ultrasonic wave intensities of 36, 72, 108, and 120 mm (24 kHz, 400 watts); moreover, the heating for US samples were also applied simultaneously. The obvious protein levels of ultrasonicated milk sample reduced as a result of the treatment, whereas butter fat content rose. Proximal analysis revealed, there was higher amount of water in U.S. milk and a decreasing trend in nonfat solid content, as well as lower pH, more lactic acid, and lower density when comparing heat treated to U.S. milk samples; color parameter of the US milk was also more white in comparison (Bermúdez et al., 2009). Utilization of ultrasound to cream milk fat globules in milk emulsions has also been studied. It was concluded that utilizing ultrasound waves at frequency of 400 kHz or 1.6 mega Hz enhances creaming process in very coarse recombined emulsion of milk and simple raw milk. Upon sonication, particle clustering and flocculation were seen in both raw milk as well as coarse milk emulsion, the quantity of which was dependent on the settings of sonication. The enhanced clustering phenomenon generated by treating milk using high-frequency ultrasound waves prior to other processing can help separate milk-fat in dairy systems. The kinetics of creaming, coalescence, and flocculation of dairy fat globules in high frequency ultrasound systems still requires more investigations. The designing of ultrasound based unit operation suited to milk and dairy sectors will only be made possible by parameters from kinetic models (Juliano et al., 2011). The effects of sonication conditions (ultrasound probe amplitude = 50, 75, and 100%, time = 5, 10, and 15 minutes, and temperature = 45, 55, and 65°C) on the characteristics of yoghurt were examined. When the sonication time, temperature, and amplitude were increased, the apparent viscosity, hardness, and lightness increased. The pH, syneresis, and flavor declined at 75% amplitude, but acidity, adhesiveness, cohesiveness, springiness, chewiness, redness, yellowness, and overall acceptability rose. Acidity, syneresis, adhesiveness, and yellowness decreased as the sonication temperature climbed. Sonication at 55°C, 10 minutes, and 75% amplitude considerably improved the qualities of yoghurt, according to the physicochemical and sensory data (Kenari and Razavi, 2021).

Free caseins reaggregate with casein micelles when sonic cavitation shear stresses increase. The ultrasonic treatment's turbulent conditions have been demonstrated to boost particle mobility and encourage the development of aggregates. Researchers investigated the impact of ultrasound on the thermal and structural properties of proteins in reconstituted whey protein concentrate. It was also studied that the structural and thermal properties of proteins in reconstituted whey protein concentrate (WPC) solutions after sonication. When WPC solutions were sonicated for up to 5 minutes, the enthalpy of denaturation was reduced. Protein aggregation enhanced the enthalpy of denaturation as a result of prolonged sonication. Although sonication had no effect on thiol content, it did cause small alterations in the protein's secondary structure and hydrophobicity. Overall, sonication had little influence on protein structure in WPC solutions, which is important for the functional features of whey protein-based dairy products to be preserved during ultrasonic processing. Microbial growth, enzymatic reactions, fermentation control, gelling processes, and the properties of rennet coagulation in milk have all been studied using low-intensity ultrasound. The use of ultrasound in cheese production lowers the time it takes for the curd to set while also increasing the curd's toughness (Chandrapala et al., 2011).

When ultrasonic/ultrasound wave is transmitted via a physical state, for example, liquids, these waves cause compression and also stretch existing molecular arrangements of liquids. This ultrasonication wave prompted stretching and compression will fluctuate usual space among the various molecules as the molecules will oscillate/vibrate around the mean positions.

Whenever the space between the fluctuating molecules surpasses even the smallest molecular-distance, which is an essential force to hold the liquid molecules together, the liquids thus breaks down and very tiny voids are formed. These voids are called cavitation-bubbles. Every collapsing bubble acts like a microreactor, in which pressures are higher than 1000 atm and the temperatures are several hundred or even thousand degrees (Sutariya et al., 2018). Ultrasound (US) creates interrupted high as well as low compression cycles by transmitting highly concentrated and high frequency sound waves inside the liquid food materials. During less pressure, certain vacuum bubbles may develop and violently breakdown throughout high-pressure cycles. The collapse causes a quick and significant increase in temperature and pressure, which leads to intense particle collisions in liquid food (Nguyen and Anema, 2010).

4.3.1 Fat size reduction

The applied ultrasound power determines the reduction in the particle size of fat. In the dairy industry, the particle size reduction is important particularly in the emulsification process. Firstly, dispersion of fat droplets due to acoustic vibration occurs into milk aqueous fractions via turbulence. Then, the breakup of fat droplets is promoted by cavitation. The reduction of fat globule due to the disintegration of the fat globule membrane and modification in its granular surface due to the interaction of casein micelle with the disrupted membrane can be achieved with ultrasonication. In a nondairy matrix, a mechanism has been proposed according to which splitting of big existing droplets to the aqueous phase is caused by ultra-sonication. Cavitation reduces the size of these droplets and a localized turbulence is produced and dispersed around the liquid (Soria and Villamiel, 2010).

4.3.2 Economical perspective

Compared to other technologies, ultrasound equipment is fairly economical. Research on low-frequencies of ultrasonic waves is much needed to decide the cost of converting existing equipment to ultrasonic adaptation, as well as the cost of adapting cooling/heating baths on an industrial scale to test target bacteria and critical enzymes for product stability (Bermudez-Aguirre et al., 2011). Depending on ultrasonic parameters (pressure temperature, amplitude, duration intensity), microbe kind and attributes (shape, size and growth phase, thickness, and wall composition), and composition of medium, power ultrasound can activate or inactivate dairy microorganisms. Only a few studies have looked at the effects of ultrasound on spore inactivation process. The log-reduction seen for the spores which were treated with intense ultrasound processing was less than those observed for vegetative/live cells of *Geobacillus stearothermophilus* in sample of reconstituted (skim) milk powder (Beatty and Walsh, 2016). Thermosonication at temperature of 70°C has shown to be highly effective as compared to isolated heat treatments in lowering (psychrotrophic) *Bacillus cereus* spore process of inactivation in skimmed milk as well as cheese slurries (Evelyn and Silva, 2015). In the food industry, ultrasound's capabilities have been proved in preservation, extraction, and processing. Ultrasound is being used to enhance a variety of processes in the food industry, and it has emerged as a very promising technology in the processing space. It becomes substantially more effective when paired with other food preservation methods. It offers various advantages over existing or traditional technologies, and when utilized together, it helps to mitigate or eliminate their weaknesses. Because of its ability to boost efficiency and minimize the amount of time necessary for various processing procedures, ultrasound has a bright future. Ultrasonic technology may complete a whole processing unit in minutes or seconds, resulting in cheaper total processing costs, greater end-product purity, no wastewater treatment, and low-energy postprocessing (Bhargava et al., 2021). Ultrasound improves the efficacy of other food preservation processes when used in combination with them. It offers a number of advantages over current technologies, and when combined with them, it can help them overcome their own constraints. When used in conjunction with other food processing technologies, ultrasound can enhance process efficiency, shorten the time necessary for various processing steps, and increase total extraction and yield rates. Because of its ability to boost production and reduce the time required for various processing procedures, ultrasound has promised a revolutionary future. Whole food processing activities can be finished in minutes or seconds, resulting in lower overall processing costs, higher end product purity, no wastewater treatment, and less postprocessing (Singla and Sit, 2021). In a recent study, researchers evaluated of fat droplet size, enzyme coagulation characteristics, and microstructure of milk as well as analyzed lipid properties. This research was conducted to evaluate the potential of an ultrasound system for effective modification of milk. For milk entering the unit at 42 or 54°C, the use of a continual US unit which was delivering 1.36 kilo watts per pass inside a 2.4 kilo watt reactor cell having parallel diaphragm plates as well as two external transducers of frequency 16/20 kHz. This process resulted in the development of substantial amounts of submicron-sized fat droplets and altered the coagulation characteristics of milk. Microscopic imaging revealed changes in milk fat droplet sizes as well as the curd matrix microstructure, revealing how the modified milk's protein-lipid and protein-protein interactions had changed. Despite the fact that particular efficiency against bacterial pathogens is required to qualify US as a thermal processing option, the system reduced germ counts at temperatures lower than pasteurization. Although additional system tuning is required, such as processing milk at temperatures over the melting point of milk fat but below the pasteurization temperature (63°C), the results of this study suggest that continuous US might be a feasible alternative for dairy food processing (Van Hekken et al., 2019).

4.3.3 Time efficient

The time parameter is very important to achieve the favorable outcome while utilizing High-Intensity Ultrasound waves on the physicochemical characters of different milk and dairy based products. With a lower than 10 minutes processing time, a trend produces while with ultrasonication for 3 minutes, enzyme activity, and pH remain unaffected. Similarly, when HIU was applied for 6 minutes, there was no change in the proximate composition of the bovine milk. However, with application of ultrasonication from 1 to 90 seconds, a nonconsistent effect was observed on the characteristics of heavy cream such as melting peak and droplet size. Some issues raised by the HIU when applied in milk include modification in color and a reduction in the chroma and lightness which may not be favorable for appearance. Also, HIU when applied to milk, it resulted in denaturation of some major proteins. A variety of factors determine color of milk such as the breed, the diet and the physiological state of the animal. However, the main contributing factor is the retinol or carotenoids in the milk. Riboflavins, tocopherols, and xanthophylls are other factors contributing to the color of dairy products. Numerous factors determine the milk color after harvesting for example the available fat, homogenization, presence of additives, and storage conditions. High value of lightness and whiteness is shown by milk as fat globules and casein micelles disperse light. Any treatment that can affect the physical structure of milk will ultimately result in changes in milk color.

4.3.4 Fermentation efficiency and biological activities

The ultrasonication/sonication technique is a novel region of research in the dairy-based industry for enhancing milk fermentation in this respect. Fermentation is the utmost time as well as resource consuming phase in the fermentation process for all fermented dairy based products, with a significant influence on product quality. Fermented dairy products, as in lactic acid producing bacteria (LABs) and probiotics, which are a key source of, essential amino acids, bioactive peptides, vitamins, minerals, fatty acids, and other physiologically active compounds for human nutrition and health, require microbial activation. Fermented milk is widely used as a foundation for functional foods, which is growing increasingly popular among health-conscious customers. Functional dairy products can be manufactured in a variety of ways, utilizing various processing techniques or different fortifications with various nutrients to enhance the use of specific product. If the permeability of the cell membranes is increased, we can speed up the delivery of oxygen and nutrients to microbes, thus, minimizing some of the ultrasound techniques negative aspects on the living microbial cells, power ultrasound is stated to increase the process of fermentation and proliferation of LAB and probiotics. Ultrasonicated microwaved milk showed 13.9%, 44.7%, and 77.3% more antioxidant activity at the end of fermentation than ultrasonicated plus autoclaved, microwaved and simply autoclaved samples, respectively. Microwaved milk samples showed protease activity that was 1.3–1.8 x(times) higher than other treatment samples after 24 hours of fermentation. Ultrasonicated microwaved milk produced considerably more exopolysaccharides, anticancerous activity and also amylase inhibitory effects after 8 hours fermentation than the other treated milks (Gholamhosseinpour et al., 2020). When compared to other treatments, microwave treatment enhanced *L. acidophilus* viability, the lactate contents and also increased protease activities, while microwave combined with ultrasound subjected samples had greater radical scavenging or antioxidant capacity, amylase-inhibition, anti-cancerous potential, better peptides concentration as well as better exopolysaccharide content. When compared to the other treatments, autoclave heating reduced the bioactive qualities of samples; hence autoclaved milks had the least of these features. When lactic acid bacteria were fermented, microwaved, or sonicated, peptide levels were higher than when they were not. It is a good idea to use these safe approaches to raise peptides in fermented foods since they contain anticancer, antioxidant, and amylase inhibitory qualities (Gholamhosseinpour et al., 2020). Normal processing (pasteurization) causes protein denaturation and precipitation because whey protein portion of milk proteins is thermolabile that break down at high temperature (i.e., over 61°C). Ultrasound is an effective replacement which can be used for pasteurization whenever precipitation process is not possible. Whey that has been treated with ultrasound achieves homogeneity, which increases stability. The study found that treating microbiological fermentation cultures with ultrasound before inoculating them in samples increases their activity and speeds up fermentation. Because the fermentation duration is reduced as well as the outcome is almost identical to pasteurization, hence, ultrasonic processing may save a lot of money. Ultrasonic technique holds a promising future in the dairy and dairy based industries. According to the findings, combining 480 W and 55°C had the best synergistic effect in terms of bacterial inactivation. As a result, when contemplating processing substitutes to thermal treatments like pasteurization, the stated parameters may be ideal for the treatment of whey. When comparison was made to the control samples, ultrasound assisted activation (84 watts/150 seconds) of dairy cultures (specific) (La-5 *L. acidophilus*, yoghurt-culture. YC-380) exhibited a 0.5 hour reduction in fermentation time with monoculture La-5. Furthermore, irrespective of culture utilized, the technique of thermo-sonication (480 watt/55°C per 8 minutes) rather than using pasteurization caused fermentation of whey having a higher number of live cells and improved sensory qualities at the end of fermentation, such as no sediment. As a result, thermo-sonication may be a useful technique to increase the quality of fermented whey beverages while simultaneously increasing microbial biomass production utilizing whey as a substrate (Barukčić et al., 2015). Improving the fermentation process without compromising the ultimate product is crucial for the dairy industry. Ultrasound has been proven in multiple investigations to enhance the textural character, syneresis, and water holding capacity of fermented milk gels. Furthermore, ultrasonic has been shown to reduce fermentation durations, boost lactose hydrolysis, and transgalactosylation in culture cells, particularly intracellular beta-galactosidase enzyme activity.

4.3.5 Functional properties

Function compounds are believed to act as antioxidants and provide the human body some protection from the damages caused by reactive-oxygen species. Numerous studies have found increased antioxidant ability in dairy products when sonication is applied. The functional properties of milk or dairy based products can be enhanced by using ultrasound processing. Increase in the bioactivity of different functional compounds was observed while utilizing ultra-sonication processing techniques. It is important as with increase in these compounds increases the anticarcinogenic and antioxidant activity. Recently, the demand for natural antioxidants has increased. Sulfur-containing amino acids, carotenoids, and vitamins A and E are responsible for anitocarconigenic and antioxidant activity in dairy products such as milk. Solubility is also another important property of dairy products. A study to analyze changes in solubility when milk is subjected to HIU pretreatment (20 kHz, 200 W) found that after 5 minutes of the treatment, there is a significant increase in the solubility from 35.78% to 80.30% due to alternations in the three dimensional structure of globular protein.

4.3.6 Protein-gels, enzyme activity, and water holding capacity

In dairy systems, sonication/ultrasounds can be used to homogenize several protein aggregates and also denature whey proteins. Ultrasonication plus high pressures as well as heat (e.g., 3.5 kg/cm^2 at 126°C approx.) lowers enzyme activity, whereas mild sonication increases it (Munir et al., 2019). MW, HPP, and US technologies can drastically modify the physical and functional characteristics of milk proteins in a number of ways. Due to whey protein aggregation, particle size rises during HPP and MW processing, but reduces in the US due to dissociation of protein clumps. The viscosity of HPP and MW treated milk rises, whereas the viscosity of US treated milk decreases. Sonication/ultrasound-induced alterations in dairy proteins can be efficiently utilized to increase the yields, physiochemical attributes as well as sensory characteristics of many milk based products. As for example, when ultra-sonicated-milk was utilized in yoghurt production, it was detected that gelling structures were more inter-connected as well as more uniform in comparison to milk which was not treated with the same method. The amplified hydrophobicity of milk protein and breakdown of the protein-aggregates due to sonication are supposed to enhance the gelling process because of the improved protein to protein interactions as well as hydrophobic interactions which leads to more firm textured curds (Munir et al., 2019). Ultra-sonication renders the different proteins present within milk protein-concentrates a little more sensitive to enzyme mediated hydrolysis owing to proteins unfolding, which increases enzyme and substrate interaction, resulting in better enzyme activity (Uluko et al., 2013). Pretreatment of goat milk in the US can improve rennet coagulation properties. Greater rennet gels were linked to smaller and more uniform particle sizes, little changes in salt balance, and moderate denaturation of whey.

Capacity of water storage (%): Ultrasonic treatment gives goat milk a high water-holding capacity (WHC) The coagulation of goat milk at various ultrasound treatment periods. The cross-linked structure within gels grew denser, resulting in firmer gels, higher G_{max}, final G_0 values, cohesiveness, and WHC, as well as a significant rise in the TSI index. Additionally, at the optimum time of ultrasonication, syneresis kinetics, and gelation time are reduced. As a result, ultrasonic pretreatments of goat milk might be a potential option for the cheese business (Zhao et al., 2014). In a study ultrasound technique was used to raise the temperature to 72°C in 6 minutes, following which it remained steady for the rest of the therapy. Traditional heating denatured about half of the whey proteins, but the US method denatured just about a fourth. Ultrasonography may significantly alter the viscosity of yogurt, whether provided before or after inoculation, if the amplitude is large enough. The volume fraction of gel fragments and the viscosity of continuous liquid whey both increase. When compared to yogurts made from conventionally heated whole milk, the qualities of yogurts made from ultrasonicated whole milk were better, with higher firmness and water-holding capacity and decreased syneresis (Riener et al., 2009).

4.3.7 Adulteration testing

The presence of fat or lactose crystals, as well as the solid component of the matrix, might explain the attenuation coefficient seen in reconstituted milk. The existence of crystals, without a doubt, changes compressibility and, as a result, attenuation. This is not the case, however, due to the presence of liquid and solids in milk. It is probable that particle size and concentration, rather than the presence of crystals and air, are to account for the matrix's higher attenuation than water. These two criteria will influence the outcome. As a result, ultrasonic attenuation may be used to assess powder reconstitution quality. Finally, research showed that the ultrasonic attenuation parameter may be utilized to detect adulteration induced by water addition or chemical treatment. The attenuation coefficient may be used to analyze the content, molecular assembly, physiological changes, and, ultimately, the quality of various reconstituted milk brands because it is such a sensitive indicator (Ali and Ahmad, 2017).

4.3.8 Mano-thermo-sonication

Data on the intensification of lethality of older heat treatments coupled with ultrasonication shows great prospect. A new combined process of microbial inactivation including heat and ultrasonication under pressure (Mano-Thermo-Sonication; MTS process) is described to be very effectives. When compared to heat treatments at the same temperature, this novel method has a much higher lethality rate. *Bacillus subtilis* spores, yeasts (*Saccharomyces cerevisiae*), and certain vegetative cells (*Aeromonas hydrophila*) all died far faster than *Bacillus subtilis* spores (between 10 and 30 times approximately). Pressure was necessary to elicit MTS lethality at temperatures over the medium's boiling point. Heat and ultrasound under pressure work together to significantly improve the lethality of heat treatments, resulting in significant time and/or temperature savings in heat operations. Microbial inactivation by MTS follows a first-order reaction kinetics pattern, similar to heat processes, allowing for the prediction of the inactivation effect. As a consequence, assuming no unwanted side effects in the processed food, Mano-Thermo-Sonication may be a more cost-effective alternative to traditional thermal procedures (Raso et al., 1994). Researchers assessed the impact/effects of "HIUS (high-intensity ultrasound)" technology with respect to quality attributes of a novel dairy based beverage, that is, chocolate and whey beverage. Two procedures were executed varying the processing time and power: HIUS.A 160 watts, 937 seconds and HIUS.B 720 watts, 208 seconds. The different methods of applying the same energy-density had a direct impact on the quality of chocolate flavored whey beverage which resulted in the formation of products with diverse physical and microstructural features. The HIUS-B method (higher power, shorter processing time) was a thermal process (maximum temperature of 71°C), whereas the HIUS-A method (lower power, longer processing time) was a nonthermal process (maximum temperature 35°C). Although the HIUS-A and HIUS-B treatments had varied impacts on fat globule size, both created a chocolate whey beverage with good kinetic and electrostatic stability. When compared to HIUS-A, the HIUS-B technique's expressive reduction in fat globules led in an apparent lightness (L*) difference. Additionally, both ultrasonic treatments conserved mineral components in the same way, indicating that the technology may be used to preserve food nutrition (Monteiro et al., 2020).

4.3.9 Limitations

Sonication has its limitations, for example, metal contamination susceptibility, reduced transmission of ultrasonic energy in bulky sample in terms of volume, poor spatiotemporal dispersal, lower energy efficiency and increased operational costs as compared to other available methods. Due to these, sonication has not been accepted widely for commercial applications. Ultrasonication's inability to be employed in a continuous process with a greater flow rate is one of its drawbacks (Sutariya et al., 2018). Cultivation conditions such as harvesting, postharvest, variety, processing, storage and a number of different factors all impact the physicochemical qualities of an agricultural product, resulting in each product's own distinct behavior. As a result of this issue, utilizing ultrasound on a large scale is challenging, and the desired results may not be obtained (Firouz et al., 2019). Intense sonication, on the other hand, produces irreversible cellular damage such as cell splitting and shearing. Lack of knowledge, understanding, and a reluctance to forsake current practices all impede the industrial adoption and commercialization of ultrasonography (Bhargava et al., 2021).

4.4 Future prospects

The use of ultrasound alone to control spoilage and pathogenic bacteria (psychrophilic, mesophilic and total coliforms) is effective, favorable results can be achieved with the combination of ultrasound with heat. However, for evaluating microbial behavior during maturation (in cheese) and during cold storage, further studies are required. While numerous studies deal with the effect of HIU application on quality parameters of dairy products, very few studies have evaluated the effect of HIU on sensory properties according to which HIU preserves the sensory characteristics of dairy products better than conventional heat treatment. While lipid oxidation and off-flavors develop in milk due to ultrasonication, these effects diminish with time. Ultrasound makes it possible to manipulate and improve the texture and quality of dairy products. The mechanism of modification in the texture of the product by ultrasound is still unclear. Therefore, further studies are needed on the effect of ultrasound on the sensory properties of dairy products. So far, the practical applications of ultrasound in the dairy industry have been the focus of research instead of its mechanism of action of bacterial cells utilized for fermented dairy based products. Dairy products or milk when subjected to ultrasonic processing improves functional properties, increases efficiency of preservation and enables the development of value added products. However, to use ultrasound for increasing the quality and quantity of functional dairy products by enzymatic hydrolysis and the use of ultrasound as a tool for producing bioactive components, further knowledge is required. Particularly, further research is needed to comprehend the interaction between sensory characters of the final product and ultra-sonication parameters. Finally, the health-promoting effect, and the effect on probiotic role in increasing shelf life should also be considered (Table 4.1).

Table 4.1 Different effects of ultrasounds on dairy-based products.

Sample	Experimental parameters	Effects of ultrasound
Anhydrous milk fat and milk fat/rapeseed oil blends	20 kilo Hz, 40 watt at 28°C. Recycling mode with a flow rate of 200 mL/min and a total volume of 500 mL samples collected at 0, 2.5, 5, 7, and 10 min of ultrasonication.	HIU had no effect on the melting of fat crystals, although it did reduce their size. Needle penetration revealed no influence on hardness.
Pasteurised full fat milk.	20 kilo Hz, 10, 30 and 50 watt at 27, 50, 30 min and 70°C. Energy in the flow = 91, 273 and 454 watts per liter, respectively.	At 30 watt and 50 C, HIU speeds up the development of acid gels. The rise in temperature caused by HIU promotes protein denaturation. Fat globule size was decreased from 3.39 to 3.89 m to 0.37–1.9 m when sonication power was more than 30 W and temperature was greater than 50 C. HIU increases the specific surface area (from 2.07 to 45.09 m^2ag 1 fat).
Pasteurized fresh bovine milk	For 3 min, a frequency of 20 kilohertz at 100% power (500 W.)	After 21 days, ultrasonication showed no influence on pH or enzyme activity when compared to control. The US did have a dispersion of 1 m particles, which might be attributed to fat globule size decrease. Denaturation of a-lacto-albumin and k-casein decrease in the US.

(continued on next page)

Chocolate milk beverage	800 watt, 19 kilo Hz. Nominal power of 400 watt energy density 0.3, 0.9, 1.8, 2.4, and 3.0 kJ/cm^3 with 13 mm diameter probe. Temperatures reached 42°C at their highest point.	After 14 d of storage, 10 min of HIU increases L*, whereas HIU decreases C*. The variables a* and b* remain unchanged. Higher HIU energy densities result in better beverage homogenization and fat globule particle size distribution. Antioxidant activity, ACE inhibitory activity, fatty acid profile, and volatile profile were all affected by HIU. Ultrasonication also boosted the retention of short and medium chain fatty acids while reducing the loss of nutritional components.
Cheeses and creams	24 kHz, 400 watt, amplitude of 100 and 50% for 0, 5, and 10 min.	Despite an increase in exudate, HIU boosted cheese yield (percent). HIU induces yellowing in cheese after 10 min. The color coordinates L*, a*, and C* remain unaltered. After 5 min of ultrasonication, the pH rose from 6.6 to 6.74, but fell after 10 min.
Fresh cream	20 kHz, 100 and 300 watt for 0, 5, 10 and 15 min. Temperature 50°C.	With HIU, the diameter of cream particles (0.296–5.867 μm) and particle size distribution were enhanced. Yellowness is reduced with HIU.
Yogurt and ice cream	Multiple ultrasonic factors used in the production of yoghurt and ice cream are reviewed.	Ultrasound helps to minimize fat droplets by increasing viscosity, whey coagulation, and water-holding capacity. Ultrasonication also reduces the size of ice crystals, allowing the freezing process to be accelerated.

4.5 Conclusion

In the light of facts discussed above it is safe to say that both microwave and sonication techniques are good options for dairy processing. Microwave technique can help prevent the production of off taste and odor, which are otherwise results of conventional thermal techniques. Microwaves are also more efficient in destroying microbes which are a threat towards dairy products. Moreover, microwave technique has also shown good results in term of gelation, another desirable effect to be used in dairy industry. Sonication technique can help mitigate the risks of adulteration, for this we have to indulge more deeply into how we can make this happen. Sonication can also increase the water holding capacity making good quality milk products. Sonication can enhance enzyme activity which is desirable in many dairy processing operations. Time is of the essence and sonication is no doubt time efficient technique. Both microwave and sonication techniques require much more refining before they can be implemented in already established dairy industries. This mean more research studies aimed at perfecting these techniques are required. The gap between research and industry should also be reduced. More public outreach and awareness programs must be set in place in order to makes these techniques widely acceptable in all sectors of dairy industry. Nonetheless, the benefits of these techniques are far too great to be ignored. So, the time is not too far when these techniques will be helping us to overcome existing challenges.

References

Ali, M., Ahmad, A., 2017. Attenuation of ultrasound in reconstituted milk. Int. J. Environ. Sci. Technol. 6, 1828–1832.

Balthazar, C.F., Santillo, A., Guimaraes, J.T., Bevilacqua, A., Corbo, M.R., Caroprese, M., Marino, R., Esmerino, E.A., Silva, M.C., Raices, R.S., 2019. Ultrasound processing of fresh and frozen semi-skimmed sheep milk and its effects on microbiological and physical-chemical quality. Ultrason. Sonochem. 51, 241–248.

Barbosa-Canovas, G., Bermudez-Aguirre, D., 2010. Other novel milk preservation technologies: ultrasound, irradiation, microwave, radio frequency, ohmic heating, ultraviolet light and bacteriocins. In: Improving the Safety and Quality of Milk. Elsevier, Woodhead Publishing, Cambridge, UK, pp. 420–450.

Barukcic, I., Jakopovic, K.L., Herceg, Z., Karlovic, S., Bozanic, R., 2015. Influence of high intensity ultrasound on microbial reduction, physico-chemical characteristics and fermentation of sweet whey. Innovative Food Sci, Emerg. Technol 27, 94–101.

Beatty, N.F., Walsh, M.K., 2016. Influence of thermosonication on *Geobacillus stearothermophilus* inactivation in skim milk. Int. Dairy J. 61, 10–17.

Bermudez, A.D., Mawson, R., Versteeg, K., Barbosa-Canovas, G.V., 2009. Composition properties, physicochemical characteristics and shelf life of whole milk after thermal and thermo-sonication treatments. J. Food Qual. 32, 283–302.

Bermudez-Aguirre, D., Mobbs, T., Barbosa-Canovas, G.V., 2011. Ultrasound applications in food processing. In: Ultrasound Technologies for Food and Bioprocessing. Springer, New York, USA, pp. 65–105.

Bhargava, N., Mor, R.S., Kumar, K., Sharanagat, V.S., 2021. Advances in application of ultrasound in food processing: a review. Ultrason. Sonochem. 70, 105293.

Bozkurt-Cekmer, H., Davidson, P.M, 2017. Microwaves for microbial inactivation-efficiency and inactivation kinetics. In: The Microwave Processing of Foods, 2nd ed. Elsevier, Woodhead Publishing, Cambridge UK, pp. 220–251.

Carrillo, L.M., Garcia-Galicia, I.A., Tirado-Gallegos, J.M., Sanchez-Vega, R., Huerta-Jimenez, M., Ashokkumar, M., Alarcon-Rojo, A.D., 2021. Recent advances in the application of ultrasound in dairy products: effect on functional, physical, chemical, microbiological and sensory properties. Ultrason. Sonochem. 73, 105467. https://doi.org/10.1016/j.ultsonch.2021.105467.33508590. In press.

Chandrapala, J., Ong, L., Zisu, B., Gras, S.L., Ashokkumar, M., Kentish, S.E., 2016. The effect of sonication and high pressure homogenization on the properties of pure cream. Innovative Food Sci. Emerg. Technol. 33, 298–307.

Chandrapala, J., Zisu, B., Palmer, M., Kentish, S., Ashokkumar, M., 2011. Effects of ultrasound on the thermal and structural characteristics of proteins in reconstituted whey protein concentrate. Ultrason. Sonochem. 18, 951–957.

Chemat, F., Khan, M.K., 2011. Applications of ultrasound in food technology: processing, preservation and extraction. Ultrason. Sonochem. 18, 813–835.

Chen, C.C., Hsieh, J.F., 2016. Microwave-assisted cross-linking of milk proteins induced by microbial transglutaminase. Sci. Rep. 6, 1–9.

Clare, D., Bang, W., Cartwright, G., Drake, M., Coronel, P., Simunovic, J., 2005. Comparison of sensory, microbiological, and biochemical parameters of microwave versus indirect UHT fluid skim milk during storage. J. Dairy Sci. 88, 4172–4182.

El-Mecherfi, K.E., Saidi, D., Kheroua, O., Boudraa, G., Touhami, M., Rouaud, O., Curet, S., Choiset, Y., Rabesona, H., Chobert, J.M., 2011. Combined microwave and enzymatic treatments for β-lactoglobulin and bovine whey proteins and their effect on the IGE immunoreactivity. Eur. Food Res. Technol. 233, 859–867.

Evelyn, E., Silva, F.V., 2015. Thermo sonication versus thermal processing of skim milk and beef slurry: modeling the inactivation kinetics of psychotropic *Bacillus cereus* spores. Food Res. Int 67, 67–74.

Firouz, M.S., Farahmandi, A., Hosseinpour, S., 2019. Recent advances in ultrasound application as a novel technique in analysis, processing and quality control of fruits, juices and dairy products industries: a review. Ultrason. Sonochem. 57, 73–88.

Gholamhosseinpour, A., Hashemi, S.M.B., Jahromi, L.R., Sourki, A.H., 2020. Conventional heating, ultrasound and microwave treatments of milk: fermentation efficiency and biological activities. Int. Dairy J. 110, 104–109.

Guzik, P., Kulawik, P., Zając, M., Migdał, W., 2021. Microwave applications in the food industry: an overview of recent developments. Crit. Rev. Food Sci. Nutr. 12, 1–20.

Jambrak, A.R., Lelas, V., Kresic, G., Badanjak, M., Brncic, S.R., Herceg, Z., Batur, V., Grcic, I., 2011. Rheological, functional and thermo-physical properties of ultrasound treated whey proteins with addition of sucrose or milk powder. Mljekarstvo 61, 79.

Jambrak, A.R., Mason, T.J., Lelas, V., Paniwnyk, L., Herceg, Z., 2014. Effect of ultrasound treatment on particle size and molecular weight of whey proteins. J. Food Eng. 121, 15–23.

Juliano, P., Kutter, A., Cheng, L.J., Swiergon, P., Mawson, R., Augustin, M., 2011. Enhanced creaming of milk fat globules in milk emulsions by the application of ultrasound and detection by means of optical methods. Ultrason. Sonochem. 18, 963–973.

Kalla, A.M., Devaraju, R., 2017. Microwave energy and its application in food industry: a reveiw. Asian J.Dairy Food Res. 36, 37–44.

Kehinde, B.A., Sharma, P., Kaur, S., 2021. Recent nano, micro and macrotechnological applications of ultrasonication in food-based systems. Crit. Rev. Food Sci. Nutr. 61, 599–621.

Kenari, R.E., Razavi, R., 2021. Effect of sonication conditions: time, temperature and amplitude on physicochemical, textural and sensory properties of yoghurt. Int. J. Dairy Technol. 74, 332–343.

Leong, T.S., Walter, V., Gamlath, C.J., Yang, M., Martin, G.J., Ashokkumar, M., 2018. Functionalized dairy streams: tailoring protein functionality using sonication and heating. Ultrason. Sonochem. 48, 499–508.

Li, S., Zhang, R., Lei, D., Huang, Y., Cheng, S., Zhu, Z., Wu, Z., Cravotto, G., 2021. Impact of ultrasound, microwaves and high-pressure processing on food components and their interactions. Trends Food Sci. Technol. 48, 641–648.

Liu, Z., Juliano, P., Williams, R.P., Niere, J., Augustin, M.A., 2014. Ultrasound improves the renneting properties of milk. Ultrason. Sonochem. 21, 2131–2137.

Madadlou, A., Mousavi, M.E., Emam-Djomeh, Z., Ehsani, M., Sheehan, D., 2009. Sonodisruption of re-assembled casein micelles at different pH values. Ultrason. Sonochem. 16, 644–648.

Martins, C.P., Cavalcanti, R.N., Couto, S.M., Moraes, J., Esmerino, E.A., Silva, M.C., Raices, R.S., Gut, J.A., Ramaswamy, H.S., Tadini, C.C., 2019. Microwave processing: current background and effects on the physicochemical and microbiological aspects of dairy products. Compr. Rev. Food Sci. Food Saf. 18, 67–83.

Monteiro, S.H., Silva, E.K., Guimaraes, J.T., Freitas, M.Q., Meireles, M.A., Cruz, A.G., 2020. High-intensity ultrasound energy density: how different modes of application influence the quality parameters of a dairy beverage. Ultrason. Sonochem. 63, 109–112.

Munir, M., Nadeem, M., Qureshi, T.M., Gamlath, C.J., Martin, G.J., Hemar, Y., Ashokkumar, M., 2020. Effect of sonication, microwaves and high-pressure processing on ACE-inhibitory activity and antioxidant potential of Cheddar cheese during ripening. Ultrason. Sonochem. 67, 105–110.

Munir, M., Nadeem, M., Qureshi, T.M., Leong, T.S., Gamlath, C.J., Martin, G.J., Ashokkumar, M., 2019. Effects of high pressure, microwave and ultrasound processing on proteins and enzyme activity in dairy systems-a review. Innovative Food Sci. Emerg. Technol. 57, 121–129.

Murthy, V.J., Kiranmai, N.S., Kumar, S., 2017. Study of dielectric properties of adulterated milk concentration and freshness. In: IOP Conference Series: Materials Science and Engineering, 225. IOP Publishing, Bristol, United Kingdom, pp. 1–7, 012285.

Nguyen, N.H., Anema, S.G., 2010. Effect of ultra-sonication on the properties of skim milk used in the formation of acid gels. Innovative Food Sci. Emerg. Technol. 11, 616–622.

O'sullivan, J., Murray, B., Flynn, C., Norton, I., 2016. The effect of ultrasound treatment on the structural, physical and emulsifying properties of animal and vegetable proteins. Food Hydrocolloids 53, 141–154.

Raso, J., Condon, S., Sala, F., Trepat, 1994. Mano-thermosonication: a new method of food preservation. Food Preserv. Combined Processes. Final report for FLAIR Concerted Action (7).

Riener, J., Noci, F., Cronin, D.A., Morgan, D.J., Lyng, J.G., 2009. The effect of thermo sonication of milk on selected physicochemical and microstructural properties of yoghurt gels during fermentation. Food Chem. 114, 905–911.

Shanmugam, A., Chandrapala, J., Ashokkumar, M., 2012. The effect of ultrasound on the physical and functional properties of skim milk. Innovative Food Sci. Emerg. Technol. 16, 251–258.

Shen, X., Shao, S., Guo, M., 2017. Ultrasound-induced changes in physical and functional properties of whey proteins. Int. J. Food Sci. Technol. 52, 381–388.

Silva, M., Zisu, B., Chandrapala, J., 2018. Influence of low-frequency ultrasound on the physico-chemical and structural characteristics of milk systems with varying casein to whey protein ratios. Ultrason. Sonochem. 49, 268–276.

Singla, M., Sit, N., 2021. Application of ultrasound in combination with other technologies in food processing: a review. Ultrason. Sonochem. 11, 105–116.

Soria, A.C., Villamiel, M., 2010. Effect of ultrasound on the technological properties and bioactivity of food: a review. Trends Food Sci. Technol. 21, 323–331.

Sutariya, S., Sunkesula, V., Kumar, R., Shah, K., 2018. Emerging applications of ultra-sonication and cavitation in dairy industry: a review. Cogent Food Agric. 4, 154–167.

Tho, P., Manasseh, R., Ooi, A., 2007. Cavitation micro streaming patterns in single and multiple bubble systems. J. Fluid Mech. 576, 191–233.

Uluko, H., Zhang, S., Liu, L., Chen, J., Sun, Y., Su, Y., Li, H., Cui, W., Lv, J., 2013. Effects of microwave and ultrasound pretreatments on enzymolysis of milk protein concentrate with different enzymes. Int. J. Food Sci. Technol. 48, 2250–2257.

Van-Hekken, D., Renye, J., Bucci, A., Tomasula, P., 2019. Characterization of the physical, microbiological, and chemical properties of sonicated raw bovine milk. J. Dairy Sci. 102, 6928–6942.

Yang, J., Huang, M., Peng, J., Shi, J., 2016. Rapid determination of the moisture content of milk powder by microwave sensor. Measurement 87, 83–86.

Yanjun, S., Jianhang, C., Shuwen, Z., Hongjuan, L., Jing, L., Lu, L., Uluko, H., Yanling, S., Wenming, C., Wupeng, G., 2014. Effect of power ultrasound pre-treatment on the physical and functional properties of reconstituted milk protein concentrate. J. Food Engg. 124, 11–18.

Zhao, L., Zhang, S., Uluko, H., Liu, L., Lu, J., Xue, H., Kong, F., Lv, J., 2014. Effect of ultrasound pretreatment on rennet-induced coagulation properties of goat's milk. Food Chem. 165, 167–174.

CHAPTER FIVE

Synergistic effect of ultrasound and microwave treatments in meat, fish, and seafood processing, preservation, and extraction

Waseem Khalid[a], Muhammad Sajid Arshad[a], Zahra Maqbool[a], G.D.T.M. Jayasinghe[b,g], B.K.K.K. Jinadasa[b,c], Scott W. Fowler[d,e], Safura Kousar[a] and Mohammad Javed Ansari[f]

[a]Department of Food Science, Government College University, Faisalabad, Pakistan
[b]National Aquatic Resources Research & Development Agency (NARA), Colombo, Sri Lanka
[c]Department of Food Science & Technology, Faculty of Livestock, Fisheries & Nutrition, Wayamba University of Sri Lanka, Gonawila (NWP), Sri Lanka
[d]School of Marine and Atmospheric Sciences, Stony Brook University, New York, USA
[e]Institute Bobby, Cap d'Ail, France
[f]Department of Botany, Hindu College Moradabad (Mahatma Jyotiba Phule Rohilkhand University Bareilly), India
[g]Universite de Pau et des Pays de l'Adour, E2S UPPA, CNRS, IPREM UMR 5254, Hélioparc, 64053 Pau, France

5.1 Introduction

There are previous works which have verified that synergism between ultrasound and microwave treatments may be a useful technology for the food industry, natural product extraction, and for the development of biomaterials (Li et al., 2020). Effective conversion of agricultural products into ready-to-eat products, which are of high quality and shelf-stable, added value can be carried out by further food processing. Food processing has single or multiple operations, and the basic notion of heat generation is used in the most conventional procedures (Singla et al., 2021). As a result, there is a high demand for substantial research and development in order to improve existing technologies like ultrasonic processing (Bindes et al., 2019), microwave processing (Rodsamran and Sothornvit, 2019), supercritical fluid extraction (Pimentel-Moral et al., 2019), control pressure drop processing (Rezzoug et al., 2005), and supercritical water extraction (Ozel et al., 2003), etc. In comparison to ultrasound and microwave treatment alone, combined ultrasound–microwave treatment may be a better method of

Ultrasound and Microwave for Food Processing: Synergism for Preservation and Extraction.
DOI: https://doi.org/10.1016/B978-0-323-95991-9.00012-6

modification. Combining ultrasound with microwave methods provides quick and efficient extraction for some compounds from meat, fish, and seaweeds. Both techniques belong to the "greener technologies." Microwave irradiation produces efficient volumetric heating of the working media without extensive energy losses compared to the conventional convection or conduction heating. Ultrasound sonication is defined as sound at a frequency that is very high which is potentially used for transmitting the ultrasound waves through the liquid (Mason et al., 2002; Van Gerven and Stankiewicz, 2009). The main benefit of combining ultrasound and microwave is that it creates a powerful energy source that can be used to promote or improve a variety of chemical processes, including synthesis, compound extraction from natural products, and sample preparation for chemical analysis (Cravotto and Cintas, 2006). One question about the synergistic ultrasonic and microwave effects is how the two technologies can be combined. There are two suggested approaches to combine these technologies. The first is having two separate reactors where one is for high quality microwave application and the second uses ultrasound waves in conjunction with a recirculating pump to transfer medium from one reactor to the other. The other approach is simply using one reactor supported by both ultrasound and microwave generators (Leonelli and Mason, 2010).

5.2 Recent technologies in the meat, fish, and seafood processing, preservation, and extraction

5.2.1 Ultrasound sonication

New "green and innovative" pasteurization techniques, such as ultrasound, typically require less water, time, energy, processing, and extraction. Ultrasound is a nonthermal method of food preparation that is advanced and gaining popularity as an alternative to, or complement to, traditional processing techniques. Food processing can benefit from the use of ultrasound for drying, homogenization, extraction, emulsification, filtration, and other operations. Ultrasound is an important tool in achieving the goal of long-term sustainability "green" chemistry and extraction (Singla and Sit, 2021). This method is well-known in the food and chemical industries for having a significant impact on the processing rate of various processes. The principle of acoustic cavitation underpins ultrasound-assisted extraction (UAE), which causes the plant to be damaged matrix cell walls, allowing bioactive compounds to be released. Ultrasound waves varying in frequency from 20 kHz to 100 MHz and intensities varying between 10 and 1000 W/cm^2

can travel as a mechanical wave through any medium, causing mechanical, physical, and chemical changes in the structure of the food. By producing phases of compressions and expansions, the motion generates two alternative cycles (Astráin-Redn et al., 2019).

Cavities form in a liquid when the liquid's tensile strength is overcome by a large negative pressure during the expansion cycle; positive pressure is created during the compression cycle, which can bind molecules. The cavities continue to grow as acoustic energy is absorbed; once they reach a critical size, they implode, releasing energy and heat into the surrounding environment, reaching temperatures high, after that, a quick cooling at a rate of more than a billion °C/s (Yu et al., 2020). For a driving frequency f, the pressure amplitude observed by the gas bubble through an acoustic field can be written as [PA = P0Pm(sin2ft)] using the trilling tonnes equation. The total partial pressure of the gas and vapor in the bubble is the total pressure inside the bubble (Khan et al., 2020).

Ultrasound is a term for high-frequency, nonaudible sound waves that have lived in nature for over a million years and with a frequency greater than 20 kHz. Since the 1880s, the piezoelectric effect has been widely utilized in the United States to create ultrasonic waves, and as technology has advanced, additional ceramic materials, such as lead zirconate, barium titanate, quartz, and lead titanate, have been used to generate waves in the United States (Kane et al., 2004). The technique developed in the United States has gradually evolved into a novel, promising, and environmentally friendly technology that may be used to replace or complement existing food preparation procedures. The "Green Food Process" is met by US technology because of its lower maintenance costs, reduced waste generation, energy, shorter processing time, solvent savings, and improved food quality. Existing food-industry unit operations have been updated as a result of the development of improved high-power ultrasonic generators, a clear example of a "technology-push" (Fu et al., 2020).

The consequences of a high-intensity ultrasonic wave on the biochemical, physical, and microbial properties of meat have piqued people's interest in recent years. Since the 1950s, ultrasonic treatments have been used in meat processing to determine the amount of fat and muscle in live animals. Real-time meat measurements help to enhance animal quality and, as a result, processed meat quality. Ultrasound can change the texture of meat by shattering the integrity of muscle tissues, especially cells, and promoting enzymatic reactions, while ultrasonic cavitation can even change connective tissue proteins (Karwowska et al., 2021). Nonthermal preservation technologies (NTPT) have shown a high potential for industrial application in improving the quality of fish and meat products and reducing microbial loads. Salmonella Enteritidis in chicken sausages can be reduced using high hydrostatic pressure processing. Similarly, UV radiation was used to inactivate E. coli in shrimp, while gamma irradiation was utilized to diminish five distinct Salmonella strains in chicken. Furthermore, these technologies produce no waste that is harmful to the environment (Rosario et al., 2021).

5.2.2 Microwave

Because of its wide variety of uses, such as drying, frying, heating, and extraction, microwave heating has piqued the interest of food producers and scientists. Raytheon Corporation built and sold the first commercial microwave oven in 1954 (Soni et al., 2020). Microwave working principle: Polar molecules inside the meal, such as water molecules, spin according to the alternating electromagnetic field as energy is absorbed. One end is positively charged, while the other is negatively charged, the water molecule is a "dipole." When exposed to an electromagnetic field, water molecules act like magnets, and these dipoles have similar orientations. Cooking heat would be generated by the rotation of these water molecules. As a result, the composition of food will influence how it heats up in a microwave oven (Bhatt et al., 2020)

Microwaves are electromagnetic waves with wavelengths and frequencies ranging from micrometers to millimeters from 300 MHz to 300 GHz and 1 m to 1 mm, respectively. Most industrial microwave systems use frequencies between 915 and 2450 MHz (2.45 GHz), whereas home appliances use 2450 MHz, which has been successfully implemented in prepackaged, frozen, ready-made, or prechilled products in pasteurization, sterilization, thawing, microwave vacuum drying, tempering, microwave freeze-drying, tempering, and baking (Li et al., 2019; Gautam, 2019).

In the microwave and radio frequency ranges, four main relaxations occur when an electric field interacts with biological tissue. Each relaxation takes place within a certain frequency range, allowing different phenomena to be identified. Microwaves and meat products interact to produce a -dispersion is a phenomenon associated to the condition of the meat and the amount of water in it (Wang et al., 2019).

Meat is inherently variable as a biological material, making the twin goals of efficiency and consistency difficult to achieve. Microwave cooking has been used to investigate the impact of various raw meats on quality. Heat-induced changes in the components of meat alter its texture, and one of the most essential aspects of muscle foods is that they are abundant in protein, with changes in connective tissue tenderizing and myofibrillar protein hardening increasing toughness. Thermal processing increases toughness by shrinking intramuscular collagen between 60 and 70°C, denaturing meat myofibrillar proteins between 40 and 50°C, and shrinking and dehydrating actomyosin between 70 and 90°C (Barbosa-Cánovas et al., 2014).

Water holding capacity and protein denaturation were significantly reduced by the mixture of microwave and air convection, allowing meat uniformity to be preserved. Microwaves can also help to prevent the formation of carcinogenic compounds during the thawing process. In comparison to other thawing methods, when microwave thawing is combined with vacuum thawing, the thawed food product has more desirable physicochemical properties. Furthermore, TBA analysis revealed that in microwaved samples, the degree of lipid oxidation is lower (Guzik et al., 2021).

Microwave food processing is always changing; the primary advantages of this technology are quick heating and good energy efficiency. Microwave treatments affect the microbial species predominance and those that are best adapted to particular environments. Some intrinsic meat components, water activity, pH, and salt concentration, as well as extrinsic factors like oxygen concentration and temperature, define the microbial species that predominate and those that are most adaptable to specific conditions (Atuonwu and Tassou, 2018).

5.3 Synergistic effect of ultrasound and microwave

Ultrasound treatment by itself is insufficient for sterilization, and to ensure food safety, it can be used in conjunction with other technologies. One of the major barriers to the implementation of ultrasonic technology in the food sector is that it requires a substantial amount of energy input. As a result, ultrasound research should be bolstered to manage the conditions or combine different technologies to avoid the negative consequences of ultrasonography (Chen et al., 2020).

Ultrasound is being used in a diverse selection of food processes because of the benefits it provides, such as improved food preservation, thermal treatment aid, and a good impact on food quality indices. It also entails processing time reduction costs by obviating the need for some downstream purification procedures. As a result, ultrasound is now employed in a wide range of food processing applications, such as freezing, drying, sterilizing, and extraction (Ekezie et al., 2017). Ultrasound and microwave technology is a cost-effective and environmentally friendly technology that enhances manufacturing efficiency and product quality while also saving money. Ultrasound's physical and chemical impacts, on the other hand, may have a deleterious impact on food. Because the "cavitation" impact produced hydroxyl radicals, which degraded ascorbic acid and reduced total phenol levels. Ultrasound-induced lipid degradation, particularly in lipid foods, not only will the secondary reaction products produce noxious odors, but they will reduce the food's nutritional safety and quality (Li et al., 2020). As a result, using a hybrid microwave-assisted ultrasonic food processing approach with a synergistic impact based on theoretical and experimental data should allow for the elimination of each methodology's flaws, and the combination technology has been extensively investigated in the food industry.

5.3.1 Synergistic effect of ultrasound and microwave on meat processing

Ultrasound has been successfully used in a variety of food-related fields to replace conventional techniques (meat tenderization, degassing, and cutting) and to assist traditional techniques (freezing, brining, thawing, and filtration). However, more research is needed in terms of optimizing process conditions and enhancing the potential for efficient use of existing processes. When paired with ultrasound, few other technologies have demonstrated the ability to efficiently manage food bacteria. A synopsis of ultrasound ideas and applications in the food business, as well as meat technology's long-term applicability and major influence, are explored as a result of the prospective benefits of ultrasound in food safety, processing, and preservation (Boateng et al., 2019).

Ultrasound applications have specifically effect on meat processing. The effects on technological properties and quality like water retention, texture, curing, color, cooking yield, marinating, thawing, freezing, and microbial inhibition are all highlighted. Following a review of the literature, it was determined that ultrasound is a valuable tool in the meat processing industry because it helps to accelerate maturation, tenderization, and mass transfer, improves equipment surface sterilization, makes mold cleaning easier, reduces cooking energy, and improves the functional properties of emulsified products (Alarcon-Rojo et al., 2015).

In the estimation of the carcass and body, chemical transcription among meat products real-time ultrasound has been widely used. When reviewed by Awad et al. (2012), intramuscular fat percentage, (sheep) carcass traits, and reflection prototypes of tissues such as internal organs, muscles, and fat in live animals were helpful in genetic improvement through low-intensity ultrasound adaptation. Because inconsistent tenderness is a major yield problem in the meat industry, and consumers appear dissatisfied with meat quality, a synergistic effect of heat and ultrasound revealed a tenderizing benefit on meat (Pagan et al., 1999).

Microwave energy is increasingly being used in a wide range of processing technologies in the food industry. Microwaves' properties of dielectric heating allowed them to be used in a variety of applications such as cooking, blanching, pasteurization, thawing, baking, and drying (Schiffmann, 2001).

Cooking, sterilization, thawing, drying, extraction of bioactive components, surimi gelation, and other seafood processing applications could all benefit from microwave energy. Microwave heating's effects on the nutritional value of seafood and its products are discussed, as well as the Future research directions for improving microwave efficiency. Microwave processing has a number of advantages over traditional processing methods. Microwave processing saves time and energy while producing products made from high-quality fish with a high nutritional value. The nutritional composition of fish is unaffected by blanching or microwave cooking. Microwave drying not only preserves the sensory qualities but also creates a porous structure that aids in the absorption of water. Microwaves combined with advanced or traditional technologies by overcoming the microwave's shortcomings have improved the fish products' quality (Viji et al., 2022). Yaqoob et al. (2020) suggested that ultrasound-extract from papaya and guava leaves improve the stability and antioxidants contents of shrimp patties. Another study suggested that gamma irradiation and *Asparagus racemosus* powder reduce microbial growth and increased the storage life (Noor et al., 2022).

5.3.2 Synergistic effect of ultrasound and microwave on meat preservation

Ultrasound, particularly high-intensity ultrasound (frequency range 18–100 kHz), is commonly used in food processing, including fermentation, cutting, crystallization, food preservation, and drying. The purpose of using high-intensity ultrasounds is to modify the medium through which the waves flow and boost the efficiency of food processing (Huang et al., 2018). Preservation, drying, filtering, extraction, and boiling are some of the processes that can be employed for this. Many food ingredients and products, on the other hand, have been shown to be thermally sensitive to physical, chemical, and microbiological alterations. Chemical loss, low production efficiency, and time and energy-intensive operations are all possible outcomes of traditional food processing methods (Chemat et al., 2011).

As a result of these flaws, new "green and innovative" pasteurization, processing, and extraction techniques have been adopted, such as supercritical fluid extraction, ultrasound-assisted processing, controlled pressure drop process, microwave processing, high pressure, extrusion, pulsed electromagnetic field, and subcritical water extraction, which typically require less time, water, and energy. Food technology under extreme or nonclassical conditions is currently a fast-developing problem in applied research and industry. Using two or more food preservation techniques at the same time to increase food safety, especially in the case of meat quality, while using lower individual treatment intensities. The synergistic strategy calls for the intelligent application of numerous preservation aspects or technologies to produce multitargeted, moderate, but consistent preservation results (Aaliya et al., 2021).

In a study published in 2021 by Aykın Dinçer on meat deterioration, it was discovered that the drying process is often used in conjunction with other techniques. Chemical and microbial deterioration can be avoided, shelf life increased, and packaging, transportation, and storage expenses decreased by drying the goods. He dried beef, chicken, and fish meats at 55 °C, 65 °C, and 75 °C using three different drying techniques: vacuum drying, ultrasonic vacuum drying, and oven drying. The ultrasonic vacuum drying technology dried meats in less time and with less energy than traditional methods.

Under normal storage circumstances, lipids, proteins, and pigments oxidize, which is a regular occurrence in meat products. During lipid oxidation, the generation of hydroperoxides encourages additional degradation processes, resulting in undesired volatile chemicals such as acids, ketones, aldehydes, and alcohols. Furthermore, during the oxidation process, physicochemical changes in amino acids and proteins affect their solubility, digestibility, bioavailability, and proteolytic activity. Microwaves and ultrasound work together to boost antioxidant efficacy while keeping adverse effects to a minimum, allowing them to be employed as a food preservation system (Pateiro et al., 2018).

Ultrasound's application in microbial inactivation and cleansing is a large topic to investigate. It has been discovered that ultrasound alone is insufficient to stop microbial development. Gram-positive bacteria, for example, have a thicker cell structure and are therefore more resistant than gram-negative bacteria. Viruses are also resistant to ultrasound due to their high depolymerization and tolerance to enzymes. As a result, using additional ultrasonic means such as heat or pressure, UV, or microwave is critical. Different dairy products and fruit juices have been found to increase their functional characteristics and quality using ultrasound (Khaire et al., 2021).

5.3.3 Synergistic effect of ultrasound and microwave on meat extraction

Meat extraction is a concentrated meat stock made primarily from beef or chicken. It is used to flavor meat, as well as to make broth for soups and other liquid-based dishes. Skeletal muscle has a different subcellular architecture than mononucleated cells, making it more complex than other tissues (Finlay-Smits et al., 2019). As a result, the interaction of several factors influences the extraction of the meat proteome such as protein localization, protein solubility, extraction method, and postmortem changes that occur as a result of the transformation of muscle into the meat. Meat scientists divide the proteome into two sections to deal with this complexity: myofibrillar and sarcoplasmic, which require separate extraction procedures because of their varying extractability and water solubilities (González-Blanco et al., 2021).

Meat accounts for a significant portion of the world's food consumption, resulting in the manufacturing of a large number of by-products in slaughterhouses each year. The antimicrobial, antioxidant, and opioid activities of the bioactive compounds isolated from these products, as well as the technological properties associated with them, could allow them to be used in food as functional products, as well as in the medical and pharmaceutical fields (Yaqoob et al., 2022). These biomolecules contain collagens, enzymes, protein hydrolysates, and polyunsaturated fatty acids. Alternative uses for meat by-products are presented as a future option for resolving issues arising from slaughterhouse waste.

Because no interspecies conversion can take place so the efficiency of quantitative extraction is required, the extraction procedure is often a critical step in speciation studies. Enzymatic digestion or sonication in an ultrasonic bath aids the extraction process, which can take several hours. The use of an ultrasonic probe is a faster alternative that has been used to speed up Arsenic extraction from soils and biological samples (Sánchez-Rodas et al., 2006). Microwave-assisted extraction (MAE) and UAE are two new techniques that have achieved significant adoption due to their adaptability in industries, ease of operation, low cost, high extraction efficiency, and little environmental effect. The UAE and MAE extraction methods have different principles, and while the input energy for both is the same; the energy supplied to the sample may be different. So, if the device's power is transformed into heat, which dissipates in the medium, it is critical to look at the processes employing the actual power/energy provided to the sample (Chutia and Mahanta, 2021).

Even though many types of pretreatments have been looked into in the past few years, there is always a need to increase efficiency, cut prices, and develop more environmentally friendly methods. Sound waves with a frequency range of 20–1000 kHz can be used to induce matrices to be disrupted by varied levels of cavitation. To obtain a suitable extraction of biologically active compounds, green extraction principles should be applied (Moreira et al., 2019). Ultrasound technology improves extraction yield by enabling solvent penetration, enhancing yields with or without solvents, when used as a pretreatment. It makes it easier to get rid of heat-sensitive compounds while reducing their exposure to heat and oxidation. In this situation, sonication pretreatment helps get more biological components out of meat and other goods (Chan et al., 2011).

Chemat et al. (2017) reported that combining ultrasound with various modalities, either simultaneously or sequentially, yielded considerable benefits above ultrasound treatment alone. The use of ultrasound and microwave radiation together can provide several benefits, including increased extraction rates and shorter processing times. MAE is a widely used method for extracting valuable chemicals from a variety of matrices. Two transport phenomena working together synergistically (mass and heat gradients) traveling in the same direction are responsible for MAE's high extraction yield. It is vital to show US cost-effectiveness, sustainability, safety, and environmental friendliness to promote increased industry using ultrasonic technology to improve process intensification on a large scale in industrial extraction.

5.4 Advantages and disadvantages of the synergistic effect of ultrasound sonication and microwave technologies

Thermal treatments make sure that food is safe and extend how long it can be stored. But they can also cause important nutrients to break down in the food (Munir et al., 2020). Ultrasound sonication is considered a nonthermal technique that has been under study over the last few years, and it appears to be quite efficient for reducing microbial populations in food products, however, enzymes cannot be inactivated (Jabbar et al., 2014; Nadeem et al., 2018; Villamiel and de Jong, 2000).

The combination of nonthermal techniques with microwaves can be an innovative way to ensure quality, safety, and stable storage ability of processed food (Rupasinghe and Yu, 2012). Microwaves are defined as a part of the electromagnetic radiation spectrum which lies in the frequency range of 300 MHz–300 GHz, 1 mm–1 m wavelengths. Combining ultrasound and microwave technology is an effective hybridization method that provides quick and efficient extractions. This combined technology is mainly used for the natural extraction process, but the consideration for the hybrid procedure has still not been appropriately utilized (Singla and Sit, 2021). The main advantages of combining ultrasound and microwave methods are the rapid sample preparation/extraction, and the cost-effective method and utilizing moderate power.

Found that when ultrasound and microwave extraction were used together, they got a higher yield of phenolic and antioxidant from brown macroalgae than when they were used separately. Ultrasound and microwave were used together to get the lipids out of vegetables and yeast, which led to a higher oil yield (Patel et al., 2019; Cravotto et al., 2008). This method also has other effects, such as mixing well, keeping food's original qualities, transferring energy and mass more quickly, reducing thermal and concentration gradients, and increasing a product's shelf life. It is an effective alternative to traditional methods (Singla and Sit, 2021). Huang et al. (2018) looked at how microwave and ultrasound work together to improve the quality of pumpkin slices during ultrasound-assisted microwave vacuum frying. They found that the application reduced the amount of oil in the pumpkin slices and made their texture better. The synergistic effects of ultrasound and microwave treatment on the physical, chemical, textural, and microstructural properties of vacuum-fried yam were studied, and it was found that the texture and color of the final products were improved by the right combination of ultrasound and microwave. It was found that when the power level is higher, the results are better. When the ultrasound and microwave powers were raised to 600W and 3000W, respectively, the frying time went from 18 minutes to 12 minutes and the amount of oil used went down by 40.29%. The biggest problem with this method is that scanning electron microscopy (SEM) analysis showed that when higher powers were used together, the structure became more broken up and porous (Chitrakar et al., 2019).

According to recent research, these hybrid procedures reduced energy usage while improving product quality and efficiency in the drying and frying cooking processes. When it comes to extraction, the combination approach yields a better yield in less time. Hybrid approaches have been found to be a good pretreatment because the structural changes in the molecules make it easier for enzymes to work later (Muñoz-Almagro et al., 2021).

The main disadvantage of the synergistic effect of ultrasound and microwave extraction is the degradation of some compounds. Wu et al. (2008) studied the degradation of phenol under combined irradiation of microwave and ultrasound which produced sono-generated radicals from the rapid thermal effect of the microwaves.

5.5 Synergistic effect of ultrasound and microwave in fish, meat, and seafood industry

Microwave dielectric heating and power ultrasonography are two of the most basic, affordable, and successful technologies in applied chemistry. This hybrid technology is one of the green strategies that greatly improves heat and mass transport, resulting in a more rapid and selective chemical reaction. This technology has a brief history but is already being used in a growing variety of applications. For the fish, meat, and seafood industries, the synergistic impact of ultrasound and microwave extraction has a very limited range of applications (Table 5.1).

Li et al. (2020) the structural and functional features of golden threadfin bream myofibrillar protein and hydrolysates were investigated after they were subjected to a combination ultrasound microwave treatment. They have also investigated the ultrasound and microwave techniques separately. They showed that the ultrasound and the ultrasound microwave combined treatment resulted in producing a partially denatured the structure and increased the susceptibility of myofibrillar proteins to the action of exogenous enzymes. By yielding a high peptide concentration and myofibrillar protein hydrolysates, this combination procedure sped up the hydrolytic stage. Furthermore, they found that myofibrillar protein peptides modified using a combination of ultrasound and microwave-assisted methods had better bioactive properties due to the mild microwave field, which promoted oriented structural modification while also inhibiting excessive denaturation

Table 5.1 Combined microwave and ultrasound techniques used in the extraction of various food products.

Food type	Purpose of use	Remarks	Reference
Golden threadfin beam	Myofibrillar protein enzymatic hydrolysis	The 300 W ultrasound couple with 100 W microwave and accelerates the hydrolytic process and yielding a high peptide concentration of myofibrillar protein	Li et al. (2020)
Brown microalgae	Extraction of bioactive compounds and antioxidants	Maximum yield of compounds achieved by the technology of combined ultrasound and microwave	Garcia-Vaquero et al. (2020)
Dried goat meat	Treatment for fatty acids	Microwave and ultrasound treatment improved the nutritional value in the terms of fatty acids	
Freshwater mussel meat	Enzymatic hydrolysis	Synergistic treatment enhanced the amino acid profile of the hydrolysate	Zhou et al. (2018)
Tilapia byproducts	Extraction of chondroitin sulfate	The optimum conditions of 50 W of ultrasound and 98 W of microwave power at 120 s were used for chondroitin sulfate extraction	Cheng et al. (2013)

and reaggregation of the sonicated myofibrillar protein. As a result, they came to the conclusion that the synergistic effects of combining ultrasonic and microwave technologies could be a feasible modification technique for producing more attractive peptides.

The biological activities of macroalgae components, such as phenolic compounds and polysaccharides, are correlated with the molecules' chemical structure which is associated with species, season, site of harvesting, macroalgae biomass, extraction procedure, and the purification method (Patel et al., 2019). The application of novel technologies singly, sequentially, or simultaneously to boost the production of bioactive chemicals and improve the efficiency of extraction methods by minimizing chemical usage, energy consumption, and waste creation is a recent trend in algae biotechnology. It was also looked at the effect of combining ultrasonic and microwave technologies on the extraction of bioactive components and antioxidants from brown macroalgae. The results revealed that using ultrasound and microwaves together enhanced the yield of chemicals extracted when compared to using ultrasound and microwave technology alone. This was the first study to look at how microwave and ultrasound-assisted technology can work together to extract high-value chemicals from macroalgae. As a result, it was proposed that more research be done to confirm the extracts' antioxidant capabilities (Garcia-Vaquero et al., 2020).

GAO et al. (2017) studied the effect of microwave/ultrasound-assisted treatments on the fatty acids of dried goat meat. Microwave treatments alone resulted in a decrease in total saturated fatty acids (SAF) and an increase in total polyunsaturated fatty acids (PUFA) when combined with ultrasound. Furthermore, the combination treatment assisted in lowering the n-6/n-3 ratio, allowing them to achieve the nutritionally acceptable levels. Ultrasound treatment alone did not significantly influence the relative content of the total SAF, monounsaturated fatty acids (MUFA) and PUFA. Overall, they found the combination of both microwave and ultrasound treatment improved the nutritional value of dried mutton in terms of fatty acids.

Cheng et al. (2013) investigated the extraction of chondroitin sulfate from Tilapia by-products with the ultrasonic-microwave combination. Their results showed the best conditions for extracting chondroitin sulfate from tilapia by-products and separating it were with using 50 W of ultrasonic power and 98 W of microwave power at 120 seconds. Under the specified conditions, the optimum solid-to-liquid medium ratio was 1:50 (g/mL), and the chondroitin sulfate extraction rate was 2.513%. The combined ultrasonic and microwave extraction technology used in this work appears to be a successful way to extract and isolate chondroitin sulfate (Cheng et al., 2013). Zhou et al. (2018) applied the ultrasound microwave combined approach to optimize the enzyme hydrolysate of freshwater mussels. The following were the ideal conditions that were considered: mussel meat liquid-to-material ratio: 10.36:1, power: 360.52 W, processing time: 6.91 min, enzyme digestion time: 4 h A degree of hydrolysis (DH) of 48.40% was discovered under these conditions. Hydrolysates with and without ultrasonic microwave pretreatment had essential amino acids of 36.62 mg/100 g and 33.40 mg/100 g, respectively. In addition, an examination of amino acid content revealed that glutamic acid and aspartic acid were the most important components for improving flavor. Hence, extracted mussel hydrolysate can be utilized as a flavor enhancer with good nutritional value.

5.6 Conclusion

It is concluded that green technologies can be suitable in processing, preservation, and extraction of different types of meat including chicken, beef, ostrich, fish, and other sea food. Nowadays, synergism of different technologies has been applied for meat and meat products processing. However, in this chapter we study the synergism effect of ultrasound and microwave. Some previous studies suggested that synergism ultrasound and microwave suppress the growth of microorganism that make meat and meat products safe for consumer because this combination improves the quality and enhance the shelf stability.

References

Aaliya, B., Sunooj, K.V., Navaf, M., Akhila, P.P., Sudheesh, C., Mir, S.A., … George, J., 2021. Recent trends in bacterial decontamination of food products by hurdle technology: a synergistic approach using thermal and non-thermal processing techniques. Food Res. Int. 147, 110514.

Alarcon-Rojo, A.D., Janacua, H., Rodriguez, J.C., Paniwnyk, L., Mason, T.J., 2015. Power ultrasound in meat processing. Meat Sci. 107, 86–93.

Astráin-Redín, L., Raso, J., Condón, S., Cebrián, G., Álvarez, I., 2019. Application of high-power ultrasound in the food industry. Sonochemical Reactions. IntechOpen.

Atuonwu, J.C., Tassou, S.A., 2018. Quality assurance in microwave food processing and the enabling potentials of solid-state power generators: a review. J. Food Eng. 234, 1–15.

Awad, T.S., Moharram, H.A., Shaltout, O.E., Asker, D., Youssef, M.M., 2012. Applications of ultrasound in analysis, processing and quality control of food: a review. Food Res. Int. 48, 410–427.

Barbosa-Cánovas, G.V., Medina-Meza, I., Candoğan, K., Bermúdez-Aguirre, D., 2014. Advanced retorting, microwave assisted thermal sterilization (MATS), and pressure assisted thermal sterilization (PATS) to process meat products. Meat Sci. 98 (3), 420–434.

Bhatt, K., Vaidya, D., Kaushal, M., Gupta, A., Soni, P., Arya, P., Sharma, C., 2020. Microwaves and radiowaves: in food processing and preservation. Int. J. Curr. Microbiol 9, 118–131.

Bindes, M.M.M., Reis, M.H.M., Cardoso, V.L., Boffito, D.C., 2019. Ultrasound-assisted extraction of bioactive compounds from green tea leaves and clarification with natural coagulants (chitosan and Moringa oleífera seeds). Ultrason. Sonochem. 51, 111–119.

Boateng, E.F., Nasiru, M.M., 2019. Applications of ultrasound in meat processing technology: a review. Food Sci. Technol. 7 (2), 11–15.

Chan, C.H., Yusoff, R., Ngoh, G.C., Kung, F.W.L., 2011. Microwave-assisted extractions of active ingredients from plants. J. Chromatogr. A 1218 (37), 6213–6225.

Chemat, F., Khan, M.K., 2011. Applications of ultrasound in food technology: processing, preservation and extraction. Ultrason. Sonochem. 18 (4), 813–835.

Chemat, F., Rombaut, N., Sicaire, A.G., Meullemiestre, A., Fabiano-Tixier, A.S., Abert-Vian, M., 2017. Ultrasound assisted extraction of food and natural products. Mechanisms, techniques, combinations, protocols and applications. a review. Ultrason. Sonochem. 34, 540–560.

Chen, F., Zhang, M., Yang, C.H., 2020. Application of ultrasound technology in processing of ready-to-eat fresh food: a review. Ultrason. Sonochem. 63, 104953.

Cheng, C.Y., Duan, W.W., Duan, Z.H., Hai, Y., Lei, X.G., Chang, H., 2013. Extraction of chondroitin sulfate from tilapia byproducts with ultrasonic-microwave synergistic. In: Advanced Materials Research, 726. Crossref: Trans Tech Publications Ltd, pp. 4381–4385 doi:10.4028/www.scientific.net/amr.726-731.4381.

Chitrakar, B., Zhang, M., Fan, D., 2019. The synergistic effect of ultrasound and microwave on the physical, chemical, textural, and microstructural properties of vacuum fried Chinese yam (Dioscorea polystachya). J. Food Process. Preserv. 43 (9), e14073.

Chutia, H., Mahanta, C.L., 2021. Green ultrasound and microwave extraction of carotenoids from passion fruit peel using vegetable oils as a solvent: optimization, comparison, kinetics, and thermodynamic studies. Innov. Food Sci. Emerg. Technol. 67, 102547.

Cravotto, G., Boffa, L., Mantegna, S., Perego, P., Avogadro, M., Cintas, P., 2008. Improved extraction of vegetable oils under high-intensity ultrasound and/or microwaves. Ultrason. Sonochem. 15 (5), 898–902.

Cravotto, G., Cintas, P., 2006. Power ultrasound in organic synthesis: moving cavitational chemistry from academia to innovative and large-scale applications. Chem. Soc. Rev. 35 (2), 180–196.

Ekezie, F.G.C., Sun, D.W., Han, Z., Cheng, J.H., 2017. Microwave-assisted food processing technologies for enhancing product quality and process efficiency: a review of recent developments. Trends Food Sci. Technol. 67, 58–69.

Finlay-Smits, P. M., & Loeffen, M. (2019). Design and deliver novel meat extract concepts.

Fu, X., Belwal, T., Cravotto, G., Luo, Z., 2020. Sono-physical and sono-chemical effects of ultrasound: primary applications in extraction and freezing operations and influence on food components. Ultrason. Sonochem. 60, 104726.

GAO, T., et al., 2017. Effects of microwave/ultrasonic assisted treatments on fatty acids of dried goat meat. Meat Res. 2017, 01.

Garcia-Vaquero, M., Ummat, V., Tiwari, B., Rajauria, G., 2020. Exploring ultrasound, microwave and ultrasound–microwave assisted extraction technologies to increase the extraction of bioactive compounds and antioxidants from brown macroalgae. Mar. Drugs 18 (3), 172.

Gautam, S. (2019). An introduction to seafood and recent advances in the processing of seafood products.

González-Blanco, L., Diñeiro, Y., Díaz-Luis, A., Coto-Montes, A., Oliván, M., Sierra, V., 2021. Impact of extraction method on the detection of quality biomarkers in normal vs. DFD meat. Foods 10 (5), 1097.

Guzik, P., Kulawik, P., Zając, M., Migdał, W., 2021. Microwave applications in the food industry: an overview of recent developments. Crit. Rev. Food Sci. Nutr. 1–20 doi:10.1080/10408398.2021.1922871.

Huang, M.S., Zhang, M., Bhandari, B., 2018. Synergistic effects of ultrasound and microwave on the pumpkin slices qualities during ultrasound-assisted microwave vacuum frying. J. Food Process Eng. 41 (6), e12835.

Jabbar, S., Abid, M., Hu, B., Wu, T., Hashim, M.M., Lei, S., …, Zeng, X., 2014. Quality of carrot juice as influenced by blanching and sonication treatments. LWT 55 (1), 16–21.

Kane, D., Grassi, W., Sturrock, R., Balint, P.V., 2004. A brief history of musculoskeletal ultrasound: from bats and ships to babies and hips. Rheumatology (Oxford) 43 (7), 931–933.

Karwowska, M., Stadnik, J., Stasiak, D.M., Wójciak, K., Lorenzo, J.M., 2021. Strategies to improve the nutritional value of meat products: incorporation of bioactive compounds, reduction or elimination of harmful components and alternative technologies. Int. J. Food Sci. Technol. 56 (12), 6142–6156.

Khaire, R.A., Thorat, B.N., Gogate, P.R., 2021. Applications of ultrasound for food preservation and disinfection: a critical review. J. Food Process. Preserv. e16091. https://doi.org/10.1111/jfpp.16091.

Khan, S.A., Dar, A.H., Bhat, S.A., Fayaz, J., Makroo, H.A., Dwivedi, M., 2020. High intensity ultrasound processing in liquid foods. Food Rev. Int. 6, 1–26.

Leonelli, C., Mason, T.J., 2010. Microwave and ultrasonic processing: now a realistic option for industry. Chem. Eng. Process. 49 (9), 885–900.

Li, S., Tang, S., Yan, L., Li, R., 2019. Effects of microwave heating on physicochemical properties, microstructure and volatile profiles of yak meat. J. Appl. Anim. Res. 47 (1), 262–272.

Li, Z., Wang, J., Zheng, B., Guo, Z., 2020. Impact of combined ultrasound-microwave treatment on structural and functional properties of golden threadfin bream (Nemipterus virgatus) myofibrillar proteins and hydrolysates. Ultrason. Sonochem. 65, 105063.

Mason, T.J., Lorimer, J.P., 2002. Applied Sonochemistry: The Uses of Power Ultrasound in Chemistry and Processing, 10. Wiley-Vch, Weinheim.

Moreira, S.A., Alexandre, E.M., Pintado, M., Saraiva, J.A., 2019. Effect of emergent non-thermal extraction technologies on bioactive individual compounds profile from different plant materials. Food Res. Int. 115, 177–190.

Munir, M., Nadeem, M., Qureshi, T.M., Gamlath, C.J., Martin, G.J., Hemar, Y., Ashokkumar, M., 2020. Effect of sonication, microwaves and high-pressure processing on ACE-inhibitory activity and antioxidant potential of Cheddar cheese during ripening. Ultrason. Sonochem. 67, 105140.

Muñoz-Almagro, N., Morales-Soriano, E., Villamiel, M., Condezo-Hoyos, L., 2021. Hybrid high-intensity ultrasound and microwave treatment: a review on its effect on quality and bioactivity of foods. Ultrason. Sonochem. 80, 105835.

Nadeem, M., Ubaid, N., Qureshi, T.M., Munir, M., Mehmood, A., 2018. Effect of ultrasound and chemical treatment on total phenol, flavonoids and antioxidant properties on carrot-grape juice blend during storage. Ultrason. Sonochem. 45, 1–6.

Noor, M.M., Arshad, M.S., Ahmad, R.S., Imran, A., Khalid, W., Suleria, H.A.R., 2022. Stability and quality improvement of shrimp patties by Asparagus racemosus and gamma irradiation. Int. J. Food Prop. 25 (1), 1328–1342.

Ozel, M.Z., Gogus, F., Lewis, A.C., 2003. Subcritical water extraction of essential oils from Thymbra spicata. Food Chem. 82 (3), 381–386.

Pagan, R., Mañas, P., Alvarez, I., Condon, S., 1999. Resistance of Listeria monocytogenes to ultrasonic waves under pressure at sublethal (manosonication) and lethal (manothermosonication) temperatures. Food Microbiol. 16, 139–148.

Pateiro, M., Barba, F.J., Domínguez, R., Sant'Ana, A.S., Khaneghah, A.M., Gavahian, M., …, Lorenzo, J.M., 2018. Essential oils as natural additives to prevent oxidation reactions in meat and meat products: a review. Food Res. Int. 113, 156–166.

Patel, A., Arora, N., Pruthi, V., Pruthi, P.A., 2019. A novel rapid ultrasonication-microwave treatment for total lipid extraction from wet oleaginous yeast biomass for sustainable biodiesel production. Ultrason. Sonochem. 51, 504–516.

Pimentel-Moral, S., Borrás-Linares, I., Lozano-Sánchez, J., Arráez-Román, D., Martínez-Férez, A., Segura-Carretero, A., 2019. Supercritical CO2 extraction of bioactive compounds from hibiscus sabdariffa. J. Supercrit. Fluids 147, 213–221.

Rezzoug, S.A., Boutekedjiret, C., Allaf, K., 2005. Optimization of operating conditions of rosemary essential oil extraction by a fast controlled pressure drop process using response surface methodology. J. Food Eng. 71 (1), 9–17.

Rodsamran, P., Sothornvit, R., 2019. Extraction of phenolic compounds from lime peel waste using ultrasonic-assisted and microwave-assisted extractions. Food Biosci. 28, 66–73.

Rosario, D.K., Rodrigues, B.L., Bernardes, P.C., Conte-Junior, C.A., 2021. Principles and applications of non-thermal technologies and alternative chemical compounds in meat and fish. Crit. Rev. Food Sci. Nutr. 61 (7), 1163–1183.

Rupasinghe, H.P.V., Yu, L.J., 2012, Emerging Preservation Methods for Fruit Juices and Beverages. In: Y. El-Samragy (Ed.), Food Additive, IntechOpen, London. doi:10.5772/32148.

Sánchez-Rodas, D., Luis Gómez-Ariza, J., Oliveira, V., 2006. Development of a rapid extraction procedure for speciation of arsenic in chicken meat. Anal. Bioanal.Chem. 385 (7), 1172–1177.

Schiffmann, R.F., 2001. Microwave processes for the food industry. In: Handbook of microwave technology for food application (pp. 299–335). CRC Press.

Singla, M., Sit, N., 2021. Application of ultrasound in combination with other technologies in food processing: a review. Ultrason. Sonochem. 73, 105506.

Soni, A., Smith, J., Thompson, A., Brightwell, G., 2020. Microwave-induced thermal sterilization-A review on history, technical progress, advantages and challenges as compared to the conventional methods. Trends Food Sci. Technol. 97, 433–442.

Van Gerven, T., Stankiewicz, A., 2009. Structure, energy, synergy, time: the fundamentals of process intensification. Ind. Eng. Chem. Res. 48 (5), 2465–2474.

Viji, P., Debbarma, J., Ravishankar, C.N., 2022. Research developments in the applications of microwave energy in fish processing: a review. Trends Food Sci. Technol.

Villamiel, M., de Jong, P., 2000. Influence of high-intensity ultrasound and heat treatment in continuous flow on fat, proteins, and native enzymes of milk. J. Agric. Food Chem. 48 (2), 472–478.

Wang, X., Muhoza, B., Wang, X., Feng, T., Xia, S., Zhang, X., 2019. Comparison between microwave and traditional water bath cooking on saltiness perception, water distribution and microstructure of grass crap meat. Food Res. Int. 125, 108521.

Wu, Z.L., Ondruschka, B., Cravotto, G., 2008. Degradation of phenol under combined irradiation of microwaves and ultrasound. Environ. Sci. Technol. 42 (21), 8083–8087.

Yaqoob, Z., Arshad, M.S., Imran, M., Munir, H., Qaisrani, T.B., Khalid, W., …, Suleria, H.A.R., 2022. Mechanistic role of astaxanthin derived from shrimp against certain metabolic disorders. Food Sci. Nutr. 10 (1), 12–20.

Yaqoob, Z., Arshad, M.S., Khan, M.K., Imran, M., Ahmad, M.H., Ahmad, S., …, Suleria, H.A.R., 2020. Ultrasound-assisted extraction of guava and papaya leaves for the development of functional shrimp patties. Food Sci. Nutr. 8 (7), 3923–3935.

Yu, H., Zhong, Q., Liu, Y., Guo, Y., Xie, Y., Zhou, W., Yao, W., 2020. Recent advances of ultrasound-assisted Maillard reaction. Ultrason. Sonochem. 64, 104844.

Zhou, Y., Yi, X., Wang, J., Yang, Q., Wang, S., 2018. Optimization of the ultrasonic-microwave assisted enzymatic hydrolysis of freshwater mussel meat. Int. J. Agric. Biol. Eng. 11 (5), 236–242.

CHAPTER SIX

Synergistic effects of sonication and microwave in juice processing

Boukaga Farmani, Somaye Mohammadkhani and Farzaneh Hadi Andabjadid
Department of Food Science and Technology, Ahar Faculty of Agriculture and Natural Resources, University of Tabriz, Tabriz, Iran

6.1 Introduction

Today, demands for nonchemical preservatives and nonsynthetic additives fresh juices have been increased among consumers. Although demands for the consumption of directly produced fresh juices from fresh products have been raised, but due to high microbial load the juices have very short shelf-stability in compared to processed juices, therefore juice processing is unavoidable (Timmermans et al., 2011).

Juices consist of bioactive compounds, nutrients, vitamins, and mineral matters with great health value that provide very suitable conditions for the growth and reproduction of different spoilage microorganisms and dangerous foodborne pathogens. Therefore, important contaminating microorganisms namely *Escherichia coli, Salmonella, Listeria monocytogenes,* and *Cryptosporidium* should be controlled in juices through different disinfection processes (Lima Tribst et al., 2009).

Traditional thermal processes (e.g., pasteurization and sterilization) are the most common antimicrobial procedures in juices processing. Currently, conventional thermal processes are still widely utilized for processing different types of fruit and vegetable juices (Dobson et al., 2017). Although conventional processes can effectively inactivate dissimilar types of microorganisms (molds, yeasts, and bacteria) and undesirable enzymes (browning and peroxidase enzymes) and increase shelf stability of juices, but it lead to quality loss in fruit juice (such as sensory deterioration, loss of nutrients, and substances with functional ability). Due to growing the consumers' awareness from health and nutrition value of foods, novel eco-friendly green technologies have been established for processing in food industry.

Ultrasound and Microwave for Food Processing: Synergism for Preservation and Extraction.
DOI: https://doi.org/10.1016/B978-0-323-95991-9.00015-1

Therefore, in recent years, green processing technologies have developed as a study attention in the field of liquid food treating. For instance, ultrasonic treatment (ultrasonication) is one of the novel nonheat technologies that can improve microbial safety and quality of liquid foods (Dolas et al., 2019). Sonication action mainly is dependent on the microorganisms and enzymes inactivation via mechanical (cavitation and shock waves) and/or chemical actions (free radicals formatted through sonolysis of water) (Khandpur and Gogate, 2015).

The request for nonadditives and/or nonpreservatives safe-to-drink juices with high quality has encouraged utilizing emerging technologies in processing different juices. Ultrasound has been combined with mild heat treatments such as thermosonication procedure for preventing food spoilage due to its synergistic and time saving effect (Knorr et al., 2004). Ultrasound combined with heat (as thermosonication) is a substitutive and reliable technique to conventional heating pasteurization (5 log microbial reduction at <100°C for few seconds or minutes). Thermo-sonication procedure is safer, more efficient and reliable than ultrasound process alone for destruction of food-borne pathogens and microbial inactivation in liquid foods (Abdullah and Chin, 2014).

6.2 Comparison of conventional method and ultrasound-microwave procedure

In present time, thermal pasteurization is widely applied for inhibition of spoilage and foodborne illnesses in various foods. This approach has negative impacts on some food attributes such as nutritional value and quality properties (Piyasena et al., 2003). The unfavorable effect of conventional heating process can decrease quality attributes such as nutrient value and sensorial and functional properties in processed foods (Barba et al., 2012; Knorr and Hendrickx, 2001). For many years, traditional pasteurization (5 log microbial reduction at <100°C for few seconds or minutes) has been used to attain high microbial stability and safety in different fruit juices (Dubrović et al., 2011).

Biocolorants (e.g., anthocyanins) and various bioactive substances are degraded during conventionally pasteurization. To minimize changes in bioactive substances, several researches investigate the ultrasonication treatment as an alternative procedure for conventionally thermal treatments in order to reduce different types of microorganisms. Also, combination of microwave and ultrasound is introduced as a promising innovative method for retention of phytochemicals (especially polyphenols and anthocyanins) as well as significant decrease of microbial load (molds, yeasts, and bacteria) during juice processing (Aadil et al., 2013; Pérez-Grijalva et al., 2018).

Microwaves are a form of electromagnetic spectrum with wavelengths of 12 to 34 cm that are commonly applied in frequencies range of 915 to 2450 MHz (Regier et al., 2016). Microwaves have no thermal energy, but its heating property is associated directly to food capability for absorbing microwave energy and transforms them into the heat. Microwave heating is generated through dipole rotation or ion conduction mechanisms. But in conventional heating method, heat is transferred from an external source to food medium (Chandrasekaran et al., 2013).

For the juices processing, the ultrasonication have some benefits in comparison to conventional procedure namely low processing time and energy consumption, high efficiency and quality of treated juices. Conventional heating such as sterilization and pasteurization are generally utilized to inactivate microorganisms and enzymes in juices. Conventional processing due to utilization of higher temperature (80°C) may lead to various undesirable changes in juices including organoleptic (nutrients, colorants and flavors) and physicochemical properties (Piasek et al., 2011; Mena et al., 2013). It also should be stated that the appearance of the treated orange juices by combinative procedure (57%) were better than traditional heating method (43%) (Samani et al., 2015).

The food-borne pathogens destruction using ultrasound need to long time, cost with low outcome. Therefore, thermosonication (combination of ultrasound and mild heat) at 37 to 75°C decrease temperature (16%) and time (55%) of process as well as negative effects on quality of juices. Thermosonication process is based on cavitation phenomenon and heat, which are included creation, expansion and collapse of microbubbles in juices. The changes of local temperature and pressure by cavitation cause cell membrane disintegration and generation of various free radicals (Koshani et al., 2015).

The results indicated that ultrasonic process alone has no considerable outcome on reduction rate of *E. coli*, only 1.3 log microbial decrease was achieved at 60 minutes. But, sonication can raise the efficiency of microwave treatment for deactivation of microorganisms. Therefore, combined treatment of combined ultrasound-microwave gives better results than combined microwave-ultrasonic. The combined action of ultrasound-microwave process causes a considerable decrease in microbial load with shorter times. This method is very valuable to process beverages of fruit juice, mainly beverages

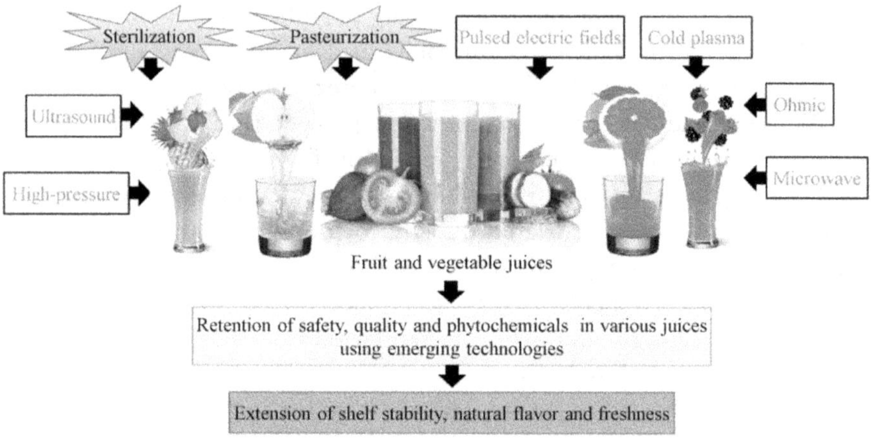

Figure 6.1 Emerging and safe and green technologies to process fruit and vegetable juices.

of orange juice which high-temperature and acid resistant microbial strains like *E. coli* are the serious concern (Kernou et al., 2021).

Conventional heating method mainly causes destruction of microorganisms through rupturing of cell membrane and nuclear and protein damages, while thermosonication is mainly related to cavitation effects on the cellular structure (Rivas et al., 2006; Anaya-Esparza et al., 2017). In combined method, rupturing cell membrane via cavitation damages of microbial structure lead to leakage of lipids, protein and nuclear and other essential compounds (Manzoor et al., 2021; Fig. 6.1).

Currently, ultrasound combined with mild heating has been reported as a green processing procedure (ultrasound plus mild heat). The technique exhibit synergistic actions in inactivating microorganisms and enzymes as well as lowering energy cost, preserving organoleptic, physicochemical and nutritional attributes of juices (Anaya-Esparza et al., 2017; del Socorro Cruz-Cansino et al., 2015). Also, the thermosonication process could decrease or even remove the degradation of heat-sensitive nutrients, bioactive substances and quality losses of juice that cause by high heating method (Abid et al., 2014a). Bottle gourd juice can treat using conventional and combined ultrasonic-microwave approaches. Combined technique has positive effects on phytochemicals and total soluble solids through the breakdown of present particles in juice (Das et al., 2020).

Microwave-ultrasound process is an emerging nonthermal technique in food industries especially for extending the shelf-stability of juices in juice industry. It is a combined technique from two different processing methods that both together have simultaneously good synergistic effects. Coupling microwave and ultrasound could increase microbial killing and enzyme inactivating effects and improve the sensory properties of the processed foods. As combined method is applied for juice processing, lower temperatures, and shorter times are required to obtain the same killing values compared to traditional processes with higher temperatures and longer times (Rani, 2020).

6.3 Synergistic effect of combined ultrasound-microwave in juice processing

Ultrasound usage alone is not operative enough as a preservation method to destroy microorganisms. Ultrasound usage with high power also can cause several negative effects on nutritional and sensory attributes of foods (Ferrario et al., 2015). Ultrasound processes used at a frequency of 42 kHz have a low influence on the destruction of *E. coli*, for instance, the maximum inactivation is 1.3 log microbial decrease of population afterward 60 minutes (Kernou et al., 2021). Low power ultrasound also has been stated to be poorly effectiveness in inactivation of bacteria. Researches display that quality factors of orange juice namely color, beta-carotene content, ascorbic acid and creation of hydroxyl radicals can be changed by ultrasonic treatment (Pingret et al., 2013).

Compared to other techniques, ultrasonic method has wide application with considerable advantages such as low energy consumption and high efficiency. Emerging technologies can be applied alone or as combined with other processing procedures. In addition, microwave can use as eco-friendly and green technology with wide practical operation in food sections (Wang et al., 2019). Microwave also has been extensively utilized to preserve types of food products (Ekezie et al., 2017).

The effect of ultrasonication via cavitation associated to the formation of free radicals by sonolysis of water molecules and collapse of microbubbles, resulting microorganism inactivation and disintegration of cell membranes, finally leakage of the intracellular matrix (Roobab et al., 2018). Several studies indicated that combined outcome of heat and ultrasound as a thermosonication method is very effective on cell livability and death of *Saccharomyces cerevisiae*. Ultrasound induces nonlethal intracellular injuries without damaging to cell membrane consequently microorganisms became more susceptible to heat injury (Wordon et al., 2012).

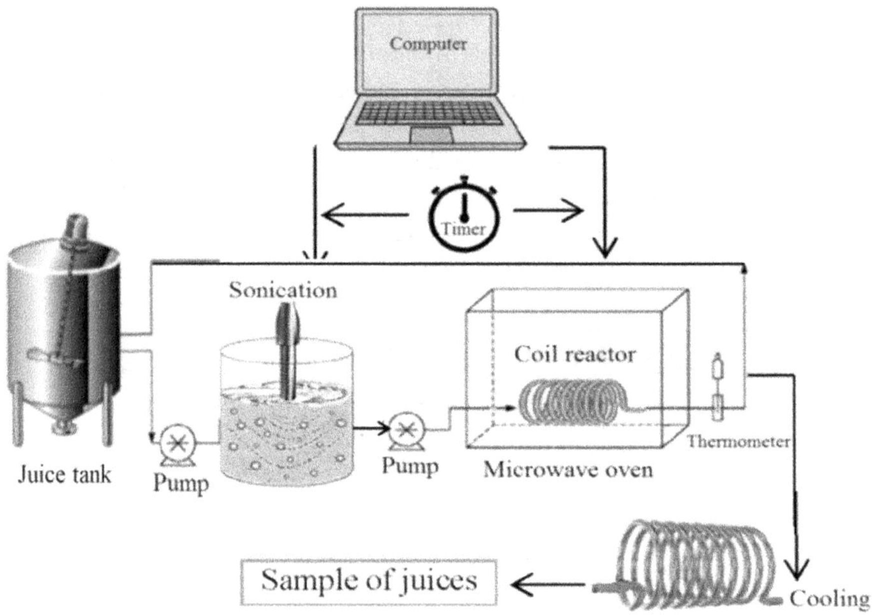

Figure 6.2 Cooperated ultrasonic and microwave for juice processing.

Fig. 6.2 illustrates the steps of juice processing with combined process mainly consist of feed tank, ultrasonication, microwave, cooling, and ultimately treated samples. The juices are pumping into ultrasound chamber followed by microwave (thermosonicated). The juices rapidly chilled then sensory attributes, physicochemical properties, and microbial safety of samples are evaluated.

Several studies have validated microwave method as a novel green preservation and antimicrobial procedure in different types of foods, especially in different liquid foods (Benlloch-Tinoco et al., 2014). From important advantages of microwave heating than conventional heating method can point out to rapid and uniform heating, lower equipment, and high energy efficiency (Chandrasekaran et al., 2013). Different researchers revealed that heating by microwave is more effective than conventionally thermal treatment for inactivating *E. coli* O157: H7 and *L. monocytogenes* in apple juice (Shamis et al., 2012). Microwave treatment has obtained popularity in food processing due to the attributes like high selectivity, good penetration, considerable reduction of heating time or shorter processing time, energy saving, high efficiency, ease, and safe operation (Kubo et al., 2019). Heating method in microwave is through electromagnetic effect, which is utilized as a substitutive to traditionally thermal processing approaches for better protection of quality and sensory attributes of foodstuffs. Microwave treatment also decrease significantly nutritional losses through inactivating different microorganisms (bacteria and yeasts) and enzymes in a shorter time compared to conventionally thermal processes (Siguemoto et al., 2018).

Ultrasonic and microwave with synergic effects are applied in extraction and process of different juices. For instance, microwave combined with ultrasound displayed a considerable replacement in processing of fruit juices and demanded by consumers and industry due to high yield. However, novel methods need to in-depth researches from different aspects such as browning reactions in jujube juices (Yu et al., 2020).

The important studies revealed reliable influence of combined ultrasonic-microwave nonthermal pasteurization on pathogenic bacterium such as *L. monocytogenes* in bottle gourd juice (vegetable juice). The model conditions was 750 W microwave at 70°C, ultrasonic with 80% amplitude for 15 minutes, resulting about 5 log *L. monocytogenes* decrease was detected. Furthermore, juice treated by combined technique displayed higher antioxidant activity, ascorbic acid and physicochemical attributes as well as storage stability in comparison with conventionally processed juice (Das et al., 2020).

Nowadays, one of the most common methods applied is thermosonication for processing different foodstuffs. In this system, a combination of heat and ultrasound is mainly utilized to treat various liquid foods. Numerous studies have showed different effects this method on the various microorganisms (yeasts and bacteria) and enzymes, which is completely dependent on the type of products. It seems that combined procedures are able to solve weak points of traditional methods (Kuldiloke, 2002). So, novel and alternative preservation processes have attracted attention of scientists and become a hot subject in food industrial field (Wang et al., 2019).

The combination of ultrasonication treatment with heating procedure has been determined to increase the microbial safety and quality values of dissimilar types of juices (vegetable and fruit juices) (Ahmed et al., 2019; Anaya-Esparza et al., 2017). The combined effect of ultrasound-microwave approach can induce synergistic actions in terms of microbial safety and quality preservation in various foods consequently as an innovative and alternative preserving technology. This combined practice in juice processing has great positive effects, at mild temperatures is capable to destroy various microbial contaminations with maintaining initial attributes of juices (Bhat et al., 2011; Rostami et al., 2018). The pretreatment effect of ultrasound can rise up microwave efficiency in destruction of *E. coli* in beverage of orange juice. A microwave process at 600 W for 30 seconds combined with ultrasound during 20 minutes (4.1 log reduction) and 30 minutes (8.0 log reduction) was more effective than when ultrasonication is performed in post-treatment in 20 minutes (2.7 log reduction) and 30 minutes (6.3 log reduction). The good efficiency of ultra-sonication as a pretreatment also was detected for decontamination at high temperatures (processed at 70, 80, 85, 90, and 95°C) on spores of *Alicyclobacillus* (*Bacillus*) *acidoterrestris* presence in apple juice (Tremarin et al., 2019). Therefore, microwave process coupled with ultrasound treatment can be a suitable replacement in processing ready-to-drink beverages such as orange juice than thermal pasteurization. This hybrid technology has high capability for usage in juice industry (Kernou et al., 2021). The enzymes and microorganisms activity cause to lose juice quality in apple juice. The ultra-sonication-microwave as a thermosonication procedure is able to deactivate the various enzymes and microorganisms (bacteria and yeasts) at lower temperature with shorter time, resulting lower losses of total phenolic compounds, flavonoids and ascorbic acid (Abid et al., 2014b).

Few scientific reports exist about effects of novel processing methods on sensory attributes. For instance, a novel and alternative process is very essential for treating rose water with minimal influence on its quality. Combined system of microwave and ultrasound is effective method for preserving nutritional value, phytochemicals and microbial safety of rose water. Effect of microwave-ultrasound method also investigated on *E. coli* in rose water (Rostami et al., 2018). The results indicated that effects of ultrasound and microwave process for destruction of *E. coli* in sour cherry juice are very different. Microwave kill microorganisms via disrupting the cell wall, while ultrasound by cavitation impacts and generated free radicals in liquid medium (Hosseinzadeh Samani et al., 2018). Investigations showed that the variations of ultrasonic power had greater effects on log microbial reduction in compared to microwave power. The results indicated a positive correlation between the ultrasonic power and inactivation rate. For example, with increase of ultrasonic power from 100 to 200 W, caused 58% increase in energy. In contrast, two times increase in temperature led to 24% reduction in the energy consumption during a log cycle deactivation. In other words, changes in temperature and energy were inversely related to each other per log cycle (Rostami et al., 2018).

Currently, ultrasound process, as a nonthermal green technique, applies widely to preserve juices without significantly negative effects on sensory properties and nutritional content than conventionally heating treatments

(Aadil et al., 2015). The ultrasound and microwave combined are known as microwave-ultrasound (microwave-sonication or thermosensation), which has been caused to increase the quality and safety attributes of dissimilar types of juices in terms of time (e.g., thermosonication [3.7 minutes] and ultrasonication [15.9 minutes]) had considerable differences to obtain the same result in juice processing (Lee et al., 2013; Ugarte-Romero et al., 2006).

6.4 Combined ultrasound-microwave effect on organoleptic attributes of juices

High temperature with short-time processes is reliable approach in food industry to reduce microbial dangers and other undesirable chances, while these methods have effects of adverse thermal degradation in food quality, sensory attributes, and bioactive compounds. Therefore, trends toward novel pasteurization and green preservation have increased in terms of microbial safety and nutritional values. From alternative and emerging techniques can point out to microwave heating with rapidly temperature come-up time. Ultrasound also another processing method becomes with great potent to expand microbial decontamination efficiency along with microwave processing. This study indicated combined ultrasound-microwave treatment had slight side effects to quality and phytochemicals than conventionally thermal processes for orange juice beverages (Kernou et al., 2021).

The safety and quality of juices are very important for consumers. So that fruit and vegetable juices are one of the important liquid nutritive products, which are high susceptible to microbial and physicochemical changes. The usage of nonthermal technologies is practicable to obtain natural and health juice drinks with negligible damages on nutritional value and organoleptic attributes. The juices processing conventionally cause some losses in bioactive compounds such as flavonoids (Igual et al., 2011) and carotenoids (Fratianni et al., 2010). The application of ultrasound alone due to the development of undesirable aromas and off-flavors has negative effect on the sensorial attributes of juices (Zinoviadou et al., 2015).

During thermosonication processing, all treated samples showed a raise in the nonenzymatic browning index in processing time that this negative influence can be related to the formation of Maillard reaction products (Manzoor et al., 2021). Next studies indicated that thermo-sonication is not able to induce Maillard reaction at $\leq 50°C$ (Dias et al., 2015). Other researchers reported that in thermosonicated samples have increased sensory attributes namely odor, taste, flavor, color, mouth feel, and overall acceptability compared to thermally pasteurized juices. But, treated samples using ultrasonic probes showed a metal taste such as processed orange juice by thermosonication, resulting proposed that probes be made from inert materials. Possibly ultrasonic probe corrosion was caused undesirable taste and mixing the corroded particles during the cavitation action with juice (Samani et al., 2015).

6.5 Combined ultrasound-microwave effect on viscosity properties and cloud retention of juices

Various enzymes such as pectin methylesterase and polygalacturonase are inactivated using thermosonication treatment. The pectine chain breakdown is catalyzed by polygalacturonase through oxygen bridge cleavage, resulting phase separation due to viscosity loss during storage period (Terefe et al., 2009; Koshani et al., 2015). For example, pectin cleavage during processing causes to reduce tomato juice consistency. Therefore, it is necessary to inactivate enzymes of pectin hydrolyzing (Verlent et al., 2004; Wu et al., 2008). Pectin methylesterase deactivation of 98.5% at 60°C and 98.9% at 65°C in tomato juice was accomplished using thermosonication treatment (Wu et al., 2008). In this process, pectin methylesterase was inactivated almost completely, but polygalacturonase was deactivated about 90% afterward 18 minutes in thermosonically tomato juice treated (20 kHz, 64 μm amplitude at 75°C) (Terefe et al., 2009). Thermo-sonically treated samples had a great apparent viscosity than conventionally thermal treated juices (Šimunek et al., 2014). Degradation rate of polygalacturonase (2.3–4 times) and pectin methylesterase (1.5–6 times) of tomato juice treated have been improved during combined heating (60–75°C) and ultrasonication (Terefe et al., 2009).

Changes of viscosity in soy-based beverage treated using ultrasound may be attributed to the synergic effects treatment time and sample protein content and pH (Bosiljkov et al., 2009). Effect of ultrasound on food viscosity can be temporary or permanent. Ultrasound's cavitation leads to shear stress, for example, in thixotropic fluids. If ultrasound energy intensity be enough changes in the molecular weight occur that it leads to temporary reduction in viscosity (Soria and Villamiel, 2010).

Ultrasound process is able to inactivate activity of enzymes like polygalacturonase and pectin methylesterase, which are capable to break down polysaccharides like pectin and consequently decreasing the viscosity of juices (Islam et al., 2014). In some juices, pectin should not be hydrolyzed by pectinase enzyme to preserve cloudiness and viscosity. In comparison with conventional procedure, including pectin and galacturonic acid of juices, which was processed using combined microwave with ultrasonic had higher compounds than control groups. The results showed that action of coupled ultrasound-microwave was very more efficiency on lower processing temperature, protection of phytochemicals, safety and nutritional quality of foods with smaller destructiveness (Guo et al., 2019).

The effective agents on almond milk cloudiness can be various components such as pectin, lipids, cellulose, proteins, and hemicellulose that cloudiness mainly is dependent on the level of turbidity and suspended particles. Thermosonication processing upgraded the cloud value (from 7.05 to 9.71) in all processed liquid samples (Manzoor et al., 2021). The cloudy increase also can be associated to the fragmentation of suspended particles during the collapse of microbubbles. Previously reported that thermosonication is able to increase significantly viscosity during longer treatment time and milder temperature (Aadil et al., 2015). Another research indicated that in grape juices treated using thermo-sonication were no significant decrease in the apparent viscosity (Ma et al., 2020).

In juices, turbidity and suspended particles called cloudiness, which was reduced significantly (from 23.50 to 8.4) in thermal pasteurization, while thermosonication increased cloudiness of juices (from 24.10 to 29.30). The results displayed that sonolysis of pectins, proteins, lipids, cellulose, and hemicellulose during cavitation mechanism of action increase cloudiness in processed juices (Aadil et al., 2015; Oladunjoye et al., 2021). Microwave non-thermal pasteurization of sour orange juice was caused significant decrease of pectin methylesterase at 70°C (75.7%) and 80°C (83.84%) with 900 W at 3 minutes. Electroporation phenomenon had main role in enzymatic degradation (Seyedabadi et al., 2020).

6.6 Combined ultrasound-microwave effect on color of juices

Juice color is a key factor that pointer to organoleptic attributes, nutritional quality, processing operations, and storage conditions. Food color displays a main role in the consumer satisfaction and acceptance due to visual character to evaluate quality of juices (Brochier et al., 2018). Color value and color variation in juices evaluate by Hunter lab scales that determining color changes are very important (Das et al., 2020). During the juices extraction from fruits and vegetables, endogenous various enzymes namely polyphenol-oxidase and peroxidase are released and should be inactivated rapidly to avoid changes of juice color. Thermosonication process is very useful in deactivating thermoresistant enzymes where it is hard to denature by conventionally thermal treatment alone, resulting overheat usage lead to very negative effects in quality of juice (Wu et al., 2008; Terefe et al., 2009; de Carvalho et al., 2015; Koshani et al., 2015). Both enzymes of polyphenol oxidase and peroxidase are causing enzymatic browning in juices that for color retention, they should be inactivated (Abid et al., 2014b). They catalyze the oxidation of phenolic substances to o-quinones, then their polymerization leads to melanin formation (dark-colored pigments) (Van Loey et al., 2001).

The results of investigations exhibited minimal revolution in color of ultrasonic-microwave treated juice than thermal pasteurization (Das et al., 2020). Juice color is mainly due to the presence of various colorants such as chlorophylls, polyphenols, and carotenoids. The minimum discoloration through degradation during processing and storage is reported in processed liquid foods using ultrasound. The positive influence of ultrasound on color stability of juices is mainly due to the effective elimination of blocked oxygen (Knorr et al., 2004). Therefore, color is a key tool for monitoring nutritional changes and quality of different juices during processing and/or storage time (Barba et al., 2012; Esteve et al., 2009). Several researches were carried out in different types of juices such as strawberry juice (Tiwari et al., 2008), grapefruit juice (Aadil et al., 2013), watermelon juice (Rawson et al., 2011), and guava juice (Cheng et al., 2007).

Thermally treated juices cause on darkening juice color, while treated juices with thermosonication lead to juice lightness (Walkling-Ribeiro et al., 2009). Lycopene (red pigment) and anthocyanin (red and blue pigments) are two different types of color pigments in juices. Thermosonicated watermelon juice with high treatment time causes raise of juice lightness. For maintaining health benefits and attraction of biocolorants in fruit and vegetable juices, thermosonication utilization for such juices can help to preserve useful bioactive compounds (Dubrović et al., 2011; Rawson et al., 2011). For example, lycopene stability of 98% to 100% in thermosonicated watermelon juice (20 kHz, 24.4 to 40 μm amplitude, 5 minutes at 25 to 30°C) (Rawson et al., 2011) and anthocyanin stability of more than 98% in thermosonicated strawberry juice (20 kHz, 60 μm amplitude, 3 minutes at 40°C) (Dubrović et al., 2011). The deactivation of various enzymes such as polyphenol oxidase (93.85%) and peroxidase (91%) in thermosonicated apple juice (60°C for 10 minutes) had significant effects on color (Abid et al., 2014b).

Emerging techniques such as ultrasound and microwave as combined presented highly inhabiting action of color changes during processing and storage than technologies alone. In usage of combined ultrasound-microwave in processing of juices, processed samples showed better retention of color pigments through inactivating polyphenol-oxidase and peroxidase enzymes, which lead to enzymatic browning reactions in different juices during processing and storage period (Liu et al., 2018). Inactivated polyphenol oxidase in almond milk using thermosonication at different temperatures (40 and 50°C) is caused considerable increase in lightness of treated samples (Manzoor et al., 2021). Time and temperature of thermo-sonication is very important in reducing polyphenol oxidase activity. Also, generated free radicals via sonolysis of water, temperature, mechanical forces, and cavitation action are responsible for the denaturation of enzymes (Fonteles et al., 2012; Manzoor et al., 2021).

Different researches have proven that thermosonication is capable to inactivate effectively various types of enzymes. With this method, many enzymes destroy completely for instance peroxidase, pectin methylesterase, protease, lipase, and polyphenol oxidase (D'amico et al., 2006; Fonteles et al., 2012). Microwave-ultrasonication process generates microbubbles which immediately collapse to produce extremely local high pressure and temperature along with shock waves and strong shear. These factors modify structure of enzymes through Van der Walls interaction or cleavage of hydrogen bond in the polypeptide chains. The cavitation action also leads to homolytic water molecules to hydrogen ($H^{\bullet}$) and hydroxyl ($^{\bullet}OH$) free radicals. Consequent changes are oxidation and inactivation of enzymes by generated free radicals (Lee et al., 2013).

In grape juice with inactivation of peroxidase and polyphenol oxidase during thermosonication process have effectively improved shelf stability and color retention of juice (Ma et al., 2020). With regard to enzymatic and nonenzymatic browning in juices, browning cause significant color changes of different types of juices. High temperatures of processing may accelerate nonenzymatic browning in juices, while in case of enzymatic browning, high temperatures effectively deactivate enzymes associated to browning in the juices consequently decrease the enzymatic browning of the juice. But in combined procedure need to ultrasonication plus mild temperature for decreasing enzymatic browning (Ma et al., 2020).

Partial sedimentation of unstable suspended particles during juice cavitation was caused to decrease color in juice thermosonication at 40 and 50°C, but increase in processed juice color at higher temperature and longer exposure time has been related to develop nonenzymatic browning products (Maillard reaction) (Oladunjoye et al., 2021). Juice color is a key case from consumers' viewpoint that pays attention to sensory quality and microbial safety of juices during processing, shelf time in storage period (Herceg et al., 2013). Enzyme inactivation (blanching) is a key step in the juices preparation. Results indicate that blanching has a positive outcome in color stability during juice processing and refrigeration storage (Pérez-Grijalva et al., 2018).

6.7 Combined ultrasound-microwave effect on microorganisms

Thermal pasteurization can destroy different microorganisms in solid or liquid foods such as molds, bacteria, and yeasts or even decrease their growth rate (Nott and Hall, 1999). Damage of protein structure by thermal shock destroy endogenous enzymes subsequently these critical conditions make for inactivation of enzymes (Islam et al., 2014).

The survival of *E. coli* in treated orange juice investigated using Weibull model. The results indicated that ultrasonication alone (42 kHz) has no considerable influence on microbial reduction rate (1.3 log cycle decrease for 60 minutes). But, ultrasonication is capable to increase the efficiency of microwave decontamination. Ultrasound utilized as pretreatment was more operative than post-treatment status. The 8 log microbial reduction was obtained when ultrasound (20 minutes) followed by microwave (900 W for 30 seconds), while 4 log cycle decrease was achieved when these arrangement reversed in processing of orange juice beverages (Kernou et al., 2021).

Investigated mechanisms for microbial inactivation by thermosonication are divided into local mechanical and chemical impacts. Formation of microbubbles in liquid medium and then collapse cause different influences like shock waves, high shear forces, damage to outer layer of different microorganisms (Zupanc et al., 2019). High local shear forces lead to serious damages to different microorganisms that along with shock waves have synergistic influences to mass transfer processes. With regard to microbubble collapses, hotspots generated with high temperatures cause serious damages to microorganisms. For example, effect on outer layer integrity, microorganisms become more susceptible to further destruction by reactive oxygen species (Yusof et al., 2016; Piyasena et al., 2003; Wigginton et al., 2012). Cavitation action also cause H_2O homolysis and generation of free radicals of $^{\bullet}OH$ and $H^{\bullet}$ as the strongest oxidants. The $^{\bullet}OH$ readily oxidize various organic compounds. The H_2O_2 also is generated from reaction between the two $^{\bullet}OH$ (Zupanc et al., 2019). Therefore, results indicated that ultrasonication is the most suitable technique for inactivating microorganisms through free radicals, local heat, and pressure and cavitation effect (Piyasena et al., 2003).

Microbial destruction by microwave method has been carried out in various juices such as grape juice, apple juice, coconut water, and milk. In this process should regard to key cases such as dielectric properties and microwave power (Salazar-González et al., 2012). Deactivation of various microorganisms and enzymes using microwaves are due to main damage to cell membrane, heat destruction, collapse of the cellular structure and generation and spread of free radicals into the fluid medium (Tajchakavit et al., 1998; Hosseinzadeh Samani et al., 2018). Other main reasons for reduction of *E. coli* count can be related to the increase horn motion in the fluid medium. So that horn motion has a direct relation on formation of microbubbles in liquid medium consequently cavitation effects (Vichare et al., 2001).

With temperature increase from 30 to 60°C, *E. coli* count decreased nearly 28.7%, while three folds enhancing treatment time from 2 to 6 minutes destroyed only 7.3% of *E. coli* count. Therefore, temperature in destruction of microorganisms was more effective than processing time. In general, temperature influence along with ultrasound (as thermosonication) on *E. coli* reduction is stronger than individual time. The microwave (352.21 W and 50°C) and ultrasound (475.13 W for 6 minutes) combined and utilized as novel and alternative processes had significantly synergistic effect on rose water treatment (Rostami et al., 2018). In another investigation evaluated the synergistic action of ultrasound and microwave procedure as nonthermal pasteurization on *E. coli* of sour cherry juice. The results displayed that temperature was more active agent to reduce the total microorganism count (Hosseinzadeh Samani et al., 2018).

Fig. 6.3 indicates the steps of microbial destruction with thermosonication treatment, microwave in heating and ultrasound in cavitation action display practical role in cell inactivation. During thermosonication, first water molecules are immediately penetrate into microbial cells, in very ultrashort time lead to swelling cells and in consequent step with rupturing

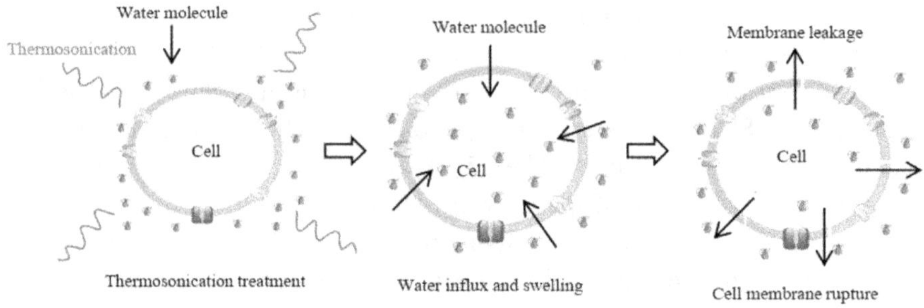

Figure 6.3 Antimicrobial mechanism of action combined ultrasonication-microwave in liquid medium.

cell membrane and cell wall under thermocavitation effects, intracellular contents leakage out.

Microorganisms can protect themselves against oxidants, but if levels of reactive oxygen species exceed their resistances in contrast to oxidative stresses disappear (Kashmiri and Mankar, 2014; Moradas-Ferreira et al., 1996). Different researchers reported that H_2O_2 has high killing effect on yeasts and bacteria as well as positive correlation occur between its concentration and lethal impact (Zupanc et al., 2019). Studies indicated that H_2O_2 toxicity is through production of oxidative species like intracellular $^{\bullet}OH$ radicals (Labas et al., 2008). Vital compounds of microorganisms such as polysaccharides, proteins, lipids, and DNA seriously damage by reactive oxygen species (Drábková et al., 2007; McDonnell and Russell, 1999). Free radicals are highly electrophilic species with short-lived period, resulting $^{\bullet}OH$ radicals attack to such compounds for oxidizing double bonds and sulfhydryl groups subsequently oxidative stress cause irreversible action for microorganisms (Flores et al., 2012). In the literature is reported that both nonradical (H_2O_2) and radical ($^{\bullet}OH$, $^{\bullet}OR$, $^{\bullet}OOR$, and $^{\bullet}ON$) reactive species attack to polysaccharides. Cleavage of glycoside backbone leads to decomposition of biopolymers, resulting main changes in function of cell constituents (Duan and Kasper, 2011). Studies also showed that nucleic acids are very susceptible to oxidative stress through reactive species. The $^{\bullet}OH$ radical inside of cells can cause a breakdown in the double helix and transformation of nitrogen bases (Flores et al., 2012; Mayer et al., 2015).

Impact of ultrasonication on bacterial inactivation is completely dependent on different operating parameters during processing samples. Experiments showed that High frequency ultrasound resulted in great destruction compared to low frequency for *Streptococcus mutans, E. coli*, and *Staphylococcus epidermidis* (Zupanc et al., 2019). Primary targets of ultrasonication effects on bacteria with gram-negative and gram-positive walls are different, outer membrane in gram-negative bacteria and cytoplasmic membrane and cell internal structure in gram-positive bacteria (Li et al., 2016). In general, due to tightly adherent the peptidoglycans layer, bacteria with gram-positive walls are more ultrasound-resistant than gram-negative (Abdullah and Chin, 2014). Additionally, proposed that hydrophobicity of bacterial surface is very important factor during ultrasonication process. Microbubbles also have hydrophobic Surface, resulting micro-bubbles highly attract to surfaces of cells. Besides, protective layers presence in bacterial cell may display a main role in their resistance to ultrasonic treatment. For example, spores of *Bacillus golbigii* have very high resistance to ultrasonication process so that need to higher decimal reduction time during lethal treatment (Kobayashi et al., 2014). Investigations indicated that ultrasonication caused serious damages to cells via ruptures on cell membrane, cell wall, vacuoles, morphological changes, shrinkage, and destruction of cells (Zupanc et al., 2019).

Experiments showed that various molds and yeasts completely inactivate at 60°C along with ultrasonication (20 kHz, 600 W, 3–9 minutes), but lower temperatures (20°C and 40°C) did not effect on such microorganisms (Jambrak et al., 2018). Other findings indicated significantly synergistic action of combined heating and ultrasonic process on destruction of bacterial spores. The most microorganisms typically have high sensitivity to sonolysis at temperatures above 50°C (Dubrović et al., 2011; Condón-Abanto et al., 2016). In general, ultrasound is able to inactive types of microorganisms mainly through cavitation action along with thermal, pressure, and chemical effects. This method can directly destroy cell wall and cell membrane of bacteria and promotes sonolysis of water molecules to formation of free radicals ($H^{\bullet}$, $O^{\bullet}$, and $^{\bullet}OH$), resulting breakdown of DNA of microorganisms and inactivation of their enzymes and ultimately killing bacteria. Ultrasound manner can effectively destroy bacteria with gram positive and gram negative walls such as *Staphylococcus aureus, Listeria innocua, Salmonella enteritidis,* and *E. coli* (Liao et al., 2018; Wu et al., 2015).

Pasteurized juices with high temperature (90°C) did not show detectably microbial growth, while thermosonication treatment in mild temperatures (50 and 60°C) indicated no detectably microbial growth such as bacteria, molds, and yeasts (especially at 60°C) (Oladunjoye et al., 2021). Complete destruction of microorganisms by pasteurization treatment could be related to destruction of cell wall, cell membrane, and nuclear components consequently microbial death (Rivas et al., 2006). Mechanisms of microbial inactivation via thermosonication could be majorly related to sonolysis (acoustic cavitation) on cell structure through generation of microbubbles and antimicrobial free radicals in juices (Anaya-Esparza et al., 2017). In juices with high acidity, induced osmotic pressure along with other reactions could increase the cavitation influence on microbial structure during thermosonication treatment through releasing intracellular components like lipids, proteins, and nuclear compounds (Wu et al., 2015; Bermúdez-Aguirre and Barbosa-Cánovas, 2012). Similar findings have obtained to inactive microorganisms presence in beverages namely pineapple, grapefruit, carrot, cranberry, apple juices using thermosonication (Martínez-Flores et al., 2015; Silva, 2016).

Pathogenic and spoilage microorganisms are main challenges in nonthermal treatments of food processing. The results indicated that ultrasonication are more effective than microwave to reduce *S. cerevisiae* in orange juice (Samani et al., 2015; Vichare et al., 2001). Increasing ultrasound frequency leads to enhancement of number the microbubbles formed in fluid medium subsequently increasing cavitation phenomenon (Vichare et al., 2001). The mechanisms deactivation of microorganisms by ultrasonication is divided to physical and chemical influences. Physical effects are effective on cell membrane and changing local temperature and pressure (5000 °K and 50,000 kPa) as well as free radicals are generated through sonochemical reaction (Piyasena et al., 2003; Valdramidis et al., 2010).

The main advantages of ultrasound treatment in juices are its high action as a novel antimicrobial technique along with minimal changes in nutritional and flavor value with increase of homogeneity in liquid medium (Soria and Villamiel, 2010). The main physical force of ultrasound for the microbial inactivation consists of cavitation action, hydrophobic surfaces of microorganisms, and microbubbles formed damage to cell wall along with intracellular cavitation formation consequently microbial inactivation (Chandrapala et al., 2012). In addition, formation of free radicals and reactive species cause cell lysis through disruption of cell walls (Sango et al., 2014). During ultrasound application in processing of different juices, microorganisms showed different behaviors. In general, spores than vegetative forms are more resistant to ultrasound cavitation (Chandrapala et al., 2012).

6.8 Combined ultrasound-microwave effect on bioactive compounds

Previous studies displayed to gain juice with high quantity, high polyphenolic substances, may use microwave treatment for blanching

(Gerard and Roberts, 2004). Microwave produces quickly uniform heat with minimal changes in anthocyanins of juices (de Ancos et al., 1999). Bioactive compounds (e.g., anthocyanins and polyphenolic substances) have anti-inflammatory, anticarcinogenic, and antioxidant actions in human healthy (Zafra-Stone et al., 2007; Dai and Mumper, 2010).

Ultrasonication of strawberry juice was caused to decrease the maximum 11% of ascorbic acid as function of treatment time and cavitation effects (10 minutes and 0.81 W/mL) (Tiwari et al., 2008). A considerable diminish reported for total phenolics in high temperature with long treatment times (>6 minutes) when processing of juices. Phenolic degradation can be attributed to oxidation function by free radicals formed ($H^{\bullet}$, ${}^{\bullet}OH$, and H_2O_2) in juice sonication (Tiwari et al., 2009). Ultrasonicated apple juices displayed a major increase (12–27%) in total carotenoid quantity (in ambient temperature) compared to control samples. Researchers attributed carotenoid increase to mechanical disruption of cell walls and bound compounds presence in food matrices (Abid et al., 2014b). Previous researches stated that in ultrasonicated grapefruit juice increased total phenolics, flavonoids, and flavonols as well as phenolic substances in cactus pear juice (Zafra-Rojas et al., 2013).

Ascorbic acid can mainly be gained from fruits and it protect against cancer and infectious diseases through keeping optimum human-immune system (Chambial et al., 2013). It has no thermal stability and easily oxidizes under conditions of overheat treatment and long heating. Therefore, mild heat is necessary to ensure high stability of ascorbic acid (Abid et al., 2014b). For instance, high ascorbic acid maintenance (about 96%) in watermelon juice occurred in mild thermosonication treatment (20 kHz, 25 μm amplitude, 29°C for 5 minutes) as well as (about 96%) for apple juice (40°C for 5 minutes) (Abid et al., 2014b; Rawson et al., 2011). In addition, eliminating dissolved oxygen during cavitation caused to stabilization of ascorbic acid in juices during thermosonication (Cheng et al., 2007).

The phytochemical content of ultrasonic-microwave processed vegetable juice has been increased significantly than raw sample, while in thermal treated juice; it has been decreased than raw juice. But combined ultrasonic-microwave and conventional treatment had not significant changes on phenolic compounds, catechin, gallic, and caffeic acid, in processed vegetable juice (Das et al., 2020). The results indicated that using process of combined ultrasonic-microwave had high retention on total phenolic and flavonoid as well as ascorbic acid contents of sugar cane juice than ultrasonic and microwave alone during treating and storage time. Bioactive compounds are used as substrate for polyphenol oxidase and peroxidase enzymes consequently inactivating these enzymes cause to maintain bioactives in processed and stored sugar cane juice. Treated samples with combined procedure also had the maximum antioxidant activity (DPPH scavenging assay) during 21 days of storage time (Zia et al., 2019). In addition, high DPPH scavenging capacity and antioxidant activity in kiwifruit puree was due to the effective destruction of polyphenol oxidase and peroxidase enzymes using thermosonication process (Benlloch-Tinoco et al., 2015).

During thermosonication process at different temperatures (30, 45, and 60°C), results displayed that bioactive compounds have been increased with cavitation action through disruption of fruit and/or vegetable particles, cell walls, and biological cells (Ghasemzadeh et al., 2015). It also was observed that thermosonication considerability increased the total antioxidant capacity and DPPH scavenging activity at 45°C (from 52.65 to 74.71%) than 30°C (from 54.13 to 68.13%) (Manzoor et al., 2021). Moreover, previous reports indicated that hydroxyl radicals scavenging activity (inhibition %) were considerably increased in at 45°C than 30°C. This might be linked to release of phenolic compounds via cavitation phenomenon (Codevilla et al., 2018). Other researches displayed that during thermosonication, free hydroxyl radical amounts had been decreased with increase of antioxidant activities, but high levels of free hydroxyl radicals for a long period can prevent antioxidant activities (Muzaffar et al., 2016; Qureshi et al., 2020). The results indicated that with degradation of antioxidant compounds (phenolic compounds) in high temperature and high power, collapse of microbubbles decrease antioxidant activity (Muñiz-Márquez et al., 2013). Degradation of ascorbic acid and phenolic compounds accelerate by hydroxyl radicals generated in cavitation action (Chen et al., 2020).

The samples treated using thermo-sonication showed high antioxidant capacity (DPPH and FRA assay) (76.03%) than control groups. The results indicated that combined thermal process with ultrasound significantly increased free radical scavenging activity through increasing antioxidant compounds through releasing blocked bioactive compounds (e.g., polyphenols) from fruit and/or vegetable particles presence in juices along with inactivating some oxidases (Ma et al., 2020). Ascorbic acid level was significantly increased in thermosonicated juices at different temperatures (40 and 50°C). Ascorbic acid stability could be related to effective removal of dissolved oxygen from juice during cavitation (Oladunjoye et al., 2021). But, ascorbic acid loss occurred in thermosonicated samples with 60°C that sonolysis action collapse of microbubbles had interfered in ascorbic acid degradation (Erkaya et al., 2015). Phenolic and carotenoid contents have been increased in thermosonicated juices at 40 and 50°C. But, both compounds have been reduced at higher temperature (60°C). In case of carotenoid, isomerization had major role. Flavonoid contents in thermosonicated juices have been increased at 40 and 50°C, but reduction of flavonoids at high temperature (60°C) could be related to excessive cavitation effects (Oladunjoye et al., 2021). Different studies related increase of processed juice antioxidant activity to decrease of free hydroxyl radical formation during juice thermosonication via releasing various bioactive compounds (Muzaffar et al., 2016; Rivas et al., 2006). It should be noted that high radical levels for a long period have a toxic influence on antioxidant activity. Therefore, this action may be effective on reduce of antioxidant values at 60°C (Qureshi et al., 2020; Oladunjoye et al., 2021).

6.9 Combined ultrasound-microwave effect on proteins of juices

Nowadays, novel and alternative nonthermal technologies are widely studied due to high efficiency, quality retention, and low cost during juice processing (Das et al., 2020). Ultrasound treatments has the ability to denature milk proteins specially whey proteins. Synergistic effect was investigated and observed higher denaturation when ultrasound treatment is combined with mild heat (Villamiel and de Jong, 2000). Proteins are compounds with high sensitive to denaturation. The local high temperature and pressure led to by ultrasound waves through formation of microbubbles consequently cavitation phenomenon can quickly promote protein denaturation. Free radicals generated by sonolysis of water also react with the residues amino acids to inactive active enzymes (Islam et al., 2014; Karaseva and Metelitza, 2006). Ultrasonication modifies protein structures through local cavitation effects, which mainly is dependent on the microbubbles collapsing consequently energy produced with high intensity. Moreover, agitation with high speed causes disruption of van der Waals and hydrogen bond in the polypeptides, thereby changing the conformational structure of the native proteins (Mawson et al., 2011; Corzo-Martínez et al., 2017).

The protein content of treated juice was estimated differently in various processing methods such as thermal treatment (22.13 μg/mg) and combined microwave-ultrasound technique (28.10 μg/mg) than raw juice (25.47 μg/mg) based on dry weight. Protein increase in this combined technique may be due to disruption of cell walls of vegetable/fruit particles. The results demonstrated that combined technique had not considerably negative effect on amino acids (cysteine, aspartic acid, lysine, histidine, valine, and proline) of processed juice thus nutritional attributes remain as intact (Das et al., 2020).

6.10 Combined ultrasound-microwave effect on shelf life of juices

On the base of ascorbic acid retention has been reported that shelf life of orange juice treated with ultrasonication was longer at compared to conventionally thermal treatment (Tiwari et al., 2009). With regard to effective removal of oxygen and stability of ascorbic acid, the predicted shelf life for ultra-sonicated orange juices ranged from 27 to 33 days compared to conventional pasteurization for 19 days (Knorr et al., 2004). The impacts of ultrasound and mild temperature (thermosonication) studied on microorganisms of orange juice. The results indicated that usage of ultrasound technique in combination with other procedures was more effective to diminish the microbial activity and consequently on shelf stability time (Valero et al., 2007).

Vegetable juice treated with combined technology indicated high shelf stability during storage period where considerable reduction of microbial growth rate obtained in juice processed and preserved for 120 days at 4°C. Significant variation in pH and molds and yeasts growth rate was not detected during storage period. Therefore, results revealed that application of combined technique in processing of juices in terms of safe shelf life, which can be utilizable as effective candidate in juice industry compared to conventional thermal treatment (Das et al., 2020). Combined treatment (ultrasonic-microwave) had considerable influence on quality (physic-chemicals and bioactive compounds), microbial safety and shelf stability of sugar cane juice for 21 days of storage at refrigerated temperature. Combined ultrasonic-microwave process is an effective emerging green technique for improving the product safety and shelf time of juice via minimizing quality variations, holding bioactive substances and falling microbial growth during storage period (Zia et al., 2019).

Microbiological stability and physicochemical quality of treated juices using microwave and ultra-sonication is very important. The studies suggested that juices treated with combined methods had a good physicochemical constancy and shelf-stability for a month. For example, in blackberry juice at the end of refrigeration storage, total polyphenols (87%) and anthocyanins (90%) had very good retention (Pérez-Grijalva et al., 2018). In contrast, conventionally treated blackberry juice indicated antioxidant activity loss (Hager et al., 2008). Ultrasonication treatment led to a significant reduction in different microorganisms (molds, yeasts, and aerobic mesophiles) in blackberry juice compared to the control sample (Pérez-Grijalva et al., 2018). The mechanical or/and physical effects of ultrasonication on microorganisms could be attributed to cavitation effects in liquid medium that induces microbubbles formation due to pressure variations (Tiwari et al., 2009). Microbubbles formed and grown ultimately collapse to shear and break cell walls and membrane structures of microorganisms (Earnshaw et al., 1995). Chemical and biological effects cause through electric discharges and free radicals generation was relative to cavitation phenomenon (Tiwari et al., 2009).

6.11 Industrial perspectives and commercial usages of combined ultrasound-microwave

The demand for ready-to-drink juices with high nutritious value and freshness in flavor and nonadditives and nonpreservatives has encouraged toward green and safe technology in juice processing. Combined microwave-ultrasound method is a novel and eco-friendly technique to replace traditional heating processes. Researches have showed that microwave-ultrasound technology is able to extent quality and stability of the different juices. The utilization of hybrid method in juice processing industry has excellent benefits in quality preservation and safe processing along with high efficiency and reliable way in destruction of food-borne pathogens (Rani, 2020).

The combined technique of microwave-ultrasound is a very effective practice for nonthermal pasteurization of juices such as bottle gourd juice and efficient enough to decrease heat-resistant bacteria like *L. monocytogenes*. This technique combined is capable to rise up for pilot plant level and even for industrial scale (Das et al., 2020).

Novel and alternative technologies such as microwave and ultrasound are eco-friendly technologies that have high efficiency and energy saving with low cost, which improves production yield, quality and microorganism safety. Disinfection and inactivation of harmful enzymes are very important for producing different juices with inhibiting loss of heat sensitive substances (Chen et al., 2020). Therefore, green processing technologies can be considered as a novel green nonthermal technique of potential, which could be utilizable in industries of juice to produce juices with high quality for commercial applications (Ma et al., 2020).

6.12 Conclusion and future prospective of combined ultrasound-microwave

Nowadays, consumers have high tendency to safe-to-drink juices with high nutritious value and microbial safety in natural taste and freshness. It is well-known that traditionally thermal processes can extend the shelf-stability and safety of various juices, but they have significantly negative effects in nutritional and quality values. Currently, many researchers utilize combined microwave-ultrasound method in different food industries as an interesting topic especially in juice industry. Combination of microwave and ultrasonication technologies, as a novel and alternative process in the juices treatment, will cause to produce fruit and vegetable juices with high antioxidant and bioactive substances as well as great microbial safety that request by consumers and juice industry. Thermosonication as mild heating can significantly improve sensory and physicochemical attributes such as color (enzymatic and nonenzymatic browning), cloudiness, bioactive components (ascorbic acid, phenolics, carotenoids, flavonoids, and antioxidant activity), and so on. The results of many research works indicated that microwave-ultrasound procedure has high potential in juice processing, storage, and shelf stability. In addition, thermosonication processing technology can decrease quality and nutritional losses in various juices. Ultimately, combined microwave-ultrasound is valuable green processing technology in food nonthermal preservation and retention of nutrients, sensory attributes, and bioactive compounds along with inhibition of food spoilage agents such as bacteria, yeasts, and molds.

References

Aadil, R.M., Zeng, X.-A., Han, Z., Sun, D.-W., 2013. Effects of ultrasound treatments on quality of grapefruit juice. Food Chem. 141, 3201–3206.

Aadil, R.M., Zeng, X.A., Zhang, Z.H., Wang, M.S., Han, Z., Jing, H., Jabbar, S., 2015. Thermosonication: a potential technique that influences the quality of grapefruit juice. Int. J. Food Sci. Technol. 50, 1275–1282.

Abdullah, N., Chin, N.L., 2014. Application of thermosonication treatment in processing and production of high quality and safe-to-drink fruit juices. Agric. Agric. Sci. Procedia 2, 320–327.

Abid, M., Jabbar, S., Hu, B., Hashim, M.M., Wu, T., Lei, S., Khan, M.A., Zeng, X., 2014a. Thermosonication as a potential quality enhancement technique of apple juice. Ultrason. Sonochem. 21, 984–990.

Abid, M., Jabbar, S., Wu, T., Hashim, M.M., Hu, B., Lei, S., Zeng, X., 2014b. Sonication enhances polyphenolic compounds, sugars, carotenoids and mineral elements of apple juice. Ultrason. Sonochem. 21, 93–97.

Ahmed, Z., Manzoor, M.F., Begum, N., Khan, A., Shah, I., Farooq, U., Siddique, R., Zeng, X.-A., Rahaman, A., Siddeeg, A., 2019. Thermo-ultrasound-based sterilization approach for the quality improvement of wheat plantlets juice. Processes 7, 518.

Anaya-Esparza, L.M., Velázquez-Estrada, R.M., Roig, A.X., García-Galindo, H.S., Sayago-Ayerdi, S.G., Montalvo-González, E., 2017. Thermosonication: an alternative processing for fruit and vegetable juices. Trends Food Sci. Technol. 61, 26–37.

Barba, F.J., Esteve, M.J., Frígola, A., 2012. High pressure treatment effect on physicochemical and nutritional properties of fluid foods during storage: a review. Compr. Rev. Food Sci. Food Saf. 11, 307–322.

Benlloch-Tinoco, M., Igual, M., Rodrigo, D., Martínez-Navarrete, N., 2015. Superiority of microwaves over conventional heating to preserve shelf-life and quality of kiwifruit puree. Food Control 50, 620–629.

Benlloch-Tinoco, M., Martínez-Navarrete, N., Rodrigo, D., 2014. Impact of temperature on lethality of kiwifruit puree pasteurization by thermal and microwave processing. Food Control 35, 22–25.

Bermúdez-Aguirre, D., Barbosa-Cánovas, G.V., 2012. Inactivation of Saccharomyces cerevisiae in pineapple, grape and cranberry juices under pulsed and continuous thermo-sonication treatments. J. Food Eng. 108, 383–392.

Bhat, R., Kamaruddin, N.S.B.C., Min-Tze, L., Karim, A., 2011. Sonication improves kasturi lime (Citrus microcarpa) juice quality. Ultrason. Sonochem. 18, 1295–1300.

Bosiljkov, T., Brnčić, M., Tripalo, B., Karlović, S., Ukrainczyk, M., Ježek, D., Rimac Brnčić, S., 2009. Impact of ultrasound-enhanced homogenization on physical properties of soybean milk. Chem. Eng. Trans. 17, 1029–1034.

Brochier, B., Mercali, G.D., Marczak, L.D.F., 2018. Effect of ohmic heating parameters on peroxidase inactivation, phenolic compounds degradation and color changes of sugarcane juice. Food Bioprod. Process. 111, 62–71.

Chambial, S., Dwivedi, S., Shukla, K.K., John, P.J., Sharma, P., 2013. Vitamin C in disease prevention and cure: an overview. Indian J. Clin. Biochem. 28, 314–328.

Chandrapala, J., Oliver, C., Kentish, S., Ashokkumar, M., 2012. Ultrasonics in food processing–food quality assurance and food safety. Trends Food Sci. Technol. 26, 88–98.

Chandrasekaran, S., Ramanathan, S., Basak, T., 2013. Microwave food processing—a review. Food Res. Int. 52, 243–261.

Chen, F., Zhang, M., Yang, C.-H., 2020. Application of ultrasound technology in processing of ready-to-eat fresh food: a review. Ultrason. Sonochem. 63, 104953.

Cheng, L., Soh, C., Liew, S., Teh, F., 2007. Effects of sonication and carbonation on guava juice quality. Food Chem. 104, 1396–1401.

Codevilla, C.F., Tischer, B., Gindri, A.L., Nogueira-Librelotto, D.R., Barin, J.S., Silva, C.D.B.D., Rolim, C.M.B., Zepka, L.Q. & Menezes, C.R.D. 2018. Cytotoxicity and antioxidant activity of goldenberry extracts obtained with high intensity ultrasound. *Ciência Rural*,48.

Condón-Abanto, S., Arroyo, C., Álvarez, I., Condón, S., Lyng, J., 2016. Application of ultrasound in combination with heat and pressure for the inactivation of spore forming bacteria isolated from edible crab (Cancer pagurus). Int. J. Food Microbiol. 223, 9–16.

Corzo-Martínez, M., Villamiel, M., Moreno, F.J., 2017. Impact of high-intensity ultrasound on protein structure and functionality during food processing. Ultrasound in Food Processing: Recent Advances. John Wiley & Sons, New York.

Dai, J., Mumper, R.J., 2010. Plant phenolics: extraction, analysis and their antioxidant and anticancer properties. Molecules 15, 7313–7352.

D'amico, D.J., Silk, T.M., Wu, J., Guo, M., 2006. Inactivation of microorganisms in milk and apple cider treated with ultrasound. J. Food Prot. 69, 556–563.

Das, M.J., Das, A.J., Chakraborty, S., Baishya, P., Ramteke, A., Deka, S.C., 2020. Effects of microwave combined with ultrasound treatment on the pasteurization and nutritional properties of bottle gourd (Lagenaria siceraria) juice. J. Food Process. Preserv. 44, e14904.

De Ancos, B., Cano, M.P., Hernandez, A., Monreal, M., 1999. Effects of microwave heating on pigment composition and colour of fruit purees. J. Sci. Food Agric. 79, 663–670.

De Carvalho, J.M., Maia, G.A., DA Fonseca, A.V.V., De Sousa, P.H.M., Rodrigues, S., 2015. Effect of processing on physicochemical composition, bioactive compounds and enzymatic activity of yellow mombin (Spondias mombin L.) tropical juice. J. Food Sci. Technol. 52, 1182–1187.

Del Socorro Cruz-Cansino, N., Ramírez-Moreno, E., León-Rivera, J.E., Delgado-Olivares, L., Alanís-García, E., Ariza-Ortega, J.A., De Jesús Manríquez-Torres, J., Jaramillo-Bustos, D.P., 2015. Shelf life, physicochemical, microbiological and antioxidant properties of purple cactus pear (Opuntia ficus indica) juice after thermoultrasound treatment. Ultrason. Sonochem. 27, 277–286.

Dias, D.D.R.C., Barros, Z.M.P., De Carvalho, C.B.O., Honorato, F.A., Guerra, N.B., Azoubel, P.M., 2015. Effect of sonication on soursop juice quality. LWT-Food Sci. Technol. 62, 883–889.

Dobson, G., Mcdougall, G.J., Stewart, D., Cubero, M.Á., Karjalainen, R.O., 2017. Effects of juice matrix and pasteurization on stability of black currant anthocyanins during storage. J. Food Sci. 82, 44–52.

Dolas, R., Saravanan, C., Kaur, B.P., 2019. Emergence and era of ultrasonic's in fruit juice preservation: a review. Ultrason. Sonochem. 58, 104609.

Drábková, M., Matthijs, H., Admiraal, W., Maršálek, B., 2007. Selective effects of H_2O_2 on cyanobacterial photosynthesis. Photosynthetica 45, 363–369.

Duan, J., Kasper, D.L., 2011. Oxidative depolymerization of polysaccharides by reactive oxygen/nitrogen species. Glycobiology 21, 401–409.

Dubrović, I., Herceg, Z., Režek Jambrak, A., Badanjak, M., Dragović-Uzelac, V., 2011. Effect of high intensity ultrasound and pasteurization on anthocyanin content in strawberry juice. Food Technol. Biotechnol. 49, 196–204.

Earnshaw, R., Appleyard, J., Hurst, R., 1995. Understanding physical inactivation processes: combined preservation opportunities using heat, ultrasound and pressure. Int. J. Food Microbiol. 28, 197–219.

Ekezie, F.-G.C., Sun, D.-W., Han, Z., Cheng, J.-H., 2017. Microwave-assisted food processing technologies for enhancing product quality and process efficiency: a review of recent developments. Trends Food Sci. Technol. 67, 58–69.

Erkaya, T., Başlar, M., Şengul, M., Ertugay, M.F., 2015. Effect of thermosonication on physicochemical, microbiological and sensorial characteristics of ayran during storage. Ultrason. Sonochem. 23, 406–412.

Esteve, M., Barba, F., Palop, S., Frígola, A., 2009. The effects of non-thermal processing on carotenoids in orange juice. Czech J. Food Sci. 27 The Effects of non.

Ferrario, M., Alzamora, S.M., Guerrero, S., 2015. Study of the inactivation of spoilage microorganisms in apple juice by pulsed light and ultrasound. Food Microbiol. 46, 635–642.

Flores, M.J., Brandi, R.J., Cassano, A.E., LABAS, M.D., 2012. Chemical disinfection with H2O2— the proposal of a reaction kinetic model. Chem. Eng. J. 198, 388–396.

Fonteles, T.V., Costa, M.G.M., De Jesus, A.L.T., De Miranda, M.R.A., Fernandes, F.A.N., Rodrigues, S., 2012. Power ultrasound processing of cantaloupe melon juice: effects on quality parameters. Food Res. Int. 48, 41–48.

Fratianni, A., Cinquanta, L., Panfili, G., 2010. Degradation of carotenoids in orange juice during microwave heating. LWT-Food Sci. Technol. 43, 867–871.

Gerard, K., Roberts, J., 2004. Microwave heating of apple mash to improve juice yield and quality. LWT-Food Sci. Technol. 37, 551–557.

Ghasemzadeh, A., Jaafar, H.Z., Juraimi, A.S., Tayebi-Meigooni, A., 2015. Comparative evaluation of different extraction techniques and solvents for the assay of phytochemicals and antioxidant activity of hashemi rice bran. Molecules 20, 10822–10838.

Guo, Z., Zhao, B., Li, H., Miao, S., Zheng, B., 2019. Optimization of ultrasound-microwave synergistic extraction of prebiotic oligosaccharides from sweet potatoes (Ipomoea batatas L.). Innovative Food Sci. Emerg. Technol. 54, 51–63.

Hager, T.J., Howard, L.R., Prior, R.L., 2008. Processing and storage effects on monomeric anthocyanins, percent polymeric color, and antioxidant capacity of processed blackberry products. J. Agric. Food Chem. 56, 689–695.
Herceg, Z., Markov, K., Sobota Šalamon, B., Režek Jambrak, A., Vukušić, T., Kaliterna, J., 2013. Effect of high intensity ultrasound treatment on the growth of food spoilage bacteria. Food Technol. Biotechnol. 51, 352–359.
Hosseinzadeh Samani, B., Khoshtaghaza, M. & Minaee, S. 2018. Modeling the simultaneous effects of microwave and ultrasound treatments on sour cherry juice using response surface methodology.
Igual, M., García-Martínez, E., Camacho, M., Martínez-Navarrete, N., 2011. Changes in flavonoid content of grapefruit juice caused by thermal treatment and storage. Innovative Food Sci. Emerg. Technol. 12, 153–162.
Islam, M.N., Zhang, M., Adhikari, B., 2014. The inactivation of enzymes by ultrasound—a review of potential mechanisms. Food Rev. Int. 30, 1–21.
Jambrak, A.R., Šimunek, M., Evačić, S., Markov, K., Smoljanić, G., Frece, J., 2018. Influence of high power ultrasound on selected moulds, yeasts and Alicyclobacillus acidoterrestris in apple, cranberry and blueberry juice and nectar. Ultrasonics 83, 3–17.
Karaseva, E., Metelitza, D., 2006. Stabilization of glucoso-6-phosphate dehydrogenase by its substrate and cofactor in an ultrasonic field. Russ. J. Bioorg. Chem. 32, 436–443.
Kashmiri, Z., Mankar, S., 2014. Free radicals and oxidative stress in bacteria. Int. J. Curr. Microbiol. App. Sci. 3, 34–40.
Kernou, O.N., Belbahi, A., Amir, A., Bedjaoui, K., Kerdouche, K., Dairi, S., Aoun, O., Madani, K., 2021. Effect of sonication on microwave inactivation of Escherichia coli in an orange juice beverage. J. Food Process Eng. 44, e13664.
Khandpur, P., Gogate, P.R., 2015. Understanding the effect of novel approaches based on ultrasound on sensory profile of orange juice. Ultrason. Sonochem. 27, 87–95.
Knorr, D., Zenker, M., Heinz, V., Lee, D.-U., 2004. Applications and potential of ultrasonics in food processing. Trends Food Sci. Technol. 15, 261–266.
Knorr, D.W., Hendrickx, M.E., 2001. Ultra High Pressure Treatments of Foods. Kluwer Academic/Plenum, Boston, MA.
Kobayashi, Y., Hayashi, M., Yoshino, F., Tamura, M., Yoshida, A., Ibi, H., Masaichi, C.-I.L., Ochiai, K., Ogiso, B., 2014. Bactericidal effect of hydroxyl radicals generated from a low concentration hydrogen peroxide with ultrasound in endodontic treatment. J. Clin. Biochem. Nutr. 54 (3), 13–86.
Koshani, R., Ziaee, E., Niakousari, M., Golmakani, M.T., 2015. Optimization of thermal and thermosonication treatments on pectin methyl esterase inactivation of sour orange juice (Citrus aurantium). J. Food Process. Preserv. 39, 567–573.
Kubo, M., Curet, S., Augusto, P., Boillereaux, L., 2019. Multiphysics modeling of microwave processing for enzyme inactivation in fruit juices. J. Food Eng. 263, 366–379.
Kuldiloke, J. 2002. Effect of ultrasound, temperature and pressure treatments on enzyme activity and quality indicators of fruit and vegetable juices.
Labas, M.D., Zalazar, C.S., Brandi, R.J., Cassano, A.E., 2008. Reaction kinetics of bacteria disinfection employing hydrogen peroxide. Biochem. Eng. J. 38, 78–87.
Lee, H., Kim, H., Cadwallader, K.R., Feng, H., Martin, S.E., 2013. Sonication in combination with heat and low pressure as an alternative pasteurization treatment–effect on Escherichia coli K12 inactivation and quality of apple cider. Ultrason. Sonochem. 20, 1131–1138.
Li, J., Ahn, J., Liu, D., Chen, S., Ye, X., Ding, T., 2016. Evaluation of ultrasound-induced damage to Escherichia coli and Staphylococcus aureus by flow cytometry and transmission electron microscopy. Appl. Environ. Microbiol. 82, 1828–1837.
Liao, X., Li, J., Suo, Y., Chen, S., Ye, X., Liu, D., Ding, T., 2018. Multiple action sites of ultrasound on Escherichia coli and Staphylococcus aureus. Food Sci. Human Wellness 7, 102–109.

Lima Tribst, A.A., De Souza Sant'ana, A., De Massaguer, P.R., 2009. Microbiological quality and safety of fruit juices—past, present and future perspectives. Crit. Rev. Microbiol. 35, 310–339.

Liu, S., Zhang, X., You, L., Guo, Z., Chang, X., 2018. Changes in anthocyanin profile, color, and antioxidant capacity of hawthorn wine (Crataegus pinnatifida) during storage by pretreatments. LWT 95, 179–186.

M. Sango, D., Abela, D., Mcelhatton, A., Valdramidis, V., 2014. Assisted ultrasound applications for the production of safe foods. J. Appl. Microbiol. 116, 1067–1083.

Ma, T., Wang, J., Wang, L., Yang, Y., Yang, W., Wang, H., Lan, T., Zhang, Q., Sun, X., 2020. Ultrasound-Combined sterilization technology: an effective sterilization technique ensuring the microbial safety of grape juice and significantly improving its quality. Foods 9, 1512.

Manzoor, M.F., Siddique, R., Hussain, A., Ahmad, N., Rehman, A., Siddeeg, A., Alfarga, A., Alshammari, G.M., Yahya, M.A., 2021. Thermosonication effect on bioactive compounds, enzymes activity, particle size, microbial load, and sensory properties of almond (Prunus dulcis) milk. Ultrason. Sonochem. 78, 105705.

Martínez-Flores, H.E., Garnica-Romo, M.G., Bermúdez-Aguirre, D., Pokhrel, P.R., Barbosa-Cánovas, G.V., 2015. Physico-chemical parameters, bioactive compounds and microbial quality of thermo-sonicated carrot juice during storage. Food Chem. 172, 650–656.

Mawson, R., Gamage, M., Terefe, N.S., Knoerzer, K., 2011. Ultrasound in enzyme activation and inactivation. Ultrasound Technologies for Food and Bioprocessing. Springer, New York, NY.

Mayer, B.K., Yang, Y., Gerrity, D.W., Abbaszadegan, M., 2015. The impact of capsid proteins on virus removal and inactivation during water treatment processes. Microbiol. Insights 8, S31441 MBI.

Mcdonnell, G., Russell, A.D., 1999. Antiseptics and disinfectants: activity, action, and resistance. Clin. Microbiol. Rev. 12, 147–179.

Mena, P., Vegara, S., Martí, N., García-Viguera, C., Saura, D., Valero, M., 2013. Changes on indigenous microbiota, colour, bioactive compounds and antioxidant activity of pasteurised pomegranate juice. Food Chem. 141, 2122–2129.

Moradas-Ferreira, P., Costa, V., Piper, P., Mager, W., 1996. The molecular defences against reactive oxygen species in yeast. Mol. Microbiol. 19, 651–658.

Muñiz-Márquez, D.B., Martínez-Ávila, G.C., Wong-Paz, J.E., Belmares-Cerda, R., Rodríguez-Herrera, R., Aguilar, C.N., 2013. Ultrasound-assisted extraction of phenolic compounds from Laurus nobilis L. and their antioxidant activity. Ultrason. Sonochem. 20, 1149–1154.

Muzaffar, S., Ahmad, M., Wani, S., Gani, A., Baba, W.N., Shah, U., Khan, A.A., Masoodi, F., Gani, A., Wani, T.A., 2016. Ultrasound treatment: effect on physicochemical, microbial and antioxidant properties of cherry (Prunus avium). J. Food Sci. Technol. 53, 2752–2759.

Nott, K.P., Hall, L.D., 1999. Advances in temperature validation of foods. Trends Food Sci. Technol. 10, 366–374.

Oladunjoye, A.O., Adeboyejo, F.O., Okekunbi, T.A., Aderibigbe, O.R., 2021. Effect of thermosonication on quality attributes of hog plum (Spondias mombin L.) juice. Ultrason. Sonochem. 70, 105316.

Perez-Grijalva, B., Herrera-Sotero, M., Mora-Escobedo, R., Zebadúa-García, J.C., Silva-Hernández, E., Oliart-Ros, R., Perez-Cruz, C., Guzmán-Gerónimo, R., 2018. Effect of microwaves and ultrasound on bioactive compounds and microbiological quality of blackberry juice. LWT 87, 47–53.

Piasek, A., Kusznierewicz, B., Grzybowska, I., Malinowska-Pańczyk, E., Piekarska, A., Azqueta, A., Collins, A.R., Namieśnik, J., Bartoszek, A., 2011. The influence of sterilization with Enbiojet® Microwave Flow Pasteurizer on composition and bioactivity of aronia and blue-berried honeysuckle juices. J. Food Compos. Anal. 24, 880–888.

Pingret, D., Fabiano-Tixier, A.-S., Chemat, F., 2013. Degradation during application of ultrasound in food processing: a review. Food Control 31, 593–606.

Piyasena, P., Mohareb, E., Mckellar, R., 2003. Inactivation of microbes using ultrasound: a review. Int. J. Food Microbiol. 87, 207–216.

Qureshi, T.M., Nadeem, M., Maken, F., Tayyaba, A., Majeed, H., Munir, M., 2020. Influence of ultrasound on the functional characteristics of indigenous varieties of mango (Mangifera indica L.). Ultrason. Sonochem. 64, 104987.

Rani, M., 2020. Thermosonication technology and its application in food industry. IJCS 8, 922–928.

Rawson, A., Tiwari, B., Patras, A., Brunton, N., Brennan, C., Cullen, P., O'donnell, C., 2011. Effect of thermosonication on bioactive compounds in watermelon juice. Food Res. Int. 44, 1168–1173.

Regier, M., Knoerzer, K., Schubert, H., 2016. The Microwave Processing of Foods. Woodhead Publishing, Sawston.

Rivas, A., Rodrigo, D., Martinez, A., Barbosa-Cánovas, G., Rodrigo, M., 2006. Effect of pef and eat pasteurization on the physical–chemical characteristics of blended orange and carrot juice. LWT-Food Sci. Technol. 39, 1163–1170.

Roobab, U., Aadil, R.M., Madni, G.M., Bekhit, A.E.D., 2018. The impact of nonthermal technologies on the microbiological quality of juices: a review. Compr. Rev. Food Sci. Food Saf. 17, 437–457.

Rostami, S., Behruzian, M., Samani, B.H., Lorigooini, Z., Hosseinabadi, T., Zareiforoush, H., Behruzian, A., 2018. Study of combined ultrasound-microwave effect on chemical compositions and E. coli count of rose aromatic water. Iranian J. Pharmaceut. Res.: IJPR 17, 146.

Salazar-González, C., San Martín-González, M.F., López-Malo, A., Sosa-Morales, M.E., 2012. Recent studies related to microwave processing of fluid foods. Food Bioprocess Technol. 5, 31–46.

Samani, B.H., Khoshtaghaza, M.H., Lorigooini, Z., Minaei, S., Zareiforoush, H., 2015. Analysis of the combinative effect of ultrasound and microwave power on Saccharomyces cerevisiae in orange juice processing. Innovative Food Sci. Emerg. Technol. 32, 110–115.

Seyedabadi, M.M., Kashaninejad, M., Jafari, S.M., Seyedabadi, E., Khojastehpour, M., 2020. Design, fabrication and evaluation of continuous fluids processing system using microwave technology. Innovative Food Technol. 7, 611–625.

Shamis, Y., Croft, R., Taube, A., Crawford, R.J., Ivanova, E.P., 2012. Review of the specific effects of microwave radiation on bacterial cells. Appl. Microbiol. Biotechnol. 96, 319–325.

Siguemoto, É.S., Gut, J.A.W., Martinez, A., Rodrigo, D., 2018. Inactivation kinetics of Escherichia coli O157: H7 and Listeria monocytogenes in apple juice by microwave and conventional thermal processing. Innovative Food Sci. Emerg. Technol. 45, 84–91.

Silva, F.V., 2016. High pressure processing pretreatment enhanced the thermosonication inactivation of alicyclobacillus acidoterrestris spores in orange juice. Food Control 62, 365–372.

Šimunek, M., Jambrak, A.R., Dobrović, S., Herceg, Z., Vukušić, T., 2014. Rheological properties of ultrasound treated apple, cranberry and blueberry juice and nectar. J. Food Sci. Technol. 51, 3577–3593.

Soria, A.C., Villamiel, M., 2010. Effect of ultrasound on the technological properties and bioactivity of food: a review. Trends Food Sci. Technol. 21, 323–331.

Tajchakavit, S., Ramaswamy, H., Fustier, P., 1998. Enhanced destruction of spoilage microorganisms in apple juice during continuous flow microwave heating. Food Res. Int. 31, 713–722.

Terefe, N.S., Gamage, M., Vilkhu, K., Simons, L., Mawson, R., Versteeg, C., 2009. The kinetics of inactivation of pectin methylesterase and polygalacturonase in tomato juice by thermosonication. Food Chem. 117, 20–27.

Timmermans, R., Mastwijk, H., Knol, J., Quataert, M., Vervoort, L., Van Der Plancken, I., Hendrickx, M., Matser, A., 2011. Comparing equivalent thermal, high pressure and pulsed electric field processes for mild pasteurization of orange juice. Part I: impact on overall quality attributes. Innovative Food Sci. Emerg. Technol. 12, 235–243.

Tiwari, B., O'donnell, C., Patras, A., Brunton, N., Cullen, P., 2009. Stability of anthocyanins and ascorbic acid in sonicated strawberry juice during storage. Eur. Food Res. Technol. 228, 717–724.

Tiwari, B., O'donnell, C., Patras, A., Cullen, P., 2008. Anthocyanin and ascorbic acid degradation in sonicated strawberry juice. J. Agric. Food Chem. 56, 10071–10077.

Tremarin, A., Canbaz, E.A., Brandão, T.R., Silva, C.L., 2019. Modelling alicyclobacillus acidoterrestris inactivation in apple juice using thermosonication treatments. LWT 102, 159–163.

Ugarte-Romero, E., Feng, H., Martin, S.E., Cadwallader, K.R., Robinson, S.J., 2006. Inactivation of Escherichia coli with power ultrasound in apple cider. J. Food Sci. 71, E102–E108.

Valdramidis, V., Cullen, P.J., Tiwari, B., O'donnell, C., 2010. Quantitative modelling approaches for ascorbic acid degradation and non-enzymatic browning of orange juice during ultrasound processing. J. Food Eng. 96, 449–454.

Valero, M., Recrosio, N., Saura, D., Muñoz, N., Martí, N., Lizama, V., 2007. Effects of ultrasonic treatments in orange juice processing. J. Food Eng. 80, 509–516.

Van Loey, A., Verachtert, B., Hendrickx, M., 2001. Effects of high electric field pulses on enzymes. Trends Food Sci. Technol. 12, 94–102.

Verlent, I., Van Loey, A., Smout, C., Duvetter, T., Hendrickx, M., 2004. Purified tomato polygalacturonase activity during thermal and high-pressure treatment. Biotechnol. Bioeng. 86, 63–71.

Vichare, N.P., Gogate, P.R., Dindore, V.Y., Pandit, A.B., 2001. Mixing time analysis of a sonochemical reactor. Ultrason. Sonochem. 8, 23–33.

Villamiel, M., De Jong, P., 2000. Inactivation of Pseudomonas fluorescens and Streptococcus thermophilus in trypticase® soy broth and total bacteria in milk by continuous-flow ultrasonic treatment and conventional heating. J. Food Eng. 45, 171–179.

Walkling-Ribeiro, M., Noci, F., Riener, J., Cronin, D., Lyng, J., Morgan, D., 2009. The impact of thermosonication and pulsed electric fields on Staphylococcus aureus inactivation and selected quality parameters in orange juice. Food Bioprocess Technol. 2, 422–430.

Wang, J., Vanga, S.K., Raghavan, V., 2019. High-intensity ultrasound processing of kiwifruit juice: effects on the ascorbic acid, total phenolics, flavonoids and antioxidant capacity. LWT 107, 299–307.

Wigginton, K.R., Pecson, B.M., Sigstam, T., Bosshard, F., Kohn, T., 2012. Virus inactivation mechanisms: impact of disinfectants on virus function and structural integrity. Environ. Sci. Technol. 46, 12069–12078.

Wordon, B., Mortimer, B., Mcmaster, L.D., 2012. Comparative real-time analysis of Saccharomyces cerevisiae cell viability, injury and death induced by ultrasound (20 kHz) and heat for the application of hurdle technology. Food Res. Int. 47, 134–139.

Wu, J., Gamage, T., Vilkhu, K., Simons, L., Mawson, R., 2008. Effect of thermosonication on quality improvement of tomato juice. Innovative Food Sci. Emerg. Technol. 9, 186–195.

Wu, T., Yu, X., Hu, A., Zhang, L., Jin, Y., Abid, M., 2015. Ultrasonic disruption of yeast cells: underlying mechanism and effects of processing parameters. Innovative Food Sci. Emerg. Technol. 28, 59–65.

Yu, Y., Cheng, X., Zhang, C., Zhang, J., Zhang, S., Xu, J., 2020. Ultrasonic and microwave treatment improved jujube juice yield. Food Sci. Nutr. 8, 4196–4204.

Yusof, N.S.M., Babgi, B., Alghamdi, Y., Aksu, M., Madhavan, J., Ashokkumar, M., 2016. Physical and chemical effects of acoustic cavitation in selected ultrasonic cleaning applications. Ultrason. Sonochem. 29, 568–576.

Zafra-Rojas, Q.Y., Cruz-Cansino, N., Ramírez-Moreno, E., Delgado-Olivares, L., Villanueva-Sánchez, J., Alanís-García, E., 2013. Effects of ultrasound treatment in purple cactus pear (Opuntia ficus-indica) juice. Ultrason. Sonochem. 20, 1283–1288.

Zafra-Stone, S., Yasmin, T., Bagchi, M., Chatterjee, A., Vinson, J.A., Bagchi, D., 2007. Berry anthocyanins as novel antioxidants in human health and disease prevention. Mol. Nutr. Food Res. 51, 675–683.

Zia, S., Khan, M.R., Zeng, X.A., Shabbir, M.A., Aadil, R.M., 2019. Combined effect of microwave and ultrasonication treatments on the quality and stability of sugarcane juice during cold storage. Int. J. Food Sci. Technol. 54, 2563–2569.

Zinoviadou, K.G., Galanakis, C.M., Brnčić, M., Grimi, N., Boussetta, N., Mota, M.J., Saraiva, J.A., Patras, A., Tiwari, B., Barba, F.J., 2015. Fruit juice sonication: implications on food safety and physicochemical and nutritional properties. Food Res. Int. 77, 743–752.

Zupanc, M., Pandur, Ž., Perdih, T.S., Stopar, D., Petkovšek, M., Dular, M., 2019. Effects of cavitation on different microorganisms: the current understanding of the mechanisms taking place behind the phenomenon. a review and proposals for further research. Ultrason. Sonochem. 57, 147–165.

CHAPTER SEVEN

Synergistic effect of sonication and microwave for inhibition of microorganism

P.V.V.P. Prudhvi[a], Sudarshanna Kar[a], Piyush Sharma[a], Jyotsana Patel[b] and Siba Prasad Nayak[c]
[a]Department of Food Process Engineering, National Institute of Technology, Rourkela, India
[b]College of Agricultural Engineering, Jawaharlal Nehru Krishi Vishwa Vidyalaya, Jabalpur, India
[c]Department of Food Engineering, National Institute of Food Technology Entrepreneurship and Management, Sonipat, India

7.1 Introduction

Food is a basic part of life whose consumption provides nutrients and energy to the body. The food may be contaminated during processing, handling and storage causing adverse effect on the health of consumers. A food contamination can be physical, chemical, or biological in nature. From farm to fork, a food contaminant may enroute the food supply chain from various ways that may cause the food to unfit for consumption. Physical contaminants include heavy metals, glasses, plastics, stones etc. Chemical contaminants include the chemicals that are used in farm as insecticides, pesticides, fumigants, or the industrial chemicals. Biological contamination refers to the spoilage due to presence of harmful microorganisms. Based on studies, World Health Organization had reported the food contamination as a global challenge and inferred that food contamination in one part of planet may affects the health of consumers on the other part. Bacterial contamination of food is of main concern due its adverse effects and vulnerability to spread. Approximately there are about more than 200 identified food borne diseases effecting most people with a complicated immune system such as children or old aged.

Yearly more than 200 million people are losing their lives due to the consumption of contaminated foods. As per the records of World Health Organization, foodborne outbreaks in United States alone results in an estimate of 48 million illnesses including 3000 deaths and 128,000 hospitalizations (Vieira et al., 2021) whereas, for the year 2015, 4362 food borne outbreaks were reported in the European Union with 17, 3892, and

Ultrasound and Microwave for Food Processing: Synergism for Preservation and Extraction.
DOI: https://doi.org/10.1016/B978-0-323-95991-9.00001-1

45, 874 cases of deaths, hospitalizations and illnesses, respectively (Bintsis, 2017). The microorganisms responsible for the foodborne illnesses include pathogenic bacteria such as the species of salmonella, Shigella, Yersinia, Clostridium, Vibrio, Escherichia coli, Listeria monocytogenes, Staphylococcus aureus, Bacillus cereus, Campylobacter jejuni, Cronobacter sakazakii, Yersinia enterocolitica, Parasites such as Cyclospora cayetanensis, Toxoplasma gondii, Trichinella spiralis, Mycotoxins like Aflatoxin, Ochratoxin A, Citrinin, Patulin, and viruses like Hepatitis A, Noroviruses (Yeni et al., 2016). Recent Corona virus outbreak in 2019 can be demonstrated as the best example of foodborne illness through viruses.

According to FAO State of Agriculture report, 2019 approximately 14% of worldwide food output is lost in the supply chain. Most of these losses in supply chain occur due to microbial spoilage. Though different conventional thermal methods are used by processors such as acid and salt treatments, roasting, hot air drying, fumigation, etc., their greater nutritional impacts a need for research in nonconventional methods of food processing. Longer treatment times to accomplish the desired thermal treatment needed to inactivate microbes deeper inside the food. Traditional heating procedures can cause unwanted changes in product quality. This problem is particularly prevalent in solid food products with intermediate moisture content (20%–50% wet basis) and low moisture content (below 20% wet basis). This is due to the fact that traditional heating is particularly nonuniform when applied to solid food products and in-package heating of liquid food materials (where only the surface is heated, and heat must conduct to the interior). A uniform temperature distribution can only be achieved if the slow heat is applied. However, in terms of food quality and energy efficiency, this increase in process time is unfavorable (Dev et al., 2012). Nonconventional thermal methods are not only energy efficient but also environment friendly which is the need of the hour. Ultrasound and microwave are increasingly being researched for their potential usage in food processing. Individually they find many usages such as preservation, sterilization, dispersion, tenderization, hydrogenation, inactivation of enzymes etc. but when combined can bring much greater reduction in microbial load with improved sensory and nutritional qualities of food.

7.2 Ultrasound treatment

7.2.1 Overview

Ultrasound (Cárcel, García-Pérez, Benedito, and Mulet) refers to sound waves with a frequency of more than 20 kHz. The application of ultrasound

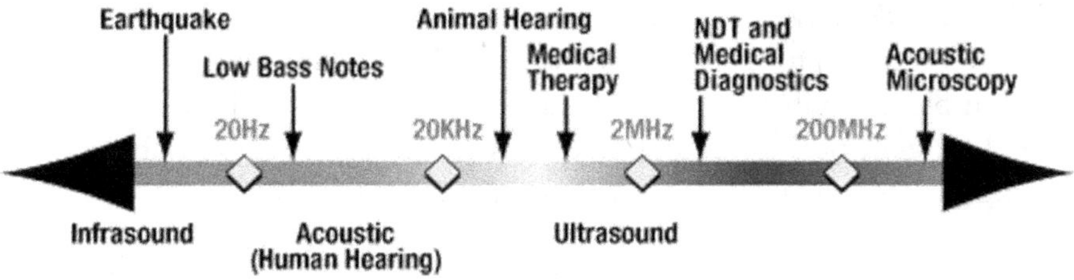

Figure 7.1 ***Frequency distribution of sound waves with application.*** (Source: www.olympus-ims.com)

in sonar, medical, nondestructive measurements and fluid are quite popular (Al-Hilphy et al., 2020). Ultrasound frequency decides its application range. Ultrasound can be broadly categorized into power and diagnostic ultrasound depending upon its frequency and power intensity. Power ultrasound has a frequency between 20 and 100 kHz and an intensity of more than 1 W/cm^2 whereas diagnostic ultrasound has a frequency more than 100 kHz and intensity less than 1 W/cm^2 (Cárcel et al., 2012). Power ultrasound causes properties alternation in material whereas diagnostic ultrasound does not induce properties alternation in the material. Ultrasound inactivates microbes by cavitation effect at normal temperature so maintain its sensorial and nutritional properties of food thus categorized into nonthermal food processing technology. Ultrasound application is mostly limited to liquid food processing as the cavitation effect is generally effective in liquids (Feng et al., 2011). Power ultrasound when passes through liquids it induces cavitation. Power ultrasound causes a series of compressions and rarefactions when passes through a liquid medium, and produces cavitation bubbles when rarefaction overcomes the force of attraction between liquid molecules. Cavitation bubbles collapse causes the mechanical and chemical effect in liquid foods thus inactivates microbes. The mechanical effect of power ultrasound is due to the generation of shear forces whereas chemical effect may be due to the generation of free radicals (Dolatowski et al., 2007). In addition to the above effects of ultrasound, it may cause the formation of hotspots (locations where the temperature is more than 4000 K and pressure around 1000 atm) and shock waves, which further adds to microbial inactivation (Fig. 7.1).

7.2.2 Ultrasound generation

Ultrasound can be produced by transducer, which converts one form of energy into sound energy. Piezoelectric, magnetostrictive and liquid-driven transducers are commonly used for ultrasound generations.

7.2.2.1 Liquid-driven transducer

Liquid food to be processed when pressed by high power pump to pass across the thin blade, it vibrates the thin blade. When vibration frequency crosses 20 kHz it generates ultrasound and thus produces cavitation bubbles. The trailing face of vibration produces cavitation in the liquid foods.

7.2.2.2 Piezoelectric transducer

The piezoelectric transducer works on the principle of electrostriction. Electrostriction is the phenomenon observed in polarized materials (quartz and barium titanate) when an electric field is applied to them. The electric field applied to polarized material induces changes in its dimension due to the alignment of polarized molecules. Dimension changes in material cause vibration and thus produce ultrasound.

7.2.2.3 Magnetostrictive transducer

Magnetostriction is the phenomenon observed in a ferromagnetic material (nickel and iron), when the magnetic field is applied to it. The magnetic field applied to ferromagnetic material in a discontinuous manner causes dimension changes, which further produces ultrasound.

Piezoelectric transducers are mostly preferred over magnetostrictive transducers in commercial-scale applications due to their scalability (magnetostrictive transducer has less power per transducer compared to piezoelectric transducers).

7.2.2.4 Ultrasonic equipment

Two types of equipment are generally used (Fig. 7.2 and Fig. 7.3).

7.2.2.5 Ultrasonic bath

Bath type transducer is used for low power application. For liquid food processing, transducer is normally provided at the bottom of liquid food filled bath to obtained a near around uniform treatment. Power of ultrasound usually kept low to prevent cavitation effect over bath wall.

7.2.2.6 Ultrasonic probe

High power application can be achieved through ultrasonic probe by attaching more than one horn to the transducer. Shapes of horn have also affect over treatment uniformity of liquid foods.

Figure 7.2 Ultrasonic bath. (Source: www.grantinstruments.com)

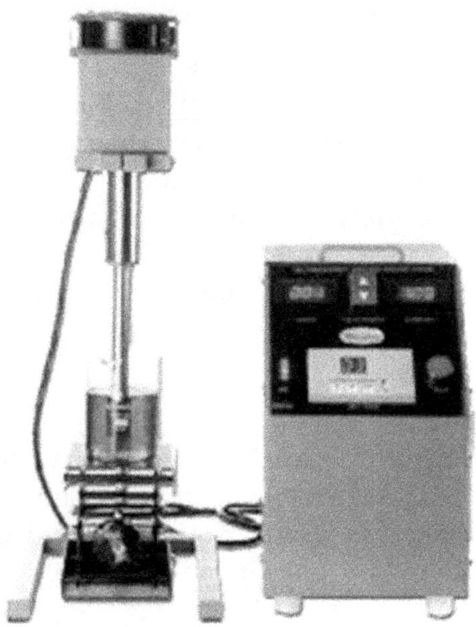

Figure 7.3 Ultrasonic probe. (Source: www.olympus-ims.com)

7.3 Principle of microbial inactivation by Ultrasound

A. Shear force due to cavitation causes disruption of cell membrane of microorganism and thus causes microbial inactivation.
B. Hotspots created due to ultrasound produces free radicals which further causes microbial inactivation. Free radical mechanism most be avoided in food processing so as to prevent detrimental chemical effect of free radicals over food.
C. Shock wave also causes microbial inactivation.
D. Temperature rises during application may cause thermal inactivation but it should be avoided.

7.4 Factors affecting inactivation of microorganisms

7.4.1 Nonultrasonic parameters

Temperature of medium, treatment medium properties (compositions and volume), strain, size and shape of microorganism, growth stage of microorganism, and vegetative microorganism or spore.

7.4.2 Ultrasound frequency

The frequency between 20 and 100 kHz is generally used for microbial inactivation, whereas frequency above 100 kHz is used for diagnostic purposes. Frequency has an inverse effect over bubble size. High cavitation bubbles produced by low frequency ultrasound cause local high pressure and temperature and thus inactivate microbes. At higher frequency (acoustic streaming) cavitation effect is no more significant and hence no microbial inactivation obtained. Frequency in the MHz range is used in medical imaging.

7.4.3 Power intensity

Power intensity and microbial inactivation has direct relations. Low power intensity causes acoustic streaming (motion and mixing in the liquids) which does not encourage bubbles formation whereas high power intensity encourages bubble formation, growth, and violent collapse of large bubbles.

7.4.4 Mode of treatment

Pulse mode saves energy and gives better microbial inactivation.

7.4.5 Temperature of medium

Temperature of medium has direct effect over microbial inactivation up to a certain limits however it should be avoided so as to maintain sensorial and nutritional value of the food. Energy released during cavitation depends on bubble growth kinetics and bubble explosion. Cavitation energy is inversely proportional to vapor pressure of the liquid food (Knorr et al., 2004). Muthukumaran et al. (2006) reported that the temperature can affect the surface tension, viscosity and vapor pressure of the liquid medium. High viscosity restricts the bubble formation. High temperature causes low viscosity of the liquid medium and thus allows violent collapse of cavitation bubbles. However temperature of the medium should be optimized so as to avoid dampening effect of higher vapor pressure at high temperature.

7.4.6 Composition of medium

A study by Wang et al. (2010) on alicyclobacilli in apple juice and buffer, suggested that the ingredients in apple juice hinder the ultrasound application and thus US found less effective in apple juice than in buffer (Wang et al., 2010).

7.4.7 Strains and shape of microorganisms

Study by Wang et al. (2010) on A. acidiphilus DSM14558^{T} and A. acidoterrestris DSM 3922 ^{T} reported that latter is more sensitive to US treatment compared to former due to different fatty acids distributions in cytoplasmic membrane, different membrane compositions and different C+G contents in DNA of the strains. Gram positive bacteria were found to be more resistant to US treatment compared to gram negative bacteria attributing to its thicker cell wall and rod shape bacteria were found to be more susceptible than to cocci due to its larger surface area (Chemat and Khan, 2011; Wang et al., 2010)

7.4.8 Vegetative microorganism or spore

Spore structure has highly resistant mechanical properties corresponding to its vegetative cells which resist the ultrasound applications (Reineke and Mathys, 2020). In additions to above, number of layers, less cytoplasm and less fluid membrane and dramatic dehydration of spore resist the different inactivation technologies.

7.5 Microbial inactivation by ultrasound

Above results indicated that 5 log reductions of microbes to satisfy USFDA standard cannot be obtained through US technology only thus require hurdle technology for better results. Thermosonication (Bintsis), manothermosonication (MTS) were nowadays emphasised to obtain 5 log reductions of micro-organisms. MTS can have additive (Baumann et al., 2005) or synergistic effects (Alvarez et al., 2003; Arroyo et al., 2012) over microbial inactivation depending upon treatment range, treatment medium and microorganism investigated. An additive effect is attributed toward the different microbial inactivation mechanisms whereas the synergistic effect was due to sensitizing phenomenon (heat treatment causes microbes to be more sensitive to ultrasound treatment). Pressure, heat and ultrasound have been reported to have synergetic effects to inactivate micro-organisms more efficiently and effectively than TS, for example, MTS (400 kPa, 59°C) required 1.4 minutes whereas TS and sonication required 3.7 minutes and 15.9 minutes respectively to get 5 log reduction of *E. coli* K12. Studies in 2011 reported that TS, MS, and TMS can inactivate spores compared to US technology only. Guerrero et al. (2001) and Lopez-Malo et al. (2005) reported the synergistic effect of TS (40–60°C and 20 kHz) in fungal spore inactivation with lower US D (Decimal reduction) values corresponding thermal D values. A study by Muñoz et al. (2012) suggested that TS and pulse light in combination (light of 200–1100 nm and applied in pulses of [100–400 μs]) was effective in obtaining 4.8 to 5.9 log CFU/mL of *E. coli* in apple juice in comparison to individual treatments. It was also reported that the hurdle sequence of the above processing technologies does not affect its effectiveness in the inactivation of *E. coli*. Another study by Palgan et al. (2011) reported that UV followed by MTS can fulfill the FDA standard of *E. coli* inactivation in fresh cranberry and grape juice blend. A study by Ross et al. (2003) over hurdle technologies reported that high pressure before ultrasonication improves its effectiveness over microbial inactivation. In another study by Gomez-Gomez et al. (2021), reported the combined effect of pulsed electric field (PEF) and high power ultrasound (HPU) and its order of application over inactivation of B. pumilus, A. niger, and *E. coli* in emulsions. PEF treatment of 152.3–176.3 kJ/kg prior to HPU application of 3 minutes was reported to be the best sequence to inactivate B. pumilus and A. niger over 1.0 and 6.6 log cycles respectively compared to its treatment as well as another sequence of HPU-PEF treatment (Table 7.1).

Table.7.1 Inactivation of microorganism by ultrasound.

Food	Microorganism	T (°C)	t (min)	F (kHz)	A (μm)	log reduction	References
Natural squeezed Apple juice	*Saccharomyces cerevisiae*	30	30	20	80	2.5	Ferrario et al. (2015)
Commercial apple juice	Alicyclobacillus acidoterrestris spores	30	30	20	80	No	Ferrario et al. (2015)
Apple juice	Total plate count Yeast and molds	20	60	25	70	1.2 1.3	Abid et al. (2014)
Apple cider	*Listeria monocytogenes*	20	20	20		1.6	Baumann et al. (2005)
Strawberry juice	*Zygosaccharomyces bailii*	20	6	20	60	-	Bevilacqua et al. (2014)
Apple juice	*Saccharomyces cerevisiae*	20	6	20	60	2.16	Bevilacqua et al. (2014)
Orange juice	*Zygosaccharomyces rouxii*	20	6	20	60	1.7	Bevilacqua et al. (2014)
Tomato juice	*Pichia fermentans*	40	10	20	61	-	Adekunte et al. (2010)
Milk	*Staphylococcus aureus, Escherichia coli*	20	12	20	60	0.8 2.2	Herceg et al. (2012)
Liquid whole egg	*Salmonella enteritidis*	55	5	20		-	Huang et al. (2006)
Grape fruit juice	TPC yeast and mold	20	30	28		0.5 0.5	Aadil et al. (2018)

Where F is frequency, A is amplitude, T is temperature, and t is time.

7.5.1 Scope of ultrasound in food industry

Ultrasound apart from obtaining microbial inactivation is now used for improvising the various food processing unit operations, that is, ultrasound-assisted freezing, extraction, drying, filtration, osmotic dehydration and rice parboiling process, etc., in food industry now days. Application of low power ultrasound (LPU) in meat product, fruits and vegetables, cereal products, fat and emulsion products, aerated foods, food gels, honey, food proteins, ultrasonic monitoring of food freezing has been extensively researched and were now applied in food industry. Applications of ultrasound in particle size determination of emulsions, impurities detection in cherry marmalades, cheese, and margarine, thickness measurement of materials, fluid velocity in pipes, and food matrix compositions, and monitoring of phase transitions in crystallization and melting has been utilized in food industry since its evolutions. Application of power ultrasound in crystallization process, emulsifications, defoaming, improving functional properties of food proteins, enzyme inactivation, freeze concentration, and freeze drying, ice cream manufacturing, frozen food thawing, and peeling has been extensively researched and are in general use in food industry now days.

7.6 Microwave treatment

7.6.1 Overview

Microwaves are the electromagnetic waves that travel at the speed of light. Its frequency varies from 300 MHz to 300 GHz. The dipolar effect existing between oxygen and hydrogen in water molecules is responsible for the rise in temperature. Moreover, the ionic interactions when microwaves pass from foods also play a key role in heating effect of microwaves. When microwaves pass from the food the polar water molecules tries to reorient themselves with oscillating electric field in due process of which they collide with each other resulting in volumetric heating of material. The dielectric characteristics and penetration depth are two of the most critical parameters that influence microwave heating and heat dispersion. The dielectric characteristics and penetration depth are two of the most critical parameters that influence microwave heating and heat dispersion (Chandrasekaran et al., 2013). Penetration with microwaves resulted into selective heating, membrane rupture, electroporation, which results in microbial damage. In selective heating preferential heating of microbes results in their damage. Electroporation theory supports the fact that proration in cell membrane which results in sweeping out of cellular materials leading to microbial death. Rupturing of membranes with high voltage happens according to cell membrane rupture theory. Disruption of internal components due to magnetic field coupling also results in microbial death (Kozempel et al., 1998).

7.6.2 Microbial inhibition mechanism

Thermal destruction is the main inactivation mechanism behind the volumetric heating during microwave assisted inhibition of microorganisms. Enzymes are the major catalysts responsible for the metabolism of most of living cells in microorganisms. On the other hand, DNA plays a major role in multiplication or replication of microorganisms. Similar to other conventional thermal mechanisms, the electromagnetic energy in the form of microwaves denatures the enzymes and DNA of microorganisms that are responsible for their growth and multiplication. In brief, the high temperature generated during volumetric heating causes irreversible effects on enzymes and DNA resulting in the reduction of active sites, inhibiting metabolic biochemical reactions required for the microbial survival, breaking hydrogen bonds between the double stranded DNA to single strands (Ahmed and Ramaswamy, 2004).

In addition to thermal mechanisms, several studies have also reported a nonthermal effect of microwaves on microbial load which is still not strongly proven (Kim et al., 2017) but some studies claim that the nonthermal effect showed successful decontamination resulted by direct interaction between specific molecules and electric field in a reaction medium (Pan et al., 2017). To prove the nonthermal effect, Shamis et al. (2011) used microwave frequency of 18GHz at sub lethal temperatures and repeated exposures on meat that resulted in reversible cellular fluid leakage and visual morphological changes in *E. coli* cells creating a significant inactivation of microbial load when compared with a single exposure. Therefore, further studies are needed to completely understand the mechanism of microbial inactivation with microwaves (Rifna et al., 2019; Sánchez-Maldonado et al., 2018). Moreover, rupture of cell due to electroporation and selective heating are also said to be the effects of microwave heating. The microorganisms attain high temperatures than that of the surroundings expediting the microbial inactivation during selective heating. Electroporation refers to the formation of spores on the cell wall of the microorganisms causing a fluid leakage and there by inhibiting their activity (Guo et al., 2017).

According to the reports by Institute of Food Technologists-Food and Drug Administration (Technologists, 2000), no microorganism was found to be resistant to the microwave heating. Also, the results of microwave treatment in the microwave treatment in an open atmosphere differs from the treatment in closed conditions as the steam generated in closed containers acts as an added advantage in microbial destruction. Therefore, to analyze the microbial effect of microwaves, a system with provision to vent off the steam generated should be used in determining D and z values.

7.6.3 Microwave-assisted microbial destruction kinetics

The nth order reaction kinetics generally describes the inactivation of microbes and enzymes which is given by:

$$\frac{dC}{dt} = -kC^n \tag{7.1}$$

Where C is the population or concentration of microorganisms, dC/dt is the rate of change in concentration, k is the constant of reaction rate and n represents the reaction order. Conventionally, the time of treatment at a constant temperature creating 90% of reduction in microbial load (*D*-value or decimal reduction time) and the increase in temperature required to reduce the *D* value by tenfold (z value or thermal resistance constant) are used in food processing to estimate the microbial thermal resistance.

$$D_{T_{ref}} = \frac{2.303}{k} \tag{7.2}$$

$$z = \frac{\Delta T}{\log D_1 - \log D_2} \tag{7.3}$$

With known thermal resistance of microorganisms, the time required for treatment can be calculated using following equation:

$$F = \int_0^t 10^{\frac{T(t)-T_R}{z}}\, dt \tag{7.4}$$

All the above-described relations can be applied for the conventional thermal treatment. But in case of microwaves where, the heating follows a nonisothermal pattern, the D value can be obtained by:

$$D = \frac{t_{\mathrm{eff}}}{\log\left(\frac{C_0}{C}\right)} \tag{7.5}$$

The t_{eff} is total effective time which is similar to F value, C_0 and C are initial and finals microbial loads, respectively. The microbial destruction kinetics with microwaves using temperature profiles were rarely reported in the literature and plays a major role in developing microwave assisted processing techniques in food processing and safety (Tajchakavit and Ramaswamy, 1997).

7.6.4 Factors effecting microwave sterilization in foods

There are many physical, thermal, and electrical properties effecting the efficient inhibition of microorganisms in food materials using microwave heating and are briefly discussed in the following sections.

7.6.4.1 Dielectric properties

Dielectric properties of materials being treated are the most important factors that affect the volumetric heating which is main reason behind microbial inhibition in foods upon the application of microwaves. They provide a complete idea on the interaction of food materials with applied electromagnetic field. Dielectric constant and dielectric loss factor mainly describe the dielectric properties of foods. The energy during microwave sterilization is mainly supplied in the form of electromagnetic waves and the materials ability to store the applied electromagnetic energy is described by the dielectric loss factor whereas, the measure of a materials ability in generating the heat through the stored electromagnetic energy is given by dielectric loss factor which is an imaginary component. Both dielectric constant and dielectric loss factor are related to each other by dielectric permittivity and can be mathematically expressed as:

$$\varepsilon^* = \varepsilon' - j\varepsilon'' \tag{7.6}$$

Where, ε^*, ε', and ε" represents the dielectric loss permittivity, dielectric constant and dielectric loss factor, respectively. J is a complex constant. Loss tangent is another component based on which the materials are classified. It is the ratio of loss factor to dielectric constant. Higher loss tangent refers to higher microwave absorption that facilitates efficient heating while materials with lower loss factor represent their transparency.

$$\text{Loss tangent} = \frac{\varepsilon''}{\varepsilon'} \tag{7.7}$$

Magnetic field interaction in food material can be obtained from the absolute permittivity of a vacuum (ε_0) which is determined by magnetic constant (μ_0) and the speed of light (C_0) as given as

$$C_0\varepsilon_0\mu_0 = 1 \tag{7.8}$$

The relative permittivity (ε_r) of a material can be obtained by dividing the absolute permittivity of material (ε_{abs}) by absolute permittivity of vacuum (ε_0). Some materials do not contain magnetic components and respond only to the electric field. As the heating of the materials under the influence of MW is mostly affected by the dielectric properties of materials, the portion of the reflected, absorbed, or transmitted energy of MW or RF can be written as a function of dielectric properties and is given below:

The proportion of reflected energy is given by:

$$P_r = \frac{\left(\sqrt{\varepsilon'} - 1\right)^2}{\left(\sqrt{\varepsilon'} + 1\right)^2} \tag{7.9}$$

The portion of the transmitted energy is as follows

$$P_{trans} = 1 - P_r \tag{7.10}$$

The power absorbed (P_{abs}) and converted to heat energy is directly related to dielectric loss factor (ε'') of material can be calculated by using the following equation

$$P_{abs} = 5.56 \times 10^{-4} f \varepsilon'' E^2 \tag{7.11}$$

Here E is the root mean square value of the electric field intensity expressed in Vcm^{-1} and f is the frequency of the electromagnetic spectrum in Hz. As seen from the above equation, power absorption or heat generation due to alternating electromagnetic field increases with an increase in dielectric loss of products. Ionic conduction and various polarization mechanisms (including dipole, electronic, atomic, and Maxwell–Wagner) all contribute to the dielectric loss factor. But in the MW ranges of food application, dipole rotation and ionic conduction are the main loss mechanisms (Tang and Chan, 2007).

The decontamination efficiency and microbial reduction are directly and highly controlled by the dielectric properties and these properties are also directly affected by other factors such as frequency, moisture content, temperature etc.

7.6.4.2 Frequency

As mentioned earlier, in order to prevent the interruptions of with communication signals the Federal Regulations for industrial applications of US allows only microwave frequencies of 915 MHz and 2450 MHz for scientific and industrial applications (Kurup et al., 2020). A 2450 MHz frequency having a wavelength of 0.122 m is mostly employed in domestic microwave ovens and industrial systems. Only large-scale industrial food processing systems uses 915MHz frequency whose wavelength is 0.328 m (Tang, 2015). The microwave frequency has a significant effect on food decontamination as the heat is generated instantly as a result of energy and materials interactions that happen in a frequency region. Food components such as lipids and ash are inert to frequency and the required frequency decides the size of magnetron, waveguide, etc.

According to a study by Ling and Wang (2017), the dielectric constant and dielectric loss factor found to be decreased as the frequency is elevated. A negative linear relationship is observed by plotting dielectric loss factor against frequency. Similar, nonlinear depression of dielectric properties was also observed by Zhang et al. (2019) where, the depression was higher at lower frequencies below 300 MHz and lower as frequency ranged from 300 to 3000 MHz. With an increase in frequency from 10 to 1800 MHz the dielectric loss factor reduced from 445.9 to 5.9 whereas, the values of dielectric constant fell to 15.6 from 45.5 (Wang et al., 2013).

7.6.4.3 Moisture content

The moisture content plays a very important role during microwave assisted thermal inhibition of microorganisms as the moisture is main reason for heat generation. In brief, the water is polar in nature and the heating effect by microwaves is based on polarity. Another important parameter known as penetration depth, which is the distance at which the applied power reduces by 36% is also influenced by the moisture content. This penetration depth plays a vital role in estimating the sample thickness to be followed during microwave heating for efficient microbial reduction. It is inversely related to moisture content. So, products with high moisture content will have lesser penetration depth. On the other hand, the high moisture significantly increases the dielectric constant and dielectric loss factor. For the food products with low moisture content, most of the moisture is bound with other food components such as proteins or carbohydrates for which, the ionic polarization is very less when compared with free water present in high moisture foods. "Moisture leveling phenomenon" is the beneficial effect resulted by the increased dielectric properties at higher moisture content as the parts containing more water absorbs more electromagnetic energy and provides faster heating rate resulting in uniform moisture of final end product.

In a study by Ling and Wang (2017), the dielectric constant experienced an elevation from 2.6 to 13.5 with an increase in moisture content from 3% to 7% when treated with microwaves at a frequency of 2450 MHz. Similar increase was also observed by Zhang et al. (2019) where the dielectric constant and dielectric loss factor during microwave treatment at 915 MHz increased from 5.2 and 0.8 to 13.8 and 4.5 with an increase in moisture content from 3% to 24%. In low moisture foods, lipids and carbohydrates involves in heat generation but are less efficient than water molecules (Sánchez-Maldonado et al., 2018).

7.6.4.4 Temperature

Both temperature and moisture content are directly related to each other in dielectric heating assisted sterilization or pasteurization of foods. The heat resistance of microorganisms varies from each other. Salmonella enterica is more resistant to heat and can survive at higher temperatures. Therefore, it should be assured that the temperature during microwave treatment has reached a value at which, the target microorganism is inhibited. However, as the temperature increases, more water is evaporated from the product being treated which reduces the dielectric properties and effects the heating rate. Hence, the microwave power level during microbial inhibition should be adjusted according to the target temperature based on the initial product temperature considering the moisture evaporation as well.

7.6.4.5 Mass

The effect of product's mass on microwave heating which is reasonable for efficient decontamination is quite intriguing. It is directly related to the microwave power absorption. This is where the term microwave power density comes into existence which is the ratio of microwave power applied to mass of the product being treated. Higher microwave power density results in faster heating and greater power absorption. For example, microwave treatment of 500 g of an agricultural commodity at 900 W for 10 seconds will provide an efficient microbial decontamination than treating 1 kg at same conditions. The higher microwave power density while treatment at 500 g is responsible for this greater effect. However, there exists a critical mass required by a magnetron depends on its capacity for efficient heating. A minimum of 250 mL of water load is required for 1 kW magnetron to facilitate good microwave power absorption and safe usage of magnetron. The dielectric properties of materials are also directly related to density.

A study by Chen et al. (2019) revealed that the bulk density and mass of product has significant effect on the dielectric heating efficiency. This study included the inactivation of salmonella enterica in three different batches of cumin seeds. The batch with lower bulk density had attained higher temperature very quick and the reduction of microbial load was under detectable levels. A study by Prudhvi et al. (2021) also reported that the microwave heating at higher power density was faster than that at the lower power density where, the treatment with lower product mass load had facilitated an efficient heating.

7.6.4.6 Geometry of product

The shape and product's geometry are also some of the important factors influencing the heating properties during dielectric heating. Nonuniform heating occurs in the products with irregular shape as they have irregular distribution of thickness. Higher heating uniformity can be observed in products that are uniformly distributed. In addition, heating in square- or rectangular-shaped products (products with corners) possess lower heating uniformity than circular or sphere-shaped objects. Centre heating in more dominant in products with no corners. Ryynänen and Ohlsson (1996) compared the microwave heating behavior of different shaped products and reported that the geometry of products being heated influenced heating uniformity. They also observed difference in temperature distribution between corners and center of product and was being varied with geometry.

7.6.4.7 Thermal properties

Apart from the above mentioned factors, the temperature profile developed during electromagnetic radiation exposure, also depends on thermo-physical properties like thermal conductivity, thermal diffusivity and specific heat of the product (Venkatesh and Raghavan, 2004). Thermal conductivity is the property of material to conduct heat or thermal energy. It varies according to the nature of product. Thermal diffusivity determines the dissipation of heat inside the material. At higher thermal diffusivity, heat transfer occurs rapidly. For example, in frozen foods, due to the high thermal conductivity of ice, the heating was efficient. It was found that increase in temperature can lead to decrease in thermal diffusivity and thermal diffusivity attains its minimum at the $\alpha-\beta$ transition (Zhang, 2016). Higher thermal conductivity and thermal diffusivity is required parameter in dielectric heating. In a dielectric material having no free electrical charges, phonon acts as carrier for thermal energy. Phonons are the wave packets generated due to the interference of displacement waves because of collective vibrations of atoms in crystal lattice. When the waves pass through a region of different refractive index, subsequent displacement waves carrying the energy of thermal motion will be scattered. As the temperature increases, scattering become more in the lattice and in many cases, this might lead to decrease in thermal conductivity at fairly high temperatures (Dante, 2015). The foods heat capacity is the influence of input heat on resultant temperature. The thermal properties of food are defined by combination of thermal conductivity k, thermal diffusivity $\grave{\alpha}$ and heat capacity c_p. Mathematically the relation can be given as:

$$'a = \frac{k}{\rho' c_p} \tag{7.12}$$

7.7 Microbial inactivation in foods using microwaves

The application of microwaves in inhibiting microorganisms has been studied by many researchers and proved a significant microbial reduction. Foods of several categories such as diary, poultry, meat, vegetables, spices, and low moisture foods have been treated under microwaves for microbial inhibition. Compared to conventional methods, the microwave treatment is rapid and hence achieves the required log reduction of bacterial and fungal count in less time thereby making the process economical and industrially feasible. The heat resistance varies from different microorganisms and therefore, different microwave pasteurization and sterilization conditions of microwave processing should be followed for efficient microbial inhibition.

For dairy products, pasteurization using heat exchangers is generally followed in industries. The replacement of this traditional method with novel microwave pasteurization of milk has been widely practiced and its effects were mainly studied on enhancement of shelf life and microbial load. Microwave assisted heat exchanger with continuous flow systems at a frequency of 2450 MHz has been developed by many researchers and studied its pasteurization effects. Moreover, domestic microwave ovens have also been used for milk pasteurization by Kudra et al. (1991) where, the inactivation levels of microbes and sensory properties in comparison with conventional treatments were reported to be similar however, the treatment times and protein denaturation were found to be minimum in microwave assisted processing. Different pathogens such as *S. faecalis*,

Lysteria monocytogens, C. jejuni, and Y. enteroclitica were successfully inactivated in milk by Choi et al. (1993a,b) using microwave pasteurization technology.

Although fruit juices are less vulnerable for microbial growth due to their pH, there are some specific microbes of concern such as *Salmonella enterica, E. Coli O157:H7,* and *Crytosporidium parvum*. Like milk, conventional plate hate exchangers are used for fruit juice high temperature short time pasteurization resulting in fouling. There also exist some enzymes that are responsible for spoilage and are more resistant to heat than microorganisms. Introduction of microwave pasteurization systems for such fruit juices inhibits both microbes and spoilage causing enzymes. According to a study by Nikdel et al. (1993), the microbial load and spoil causing enzymes inactivation in orange juice was found to be enhanced by using microwave technology when compared with conventional plate heat exchanger system. Canumir et al. (2002) and Tajchakavit et al. (1998) reported the enhanced inactivation of *saccharomyces cerevisiae, Lactobacillus plantarum* and *E. coli* in apple juice. The Z values of inactivation of *E. coli* were *Saccharomyces cerevisiae*, and *Lactobacillus plantarum* were 8.4 and 10.9°C lesser than that in a conventional plate heat exchanger system.

Microwaves were also said to have more efficient in egg pasteurization in very less time. But the low microwave power was recommended for egg pasteurization to avoid egg bursting at high power levels. Egg shells were found to possess an excellent transparency to microwaves and when the egg yolk reaches 61°C the whole eggs were reported to get pasteurized (Dev et al., 2008). 10^7 cells/100 μL of Salmonella Typhimurium inoculated on whole egg were reduced to 0.08 CFU/mL under microwave treatment for 15s (Shenga et al., 2010). Application on meat products such as salmon resulted in reduction of *E. coli, S. enteritidis,* and *Enterococcus spp.* to undetectable levels with microwave does of 430 kJ/g (Zeinali et al., 2015). Microwave treatment for 60 seconds after the temperature reached a value greater than 74°C on chicken meat resulted in complete inhibition of *Listeria monocytogenes*. Similar results were also reported for *E. coli* o15:H7 on beef (Jamshidi, Seifi, & Kooshan, 2010).

The increase in treatment time increased the reduction in microbial load of carrot where, 1.70, 1.50, and 2.1 log reductions of total aerobic count, yeast, and molds were observed, respectively (Yaghmaee and Durance, 2007). In a study, the increase in microwave power level expedited the inactivation rates of Salmonella Enteritidis in potatoes. Similarly, microwave treatment at 700 and 350 W for 50 seconds resulted in 2.07 and 1.70 log reductions of Salmonella enterica (Valero et al., 2014). The reduction in D values for L. monocytogenes inhibition in kiwi fruit was reported with an increase in microwave power level (Lu et al., 2011). The application of microwaves in decontamination of low moisture foods like spices has also shown promising results. For example, microwave treatment when applied on black pepper for 12.5 minutes at 663 W microwave power with 40 W/g intensity resulted the desired microbial load reduction as specified by International Commission on Microbiological Specifications for Foods (ICMSF) and the European Spice Association (Jamshidi et al.) that is, 4.02 and 4.11 log reduction of total yeast and mold (TYM), total mesophilic aerobic bacteria (TMAB), respectively and salmonella, shigella, *E. coli*, total coliforms (TC) to undetectable levels with no sample deterioration (Jeevitha et al., 2016).

7.7.1 Effects of Microwave sterilization on final product quality

The microwave treatment may have significant effect on the quality characteristics of end products such as nutritional profile, degradation of bioactive compounds, activity of enzymes, antioxidant properties, changes in color and texture. Despite of the high temperatures attained by the product under microwaves, many studies have proved that the effect on food quality can be reduced to minimum due to the rapid microbial inactivation mechanism. For example, the loss of anthocyanin content in aronia was ranged between 66.1% and 99.8% during conventional treatment at 100°C whereas, the application of microwaves reduced it to 39.7% (Piasek et al., 2011). The microwave treatment of strawberry puree also resulted in comparatively lower degradation of polyphenols, Vitamin C, and anthocyanins than conventional treatment for 15 minutes at 90°C where, the microwave treatment was carried out at same temperature for 10 seconds (Marszałek et al., 2015). Degradation of lycopene and ascorbic acid content of grapes were also found to fall below 13.52% and 6.83% under microwave treatment (Lu et al., 2011) proving that application of microwaves is the best alternative to conventional methods in lowering the degradation of antioxidant and bioactive compounds activity.

There are also several enzymes in addition to the microorganisms that are responsible for the spoilage in foods. The microwave treatment for sterilization also found to facilitate the reduction of spoilage causing enzymes in foods along microorganisms. Upon the application of microwave treatment, the lipase activity was reduced by 60% and 100% at 45 and 60°C in wheat germs (Chen et al., 2017). Microwave sterilization of oil palm fruit additionally reduced the lipolysis enzymatic activity (Umudee et al., 2013). Peroxidase and polyphenol oxidase are also the most common enzymes in food products that are responsible for food spoilage. Sterilization of strawberry puree with microwaves by observed the 98% and 100% degradation of polyphenol oxidase and peroxidase enzymes, respectively at 120°C (Marszałek et al., 2015).

The effects on color and texture due to the application of microwaves for sterilization were also proved to be nonsignificant. Several studies on grape (Lu et al., 2011), jalapeno pepper and coriander reported the negligible changes in product firmness (De La Vega-Miranda et al., 2012). The short-time application is the main reason behind the negligible or insignificant effects on texture where, longer exposure of microwaves might cause major changes in texture. On the other hand, the color changes due to application of microwave sterilization depend on the type of product being used. For example, application on red colored vegetables shows lower changes as the red color is the function of lycopene which remains unaffected by microwaves However, for green vegetables the change in color especially in L value representing lightness may be observed. Retention of original structure in microwave sterilized macaroni and cheese was noticed when comported with retorted product. The color and texture retention of microwave treated peas was also noticed in comparison with canned product.

7.8 Synergistic effects ultrasound and microwaves on microorganisms

Kernou et al. (2021) investigated the impact of combined treatment of microwave and ultrasound on *E. coli* in orange juice. Results obtained shows that there is little contribution of ultrasound alone in killing *E. coli* as only 1.3 log reductions are obtained even after 1 hour of sonication but at the same time if ultrasound is coupled with microwaves and then juice is treated then improved efficiency is observed. 8.0 log reductions are obtained when treatment of microwave (900 W/30 s) is given with ultrasound for 20 minutes. The author concludes with greater applications of the joint treatment as a HTST process where safety is obtained along with quality.

A research was conducted by Wang et al. (2010) to inactivate vegetative cells of different *Alicyclobacillus strains* by giving a combination of MW-UV treatment. Different treatment combinations were tried, and an optimized and effective treatment was sought to obtain. Treatment with microwave power ranging from 300 to 900 W and sonication of 400W for period b/w 1 minute to 30 minutes is given. The optimized parameter was found at ultrasound of 400 W for 20 minutes and microwave for 10 minutes at 900 W. Overall study concludes lauding with efficacy of combined treatment and it potential applications in beverages industry.

A similar study on phalsa drink was carried out by Nadeem et al. (2021), who looked at the cumulative effects of MW-US treatment on phalsa drink. High quality juice with greater phytochemical and physiochemical stability was obtained. Improved properties of juice with greater shelf life and storage without chemical preservation show the potential benefits of combined treatment.

Another study was conducted on blackberry juice by Pérez-Grijalva et al. (2018) where US and MW impact on bioactive and antioxidant activity was observed. It was concluded that superior quality of juice can be obtained when treatment of MW and US in combination is applied.

7.9 Effect on quality parameters of food

7.9.1 Color

Color is one of the most perceived and sought quality of foods that determines its consumer acceptability. Differences in perception can be determined analytically based on color difference values, that is, very distinct, distinct and minor difference based on values greater than 3, between 1.5 and 3, and less than 1.5 (Tiwari et al., 2008).

Ultrasound processing generally enhances the color as its exposes different pigments such as carotenoids, betalins, anthocyanins, etc. Cruz et al. (2007) reported enhancement in samples of watercress. A reduction of 5% anthocyanin was found at higher amplitude levels and treatment periods. The phenolic compounds identified were increased in red wine after being exposed to weak ultrasonic irradiation (Masuzawa et al., 2000).

Microwave processing though impact the color of food slightly but overall the loss of color is less than conventional thermal processing. Yongsawatdigul and Gunasekaran (1996) reported that microwave dried cranberries are superior in color.

Combined treatment of US and Microwave also showed an increase in pigments responsible for color, that is, anthocyanins. Nadeem et al. (2021) reported an increase in anthocyanins. Similar results were noted by Sharma (2012) for jamun.

7.9.2 Bioactive and phenolic compounds

Treatment with US and MW leads to transformation of old compounds to new ones. Šimunek et al. (2013) noted a change in aroma profile when high power sonication is applied to apple juice. Bhat et al. (2011) researched about sonicated - kasturi lime and found that bioactive compounds got enhanced upon treatment. A similar result was obtained by Zafra-Rojas et al. (2013) where they noticed an increase in phenolic compounds and bioactive compounds. Kammoun Bejar et al. (2011) found increment in phenolic compounds for both orange peel and leaves after treating with microwaves. A positive impact on DPPH (antioxidant activity) when treated with MW suggests that as the amount of time passed, the oil's DPPH free radical scavenging ability improved dramatically (Karrar et al., 2020). Two times increment in phenolic content was obtained when sonication was performed for mulberry juice (Nguyen and Nguyen, 2018).

7.9.3 Nutritional properties

The total protein content of microwave dried wheat product resulted in insignificant changes in the protein content. Research carried by Velu et al. (2006) on maize on maize showed an insignificant reduction in protein. Degradation of ascorbic acid or Vitamin C is one of the leading concerns in food industry. Khraisheh et al. (2004) reported 75% retention in ascorbic acid content for microwave treated samples of potatoes. Karatas and Kamışlı (2007) evaluated variations of Vit. A, C, E of apricot samples treated with different techniques and concluded that microwave treatment retained maximum vitamins. Ultrasound treatment shows minimal effect on ascorbic acid content. Tiwari et al. (2008) reported 5% maximum degradation of ascorbic acid in orange juice. Combined effects of ultrasound and microwave on ascorbic acid were studied by Nadeem et al. (2021) over different combinations with variation in processing time. Increment in ascorbic acid content was noticed till 60 minutes after which eventually degradation was noticed which might be due to formation of hydroxyl radicals (Pétrier et al., 2007).

7.10 Conclusions

Food contamination with microorganisms has become a major concern there is a strong need for an alternative technology or method to ensure the complete inhibition of microorganisms. The existing conventional methods have some serious effects on the end product's quality. Application of microwaves and ultrasound technologies in food decontamination has been practiced from several years. The efficient microbial inhibition of both the technologies with minimal damage to the final product quality has brought some light on the synergism of two technologies for enhanced microbial inhibition. Although the synergism has been proved to be efficient, there exist some limitations associated with each technology individually.

The heating nonuniformity in the microwaves is a major limitation in microwave assisted sterilization process. The accurate temperature profile during microwave heating is not possible and the temperature measurement at fewer predetermined locations doesn't assure the accuracy. The commercialization of the microwave assisted microbial inhibition systems at an industrial level is also limited. Whereas for ultrasound, it is limited to liquid foods and its process non uniformity restricts its further popularizations. Commercialization of the technology requires extensive study over microbial and enzyme inactivation kinetics. Ultrasound itself is not sufficient to obtain FDA standard of microbial inactivation and thus, more research works must be carried to address all the mentioned limitations and synergistic effects of hurdle technologies and its additives needs to be studied extensively.

References

Aadil, R.M., Zeng, X.A., Han, Z., Sahar, A., Khalil, A.A., Rahman, U.U., ... Mehmood, T., 2018. Combined effects of pulsed electric field and ultrasound on bioactive compounds and microbial quality of grapefruit juice. J. Food Process. Preserv. 42 (2), e13507.

Abid, M., Jabbar, S., Hu, B., Hashim, M.M., Wu, T., Wu, Z., ... Zeng, X., 2014. Synergistic impact of sonication and high hydrostatic pressure on microbial and enzymatic inactivation of apple juice. LWT-Food Sci. Technol. 59 (1), 70–76.

Adekunte, A., Tiwari, B., Cullen, P., Scannell, A., O'donnell, C., 2010. Effect of sonication on colour, ascorbic acid and yeast inactivation in tomato juice. Food Chem. 122 (3), 500–507.

Ahmed, J., Ramaswamy, H.S., 2004. Microwave pasteurization and sterilization of foods. Food Sci. Technol.-N. Y.-Marcel Dekker- 167, 691.

Al-Hilphy, A.R., Al-Temimi, A.B., Al Rubaiy, H.H.M., Anand, U., Delgado-Pando, G., Lakhssassi, N., 2020. Ultrasound applications in poultry meat processing: a systematic review. J. Food Sci. 85 (5), 1386–1396.

Alvarez, I., Mañas, P., Sala, F., Condón, S., 2003. Inactivation of Salmonella enterica serovar Enteritidis by ultrasonic waves under pressure at different water activities. Appl. Environ. Microbiol. 69 (1), 668–672.

Arroyo, C., Cebrián, G., Pagán, R., Condón, S., 2012. Synergistic combination of heat and ultrasonic waves under pressure for Cronobacter sakazakii inactivation in apple juice. Food Control, 25(1), 342–348.

Baumann, A.R., Martin, S.E., Feng, H., 2005. Power ultrasound treatment of Listeria monocytogenes in apple cider. J. Food Prot. 68 (11), 2333–2340.

Bevilacqua, A., Speranza, B., Campaniello, D., Sinigaglia, M., Corbo, M.R., 2014. Inactivation of spoiling yeasts of fruit juices by pulsed ultrasound. Food Bioprocess Technol. 7 (8), 2189–2197.

Bhat, R., Kamaruddin, N.S.B.C., Min-Tze, L., Karim, A., 2011. Sonication improves kasturi lime (Citrus microcarpa) juice quality. Ultrason. Sonochem. 18 (6), 1295–1300.

Bintsis, T., 2017. Foodborne pathogens. AIMS Microbiol. 3 (3), 529.

Canumir, J.A., Celis, J.E., de Bruijn, J., Vidal, L.V., 2002. Pasteurisation of apple juice by using microwaves. LWT-Food Sci. Technol. 35 (5), 389–392.

Cárcel, J., García-Pérez, J.V., Benedito, J., Mulet, A., 2012. Food process innovation through new technologies: use of ultrasound. J. Food Eng. 110 (2), 200–207.

Chandrasekaran, S., Ramanathan, S., Basak, T., 2013. Microwave food processing—a review. Food Res. Int. 52 (1), 243–261.

Chemat, F., Khan, M.K., 2011. Applications of ultrasound in food technology: processing, preservation and extraction. Ultrason. Sonochem. 18 (4), 813–835.

Chen, L., Wei, X., Irmak, S., Chaves, B.D., Subbiah, J., 2019. Inactivation of Salmonella enterica and Enterococcus faecium NRRL B-2354 in cumin seeds by radiofrequency heating. Food Control 103, 59–69.

Chen, Z., Li, Y., Wang, L., Liu, S., Wang, K., Sun, J., Xu, B., 2017. Evaluation of the possible non-thermal effect of microwave radiation on the inactivation of wheat germ lipase. J. Food Process Eng. 40 (4), e12506.

Choi, H., Marth, E., Vasavada, P., 1993a. Use of microwave energy to inactivate Yersinia enterocolitica and Campylobacter jejuni in milk. Milchwissenschaft 48 (3), 134–136.

Choi, K., Marth, E., Vasavada, P., 1993b. Use of microwave energy to inactivate Listeria monocytogenes in milk. Milchwissenschaft 48 (4), 200–203.

Cruz, R.M., Vieira, M.C., Silva, C.L., 2007. Modelling kinetics of watercress (Nasturtium officinale) colour changes due to heat and thermosonication treatments. Innovative Food Sci. Emerg. Technol. 8 (2), 244–252.

Dante, R.C., 2015. Handbook of Friction Materials and Their Applications. Woodhead Publishing, Sawston, United Kingdom.

De La Vega-Miranda, B., Santiesteban-López, N., López-Malo, A., Sosa-Morales, M., 2012. Inactivation of Salmonella Typhimurium in fresh vegetables using water-assisted microwave heating. Food Control 26 (1), 19–22.

Dev, S., Birla, S., Raghavan, G., Subbiah, J., 2012. Microbial Decontamination of Food by Microwave (MW) and Radio Frequency (RF) *Microbial Decontamination in the Food Industry*. Elsevier, Amsterdam, Netherlands, pp. 274–299.

Dev, S., Raghavan, G., Gariepy, Y., 2008. Dielectric properties of egg components and microwave heating for in-shell pasteurization of eggs. J. Food Eng. 86 (2), 207–214.

Dolatowski, Z.J., Stadnik, J., Stasiak, D., 2007. Applications of ultrasound in food technology. Acta Scientiarum Polonorum Technologia Alimentaria 6 (3), 88–99.

Feng, H., Barbosa-Cánovas, G.V., Weiss, J., 2011. Ultrasound Technologies for Food and Bioprocessing, 1. Springer, Berlin, Germany.

Ferrario, M., Alzamora, S.M., Guerrero, S., 2015. Study of the inactivation of spoilage microorganisms in apple juice by pulsed light and ultrasound. Food Microbiol. 46, 635–642.

Gomez-Gomez, A., Brito-de la Fuente, E., Gallegos, C., Garcia-Perez, J., Benedito, J., 2021. Combined pulsed electric field and high-power ultrasound treatments for microbial inactivation in oil-in-water emulsions. Food Control 130, 108348.

Guo, Q., Sun, D.-W., Cheng, J.-H., Han, Z., 2017. Microwave processing techniques and their recent applications in the food industry. Trends Food Sci. Technol. 67, 236–247.

Guerrero, S., López-Malo, A., Alzamora, S., 2001. Effect of ultrasound on the survival of Saccharomyces cerevisiae: influence of temperature, pH and amplitude. Innov. Food Sci. Emerg. Technol. 2 (1), 31–39.

Herceg, Z., Režek Jambrak, A., Lelas, V., Mededovic Thagard, S., 2012. The effect of high intensity ultrasound treatment on the amount of Staphylococcus aureus and Escherichia coli in milk. Food Technol. Biotechnol. 50 (1), 46–52.

Huang, E., Mittal, G., Griffiths, M., 2006. Inactivation of Salmonella enteritidis in liquid whole egg using combination treatments of pulsed electric field, high pressure and ultrasound. Biosystems Eng. 94 (3), 403–413.

Jamshidi, A., Seifi, H.A., Kooshan, M., 2010. The effect of short-time microwave exposures on Escherichia coli O157: H7 inoculated onto beef slices. Afr. J. Microbiol. Res. 4 (22), 2371–2374.

Jeevitha, G.C., Sowbhagya, H.B., Hebbar, H.U., 2016. Application of microwaves for microbial load reduction in black pepper (Piper nigrum L.). J. Sci. Food Agric. 96 (12), 4243–4249.

Kammoun Bejar, A., Kechaou, N., Boudhrioua Mihoubi, N., 2011. Effect of microwave treatment on physical and functional properties of orange (Citrus sinensis) peel and leaves. J. Food Process. Technol. 2 (2), 109.

Karatas, F., Kamışlı, F., 2007. Variations of vitamins (A, C and E) and MDA in apricots dried in IR and microwave. J. Food Eng. 78 (2), 662–668.

Karrar, E., Sheth, S., Wei, W., Wang, X., 2020. Effect of microwave heating on lipid composition, oxidative stability, color value, chemical properties, and antioxidant activity of gurum (Citrulluslanatus var. Colocynthoide) seed oil. Biocatal. Agric. Biotechnol. 23, 101504.

Kernou, O.N., Belbahi, A., Amir, A., Bedjaoui, K., Kerdouche, K., Dairi, S., ... Madani, K., 2021. Effect of sonication on microwave inactivation of Escherichia coli in an orange juice beverage. J. Food Process Eng. 44 (5), e13664.

Khraisheh, M., McMinn, W., Magee, T., 2004. Quality and structural changes in starchy foods during microwave and convective drying. Food Res. Int. 37 (5), 497–503.

Kim, J.E., Choi, H.-S., Lee, D.-U., Min, S.C., 2017. Effects of processing parameters on the inactivation of Bacillus cereus spores on red pepper (Capsicum annum L.) flakes by microwave-combined cold plasma treatment. Int. J. Food Microbiol. 263, 61–66.

Knorr, D., Zenker, M., Heinz, V., Lee, D.-U., 2004. Applications and potential of ultrasonics in food processing. Trends Food Sci. Technol. 15 (5), 261–266.

Kozempel, M.F., Annous, B.A., Cook, R.D., Scullen, O., Whiting, R.C., 1998. Inactivation of microorganisms with microwaves at reduced temperatures. J. Food Prot. 61 (5), 582–585.

Kudra, T., Van, d. V., F.R., Raghavan, G., Ramaswamy, H., 1991. Heating characteristics of milk constituents in a microwave pasteurization system. J. Food Sci. 56 (4), 931–934.

Kurup, A.H., Deotale, S., Rawson, A., Patras, A., 2020. Thermal Processing of Herbs and Spices. Herbs, Spices and Medicinal Plants: Processing, Health Benefits and Safety. WILEY, Newjersey, pp. 1–21.

Ling, B., Wang, S., 2017. Dielectric properties of pistachio kernels as influenced by frequency, temperature, moisture and salt content. Paper presented at the 2017 ASABE Annual International Meeting.

Lopez-Malo, A., Palou, E., Jiménez-Fernández, M., Alzamora, S. M., Guerrero, S., 2005. Multifactorial fungal inactivation combining thermosonication and antimicrobials. J. Food Eng., 67 (1–2), 87–93.

Lu, Y., Turley, A., Dong, X., Wu, C., 2011. Reduction of Salmonella enterica on grape tomatoes using microwave heating. Int. J. Food Microbiol. 145 (1), 349–352.

Marszałek, K., Mitek, M., Skąpska, S., 2015. Effect of continuous flow microwave and conventional heating on the bioactive compounds, colour, enzymes activity, microbial and sensory quality of strawberry purée. Food Bioprocess Technol. 8 (9), 1864–1876.

Masuzawa, N., Ohdaira, E., Ide, M., 2000. Effects of ultrasonic irradiation on phenolic compounds in wine. Japan. J. Appl. Phys. 39 (5S), 2978.

Muñoz, A., Caminiti, I.M., Palgan, I., Pataro, G., Noci, F., Morgan, D.J., . . . Lyng, J.G., 2012. Effects on Escherichia coli inactivation and quality attributes in apple juice treated by combinations of pulsed light and thermosonication. Food Res. Int. 45 (1), 299–305.

Muthukumaran, S., Kentish, S.E., Stevens, G.W., Ashokkumar, M., 2006. Application of ultrasound in membrane separation processes: a review. Rev. Chem. Eng. 22 (3), 155–194.

Nadeem, M., Ghaffar, A., Hashim, M.M., Murtaza, M.A., Ranjha, M.M.A.N., Mehmood, A., Riaz, M.N., 2021. Sonication and microwave processing of phalsa drink: a synergistic approach. Int. J. Fruit Sci. 21 (1), 993–1007.

Nguyen, C.L., Nguyen, H.V., 2018. The quality of mulberry juice as affected by enzyme treatments. Beverages 4 (2), 41.

Nikdel, S., Chen, C.S., Parish, M.E., MacKellar, D.G., Friedrich, L.M., 1993. Pasteurization of citrus juice with microwave energy in a continuous-flow unit. J. Agric. Food Chem. 41 (11), 2116–2119.

Palgan, I., Caminiti, I., Muñoz, A., Noci, F., Whyte, P., Morgan, D., ... Lyng, J., 2011. Effectiveness of high intensity light pulses (HILP) treatments for the control of Escherichia coli and Listeria innocua in apple juice, orange juice and milk. Food Microbiol. 28 (1), 14–20.

Pan, Y., Sun, D.-W., Han, Z., 2017. Applications of electromagnetic fields for nonthermal inactivation of microorganisms in foods: an overview. Trends Food Sci. Technol. 64, 13–22.

Pérez-Grijalva, B., Herrera-Sotero, M., Mora-Escobedo, R., Zebadúa-García, J.C., Silva-Hernández, E., Oliart-Ros, R., ... Guzmán-Gerónimo, R., 2018. Effect of microwaves and ultrasound on bioactive compounds and microbiological quality of blackberry juice. LWT 87, 47–53.

Pétrier, C., Combet, E., Mason, T., 2007. Oxygen-induced concurrent ultrasonic degradation of volatile and non-volatile aromatic compounds. Ultrason. Sonochem. 14 (2), 117–121.

Piasek, A., Kusznierewicz, B., Grzybowska, I., Malinowska-Pańczyk, E., Piekarska, A., Azqueta, A., ... Bartoszek, A., 2011. The influence of sterilization with EnbioJet® Microwave Flow Pasteurizer on composition and bioactivity of aronia and blue-berried honeysuckle juices. J. Food Compos. Anal. 24 (6), 880–888.

Prudhvi, P., Deepika, S., Sutar, P., 2021. Modeling moisture and solids transfer kinetics during a novel microwave assisted water absorption-desorption process of dry red gram (Cajanus cajan L.) splits. J. Food Eng., 110891.

Reineke, K., Mathys, A., 2020. Endospore inactivation by emerging technologies: a review of target structures and inactivation mechanisms. Annu. Rev. Food Sci. Technol. 11, 255–274.

Rifna, E., Singh, S.K., Chakraborty, S., Dwivedi, M., 2019. Effect of thermal and non-thermal techniques for microbial safety in food powder: Recent advances. Food Res. Int. 126, 108654.

Ross, A.I., Griffiths, M.W., Mittal, G.S., Deeth, H.C., 2003. Combining nonthermal technologies to control foodborne microorganisms. Int. J. Food Microbiol. 89 (2–3), 125–138.

Ryynänen, S., Ohlsson, T., 1996. Microwave heating uniformity of ready meals as affected by placement, composition, and geometry. J. Food Sci. 61 (3), 620–624.

Sánchez-Maldonado, A.F., Lee, A., Farber, J.M., 2018. Methods for the control of foodborne pathogens in low-moisture foods. Annu. Rev. Food Sci. Technol. 9, 177–208.

Shamis, Y., Taube, A., Mitik-Dineva, N., Croft, R., Crawford, R.J., Ivanova, E.P., 2011. Specific electromagnetic effects of microwave radiation on Escherichia coli. Appl. Environ. Microbiol. 77 (9), 3017–3022.

Sharma, D. (2012). Processing of Jamun (Syzygium cumini L.) into Value-added Products. Ph. D. thesis. Division of Food Science Technology, Sher-e-Kashmir University of Agricultural Sciences and Technology, Jammu, India.

Shenga, E., Singh, R., Yadav, A., 2010. Effect of pasteurization of shell egg on its quality characteristics under ambient storage. J. Food Sci. Technol. 47 (4), 420–425.

Šimunek, M., Režek Jambrak, A., Petrović, M., Juretić, H., Major, N., Herceg, Z., ... Vukušić, T., 2013. Aroma profile and sensory properties of ultrasound-treated apple juice and nectar. Food Technol. Biotechnol. 51 (1), 101–111.

Tajchakavit, S., Ramaswamy, H., 1997. Continuous-flow microwave inactivation kinetics of pectin methyl esterase in orange juice. J. Food Process. Preserv. 21 (5), 365–378.

Tajchakavit, S., Ramaswamy, H., Fustier, P., 1998. Enhanced destruction of spoilage microorganisms in apple juice during continuous flow microwave heating. Food Res. Int. 31 (10), 713–722.

Tang, J., 2015. Unlocking potentials of microwaves for food safety and quality. J. Food Sci. 80 (8), E1776–E1793.

Tang, J., Chan, T., 2007. Microwave and radio frequency in sterilization and pasteurization applications. In: Heat Transfer in Food Processing. WIT Press, Southampton, pp. 101–142.

Technologists, I.o.F., 2000. Kinetics of Microbial Inactivation for Alternative Food Processing Technologies. Institute of Food Technologists, Chicago, United States.

Tiwari, B., Muthukumarappan, K., O'donnell, C., Cullen, P., 2008. Effects of sonication on the kinetics of orange juice quality parameters. J. Agric. Food Chem. 56 (7), 2423–2428.

Umudee, I., Chongcheawchamnan, M., Kiatweerasakul, M., Tongurai, C., 2013. Sterilization of oil palm fresh fruit using microwave technique. Int. J. Chem. Eng. Appl. 4 (3), 111.

Valero, A., Cejudo, M., García-Gimeno, R., 2014. Inactivation kinetics for Salmonella Enteritidis in potato omelet using microwave heating treatments. Food Control 43, 175–182.

Velu, V., Nagender, A., Rao, P.P., Rao, D., 2006. Dry milling characteristics of microwave dried maize grains (Zea mays L.). J. Food Eng. 74 (1), 30–36.

Venkatesh, M., Raghavan, G., 2004. An overview of microwave processing and dielectric properties of agri-food materials. Biosystems Eng. 88 (1), 1–18.

Vieira, K.C.d.O., Silva, H.R.A.d., Rocha, I.P.M., Barboza, E., Eller, L.K.W., 2021. Foodborne pathogens in the omics era. Crit. Rev. Food Sci. Nutr. 62 (24), 1–16.

Wang, J., Hu, X., Wang, Z., 2010. Kinetics models for the inactivation of Alicyclobacillus acidiphilus DSM14558T and Alicyclobacillus acidoterrestris DSM 3922T in apple juice by ultrasound. Int. J. Food Microbiol. 139 (3), 177–181.

Wang, J., Zhao, G., Liao, X., Hu, X., 2010. Effects of microwave and ultrasonic wave treatment on inactivation of Alicyclobacillus. Int. J. Food Sci. Technol. 45 (3), 459–465.

Wang, Y., Zhang, L., Gao, M., Tang, J., Wang, S., 2013. Temperature-and moisture-dependent dielectric properties of macadamia nut kernels. Food Bioprocess Technol. 6 (8), 2165–2176.

Yaghmaee, P., Durance, T., 2007. Efficacy of vacuum microwave drying in microbial decontamination of dried vegetables. Drying Technol. 25 (6), 1099–1104.

Yeni, F., Yavaş, S., Alpas, H., Soyer, Y., 2016. Most common foodborne pathogens and mycotoxins on fresh produce: a review of recent outbreaks. Crit. Rev. Food Sci. Nutr. 56 (9), 1532–1544.

Yongsawatdigul, J., Gunasekaran, S., 1996. Microwave-vacuum drying of cranberries: Part I. Energy use and efficiency. J. Food Process. Preserv. 20 (2), 121–143.

Zafra-Rojas, Q.Y., Cruz-Cansino, N., Ramírez-Moreno, E., Delgado-Olivares, L., Villanueva-Sánchez, J., Alanís-García, E., 2013. Effects of ultrasound treatment in purple cactus pear (Opuntia ficus-indica) juice. Ultrason. Sonochem. 20 (5), 1283–1288.

Zeinali, T., Jamshidi, A., Khanzadi, S., & Azizzadeh, M. (2015). *The effect of short-time microwave exposures on Listeria monocytogenes inoculated onto chicken meat portions.*Paper presented at the Veterinary Research Forum.

Zhang, J., Li, M., Cheng, J., Wang, J., Ding, Z., Yuan, X., ... Liu, X., 2019. Effects of moisture, temperature, and salt content on the dielectric properties of Pecan kernels during microwave and radio frequency drying processes. Foods 8 (9), 385.

Zhang, Z.-X., 2016. Rock Fracture and Blasting: Theory and Applications. Butterworth-Heinemann, Amsterdam, Netherlands.

CHAPTER EIGHT

Synergistic effect of sonication and microwave for deactivation of enzymes

Janifer Raj Xavier[a] and Mohammad Javed Ansari[b]
[a]Defence Food Research Laboratory, Defence Research and Development Organization, Mysore, Karnataka, India
[b]Department of Botany, Hindu College Moradabad (Mahatma Jyotiba Phule Rohilkhand University Bareilly), India

8.1 Introduction

Food is the important source of nutrients and provides thermodynamic energy to living beings on earth. Stabilization of foods from perishable to nonperishable forms is made possible due to advancements in food processing technologies. Moreover, postharvest losses are reduced as well as convenience and choice of foods are being enabled to customers by various kinds of food processing methods. Enzymes are naturally occurring substances capable of increasing the rate of specific biochemical reactions in living systems without being used up during the process. Their specificity, efficacy, safety, suitability and catalytic capability are attributed to important applications in various industrial processes including food industries. Endogenous enzymes play a significant role on flavor, texture, freshness and nutritional quality of foods. Milk, egg and wheat contain about 20, 19, and 55 enzymes, respectively. Inactivation of these enzymes by thermal and nonthermal methods is essential for improving storage stability of processed foods. Recently interest for nonthermal methods and combination treatments involving heat and chemicals for inactivation of endogenous enzymes such as polyphenol oxidase, peroxidase, pectinases, amylases, lipoxygenase, lacto peroxidase, etc. has gained interest due to their ability to reduce thermal abuse and preserve valuable nutrients during processing. Heat-induced undesirable changes on flavor, color, and nutrient content of the final product could be prevented by use of nonthermal methods for food processing (Thirumdas and Annapure, 2020). However, reports are scarce for efficient enzyme inactivation by exclusive use of nonthermal methods. Hence, combination of mild thermal and upcoming nonthermal methods

DOI: https://doi.org/10.1016/B978-0-323-95991-9.00009-6

to avoid thermal abuse of product while achieving enzyme inactivation is being welcomed. Enzyme inactivation is not a completely understood topic; they being proteinaceous molecules are capable of losing their structure and conformity due to denaturing effect of applied external forces such as heat, electrical pulses etc., by unfolding active quaternary structure of enzymes (Galanakis, 2021).

Li et al. (2021) reviewed impact of innovative food processing methods such as ultrasound, microwave and high-pressure on food matrices and, the expected modifications as well as necessary adjustments to control the level of interaction between treatments and properties of the food processed for minimizing loss of nutrients, flavor and other components. Nonthermal food processing technologies are being used along with conventional thermal treatments termed as hurdle and considered milder or gentler treatments. Enzyme and microbial activation of foods using heat along with high pressure or ultrasound were reported (Sala et al., 1995). Leonelli and Mason (2010) elaborated use of interactive effects of microwave and ultrasound during pilot scale and industrial applications. Combination of nonthermal food processing methods such as cold plasma, high pressure processing and ultra violet light with microwave assisted ultra sound processing by novel mixing of technologies were also described. Electromagnetic radiations from microwave source and energy for causing cavitations derived from ultrasound treatments for utilization in food processing were also explored.

Iqbal et al. (2019) elaborated use of ultrasound in combination with microwave to protect quality, increase shelf stability and enhanced consumer acceptance of processed foods. Cavitation of tissues caused by ultrasound treatment is considered important for inactivation of enzymes. Mechanism of action has been described as cell lysis due to vibrations caused by ultrasound followed by formation of bubbles filled with high energy which eventually explode resulting in changes in conformation of enzymes available in treated tissues. Microwave is a process where electromagnetic energy is transferred as thermal energy, in noncontact route by methods such as ionic conduction and dipole rotation. Mechanical rupture of tissues on heating moisture available inside cells aid in enhanced absorption of energy and reduces thermal gradients (Feng et al., 1999; Souraki et al., 2009). Therefore, environment friendly ultrasound technology in combination with microwave is a promising technology to inactivate enzymes with minimum damage to raw materials and finished food products. Selected examples of enzyme inactivation using sonication and microwave treatments and its synergistic effects in food matrixes and model systems are presented and discussed in the present chapter.

8.2 Sonication and microwave: Mechanism of action/enzyme inactivation

8.2.1 Sonication

Use of sound energy for various applications such as enzyme and microbial inactivation, extraction of bioactive molecules, shelf life extension or preservation, pasteurization, etc. in food processing sector is a recent trend due to its nonthermal nature. It is praised for its simple nature, cost effective as well as power saving (Radhakrishnan and Tiwari, 2021). Increased heat and mass transfer, rupturing of macromolecules by disruption of polymers, and aggregate disintegration aid in achieving desirable result by use of sonication technology (Bhaskaracharya et al., 2009). Inactivation of enzymes such as peroxidase and polyphenol oxidase in dehydrated ginger (*Zingiberofficinale*) slices (Osae et al., 2019), polyphenol oxidase in carrot (*Dacus carota*) and mushroom (*Agaricus* sp.) (Gamboa-Santoset al., 2012; Cheng et al., 2013), peroxidase and polyphenol oxidase causing pericarp browning of litchi (*Litchi chinensis*) fruit and pakchoi (*Brassica chinensis* L.) (Khademi et al., 2019; Zhang et al., 2019), polygalacturonase, β-Galactosidase (β-Gal), pectin methylesterase in apricots during postharvest storage (Osae et al., 2019), lipase and peroxidase inactivation of wheat (*Triticum* sp.) bran, polyphenol oxidase inactivation of strawberry (*Fragaria* × *ananassa*) juice (Xu et al., 2022), and milk (Munekata et al., 2021) were reported earlier. Cameron et al. (2009) also suggested that ultrasonication is an effective means of pasteurization in raw milk; however, it should be combined with an adequate heat treatment to achieve inactivation of alkaline phosphatase and lactoperoxidase.

Diagnostic or high-frequency (2–20 MHz and 0.1–1 W/cm^2) and power or high-intensity (20–100 kHz and 10 to 1000 W/cm^2) ultrasound waves are the two major categories of ultrasonic waves. Power or high-intensity finds wider application in the field of food processing and preservation viz. crystallization control, emulsion stability, enzyme inactivation, enhanced dehydration, extraction and filtration (Zheng and Sun, 2006). During sonication treatment, sound waves are used to homogenize or disturb a biological medium or chemical solution and impart chemical and physical changes by induction of cavitations (indirect) and sponge (direct) effect.

Power and frequency of ultrasound could be varied that is, sonication, thermosonication, manosonication, and manothermosonication based on factors such as duration, temperature and pH (Islam et al., 2015).

8.2.2 Microwave

Heating effect of microwave radiation is attributed to dipole effect of electromagnetic radiation emitted at frequencies between 300 MHz and 300 GHz from magnetron. Federal Communications Commission (FCC) of USA allocated microwave radiation of frequency band 915 MHz and 2450 MHz for industrial and domestic usage, respectively. Dielectric properties of treated material plays an important role on the effectiveness of microwave effect in terms of distribution and heating effect imparted as heating occurs through depolarization and conduction (Tang, 2002). Enzymatic inactivation is found to take place in a rapid manner during microwave heating in comparison to traditional heating methods. Activity, stability and selectivity of enzymes are based on their structure and conformation; intermolecular friction as well as molecular rotation of dipolar substances, heating of core and release of heat takes place on treatment of living tissues during food processing applications. Therefore, hydration or aqueous state and availability of polar molecules in media are the significant parameters for successful microwave effect. Aqueous media enables microwave effect on all its components while nonaqueous media enables focused microwave effect on polar compounds only.

Thermal and nonthermal effects are attributed to microwave radiation. Enzyme inactivation by use of microwave is due to direct transfer of energy from electromagnetic radiation to polar domains of poly peptides of protein macro molecule, which leads to modification of structure, enhanced flexibility and change in functional property (Porcelli et al., 1997). Hence, thermal effect caused by microwave energy, subsequent polarization of polar molecules results in enzyme inactivation. Furthermore, nonthermal effects of microwave on thermophilic enzymes wherein change in conformation of enzyme structure without affecting function were also reported. Physical (increase in internal pressure of cells) and chemical changes of treated material were reported (Quenzer and Burns, 1981) while nonuniform heating due to improper dissipation of electromagnetic energy has been recorded as an inherent problem during microwave assisted food processing (Vadivambal and Jayas, 2007). Inactivation of enzymes such as polyphenol oxidase and pectinmethylesterase in processed avocado (*Perseaamericana*) puree (Zhou et al., 2016), polyphenol oxidases during blanching of mushrooms (*Agaricus* sp.) (Devece et al., 1999), peroxidase and polyphenol oxidase in red beet

(*Beta vulgaris* L. var.*conditiva*) (Latorre et al., 2012), polyphenol oxidase in kiwifruit (*Actinidia deliciosa*) puree (Benlloch-Tinoco et al., 2013), mushroom tyrosinase and polyphenol oxidase present in coconut water (*Cocus nucifera*) (Cavalcante et al., 2021) using microwave radiation and their effectiveness were reported earlier.

Enzyme inactivation in during processing of foods is most essential to maintain quality during consumption and storage. Enzyme activity is based on its structure, conformational stability and availability of substrate; chemical bonds viz. disulfide links, hydrogen and hydrophobic bonds as well as van der Waals forces ensure primary, secondary and tertiary structure of enzymes, respectively. Energy inputs in form of heat, sound waves, high pressure, etc. causes imbalance between various forces holding the desired active conformation of enzyme (Pedro et al., 2018). Traditional methods of enzyme inactivation depend on thermal treatments which affects nutritional and functional property of developed products. Alternative nonthermal food processing technologies alone or in combination with mild heat treatments have gained importance as promising processing tools especially for applications related to enzyme inactivation. Ordonezez et al. (1984) explored combination of ultrasound with other thermal treatments, that is, thermosonication for enzyme and microbial inactivation and results were promising for combination method in comparison to individual treatments. Enzymes such as pectin esterases and polygalactouronases cause cloud instability, polyphenol oxidases and peroxidases affect color and nutritive quality by activating browning reactions, lipoxygenase causes off flavor and destruction of chlorophyll in processed food products, if inactivated during processing (Balaban and Spilimbergo, 2012).

Raviyan et al. (2005) described the enhanced efficiency of enzyme inactivation on combined use of ultrasound along with thermal processes. Thermosonication and its application in foods for enzyme inactivation during processing of apple (*Malus domestica*) juice (Abid et al., 2014), tomato (*Solanum lycopersicum*) juice (Wu et al., 2008), coconut water (*Cocus nucifera*) (de Matos Ribeiro et al., 2017), elderberry (*Sambucus* sp.) puree (Etzbach et al., 2019) and other fruit and vegetable products (Silva and Sulaiman, 2017) has been reported. Heating capacity of microwave is considered of better-quality and synergistic action of ultrasound and microwave energy could lead to increased or intensified processes such as chemical changes (Martina et al., 2017). "Synergistic effect" is created by "heating effect" of microwave and "vibration effect" of ultrasound when applied together" as quoted by Su et al. (2018) (Fig. 8.1). Combination effect of microwave and ultrasound and its usage as nonconventional food processing applications

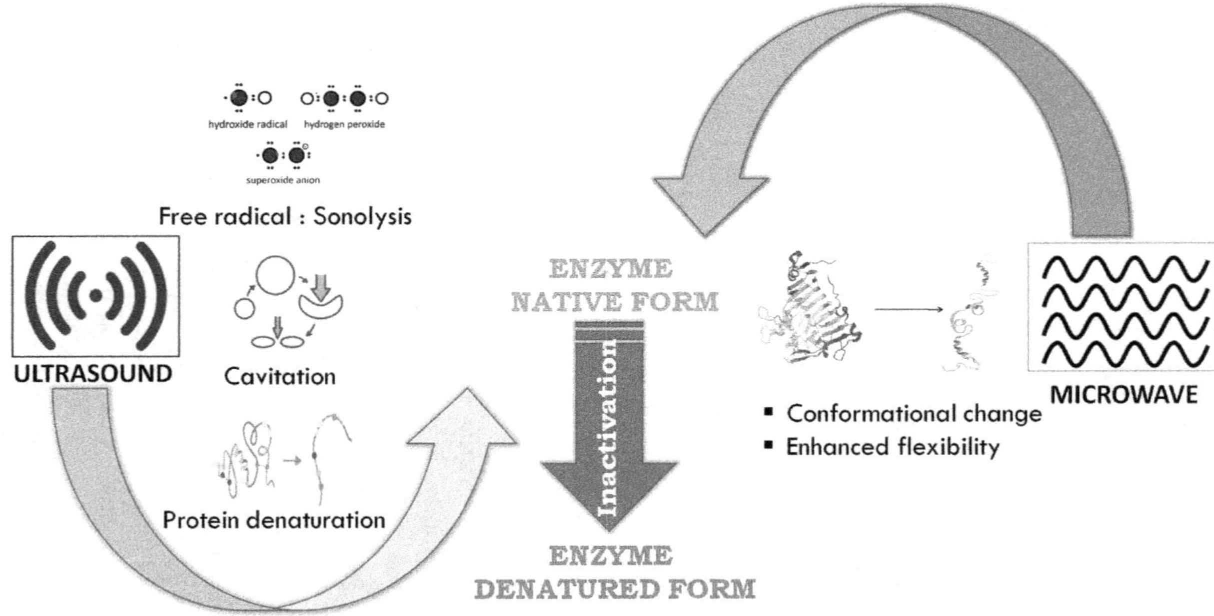

Figure 8.1 Possible mechanism for synergistic action of sonication and microwave on enzyme inactivation.

such as quick frying, improved food texture and inactivation of enzymes responsible for loss of quality and nutrients are being reported in recent past (Awad et al., 2012). Nonconventional dairy processing using of microwave, ultra sound and high pressure and their effects on enzymes and proteins were elaborated by Munir et al. (2019). Insights were provided for using these nonthermal methods alone or in combinations were also provided for obtaining shelf stable healthy milk and related products.

8.3 Individual to combination processing techniques: A road map to process intensification

8.3.1 Inactivation of endogenous enzymes using sonication and microwave

8.3.1.1 Color deteriorating/browning enzyme

Color is the most important sensory attribute of food pertaining to appearance and, first quality parameter observed by the consumer to accept or reject the food. Natural pigments such as chlorophyll, carotenoids, and oxidisable-phenolics compounds present in plant and animal foods undergo desirable and undesirable changes during food processing. Enzymatic browning is referred to the oxidation of phenol rich compounds to dark melanin pigments mediated by enzymes such as phenolase, peroxidase (donor hydrogen-peroxide oxidoreductase; EC no. 1.11.1.7), tyrosinase or polyphenol oxidase (monophenolL-dopa:oxygenoxidoreductase; EC 1.14.18.1) based on available substrates (Zeece, 2020). Browning reactions occur during operations such as hot air drying, freeze drying, freezing, thawing, etc. and requires related enzymes, oxygen, and copper as a catalyst. Phenolic compounds and browning causing enzymes are present in different cell compartments when tissues are intact. However, during food processing operations such as peeling, size reduction, pulping, etc. oxidation reactions are triggered by enzymes which result in formation of brown colored pigments. Enzymatic reaction in presence of oxygen produces highly reactive quinones on further condensation forms dark colored melanin pigments (Drogoudi et al., 2016; Jiang et al., 2016) (Figs. 8.2A and B, Source: www.brendaenzymes.com). Thermostable polyphenolase catalyzed oxidation of substrate results in production of hydrogen peroxide which in turn leads to change in taste and flavor of food (Koblitz, 2008). Improvement in color quality of processed foods especially involving fruits, vegetables, dairy and seafood could be guaranteed by newer processing methods to inactivate enzymes which are responsible for degradation of color as well as nutrients (Chemat et al., 2017).

(A)

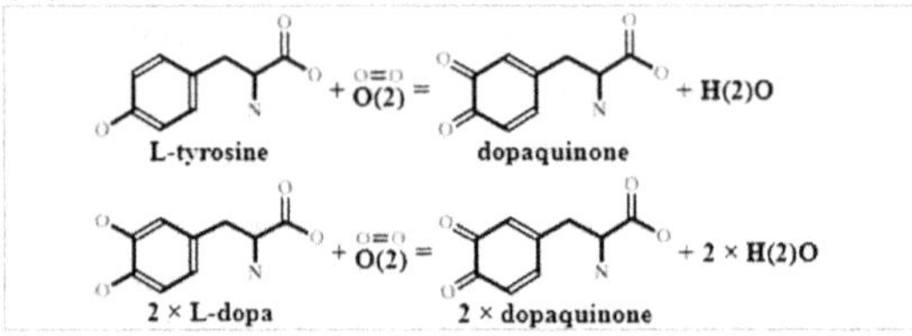

EC 1.14.18.1 Tyrosinase Reaction:(1) L-tyrosine + O(2) = dopaquinone + H(2)O. (2) 2 L-dopa + O(2) = 2 dopaquinone + 2 H(2)O

2 phenolic donor + H2O2 = 2 phenoxyl radical of the donor + 2 H2O

2 phenolic donor + HO – OH = 2 phenoxyl radical of the donor + 2 H_2O

(B)

EC 1.11.1.7 Lactoperoxidase Reaction: 2 phenolic donor + H(2)O(2) = 2 phenoxyl radical of the donor + 2 H(2)O

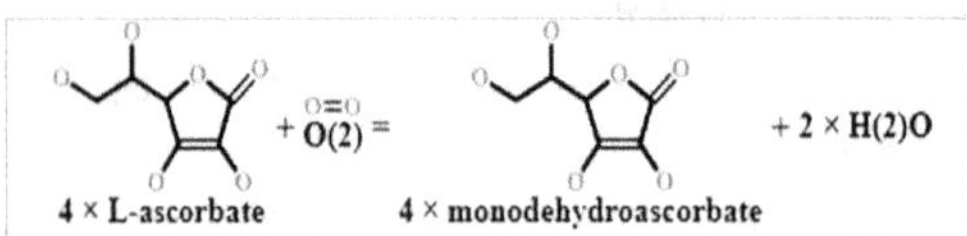

EC 1.10.3.3 L-ascorbate oxidase Reaction: 4 L-ascorbate + O(2) = 4 monodehydroascorbate + 2 H(2)O

EC 1.10.3.1 Catechol oxidase Reaction:2 catechol + O2 = 2 1.2-benzoquinone + 2 H2O

Figure 8.2 Mechanism of action–color deteriorating/browning enzyme.

Peroxidase is considered as a marker for completion of blanching operation and 90% reduction during food processing operations is essential to maintain quality while storage (Bahçeci et al., 2005). Cruz et al. (2006) explained mechanism of action of peroxidase inactivation takes place due to changes in apozyme conformation or prosthetic group disintegration. Polyphenol oxidase, an enzyme capable of catalyzing oxidoreduction reactions between phenols and its suitable substrates was inactivated by synergistic action of ultrasound and microwave 60°C at temperature for 60 min (Cheng et al., 2013). Zhang et al. (2017) reported inactivation of enzymes such as polyphenol oxidase, ascorbic acid oxidase and peroxidase in daylilies (*Hemerocallis* sp.) using low intensity ultrasound of less than 1 W/cm^2 in combination with higher temperature of more than 60°C. Microwave vacuum drying of daylilies after above mentioned pretreatment inactivated enzymes, retained nutrients and preserved color in comparison to traditional thermal treatments. Pigments such as chlorophyll and carotenoids and nutritional parameters were enhanced while enzymatic and nonenzymatic browning reactions were under check. Peroxidase showed higher activation energy and D values indicating the ability of enzyme to resist pretreatment while polyphenol oxidase and ascorbic acid oxidase were inactivated.

Microwave pasteurization for three minutes followed by 30 minutes of ultrasonication at 28 $\pm$ 2 kHz frequency in an ultrasonic bath at temperature of 60°C inactivated lipoxygenase enzyme during processing of soymilk. Inactivation rate constant for LOX was 0.3232 min^{-1} while activation energy reduced from 53.47 to 3.07 kJ mol^{-1} after microwave treatment for a period of 5 minutes (Kumar et al., 2021). Polyphenol oxidase of blueberry (*Vaccinium* sect.*Cyanococcus*) juice was inactivated using mild heat treatment along with ultrasound procedure (Zhu et al., 2017).

8.3.1.2 Texture and consistency modifying enzyme

Texture is an important parameter which plays a critical role in consumer acceptance of food. International Standards Organization (ISO) defined food texture as "All the rheological and structure (geometrical and surface) attributes of a food product perceptible by means of mechanical, tactile, and where appropriate, visual and auditory receptors" (International Standards Organization, ISO, 1994; Bolhuis et al., 2020). Texture is strongly related to structure of foods while former is considered as a sensory quality and latter is mechanical property (Chen and Rosenthal, 2015). Interrelation between texture and structure of foods was elaborated by James (2018) as "group of physical characteristics arising from the structural elements of the food, primarily sensed by the feeling of touch, related to the deformation, disintegration, and flow of the food under a force, and measured objectively by functions of mass, time, and distance". Fellows et al. (2017) described textural effects of foods caused by type and quantity of structural carbohydrates such as starch, cellulose, hemicelluloses, and pectins along with nutritional components such as moisture, protein, and fat.

Pectinases are enzymes which play a major role in degradation or depectinization of complex pectins, on maceration of fruit pulp resulting in increased juice yield, run-off and reduction in viscosity. Thus a mixture of pectinases is used to get desired liquefaction by reduction of viscosity during processing of fruit juices and avoid formation of haze during cold storage. Pectinases belong to categories such as (1) Polygalacturonases, (2) Pectin lyases, and (3) Pectin methylesterases. Galactouronic acid residues linked by

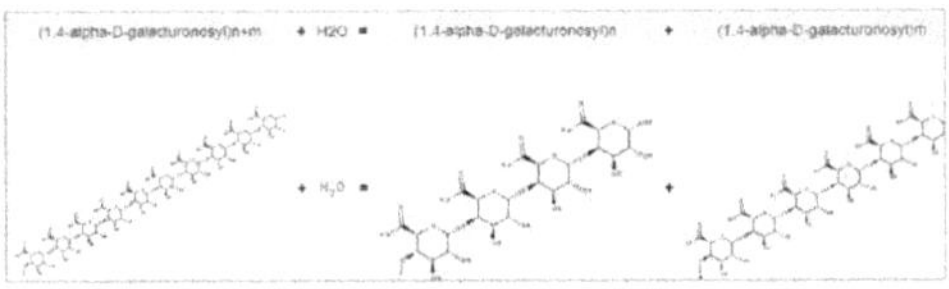

EC 3.2.1.15 endo-polygalacturonase Reaction: (1,4-alpha-D-galacturonosyl)n+m + H2O = (1,4-alpha-D-galacturonosyl)n + (1,4-alpha-D-galacturonosyl)m

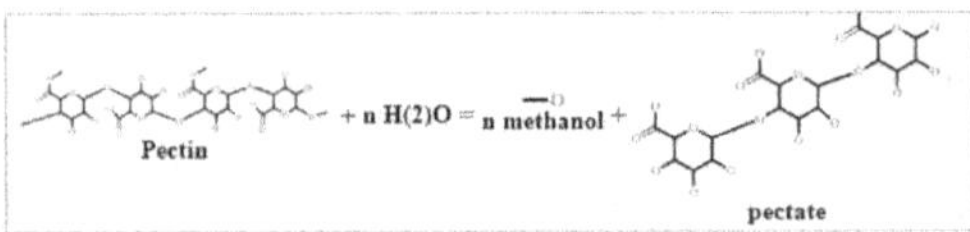

EC 3.1.1.11 Pectinesterase Reaction: Pectin + n H(2)O = n methanol + pectate

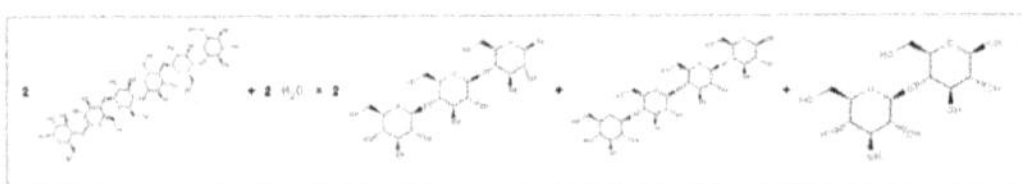

EC 3.2.1.91 cellulose 1,4-beta-cellobiosidase Reaction: cellohexaose + 2 H2O = cellotriose + cellotetraose + cellobiose

Figure 8.3 Mechanism of action–texture and consistency modifying enzyme.

alpha-1,4-glycosidic linkages form the main chain of pectin molecule are hydrolyzed in random fashion by the action of exo- and endopolygalacturonases (EC3.2.1.15). Nonreducing ends of galactouronic acid residues of pectin molecule are cleaved by pectin lyases and release oligomers. Pectate lyases (EC 4.2.2.7) and pectin lyases (EC 4.2.2.10) act upon the main chain of pectin. Pectin methyl esterase (EC 3.1.1.11) act upon the methyl group of side chains of pectin polymer releasing methanol and also eases the action of polygalactouronases by removing the esterified side chains (Rastogi and Bhatia, 2019) (Fig. 8.3, Source: www.brendaenzymes.com).

Polysaccharides such as cellulose and hemicelluloses fractions tend to be affected less by food processing operations in comparison to pectic substances which are easily subjected to modification by physical, chemical and enzymatic treatments. Combined action of pectin methylesterases and polygalactouronase is responsible for desirable textural properties of final processed product. Cloud stability and consistency of fruit juices are decided by composition of texture determining polymers such as pectin, starch, cellulose, etc. (Van Dijk and Tijskens, 2000). Cinnamic acid and tyrosine containing proteins of plant cell walls oxidize in presence of polyphenol oxidase and results in cross linking of cell wall material which lead to modifications in texture of foods (Van Buggenhout et al., 2009). Moreover, integrity of cell wall and structure related enzymes of fresh foods play an important role in enhancing storage life of foods especially fresh fruits and vegetables. Action of ultrasound and microwave as well as their synergistic effect on cell wall enzymes and their modification has been reported in recent investigations on plum (*Prunus domestica*) (Chen and Zhu, 2011), apple (*M. domestica*) (Pieczywek et al., 2017), strawberry (*Fragraria* sp.) (Cao et al., 2010), and tomato (*S. lycopersicon*) (Zhang et al., 2020). Combined action of ultrasound with microwave treatment or lyophilisation and effect on texture related enzymes has been studied during dehydration of fruits and vegetables (Yildiz and Izli, 2019; Aydar, 2020).

Soluble fiber component is composed of pectin in fruit and vegetable based juices and is responsible for cloud stability of the final product (Balestrieri et al., 1990). In citrus based juices, cloud stability is a vital parameter and loss of cloud stability leads to poor quality of juice. Cloud stability of juices also contributes to other sensory parameters such as texture, aroma, color and appeal to the consumer. Stability of cloud, the solid component of biphasic solution is disintegrated due to activity of pectin hydrolyzing enzymes. Pectin methyl esterase plays a key role in forming large aggregates of pectin chains and aid in cross linking aggregated polymers which causes precipitation of solid phase and separation of liquid phase resulting in cloud instability. Consequently, inactivation of this enzyme is crucial for maintaining quality of the fruit juices produced from citrus (*Citrus* sp.) and tomato (*S. lycopersicon*) (Cameron et al., 1999).

Action of pectin methyl esterases on pectin polymers releases demethoxylated pectin which is the suitable substrate for another texture influencing enzyme, polygalactouronase which readily acts on galactouronic acid residues causing changes related to viscosity during food product development. Detailed studies are being taken up in recent years, on the effect of texture modifying enzymes, during development of tomato-based products with different viscosity levels such as sauces, juices, ketchups, etc. (Verlent et al., 2006). Thermal treatments viz. hot break and low break are being used time immemorial to control enzymes such as pectin methyl esterases and polygalactouronases for viscosity modification of final products. However, thermal treatments compromise quality by degradation of lycopene and other heat labile nutrients of tomato. Viscosity modifications by enzyme inactivation of pectinases were reported for pectin rich fruit and vegetable products such as apple (*M. domestica*), guava (*Psidium guajava*), tomato (*S. lycopersicon*), etc. (Goodman et al., 2002).

Sonication assisted by thermal treatments could act as possible alternative in place of sole thermal treatment to inactivate texture modifying enzymes at a comparatively lower temperature. Ultrasonic treatment of fruits and vegetable juices for inactivation of enzymes, responsible for modification of textural properties was reported earlier (Costa et al., 2013). Original state of pectin molecules could be maintained by inactivating enzymes using thermal sonication procedure using 60°C for 60 minutes inactivated pectin methyl esterase up to 91% level in tomato juice and resulted in high viscosity juice. Thermal treatment up to 60°C and ultrasound effect showed inactivation of pectin methyl esterase and peroxidase while reduction in temperature does not inactivate enzymes (Wu et al., 2008). Grapefruit treated using combination of heat up to 60°C for 60 minutes along with ultrasound frequency of 28 kHz and power 420 W inactivated pectin methyl esterase up to 91% and stabilized juice cloud. Enzymes polyphenol oxidase and perodidase were also inactivated by the same treatment up to 90% and 89%, respectively (Adil et al., 2015).

Cellulase catalyzes breakdown of glucose polymer into simpler glucose units and is composed of three enzyme groups such as (1) endo-(1,4)-β-D-glucanase (2) exo-(1,4)-β-D-glucanase, and (3) β-glucosidases. Endoglucanase (EC 3.2.1.4) acts in random manner upon internal less crystalline regions of cellulose polymer and yields smaller chains of cellulose which would be acted upon by other cellulases for further breakdown. Endoglucanase (EC 3.2.1.91) catalyses breakdown of crystalline cellulose polymer into smaller oligosaccaharides by acting upon ends of giant polymer. β-Glucosidase (EC 3.2.1.21) does not act upon crystalline and amorphous forms of cellulose but catalyses breakdown of cello oligosaccharides from nonreducing ends into glucose molecules (Collins et al., 2005). Hemicellulases are enzymes available in endo and exo forms for degradation of polymers from within and reducing ends, respectively along with oligomer-hydrolyzing enzymes. Endohemicellulase acts upon polysaccharides and have very less activity on short oligomers. Endo-β,1-4 xylanases are specific for xylan while endo-β,1-4 mannosidases and β-D-glucosidases act on mannan and glucan polymers, respectively. Exohemicellulases such as β-xylosidase, β-glucosidase and mannobiosidase act upon the reducing ends of xylan, glucan and mannan, respectively. Amylases belong to glycoside hydrolase (GH) family and classified as a-amylases (GH 13) and β-amylases (GH 14) and are explored extensively for starch hydrolysis (Henrissat and Bairoch, 1993). Tateishi (2008) reported action of β-galactosidase and arabinylfuronidases on hydrolysis of β-1,4-galactan polysaccharides and arabinosyl residues available in cell wall and resultant modifications in textural consistency of food product. Zhang et al. (2005) described degradation of anthocyanin pigment by anthocyanase, a special type of β-glucoside glucohydrolase into phenolic aglycones called anthocyanidins and affect color quality of food products.

8.3.1.3 Off-flavor causing enzymes

Off-flavors and taints are considered to be of importance with respect to quality of foods, and acceptability of foods by consumers is influenced by their flavor and aroma (Whitfield and Shaw, 1985). Widespread reason for food rejection has been attributed to off-flavors in processed foods (Reineccius, 1991). Exogenous (packaging material, microbial contamination, air, and water) and endogenous factors (enzymatic reaction, oxidation of macromolecules) causing common off-flavors were elaborated (Nijssen, 1991). Identification and quantification of molecules responsible for development of undesirable flavors and taints was reviewed in detail (Ridgway et al., 2010).

Off-flavor development is closely associated with deterioration of nutritional quality of processed foods, and enzymes such as lipoxygenase oxidizing polyunsaturated fatty acids and peroxidase oxidizing peroxides of polyphenols along with production of hydrogen peroxide radicals (Ludikhuyze et al., 1998) and brown melanin pigments (Vernwal et al., 2006), respectively. Hydroperoxylipases act up on13-hydro-peroxy lipids and produce off-flavor generating aldehydes such as 6-hexanal and hexenal (Feussner and Wasternack, 2002). Sulfhydryl proteins present in specific group of plants such as crucifers (cauliflower (*Brassica oleracea* var. *botrytis*), cabbage (*B. oleracea* var. *capitata*), and broccoli (*B. oleracea* var. *italica*) are hydrolyzed by actinidin which acts up on carboxylic acid group of lysine, liberating allergenic and off-flavor causing amines and carboxylic acids (Katsaros et al., 2009). Hydrolysis of aryl and alkyl β-D-glucoside compounds by β -glucoside glucohydrolase (glucohydrolase; EC 3.2.1.21) produces β-D-glucose. Action on various glucoside compounds in fruits leads to production of flavor and off-flavor producing molecules in fruit juices (Maicas and Mateo, 2005).

Lipids are broken into fatty acids and glycerol in aqueous or solvent phase by catalyzing action of lipases (triacylglycerol hydrolases EC 3.1.1.1). Enzyme action of lipase takes place in two steps: (1) acylation and (2) deacylation, oxyanion formation takes place in former step while release of fatty acid from oxyanion group and release of enzyme for further catalysis takes place in latter step (Casas-Godoy et al., 2012) (Fig. 8.4, Source: www.brendaenzymes.com). Liopxygenase pathway which leads to an array

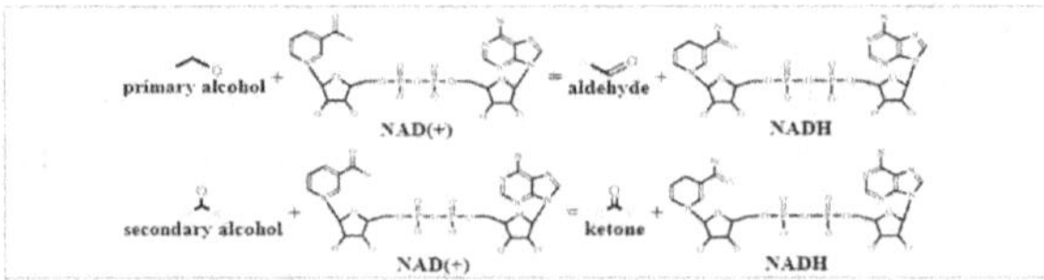

EC 1.1.1.1 Alcohol dehydrogenase Reaction:(1) A primary alcohol + NAD(+) = an aldehyde + nadh
(2) A secondary alcohol + NAD(+) = a ketone + nadh.

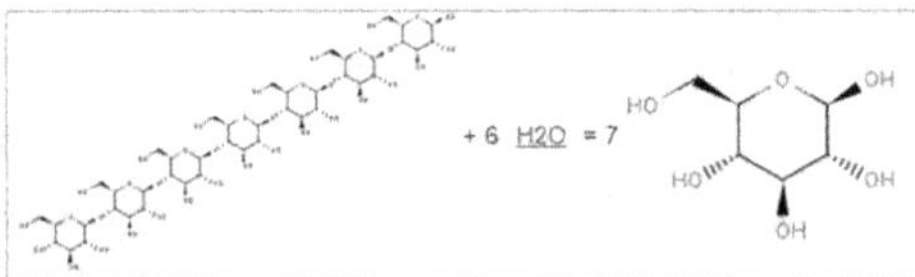

EC 3.2.1.21 beta-glucosidase Reaction: Hydrolysis of terminal, non-reducing beta-D-glucosyl residues with release of beta-D-glucose

Figure 8.4 Mechanism of action–off-flavor causing enzymes.

of flavor causing food components involving enzymes viz. lipoxygenase (EC 1.3.11.12), hydroperoxide isomerase (EC 4.2.1.92), and alcohol dehydrogenase (EC 1.1.1.1) has been elaborated (Feussner and Wasternack, 2002). Long chain (>C12) fatty acids does not impact flavor as influenced by short (C4) and medium chain fatty acids (C12). Hydroperoxide intermediates generated by action of lipoxygenase on polyunsaturated fatty acids are acted up on hydroperoxide isomerase, and oxidation of hydrocarbon skeleton leads to production of aldehydes, ketones and ω-fatty acids and reduced further into alcohols by action of alcohol dehydrogenase. Sharp and tangy flavor leading to formation of rancid notes and soapy flavor in fresh and processed foods containing lipids are due to production of short and medium chain fatty acids, respectively due to action of lipases (Nelson, 1972).

Secondary metabolite pathways lead to synthesize a variety of compounds in living systems and are compartmentalized in cells. Action of enzymes on such precursors leads to their activation and generation of flavors. Glycosidases hydrolyzes glycosides (carbohydrate-aglycone) and leads to production of volatile aromatic compounds. Polyphenols, sulfur compounds and alcohols may be characterized by attachment of proteins through specific bonds and remain inactive till hydrolyzed by action of enzymes. Glycoside concentration builds up in mature ripen fruits in form of O-alpha-L-rhamnopyranoside and O-alpha-L-rabinofuranoside which on action of enzymes generate aroma compounds. Action of enzymes such as aldehydrolyase, lipoxygenases, esterases, etc. was reported during storage of processed apple (*M. domestica*) juice (Jenniskens et al., 1991). Therefore, inactivation of off-flavor causing enzymes is an essential prerequisite to safeguard quality of food product and avoid rejection by consumers. Temperature has been used from time immemorial to inactivate enzymes, by causing damage to enzyme confirmation and structure, which compromised nutritional quality of processed foods.

Reports on synergistic action of ultrasound and microwave or other thermal treatments to inactivate major enzymes responsible for flavor development are scarce. Mahn et al. (2020) reported ultrasound assisted blanching treatment of broccoli at 60°C for 4 minutes inactivated myrosinase effectively in comparison to thermal treatments as well as preserved sulforaphane, an isothiocyanate documented as a potent anti-cancer agent in vegetative parts which could be used for extraction during volarization of broccoli processing waste. Ultrasound treatment at 20 kHz with mild heat inactivated lipoxygenase in soybean (*Glycine max*) (Lopez et al., 1994). Ultrasound assisted osmodehydration as well as osmosonication as pretreatments inactivated polyphenol oxidase and enhanced quality parameters of dehydrated ginger (*Z. officinale*) (Osae et al., 2019).

Significant reduction of decimal point of peroxidase from 65 minutes to 10 minutes time during horseradish (*Armoracia rusticana*) processing was reported on high ultrasound treatment using frequency up to 60 kHz at 80°C (De Gennaro et al., 1999). Grintsevich and Metelitsa (2002) elaborated pH of processing medium and its effect on inactivation of horseradish peroxidase during ultrasound treatment of 1 W/cm^2 for 1–2 hours at 35.5–55°C. Reduction in pH enhanced inactivation of enzyme which may be attributed to the increased production of harmful free radicals in medium. pH values less than 5.0 enhanced attack of free radical on active sites of enzyme leading to its inactivation. Additives such as sucrose and glycerol improved horseradish (*A. rusticana*) peroxidase inactivation efficiency of heat assisted and heat and pressure assisted sonication. Lower pH also favored inactivation of peroxidase (Lopez and Burgos, 1995). Thermosonication of watercress (*Nasturtium officinale* W.T.Aiton) in temperature range 40–92.5°C, 20 kHz and 50% power inactivated peroxidase after increase in temperature of 85.0°C. Blanching treatments assisted with heat and ultrasound effectively inactivated peroxidase in comparison to traditional blanching methods (Xin et al., 2015). High power ultrasound of 360 W and at 20 kHz, combined with heat treatment up to 40.0°C for 2 minutes time inactivated lactoperoxidase in milk samples showing synergistic action of heat and sound on enzyme inactivation (Ertugay et al., 2003). Increase in duration of sonication of frequency 22 kHz inactivated off-flavor causing enzymes such as lactoperoxidase, peroxidase and horseradish peroxidase (Gebicka and Gebicki, 1997).

8.4 Conclusions

Growing demands of consumer for healthier and nutritious food products as well as newer regulations of food processing have led to emergence of green processing technologies, which are environmental friendly and produce high quality processed foods. Recent trends in food processing show use of combination production technologies involving thermal and nonthermal processes. Ultrasound and microwave technology for inactivation of endogenous enzymes in foods to improve their storage stability is showing great promise. Though adequate developments in these recent technologies are being reported for independent applications of sonication and microwave, detailed studies on synergistic action during use of combination of these technologies in food processing applications are need of the hour. Future research should also focus on determining the optimum combination of microwave assisted sonication treatments for enzyme inactivation by determining action of various active sites for different enzymes. Industrial scale-up depends up on clear understanding of the technology as well as economical gains. Further studies are essential to develop pilot scale processing plants and customized equipment that support large scale processing applications using newer technologies. Furthermore, impact of recent technology on safety of food products need to be assessed and reported to gain trust of customers. Effect of cavitations caused in food systems due to ultrasound as well as safety on use of electromagnetic radiation in form of microwave treatment as well as their combined effects still needs to be investigated in detail. Research priorities in the above mentioned subjects in future would pave way for use of alternative green technologies as processing tools as well as their industrial scale-up applications.

References

Aadil, R.M., Zeng, X.A., Zhang, Z.H., Wang, M.S., Han, Z., Jing, H., Jabbar, S., 2015. Thermosonication: a potential technique that influences the quality of grapefruit juice. Int. J. Food Sci. Technol. 50 (5), 1275–1282.

Abid, M., Jabbar, S., Hu, B., Hashim, M.M., Wu, T., Lei, S., … Zeng, X., 2014. Thermosonication as a potential quality enhancement technique of apple juice. Ultrason. Sonochem. 21 (3), 984–990.

Awad, T.S., Moharram, H.A., Shaltout, O.E., Asker, D.Y.M.M., Youssef, M.M., 2012. Applications of ultrasound in analysis, processing and quality control of food: a review. Food Res. Int. 48 (2), 410–427.

Aydar, A.Y., 2020. Investigation of ultrasound pretreatment time and microwave power level on drying and rehydration kinetics of green olives. Food Sci. Technol. 41, 238–244.

Bahçeci, K.S., Serpen, A., Gökmen, V., Acar, J., 2005. Study of lipoxygenase and peroxidase as indicator enzymes in green beans: change of enzyme activity, ascorbic acid and chlorophylls during frozen storage. J. Food Eng. 66 (2), 187–192.

Balaban, M.O., Ferrentino, G., Spilimbergo, S., 2012. Dense phase CO2 (DPCD) for microbial decontamination of food. In: Microbial Decontamination in the Food Industry. Woodhead Publishing, Sawston, Cambridge, pp. 665–697.
Balestrieri, C., Castaldo, D., Giovane, A., Quagliuolo, L., Servillo, L., 1990. A glycoprotein inhibitor of pectin methylesterase in kiwi fruit (Actinidachinensis). Eur. J.Biochem. 193, 183–187.
Benlloch-Tinoco, M., Igual, M., Rodrigo, D., Martínez-Navarrete, N., 2013. Comparison of microwaves and conventional thermal treatment on enzymes activity and antioxidant capacity of kiwifruit puree. Innovative Food Sci. Emerg. Technol. 19, 166–172.
Bhaskaracharya, R.K., Kentish, S., Ashok kumar, M., 2009. Selected applications of ultrasonics in food processing. Food Eng. Rev. 1 (1), 31–49.
Bolhuis, D.P., Forde, C.G., 2020. Application of food texture to moderate oral processing behaviors and energy intake. Trends Food Sci. Technol.
Cameron, M., McMaster, L.D., Britz, T.J, 2009. Impact of ultrasound on dairy spoilage microbes and milk components. Dairy Sci. Technol. 89 (1), 83–98.
Cameron, R.G., Baker, R.A., Buslig, B.S., Grohmann, K., 1999. Effect of juice extractor settings on juice cloud stability. J. Agric. Food Chem. 47 (7), 2865–2868.
Cao, S., Hu, Z., Pang, B., Wang, H., Xie, H., Wu, F., 2010. Effect of ultrasound treatment on fruit decay and quality maintenance in strawberry after harvest. Food Control 21 (4), 529–532.
Casas-Godoy, L., Duquesne, S., Bordes, F., Sandoval, G., Marty, A., 2012. Lipases: an overview. Methods Mol. Biol. 861, 3–30.
Cavalcante, T.A.B.B., dos Santos Funcia, E., Gut, J.A.W., 2021. Inactivation of polyphenol oxidase by microwave and conventional heating: Investigation of thermal and non-thermal effects of focused microwaves. Food Chem. 340, 127911.
Chemat, F., Rombaut, N., Meullemiestre, A., Turk, M., Perino, S., Fabiano-Tixier, A.S., Abert-Vian, M., 2017. Review of green food processing techniques. Preservation, transformation, and extraction. Innovative Food Sci. Emerg. Technol. 41, 357–377.
Chen, J., Rosenthal, A., 2015. *Modifying Food Texture: Volume 2: Sensory Analysis, Consumer Requirements and Preferences*. Woodhead Publishing, Sawston, Cambridge, pp. 3–24 ISBN 9781782423331 https://doi.org/10.1016/B978-1-78242-333-1.00001-2.
Chen, Z., Zhu, C., 2011. Combined effects of aqueous chlorine dioxide and ultrasonic treatments on postharvest storage quality of plum fruit (Prunussalicina L. Postharvest Biol. Technol. 61 (2-3), 117–123.
Cheng, X.F., Zhang, M., Adhikari, B., 2013. The inactivation kinetics of polyphenol oxidase in mushroom (Agaricusbisporus) during thermal and thermosonic treatments. Ultrasonicssonochemistry 20 (2), 674–679.
Collins, T., Gerday, C., Feller, G., 2005. Xylanases, xylanase families and extremophilicxylanases. FEMS Microbiol. Rev. 29 (1), 3–23.
Costa, M.G.M., Fonteles, T.V., de Jesus, A.L.T., Almeida, F.D.L., de Miranda, M.R.A., Fernandes, F.A.N., Rodrigues, S., 2013. High-intensity ultrasound processing of pineapple juice. Food Bioprocess Technol. 6 (4), 997–1006.
Cruz, R.M., Vieira, M.C., Silva, C.L., 2006. Effect of heat and thermosonication treatments on peroxidase inactivation kinetics in watercress (Nasturtium officinale). J. Food Eng. 72 (1), 8–15.
De Gennaro, L., Cavella, S., Romano, R., Masi, P., 1999. The use of ultrasound in food technology I: inactivation of peroxidase by thermosonication. J. Food Eng. 39 (4), 401–407.
de Matos Ribeiro, M., Valdramidis, V.P., Nunes, C.A., de Souza, V.R., 2017. Synergistic effect of thermosonication to reduce enzymatic activity in coconut water. Innovative Food Sci. Emerg. Technol. 41, 404–410.
Devece, C., Rodríguez-López, J.N., Fenoll, L.G., Tudela, J., Catalá, J.M., de los Reyes, E., García-Cánovas, F., 1999. Enzyme inactivation analysis for industrial blanching applications:

comparison of microwave, conventional, and combination heat treatments on mushroom polyphenoloxidase activity. J. Agric. Food Chem. 47 (11), 4506–4511.

Drogoudi, P., Pantelidis, G.E., Goulas, V., Manganaris, G.A., Ziogas, V., Manganaris, A., 2016. The appraisal of qualitative parameters and antioxidant contents during postharvest peach fruit ripening underlines the genotype significance. Postharvest Biol. Technol. 115, 142–150.

Ertugay, M.U.S.T.A.F.A., Yuksel, Y., Sengul, M., 2003. The effect of ultrasound on lactoperoxidase and alkaline phosphatase enzymes from milk. Milchwissenschaft 58, 593–595.

Etzbach, L., Pfeiffer, A., Schieber, A., Weber, F., 2019. Effects of thermal pasteurization and ultrasound treatment on the peroxidase activity, carotenoid composition, and physicochemical properties of goldenberry (Physalisperuviana L.) puree. LWT 100, 69–74.

Fellows, P.J., 2017. 1 - Properties of food and principles of processing. In: Fellows, P.J. (Ed.), Woodhead Publishing Series in Food Science, Technology and Nutrition,Food Processing Technology. Woodhead Publishing, Sawston, Cambridge, pp. 3–200 ISBN 9780081019078 https://doi.org/10.1016/B978-0-08-100522-4.00001-8.

Feng, H., Tang, J., Cavalieri, R., 1999. Combined microwave and spouted bed drying of diced apples: effect of drying conditions on drying kinetics and product temperature. Invited contribution to the special Hall Issue. Dry Technol. 17, 1981–1998.

Feussner, I., Wasternack, C., 2002. The lipoxygenase pathway. Annu. Rev. Plant Biol. 53 (1), 275–297.

Galanakis, C.M., 2021. Functionality of food components and emerging technologies. Foods 10 (1), 128.

Gamboa-Santos, J., Montilla, A., Soria, A.C., Villamiel, M., 2012. Effects of conventional and ultrasound blanching on enzyme inactivation and carbohydrate content of carrots. Eur. Food Res. Technol. 234 (6), 1071–1079.

Gębicka, L., Gębicki, J.L, 1997. The effect of ultrasound on heme enzymes in aqueous solution. J. Enzym Inhib. 12 (2), 133–141.

Goodman, C.L., Fawcett, S., Barringer, S.A, 2002. Flavor, viscosity, and color analyses of hot and cold break tomato juices. J. Food Sci. 67 (1), 404–408.

Grintsevich, E.E., Metelitsa, D.I, 2002. Kinetics of ultrasonic inactivation of peroxidase in aqueous solutions. Russ. J. Phys. Chem. A 76 (8), 1368–1373.

Henrissat, B., Bairoch, A., 1993. New families in the classification of glycosyl hydrolases based on amino acid sequence similarities. The Biochemical Journal, 293 (Pt 3) (Pt 3), 781–788. https://doi.org/10.1042/bj2930781.

Iqbal, A., Murtaza, A., Hu, W., Ahmad, I., Ahmed, A., Xu, X., 2019. Activation and inactivation mechanisms of polyphenol oxidase during thermal and non-thermal methods of food processing. Food Bioprod. Process. 117, 170–182.

Islam, M.N., Zhang, M., Liu, H., Xinfeng, C., 2015. Effects of ultrasound on glass transition temperature of freeze-dried pear (Pyruspyrifolia) using DMA thermal analysis. Food Bioprod. Process. 94, 229–238.

James, B., 2018. Oral processing and texture perception influences satiation. Physiol. Behav. 193, 238–241.

Jenniskens, L.D., Voragen, A.J., Pilnik, W., Posthumus, M.A, 1991. Effects of the treatment of apple pulp with liquefying enzymes on the aroma of apple juice. Lebensmittel-Wissenschaft+ Technologie 24 (1), 86–92.

Jiang, J., Xiong, Y.L, 2016. Natural antioxidants as food and feed additives to promote health benefits and quality of meat products: a review. Meat Sci. 120, 107–117.

Katsaros, G.I., Katapodis, P., Taoukis, P.S, 2009. Modeling the effect of temperature and high hydrostatic pressure on the proteolytic activity of kiwi fruit juice. J. Food Eng. 94 (1), 40–45.

Khademi, O., Ashtari, M., Razavi, F., 2019. Effects of salicylic acid and ultrasound treatments on chilling injury control and quality preservation in banana fruit during cold storage. ScientiaHorticulturae 249, 334–339.

Koblitz, M.G.B., 2008. Bioquímica de Alimentos-teoria e aplicaçõespráticas. Guanabara Koogan, Rio de Janeiro, p. 242.

Kumar, R., Kumar, A., Jayachandran, L.E., Rao, P.S, 2021. Sequential Microwave–Ultrasound assisted extraction of soymilk and optimization of extraction process. LWT 151, 112220.

Latorre, M.E., Bonelli, P.R., Rojas, A.M., Gerschenson, L.N, 2012. Microwave inactivation of red beet (Beta vulgaris L. var.conditiva) peroxidase and polyphenoloxidase and the effect of radiation on vegetable tissue quality. J. Food Eng. 109 (4), 676–684.

Leonelli, C., Mason, T.J., 2010. Microwave and ultrasonic processing: now a realistic option for industry. Chem. Eng. Process. 49 (9), 885–900.

Li, S., Zhang, R., Lei, D., Huang, Y., Cheng, S., Zhu, Z., … Cravotto, G., 2021. Impact of ultrasound, microwaves and high-pressure processing on food components and their interactions. Trends Food Sci. Technol. 109, 1–15. doi:10.1016/j.tifs.2021.01.017.

Lopez, P., Burgos, J., 1995. Peroxidase stability and reactivation after heat treatment and manothermosonication. J. Food Sci. 60 (3), 451–455.

Ludikhuyze, L., Indrawati, Van den Broeck, I., Weemaes, C., Hendrickx, M., 1998. Effect of combined pressure and temperature on soybean lipoxygenase. 2. Modeling inactivation kinetics under static and dynamic conditions. J. Agric. Food Chem. 46 (10), 4081–4086.

Mahn, A., Quintero, J., Castillo, N., Comett, R., 2020. Effect of ultrasound-assisted blanching on myrosinase activity and sulforaphane content in broccoli florets. Catalysts 10 (6), 616.

Maicas, S., Mateo, J.J., 2005. Hydrolysis of terpenyl glycosides in grape juice and other fruit juices: a review. Appl. Microbiol. Biotechnol. 67 (3), 322–335.

Martina, K., Tagliapietra, S., Barge, A., Cravotto, G., 2017. Combined microwaves/ultrasound, a hybrid technology. In: Colmenares, J., Chatel, G. (Eds.), Sonochemistry. Topics in Current Chemistry Collections. Springer, Cham. 175–201. https://doi.org/10.1007/978-3-319-54271-3_7.

Munekata, Paulo, E.S., Wang, M., Zhou, J., Lorenzo, J.M., Barba, FJ., 2021. Ultrasound as a preservation technique. In: Design and Optimization of Innovative Food Processing Techniques Assisted by Ultrasound. Academic Press, United States, pp. 39–54.

Munir, M., Nadeem, M., Qureshi, T.M., Leong, T.S., Gamlath, C.J., Martin, G.J., Ashok kumar, M., 2019. Effects of high pressure, microwave and ultrasound processing on proteins and enzyme activity in dairy systems—a review. Innovative Food Sci. Emerg. Technol. 57, 102192.

Nelson, J.H., 1972. Enzymatically produced flavors for fatty systems. J. Am. Oil Chem. Soc. 49 (10), 559–562.

Nijssen, B., 1991. Off-Flavors. In: Maarse, H. (Ed.), Volatile Compounds in Foods and Beverages. Marcel Dekker, Inc, New York, pp. 689–735.

Ordonez, J.A., Sanz, B., Hernandez, P.E., Lopez-Lorenzo, P., 1984. A note on the effect of combined ultrasonic and heat treatments on the survival of thermoduric streptococci. J. Appl. Bacteriol. 56 (1), 175–177.

Osae, R., Zhou, C., Xu, B., Tchabo, W., Tahir, H.E., Mustapha, A.T., Ma, H., 2019. Effects of ultrasound, osmotic dehydration, and osmosonication pretreatments on bioactive compounds, chemical characterization, enzyme inactivation, color, and antioxidant activity of dried ginger slices. J. Food Biochem. 43 (5), e12832.

Pieczywek, P.M., Kozioł, A., Konopacka, D., Cybulska, J., Zdunek, A., 2017. Changes in cell wall stiffness and microstructure in ultrasonically treated apple. J. Food Eng. 197, 1–8.

Porcelli, M., Cacciapuoti, G., Fusco, S., Massa, R., d'Ambrosio, G., Bertoldo, C., … Zappia, V., 1997. Non-thermal effects of microwaves on proteins: thermophilic enzymes as model system. FEBS Lett. 402 (2-3), 102–106.

Quenzer, N.M., Burns, E.E., 1981. Effects of microwave, steam and water blanching on freeze-dried spinach. J. Food Sci. 46 (2), 410–413.

Radhakrishnan, M., Tiwari, B., 2021. Application of ultrasound in fruit and vegetable processing. In: Innovative Food Processing Technologies, 2021. Elsevier, Amsterdam, Netherlands, pp. 494–503 ISBN 9780128157824. https://doi.org/10.1016/B978-0-12-815781-7.00025-1.

Rastogi, H., Bhatia, S., 2019. Future prospectives for enzyme technologies in the food industry. In: Enzymes in Food Biotechnology. Academic Press, United States, pp. 845–860.

Raviyan, P., Zhang, Z., Feng, H., 2005. Ultrasonication for tomato pectinmethylesterase inactivation: effect of cavitation intensity and temperature on inactivation. J. Food Eng. 70 (2), 189–196.

Reineccius, G., 1991. Off-flavors in foods. Crit. Rev. Food Sci. Nutr. 29 (6), 381–402.

RidgwayK., S.P.D.L, Smith, R.M., 2010. Analysis of food taints and off-flavours: a review. Food Additives Contaminants A 27 (2), 146–168. doi:10.1080/19440040903296840.

Sala, F.J., Burgos, J., Condon, S., Lopez, P., Raso, J., 1995. Effect of heat and ultrasound on microorganisms and enzymes. In: New Methods of Food Preservation. Springer, Boston, MA, pp. 176–204.

Souraki, B.A., Andre´s, A., Mowla, D., 2009. Mathematical modeling of microwave-assisted inert medium fluidized bed drying of cylindrical carrot samples. ChemEng Process 48, 296–305.

Su, Y., Zhang, M., Adhikari, B., Mujumdar, A.S., Zhang, W., 2018. Improving the energy efficiency and the quality of fried products using a novel vacuum frying assisted by combined ultrasound and microwave technology. Innovative Food Sci. Emerg. Technol. 50, 148–159.

Sulaiman, A., Farid, M., Silva, F.V., 2017. Quality stability and sensory attributes of apple juice processed by thermosonication, pulsed electric field and thermal processing. Food Sci. Technol. Int. 23 (3), 265–276.

Tang, J., Feng, H., Lau, M., 2002. Microwave heating in food processing. In: Young, X, Tang, J (Eds.), Advances in Bioprocessing Engineering. World Scientific, Singapore.

Tateishi, A., 2008. β-Galactosidase and α-L-arabinofuranosidase in cell wall modification related with fruit development and softening. J. Japanese Soc. Horticult. Sci. 77 (4), 329–340.

Thirumdas, R., Annapure, U.S, 2020. Enzyme inactivation in model systems and food matrixes by cold plasma. In: Advances in Cold Plasma Applications for Food Safety and Preservation. Academic Press, United States, pp. 229–252.

Vadivambal, R., Jayas, D.S, 2007. Changes in quality of microwave-treated agricultural products—a review. Biosystems Eng. 98 (1), 1–16.

Van Buggenhout, S., Sila, D., Duvetter, T., Van Loey, A.M., Hendrickx, ME., 2009. Pectins in processed fruits and vegetables: part III-Texture engineering. Compr. Rev. Food Sci. Food Saf. 8 (2), 105–117.

Van den Broeck, I., Ludikhuyze, L.R., Van Loey, A.M., Hendrickx, ME., 2000. Inactivation of orange pectinesterase by combined high-pressure and -temperature treatments: a kinetic study. J. Agric. Food Chem. 48 (5), 1960–1970.

Verlent, I., Hendrickx, M., Rovere, P., Moldenaers, P., Loey, A.V, 2006. Rheological properties of tomato-based products after thermal and high-pressure treatment. J. Food Sci. 71 (3), S243–S248.

Vernwal, S.K., Yadav, R.S.S., Yadav, K.D.S, 2006. Purification of a peroxidase from Solanum-melongena fruit juice. Indian J BiochemBiophys 43 (4), 239–243.

Wu, J., Gamage, T.V., Vilkhu, K.S., Simons, L.K., Mawson, R., 2008. Effect of thermosonication on quality improvement of tomato juice. Innovative Food Sci. Emerg. Technol. 9 (2), 186–195.

Xin, Y., Zhang, M., Xu, B., Adhikari, B., Sun, J., 2015. Research trends in selected blanching pretreatments and quick freezing technologies as applied in fruits and vegetables: a review. Int. J. Refrig. 57, 11–25.

Xu, B., Sylvain Tiliwa, E., Yan, W., Roknul Azam, S.M., Wei, B., Zhou, C., Ma, H., Bhandari, B., 2022. Recent development in high quality drying of fruits and vegetables assisted by ultrasound: A review. Food Res. Int. (Ottawa, Ont.) 152, 110744. https://doi.org/10.1016/j.foodres.2021.110744.

Yildiz, G., İzli, G., 2019. Influence of microwave and microwave-convective drying on the drying kinetics and quality characteristics of pomelo. J. Food Process. Preserv. 43 (6), e13812.

Zeece, M., 2020. Introduction to the Chemistry of Food. Academic Press, United States.

Zhang, L.F., Wang, P., Sun, X.Y., Chen, F.S., Lai, S.J., Yang, H.S., 2020. Calcium permeation property and firmness change of cherry tomatoes under ultrasound combined with calcium lactate treatment. UltrasonSonochemistry 60, 104784. https://doi.org/10.1016/j.ultsonch.2019.104784.

Zhang, X.T., Zhang, M., Devahastin, S., Guo, Z., 2019. Effect of combined ultrasonication and modified atmosphere packaging on storage quality of pakchoi (*Brassica chinensis* L.). Food Bioprocess Technol. 12 (9), 1573–1583.

Zhang, Z., Niu, L., Li, D., Liu, C., Ma, R., Song, J., Zhao, J., 2017. Low intensity ultrasound as a pretreatment to drying of daylilies: Impact on enzyme inactivation, color changes and nutrition quality parameters. UltrasonicsSonochemistry 36, 50–58.

Zhang, Z., Pang, X., Xuewu, D., Ji, Z., Jiang, Y., 2005. Role of peroxidase in anthocyanin degradation in litchi fruit pericarp. Food Chem. 90 (1-2), 47–52.

Zheng, L., Sun, D.W., 2006. Innovative applications of power ultrasound during food freezing processes—a review. Trends Food Sci. Technol. 17 (1), 16–23.

Zhou, L., Tey, C.Y., Bingol, G., Bi, J., 2016. Effect of microwave treatment on enzyme inactivation and quality change of defatted avocado puree during storage. Innovative Food Sci. Emerg. Technol. 37, 61–67.

Zhu, J., Wang, Y., Li, X., Li, B., Liu, S., Chang, N., … Meng, X., 2017. Combined effect of ultrasound, heat, and pressure on Escherichia coli O157: H7, polyphenol oxidase activity, and anthocyanins in blueberry (Vacciniumcorymbosum) juice. Ultrasonicssonochemistry 37, 251–259.

CHAPTER NINE

Sonication microwave synergistic extraction of bioactive compounds from plant source

Md. Anisur Rahman Mazumder[a], Juwel Rana[b], Md. Fahad Jubayer[c], Thottiam Vasudevan Ranganathan[d] and Mohammad Javed Ansari[e]

[a]Department of Food Technology and Rural Industries, Bangladesh Agricultural University, Mymeningh, Bangladesh
[b]Department of Nutrition and Food Engineering, Daffodil International University, Dhaka, Bangladesh
[c]Department of Food Engineering and Technology, Sylhet Agricultural University, Sylhet, Bangladesh
[d]Department of Food Processing Technology, School of Agriculture and Bioscience, Karunya Institute of Technology and Sciences, Coimbatore, Tamil Nadu, India
[e]Department of Botany, Hindu College Moradabad (Mahatma Jyotiba Phule Rohilkhand University Bareilly), India

9.1 Introduction

Nowadays, potential challenges are the key drivers to move innovative research, as well as technology diffusion. As a consequence of the advancement of numerous technologies and industrial practices, consumer perception is changing. There is an increasing demand among the consumers for such products that provide health benefits, bring the message of no side effects while using, and cause no harm to the environment. To achieve these goals, businesses or related stakeholders must design products that are both healthy and environmental friendly (Palma et al., 2021).

The majority of bioactive compounds are sourced from plants; to be more specific, they are obtained from medicinal plants. Plants often possess only a modest number of active compounds, but their great value justifies the creation of a high performance technique in most circumstances. To extract bioactive chemicals from plants with minimal activity loss and high purity, new extraction methods have been developed (Vardanega et al., 2014). Extraction is an essential stage in the recovery of bioactive substances, and it can be accomplished using different approaches like application of solvents for extraction, mechanical force for expelling the extracts, application of supercritical fluids as well as microwave extraction. These techniques have constraints, such as an excess usage of solvents in the case of solvent extraction, unsatisfactory yield while applying mechanical pressing, high capital cost of supercritical fluid extraction, and the requirements of an

Ultrasound and Microwave for Food Processing: Synergism for Preservation and Extraction.
DOI: https://doi.org/10.1016/B978-0-323-95991-9.00013-8

aqueous medium in microwave aided extraction. In comparison, sonication or ultrasonic extraction offers benefits such as using fewer time and energy, extracting at a lower temperature, and retaining the extract's quality (Kumar et al., 2021). This method provides a number of environmental and economic benefits because it is found to decrease the application of solvents and chemical additives, consume less time, and allows for the separation and recycling of constituents (Palma et al., 2021). Ultrasonic waves produce disturbances in plant tissues, due to the physiological forces created by cavitation, and hasten the release of extractable compounds into the solvent medium, thus enhancing mass transfer (Kumar et al., 2021).

Ultrasound has been implemented satisfactorily to extract polyphenols, carotenoids, scents, and polysaccharides from plant matrix which contains both the plant and plant by-products. Different factors like time, temperature, power, frequency, duty cycle, solvent type, and liquid solid ratio must be controlled carefully and precisely for an efficient extraction (Vilkhu et al., 2008).

9.2 History and recent application of ultrasound

9.2.1 History

The research on acoustics goes back to the seventeenth century when key scientists like Boyle, Hooke, and Newton initiated their research on sound waves and their nature and put forth their hypothesis (Thompson, 2000; Garner and Hunt, 1980). Sir Isaac Newton presented the historical breakthroughs of ultrasound when he proposed his concept of sound waves in 1687. Sound is made up of pressure waves that are compressed and expanded as they travel through a medium such as a gas, liquid, or solid. Newton was interested in sound transmission and is said to be the first to define the links between sound speed and the related quantifiable medium variables, namely pressure and density. Newton proposed in his Principia that the speed of sound is proportional to the square root of the elastic force to transmission medium density ratio (Cravotto and Cintas, 2011). The extraction process has been applied after the advent of fire. The main goal of extraction is to get the bioactive plant material from inside the cells. Egyptians, Chinese, Greeks, Arabs, and Romans all historically employed extraction and distillation processes for fragrances, aesthetics, and food. Currently, manufacturing line of every field, including the food, medical, cosmetic, pharmaceutical, or renewable sectors apply one or the other method of extraction. Due to the high energy expenditure and the need

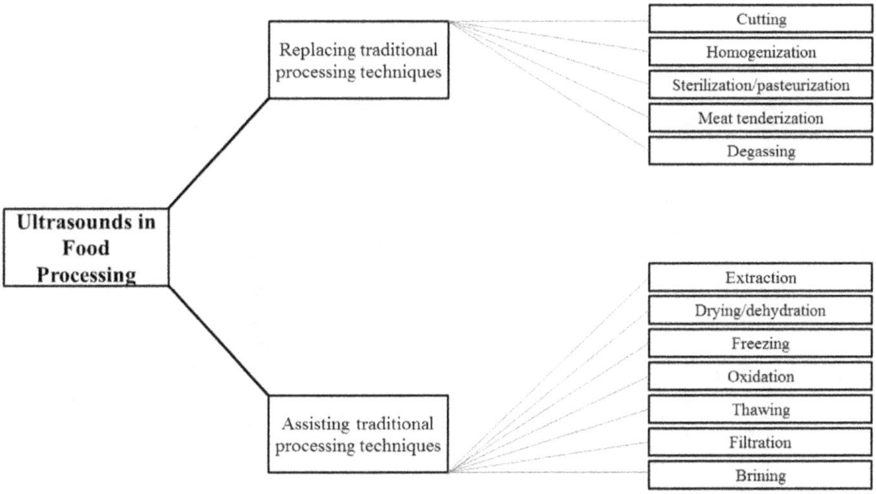

Figure 9.1 Summarized application of ultrasound in food processing (Tao and Sun, 2015).

to reduce carbon emissions, the food sectors are being pushed to create innovative energy-saving solutions. Given the industrial need, ultrasound might be a new tool for extracting intracellular bioactive plant components (Chemat and Strube, 2015).

9.2.2 Recent application of ultrasound

Several applications of ultrasound in food processing have been developed in recent years, including product quality control and quality assurance, extraction, drying, filtration, emulsification, homogenization, and so on. This section of the chapter summarizes some of these applications (Fig. 9.1). Table 9.1 lists the many procedures where ultrasound can be used.

9.3 Classification of bioactive compounds

Bioactive metabolites may be found in abundance in plants. A cell's metabolites may be grouped into two categories: primary metabolites (PMs) and secondary metabolites (SMs) (Sharma et al., 2019). Carbohydrates, amino acids, proteins, and lipids are all included in PMs, which promote growth and development. Compounds known as SMs are thought to aid plants in surviving and thriving in their natural habitats by helping them to communicate with their immediate environment. SMs, which give plants their color,

Table 9.1 Recent application and use of ultrasound in different food processing operations (Adopted and modified from Singla and Sit, 2021).

Application	Principle	Products	Advantages
Emulsification	Cavitation phenomenon	Mayonnaise	Develops rheological property; improves emulsions stability and fewer time consumed (Sarheed et al., 2020).
Filtration	Vibrations	Liquid foods, e.g., juice	Increase membrane penetration and less time required (Naji et al., 2020).
Freezing	Uniform and steady heat transfer	Fruits and vegetables, e.g., potato	Assists in the preservation of the microstructure; amended diffusion mechanism; Aids in quick freezing, rapid reduction in temperature and the development of small crystals (Zhu et al., 2020).
Drying	Uniform and steady heat transfer	Dehydrated food products	Increases the rate of mass transfer with a simultaneous reduction in drying time and an improvement in the organoleptic properties (Zhang and Abatzoglou, 2020).
Thawing	Uniform and steady heat transfer	Frozen products, e.g., bighead carp	Decrease the thawing time; prevent degradation of lipids and improved of product quality (Li et al., 2020).
Foaming	Dispersion of gas bubbles	Protein, e.g., pea	Increased the potentiality for foaming and decreased the stability of foam capacity to retain water (Xiong et al., 2018).

(continued on next page)

Degassing/deaeration	Agitation	Carbonic beverages	Decreased overflow of liquor and less time required for processing (Matsuura et al., 1994).
Cooking	Uniform and steady heat transfer	Fruits and vegetables, e.g., sweet potato	Enhanced stability of nutrients; increased heat transfer rate; enhanced the organoleptic properties and increased tenderization (Su et al., 2018).
Cutting	Cavitation phenomenon	Bread and cheese	Increased the precision in cutting; less time consumed and reduced product loss (Liu et al., 2019).
Extraction	Diffusion	Bioactive compounds	Enhance of extraction performance; time reduced for extraction and less solvent required (Dzah et al., 2020).
Rehydration	Absorption	Mushroom slices	Reduce the rehydration time of products (Zhao et al., 2018).
Size reduction	Size reduction	Milk fat	Reduce the fat globule size and increasing of surface area (Sutariya et al., 2018).
Aroma preservation	Encapsulation	Cheese	Enhanced aroma preservation and improvement of mass transfer (Mongenot et al., 2000).
Inactivation of enzyme	Enzyme inactivation	Inactivate peroxidase, e.g., tomato	Increased the shelf life (Ercan and Soysal, 2011).

flavor, and scent, are responsible for most of the plant's bioactive chemicals. These kind of molecules are defined as secondary plant metabolites that have an impact on human and animal health's and may influence the growth and development of cells in a positive or negative way. Bioactive chemicals can be found in every plant material parts, for example, roots, fruits, leaves, herbs, and grains. Some plants have medical and therapeutic advantages, while others have poisonous and harmful consequences (Azmir et al., 2013).

Although the classification of bioactive compounds into various categories and subcategories is still ongoing, they can be divided into three major types: (1) terpenes and terpenoids; (2) alkaloids, and (3) phenolic compounds are the three types of chemicals. Fig. 9.2 depicts the classification and structure of typical bioactive chemicals found in plants, along with examples.

9.3.1 Terpenes and terpenoids

Terpenes, also known as terpenoids, are perhaps the most varied group of bioactive molecules found in nature. This class is thought to include up to 40,000 distinct compounds (Lu et al., 2016). Monoterpenes, sesquiterpenes, diterpenes, triterpenes, tetraterpenes, polyterpenes, and others may be classified depending on the number of carbon molecules and the inclusion of isoprene residues (IPR). Terpenes are secondary metabolites; their role in plant development is well understood. For example, gibberellins operate as phytohormones, sterols act as cell stabilizers, and carotenoids have antiphotooxidant activity. Carotenes and oxycarotenoids, xanthophylls are the most well-known members of this class of chemicals. Carotenoids, which include β-carotene, lutein, lycopene, and zeaxanthin, are lipid soluble pigments found in vegetables and fruits that give them a yellow, orange, or sometimes red appearance (Charles, 2013). The most significant terpenes are thought to be volatile monoterpenes and sesquiterpenes, together referred to as essential oils (EOs). These are found mostly in herbs and spices, an in certain fruits, and provide a distinct aroma to them. Peppermint (*M. piperita*), basil (*O. basilicum*), lemon (*Citrus limon*), cinnamon (*C. zeylanicum*), and rosemary (*R. officinalis*) are all plants that have a high concentration of essential oils. EOs is familiar bioactive chemicals that have a variety of applications in aromatherapy, microbiology, and agriculture. It has been demonstrated that terpenes possesses a variety of beneficial pharmacological and therapeutic effects, including anti-inflammatory, chemoprotective, antitumor, antimicrobial, antimalarial, antiviral, and hepatoprotective effects (Charles, 2013).

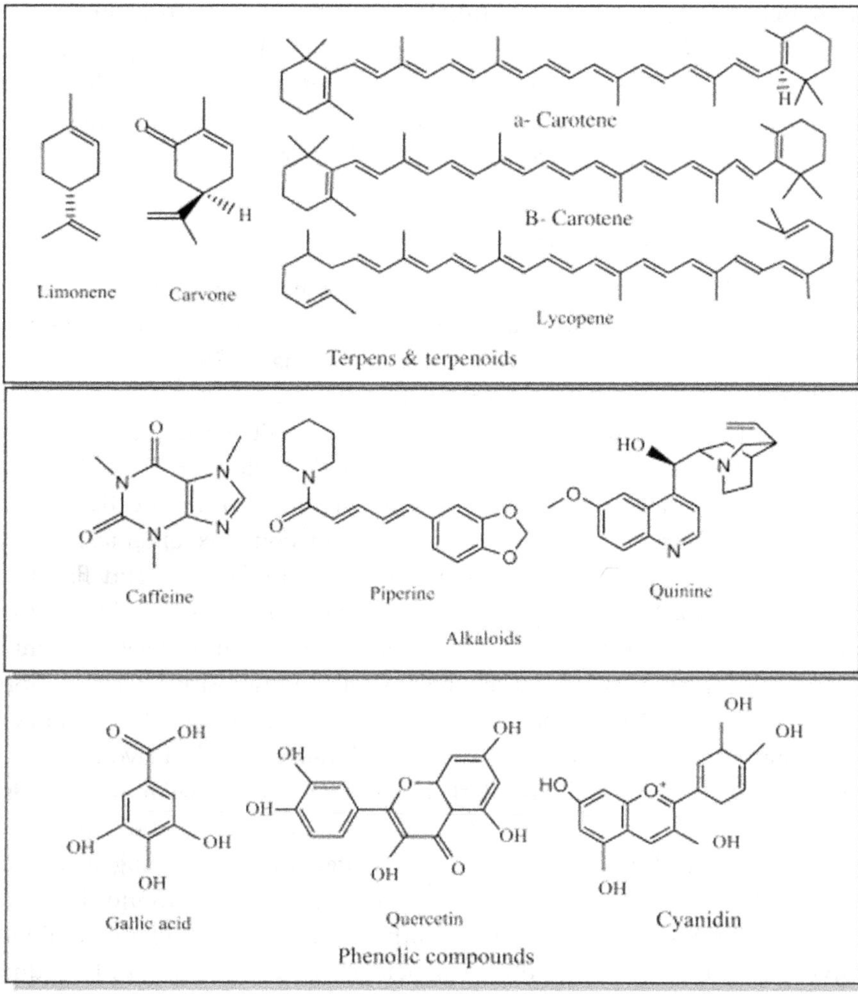

Figure 9.2 Structure and classification of bioactive compounds (Khezerlou and Jafari, 2020).

9.3.2 Alkaloids

Alkaloids are the next family of bioactive compounds often found in plants. They are characterized as cyclic compounds containing nitrogen in a reduced form. Alkaloids are often divided into three types such as true alkaloids, protoalkaloids, and pseudoalkaloids. Alkaloids are thought to be present in around 14–20% of plant species (Antolak and Kregiel, 2017). Plants belonging to the groups Solanaceae, Papaveraceae, Ranunculaceae, Erythroxylaceae, Rubiaceae, and Campanulaceae are the primary representatives. Alkaloids are regarded a growth regulator in plants, operate as a reserve chemical, and play a significant role in plants (Khatoon, 2016). Alkaloids have diversity in their biological activities, including anti-inflammatory, anticancer, depressive, antiviral, antihypertensive, antimalarial, and antibacterial effects (De Sousa Falco et al., 2008).

9.3.3 Phenolic compounds

The phrase phenolic compounds refers to substances that include single or more hydroxylated aromatic rings and are further classified into phenolic acids, flavonoids, stilbenes, ellargic acids, and lignans. Phenolic chemicals are abundant and comprise a significant category of compounds found in plants. The polyphenol family has around 8000 structurally distinct chemicals that are found in a variety of foods including fruits, flowers, vegetables, seeds, and leaves (Antolak and Kregiel, 2017). The flavonoids subgroup includes a large range of simple and complex chemicals, such as anthocyanins, flavanones, flavonols, chalcones, isoflavones, and flavones (Zhang and Tsao, 2016). Polyphenols are antioxidants that may influence the redox state of cells by directly quenching oxidative stress and chelating agents. Additionally, they stimulate the expression of redox-sensitive transcription factors for antioxidative enzymes like as catalase, superoxide dismutase, and glutathione peroxidase (Upadhyay and Dixit, 2015). Flavonoids are another class of phenolic chemicals that includes flavonols, flavones, flavanols, flavanones, and anthocyanins, isoflavones (Jakobek et al., 2007; Nijveldt et al., 2001). Catechin, epicatechin, epigallocatechin, epicatechingallate, and epigallocatechin are the most abundant flavanols. They are found in green tea, berries, apples, cocoa, and grapes. Anthocyanins, which include cyanidin, malvidin, delphinidin, peonidin, and pelargonidin, are commonly found in fruits, imparting their distinctive color and health benefits (Castañeda-Ovando et al., 2009). Fig. 9.3 illustrates different phenolic compounds.

9.4 Factors affecting the sonication process

The extraction of phytochemical substances is influenced by a variety of elements, including the raw resources, the organic solvent, the temperature, pH, and the technology used, such as ultrasonic power, frequency, and so on. For successful and better performance of ultrasound-assisted extraction (UAE), the above variables should be addressed carefully. These variables greatly impact the process. The efficiency of this process is dependent upon

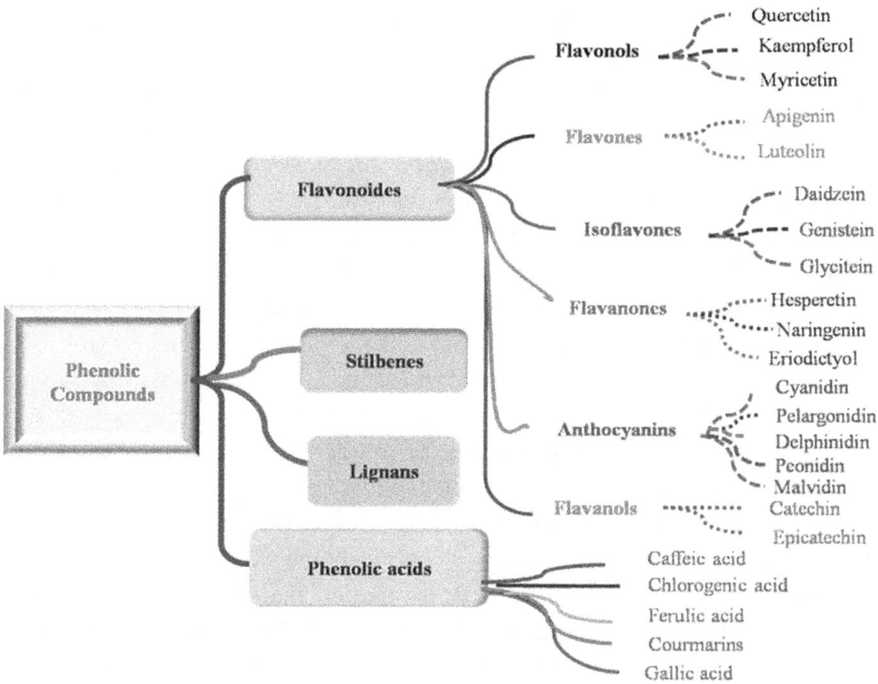

Figure 9.3 Different phenolic compounds.

the method of extraction followed. Furthermore, another noteworthy point to be considered is that the process developed should be ecofriendly and sustainable, while simultaneously being energy efficient.

9.4.1 Ultrasonic power

The production of bioactive chemicals is affected by ultrasonic power. The output of UAE extraction rises with increasing capacity or system performance and then decreases after attaining its peak (Raza et al., 2017; Rami et al., 2015; Maran and Priya, 2014; Chen et al., 2011). The power produced during the UAE was also given as an amplitude percentage ranging from 0 to 100% (Palma and Barroso, 2002). When the system was operated between the ranges of 30% and 80% of the maximum power, the ultrasonic intensity was found to vary linearly with the output power (Xu et al., 2014). Between 20 and 700 W of power is normally applied for the extraction of bioactives from fruit and vegetable by-products, the wattage being dependent on the nature of the cellular matrix and the component to be extracted (Samaram et al., 2015; Zhang et al., 2008).

Hydrodynamic force also rises when power is increased, resulting in tissue damage and an increase in the yield (Samaram et al., 2015). In rare situations, the high ultrasonic intensity may also cause the bioactive chemical to deteriorate, hence reducing the yield. The combined impact of power, temperature, and duration of sonication on the yield of total phenolic compounds in the wasted coffee grounds indicated that an increase in the power from 100 to 200 W improved the yield. However further increase in power to over 250 W lowered the yield (Al-Dhabi et al., 2017). This is in variance with the observation made by González-Centeno et al. (2014), who observed a linear correlation between the power applied and the yield of polyphenols, total flavonoid, and total antioxidant. Power has an influence on yield in addition to other extraction factors such as time, temperature, and the solvent pH. When the processing time was shorter, the yield increased with an increase in the power applied. However increase in the time of exposure was found to decrease the yield, irrespective of the power applied (Kazemi et al., 2019). A comparison between the continuous and pulsed mode of ultrasound assisted extraction revealed that there was a marginal increase in the yield of antioxidants (24% and 22% respectively) and a reduction in the extraction time (90% and 87%) on applying the continuous mode. However, considering the economics of the process, pulsed mode could be the method of choice (Pan et al., 2012a).

9.4.2 Frequency of ultrasonics

Another factor that influences the extraction performance is the frequency applied. The UAE uses 20–120 kHz for bioactive chemicals from fruit and vegetable by-products. Increasing the ultrasonic frequency or intensity successfully decreases or diminished the liquid cavitation bubbles. The high frequency inhibits the compression-rarefaction cycle, making it harder to develop cavitation bubbles. Size of the cavitation bubble and mass flow are linked to frequency selection (Esclapez et al., 2011). Thus, choosing the right frequency is critical for ultrasonic extraction. Extreme shear forces are produced by low-frequency high-intensity ultrasound, whereas high-frequency low-power density creates enormous numbers of reactive compounds.

Pectin was extracted from the peel of pomegranate, passion fruits, mango, oranges, sisal waste, grapefruit, and antioxidant from pomegranate extract at a persistent frequency of 20 kHz, while mallic and tartaric acid were extracted from grape seed at a constant frequency of 24 kHz

(Kumar et al., 2021). The influence of altering frequency (19 kHz, 25 kHz, 40 kHz, and 300 kHz) on the UAE of vegetable oil obtained from soybean germ, concluding that the greatest yield was achieved at 19 kHz (Cravotto et al., 2008). Similarly, González-Centeno et al. (2014) evaluated the impact of altering the frequency (40 kHz, 80 kHz, and 120 kHz) on the extraction of phenolic compounds from grapefruit and pomegranate by using the response surface approach. He also concluded that 40 kHz was the most effective and yield was increased at low and high frequencies but decreased at intermediate frequencies.

9.4.3 Solvent used in ultrasonic extraction

Water is often utilized as a solvent in ultrasonic processes. Additionally, ethanol, acetone, methanol, and hexane are utilized in the extraction process, with ethanol being the preferred solvent owing to its safety. Acidified water is utilized as the solvent for the extraction of pectin. Mineral acid or citric acid has been used to acidify the water. Hydrochloric acid (for durian rind, grapefruit peel) and nitric acid were used to adjust the pH of deionized water and examine their influence on the extraction of pectin (for mango peel, passion fruit peel). Alcohols and acetones containing varying amounts of water have been extensively utilized to extract phenolic chemicals from plant sources. Ethanol has been shown to have the greatest affinity for phenolic compounds in a variety of examined systems, and so is the preferred solvent for extracting polyphenols from waste of fruits and vegetables (Rami et al., 2015). Citric acid has been used to extract pectin from grape pomace, orange peel, banana peel, and sunflower head. Citric acid was used instead of mineral acid to extract pectin from grape pomace and banana peel, respectively, and the combination of citric acid and ultrasound was shown to be very effective for pectin extraction (Minjares-Fuentes et al., 2014). Additionally, ethanol is recommended as a solvent owing to its cheap, renewable supply, and GRAS (generally regarded as safe) status (Rodrigues et al., 2015). The concentration of ethanol is a critical component impacting yield in the UAE. Changes and increase in ethanol concentration enhances the extraction of phenolic compounds until the concentration reaches its maximum, after which the yield decreases. Such a trend has been found in the extraction of flavonoids from grapefruit solid waste (Garcia-Castello et al., 2015), hawthorn seed extraction (Garcia-Castello et al., 2015), and hawthorn seed extraction (Garcia-Castello et al., 2015; Pan et al., 2012b).

9.4.4 Liquid to solid ratio

The concentration of the solvent was expressed as the liquid to solid ratio (LSR) or the solid to liquid ratio (SLR). UAE of compounds is found to increase with an increase in LSR upto a certain point, after which the yield is found to decrease. The extracting solution has a high viscosity at low LSR, making turbulence more difficult owing to the mutually opposing factors of a negative pressure of the rarefaction cycle and the stronger cohesive force of the viscous solution. When the LSR is increased, the ultrasound intensity is increased, resulting in increased fragmentation, erosion, and pore development, hence enhancing the yield. Increased area of contact between the substance and the solvent may also boost yield. The reduction in yield at a further very high LSR could possibly be attributed to increased cavitation, which causes degradation of the desirable solute (Maran and Priya, 2014; Samaram et al., 2015). When the LSR was increased from 20:1 to 30:1, the extraction of pectin from sisal waste enhanced, but subsequently fell when the LSR was continued to increase to 40:1 (Maran and Priya, 2014).

9.4.5 Temperature of UAE

Temperature possesses has both positive and negative effect on the extraction process. High temperatures may disturb the interaction of the matrix and the solvent. Yet, because high temperature improves solvent diffusion rates (Richter et al., 1996), it enhances UAE yield; however, raising the temperature further diminishes yield, similar to the impact of increasing the power. Temperature improves the UAE yield because to its dual influence on both the solute and the solvent. Higher temperature improves the dissociation characteristic and accessibility of the solute in the solvent. The tissue matrix's solvent diffusion coefficient improves as the solvent viscosity is reduced. Increased temperature reduces yield owing to the decreased cavitation action. Al-Dhabi et al. (2017) noticed this tendency in the extraction of phenolic compounds from discarded spent coffee grounds in the UAE, where the yield rose as the temperature increased from 30°C to 45°C and dropped as the temperature went above 45°C. Similar results have been reported for pectin extraction from pomegranate peel (Moorthy et al., 2015), polysaccharide extraction from rambutan fruit peel (Maran and Priya, 2014), and grapefruit peel extraction (Moorthy et al., 2015; Xu et al., 2014).

9.4.6 Solvent pH

pH is a critical feature impacting the quantity and characteristics of bioactive compounds extracted in the United Arab Emirates. The separation of

Table 9.2 The optimum pH for the extraction of pectin by ultrasonic assisted extraction (UAE).

Sample name	Optimum pH	Extracted compound
Pomegranate peel	1.2	Pectin (Moorthy et al., 2015)
Orange peel	1.5	Pectin (Hosseini et al., 2019)
Eggplants peel	1.5	Pectin (Kazemi et al., 2019)
Jackfruit peel	1.5	Pectin (Moorthy et al., 2017)
Banana peel	3.2	Pectin (Ponmurugan et al., 2017)
Sunflower head	3.2	Pectin (Ponmurugan et al., 2017)

polysaccharides (pectin) from fruits and vegetable waste and by-products has been inspected in the pH range of 1–5. When the pH is kept low, pectin recovery is increased. The extremely acidic media promotes cell wall disintegration, solubilization and breakdown of pectin, increasing dispersion into the surrounding fluid and facilitating recovery. Pectin clumps at high pH values, limiting its release into the extraction media. Furthermore, the pH level influences the degree of esterification (DE) of the extracted pectin. At low pH, DE is minimal owing to pectin desertification under very acidic conditions. The linear relationship between pH and total phenolic content was significant considerably. Similarly, the optimal pH for extraction of pectin is between 1 and 3. The optimal pH for pectin extraction from various peel sources is listed in Table 9.2.

9.4.7 Time of UAE

The time of UAE has been widely investigated and reported on in terms of its influence on the production of bioactive components of fruits and vegetables by-products. Increased sonication duration initially enhances yield. However, yield declines when time are increased further; this impact is analogous to increased power and temperature. When the sonication period is increased initially, the cavitation action of the ultrasound accelerates the swelling, fragmentation, hydration, and pore development of the tissue matrix of the plant from which the solute is supposed to be obtained. Every one of these qualities enhances the solute and extraction medium's exposure and allows them to be discharged into the solvent. Ultrasound exposure over an extended period of time damages the solute structurally and reduces the extraction yield. This trend has been observed in the extraction of antioxidant polysaccharides from sijiaoling stems (Raza et al., 2017), pomegranate peel (Moorthy et al., 2015), jackfruit peel (Moorthy et al., 2017), sunflower heads (Ponmurugan et al., 2017), banana peel

(Maran et al., 2017), grapefruit peels (Xu et al., 2014). Pectin from eggplant peel (Kazemi et al., 2019), grape pomace (Minjares-Fuentes et al., 2014), grapefruit seed (Ghafoor et al., 2009), polyphenols, and anthocyanin from jabuticaba peel (Rodrigues et al., 2015), and flavonoids content and antioxidant activity from grape pomace. Unlike pectin from sour peels, which has a negative linear impact on the yield (Hosseini et al., 2019), phenolic components from coconut (*Cocos nucifera*) shell powder possess a significant and positive linear effect on productivity in the time range of 20 to 60 minutes (Rodrigues et al., 2008). At high temperatures and a larger solvent to solid ratio, the yield of oil extracted from papaya seed increased over time (Samaram et al., 2015). The optimal sonication duration for extracting pectin has been reported to be between 10 and 60 minutes, but the optimal sonication time for extracting phenolic compounds has been found to be between 10 and 90 minutes, depending on the fruit or vegetable by-product.

9.5 Types of ultrasound equipment

Ultrasound equipment is composed of a power generator, and an emitter, a transducer(s) which are responsible for physically transmitting the ultrasonic sound waves to the media (Povey and Mason, 1998). The emitter is the critical component that controls the kind of ultrasound. The major functions of ultrasound equipment are to transmit the ultrasonic sound waves to the system by physical means. As a nutshell, the ultrasonic instruments utilized in the UAE may be classified generally into two categories: ultrasound bath and ultrasonic probe mode. Fig. 9.4 illustrates a schematic diagram of ultrasound equipment (ultrasound bath and ultrasound probe).

9.5.1 Ultrasound bath

An acoustic sounder creates an excitation signal that matches the modal frequencies of the tuning assembly in ultrasound bath systems. Piezoelectric transducers convert electrical energy to mechanical vibrations, and many transducers are often installed underneath an ultrasonic source made of stainless steel. Few tanks are still equipped with thermostatically controlled heaters. While the majority of commercial ultrasonic baths provide ultrasonic levels that are acceptable for cleaning, solvent degassing, and removing metals and metallic ions, organic pollutants from environment. However, it

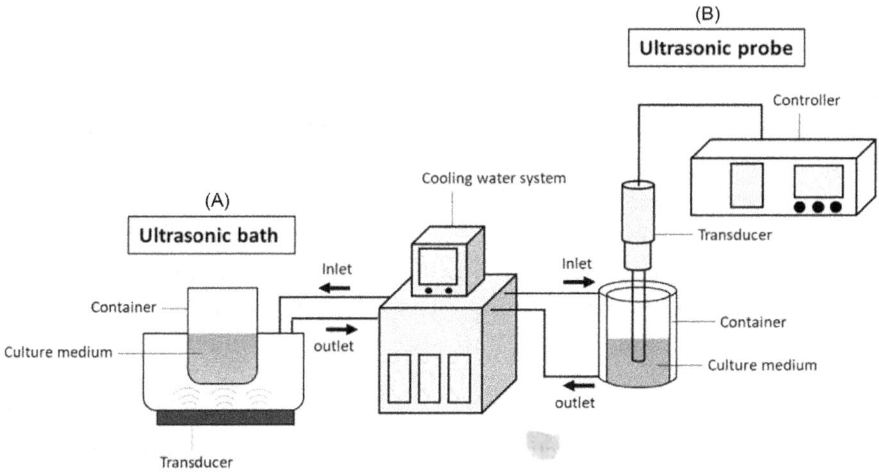

Figure 9.4 Ultrasound Equipment: (a) ultrasound bath and (b) ultrasound probe (Behzadnia et al., 2019).

is less successful at extracting complex analyses (Mohammed and Alhajhoj, 2019). The force should be sufficient to create cavitation inside the bath to initiate the extraction (Astráin-Redn et al., 2019). The vessel position inside the ultrasound bath is a crucial element in affecting extraction performance. The extraction tank must be positioned directly above the transducer (Kulkarni and Rathod, 2014). Ultrasound baths offer the benefit of being simple to operate and economically viable, they are not suited for chemical reactions owing to the reaction's limited repeatability, since the intensity of ultrasonic system is greatly dampened by solvent used and laboratory glassware (Sabnis et al., 2020). While the ultrasonic bath may produce high extraction yields through indirect extraction (Vinatoru, 2001), it can extract just a limited number of bioactive chemicals, while direct extraction can yield a significant quantity of extracts.

9.5.2 Ultrasound probe type

The main part of an ultrasonic probes type is a single transducer which is connected to the control panel through a wire, which allows for frequency, time, and temperature control of the system. The sample is then put in a distinct stainless-steel (SS) tank and subjected to ultrasonic treatment. The efficiency of a probe type sonicator is higher than ultrasound bath, which may supply up to 100 times energy to the extraction medium (Ranjha et al., 2021). Thus, enhanced performance is envisaged, owing primarily to the fact that this ultrasonic probe can be immersed directly to the solution during sonication process, increasing the interfacial interaction with the substance and thus reducing mass transfer. This system is often used to sonicate materials with a tiny volume. The ultrasonic intensity has been conveyed solely just at the tip of the probe, which makes it more potent than ultrasonic bath. Temperatures of the materials may rapidly rise throughout sonication, as a result, the configuration of the equipment must be carefully considered. Initially, an open type beaker is immersed directly in the ultrasonic bath, and subsequently the influence of acoustic probes on repeatability of the data was explored (Weissler et al., 1950). With advancements in ultrasound and ultrasonic generator technology, piezoelectric transducer may directly penetrate into the ultrasonic solvent. While this technology facilitates mass transfer, it also increases sample destruction, owing mostly to the deterioration of the ultrasonic probe (Yao, 2016).

9.6 Mechanism of extraction

Extraction is a common practice in the food industry. The primary goal of extraction is to liberate target chemicals from a variety of plant matrices. Ultrasound is a sequence of compressive and rarefaction oscillations that may be transmitted throughout a solid, liquid and gaseous media, resulting in displacement and the dislodging of molecules. Passing of high-intensity sound wave causes a negative pressure due to deformation. High intensity sound wave overcomes all forces of attraction that keep the molecules apart, separating them and causing cavitation bubbles to form. These bubbles expand during coalescence phase and then collapse during compression phase, resulting in hot spots and severe local conditions. Temperatures may rise to 5000 K and pressures can get to 1000 atm. These hot spots accelerate biological processes nearby (Chemat et al., 2017; ŠicŽlabur et al., 2015; Rosello-Soto et al., 2015).

Transducers were employed in the UAE, since they are the primary component of ultrasonic equipment, as they are able to convert electrical or mechanical energy into sound waves patterns. During the UAE process, the sound wave travels through the tank filled with medium until transducers cause acoustic resonance, resulting in the formation of compressive and rarefaction (Fu et al., 2020). Sonication-induced cavitation and implosion tear the cell wall and increase the number of ruptured cells. Rapid fragmentation results in the solubility of the active constituents in the solvent as a result of a smaller particle size, an increased surface area and an increase in diffusion

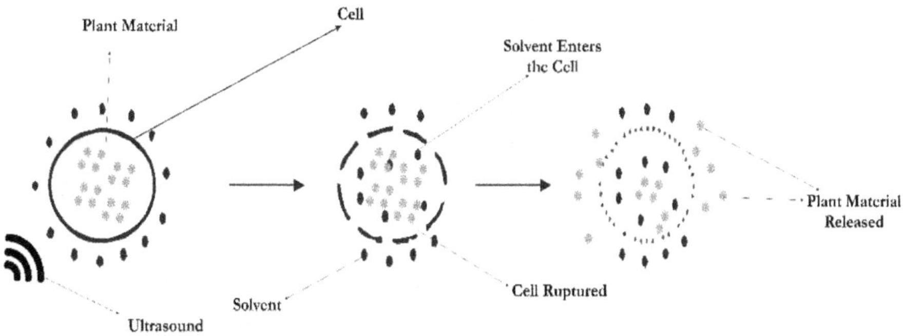

Figure 9.5 Mechanism extraction of ultrasound (Ranjha et al., 2021).

rates in the solid matrix's boundary layer. Cell disruption by sonication ensures free access of solvent to the cellular material, thus enhancing the yield (Gallo et al., 2018). Fig. 9.5 illustrates the technique of extraction by a UAE process.

9.7 Influence of treatment conditions on extraction

The ultrasound aided extraction is affected by the treatment conditions. Temperature, ultrasonic frequency, time of extraction, type of solvent or medium not only affects the extraction efficiency, but also composition of extracts.

9.7.1 Influence of temperature

Temperature is an important component in the conventional extraction process since it can promote solvent diffusion and penetration into the sample matrix (Fu et al., 2020). While increasing the temperature may result in improved extraction, it may also cause degradation of plant components (Flórez et al., 2014). The number of cavitation nuclei is affected by temperature. The tension and vapor pressure within the cavity increase as the temperature rises from 10 to 50°C, resulting in a reduced Pmax (maximum pressure) and diminished sonochemical effects. When adjusting the temperature, several considerations must be made such as boiling point of the reagent being employed and the boiling point of the target component. If the temperature of the solvent is greater than its boiling point, there is a possibility of an uneconomic extraction. Additionally, a variety of phytoconstituents are vulnerable to a greater temperature, since the critical level is surpassed, the component may begin deteriorating.

9.7.2 Influence of frequency

Frequency has also an impact on the sonication process. Loops are shortened when a high frequency is used therefore; insufficient time to generate sufficient negative pressure inhibits the creation of bubbles (Lavilla and Bendicho, 2017). Thus, as the ultrasonic frequency rises, the creation of cavitation bubbles decreases. This is because the rarefaction time isn't long enough for bubbles to form and disrupt the liquid. It's important to remember the optimum frequency setting when changing the temperature. Increasing the frequency of the extraction process may result in an uneconomical extraction method. Additionally, increased frequency may result in phytoconstituent degradation. A link between sonication frequency and catechin degradation was identified in a research work (Zhu et al., 2018). The authors hypothesized that a greater frequency would result in catechin breakdown.

9.7.3 Influence of time

Extraction is influenced by the passage of time. While a longer time period provides for better extraction, beyond a certain point, the target bioactive component may begin to degrade. While modifying the temperature, it is important to keep in mind that extending the time duration may result in an uneconomic extraction. Additionally, prolonged exposure to phytoconstituents may result in their breakdown (Safdar et al., 2016).

9.8 Impact of sonication technology on bioactive compounds and biological activities

Nowadays consumers are increasingly choosing meals that contain biomolecules that are not only good for vital metabolic functions but also offer long-term health advantages. It has long been recognized that the link between health and diet is interrelate. There is mounting evidence that such bioactive chemicals are directly linked to human health, wellbeing, and sickness management (Lobo et al., 2010). The majority of active chemicals is contained inside the cells and must be released during the extraction procedure. Decoction, maceration, solvent extraction, hydrodistillation, steam distillation, and Soxhlet extraction procedures, as well as high hydrostatic pressure extraction, high pressure process, and pulse electric field process, have all used to extract polyphenolic chemicals from different plant materials (Wen et al., 2018). All of these procedures have certain disadvantages, including limited extraction rates, increased time and solvent consumption, and the need for high temperatures, which may result in the destruction of polyphenolic compounds (Lianfu and Zelong, 2008). Furthermore, these procedures are difficult to execute correctly in order to get the appropriate crushing impact (Vilkhu et al., 2008). To address these limitations, innovative strategies for extracting bioactive chemicals from plant materials have been developed that do not result in nutritional degradation or loss. Ultrasound may be employed as a green, beneficial, and alternative technology for improving solvent extraction of bioactive substances. Due to its great efficiency and ease of use, ultrasonic technology has garnered increased attention. Additionally, US extraction may be employed to achieve higher yields at lower temperatures, shorter processing times, lesser organic solvent uses and energy input, less waste, and environmental contamination. Additionally, US equipment is more user-friendly than previous approaches (Chemat et al., 2017).

9.8.1 Extraction of bioactive components from cash crops

Cash crops are grown with the intention of selling them on the market or exporting them for profit. Plants produce phytochemicals, which are biologically active compounds. Plant components such as leaves, seeds, seed coats, roots, fruit pulps, barks, and flowers can all be used to extract phytochemicals. Cash crops have a diverse array of phytochemicals, including polyphenols, terpenes, proteins, and pigments that are widely employed in the culinary, medicinal, and aesthetic sectors (Joana Gil-Chávez et al., 2013). The rising interest in such bioactive compounds has boosted the need for more efficient extraction techniques that can extract the maximum quantity of bioactive compounds at a minimum time period and with lowest feasible cost (Esclapez et al., 2011). High-intensity ultrasound offers the benefits of being inexpensive, repeatable, simple to use, and effective. It may be a simple industry-related technique for increasing the extraction of bioactive components from food (Soria and Villamiiel, 2010; Vilkhu et al., 2008). The process is referred to as the cavitation effect, and it consists of four phases. To begin, when ultrasound is administered to the plant substrate, cavitation bubbles form at the surface. Second, the bubble explodes, thus releasing the micro-pressure jet's and hence elevating the surface temperature. Disruption of the matrix surface allows direct interaction between the cellular materials and the solvent. Lastly, the bioactive substance is liberated from the solvent and is recovered in a suitable medium. Ultrasound may facilitate mass transfer in this manner. As a result, this technique is applicable to any liquid media, and its advantage over previous approaches is that it is not solvent-dependent. The UAE approach has a lot of potential in food processing and manufacturing since it may considerably increase extraction rate while reducing extraction time (Chemat et al., 2011).

9.8.2 Extraction of bioactive components from fruits and vegetables

Along with rising consumer demand for bioactive substances, fruits, and vegetable crops are gaining greater attention. Numerous studies have indicated that bioactive chemicals found in horticulture crops have considerable beneficial impacts on human health, which will also stimulate increased consumption of horticultural crops. Fruits, vegetables, and flowers are the most common horticultural crops. The bioactive components found in fruits and vegetables have been shown to be resistant to a variety of illnesses, which has been associated with the presence of multiple antioxidants. There has been a greater focus on identification and extraction of bioactive compounds from fruits. These include compounds such as polysaccharides and polyphenols. These bioactive chemicals have been isolated from the flesh, peel, seeds, and bark of many fruits. The extraction factors, notably the duration and temperature of the extraction, have a profound influence on the rate of separation of phytochemicals. It is discovered that ultrasonic extraction of pectin from horticultural crops considerably increased yields as compared to the usual approach (Wang et al., 2016). Pectin from mango peel was compared to laboratory-grade citrus pectin in terms of molecular composition, rheological qualities, and emulsifying characteristics. The degree of methoxylation and average molecular weight of the mango peel pectin were significantly higher than those of the standard extraction samples.

Garlic contains a lot of plant proteins that have chemotherapeutic, antibacterial, antioxidant, and many other health benefits, thus it is popular in the world of functional foods. Processing garlic with high-intensity ultrasound is considered to break down chemical bonds between polysaccharides and amino acids in cell membranes, increase the hydrophobicity of the surface, and free the protein from tissue (Harnedy et al., 2015). Nevertheless, hydrolysis of garlic powder by enzymatic treatment may boost angiotensin-converting enzyme inhibitory activity (ACE). Ma et al. (2015) studied the effects of single-frequency ultrasonic pretreatment and appropriate concurrent acoustic pretreatment on the production of ACE inhibitors from garlic hydrolysates. The findings indicated that when the ideal ultrasonic settings were used, garlic hydrolysate's ACE inhibitor action was greatly enhanced. Tian et al. (2012) extracted A. bisporus polysaccharide (ABPS) from mushroom the United Arab Emirates. The molecular weight of this polysaccharide is 158 kDa, and it is mostly made up of glucose with a -type glycosidic link. Additionally, the findings suggested that this polysaccharide had significant antioxidant properties. Chen et al. (2015) successfully used ultrasonic to extract anthocyanins and total phenolic compounds from sugar beet molasses. The findings demonstrated a considerable link between the biological activity of extracts and their chemical compositions. UAE obtains a larger yield of carnosic acid from rosemary when ethanol is used as the solvent, but methanol yields a higher yield of rosmarinic acid (Jacotet-Navarro et al., 2015). Additionally, the extracts have significant antioxidant activity and are employed in a variety of applications, such as antioxidants, preservatives, and freshness-retention agents. Ultrasonic aided extraction to extract phenolic antioxidants from green tea (Yan et al., 2011). In comparison to typical shaken extraction procedures, ultrasonic extraction yields extracts with a reduced caffeine level while maintaining their antioxidant potential. The following findings demonstrate a strong correlation between the physicochemical qualities and structural properties of bioactive substances. As a result, the link between structure and function should be clarified in order to facilitate future research. Table 9.3 shows the **o**ptimized conditions for extracting polyphenolic compounds from fruit and vegetable waste materials.

9.8.3 Extraction of bioactive compounds from medicinal plants

For ages, traditional medicine has been utilized to prevent and treat ailments. Many plant and its different parts are used to prevent and treat a variety of diseases, as we all know. The isolation, purification, and extraction of active constituents in plants have been a crucial challenge and a major theme in the evolution of medicine. Currently, traditional extraction processes such as decoction, reflux extraction, percolation, and dipping are used extract the bulk of therapeutic chemical components. Nonetheless, these traditional procedures need heating, a prolonged extraction time, a poor extraction rate, and a low impurity clearing rate, among other requirements. To address these shortcomings, several investigations on the extraction of phytochemical components from medicinal plants using ultrasound have been studied. In compared to conventional approaches, bioactive compounds were extracted

Table 9.3 Optimized conditions for extracting phenolic/flavonoids/antioxidants from fruit and vegetable waste materials.

Waste materials	Optimum conditions for UAE/major outcomes
Mango peel	Ultrasound has no negative impact of the total phenolic compounds (Guandalini et al., 2019).
Banana peel	T: 30; t: 5; P: 150W; SSR = 8:100; AC: 60% (Vu et al., 2017)
Coffee ground spent	P: 244 W; T: 40; t: 40; LSR = 17:1 (Al-Dhabi et al., 2017).
Malt Peel	P: 70; t: 35; LSR = 40:1 (Nishad et al., 2019).
Grape pomace	P: 150; t: 25; F: 40 (González-Centeno et al., 2014).
Grape seed	T: 56.03; t: 29; EC: 53.15 for maximum total phenolic components (Ghafoor et al., 2009).
Grape fruit solid waste	T: 25; t: 55; EC: 40 g/10 g (Garcia-Castello et al., 2015)
Hawthorn seed	T: 91; t: 37; EC: 72%. LSR = 18:1 (Pan et al., 2012a).
Black chokeberry by products from tea industry	T: 70; t: 80; P: 206 for total phenolic compounds; T: 70; t: 75; P: 210 for flavonoids; T: 70; t: 45.6; P: 216 for anthocyanin; T: 70; t: 89.7; P: 199 for proanthocyanidins (Ramic et al., 2015).
Black chokeberry waste	T: 20-70; t: 0-240; P: 0-100 W; EC: 0-50 for high temperature improved the extraction (Galván D'Alessandro et al., 2014).
Olive leaf extract	t: 60; EC; 50%; LSR = 20:1 (Şahin and Şamlı, 2013).
Coconut shell powder	T: 30; t: 15, pH: 6.5; LSR = 50:1 (Rodrigues et al., 2008).
Jabuticaba peels	T: 10; pH: 1; EC: 46 (Rodrigues et al., 2015).
Pomegranate peel	T: 25; F: 20; LSR = 50:1 (Continuous UAE increased yield by 24% and time reduction by 90%) (Pan et al., 2012b).

T means Temperature (°C); t means sonication time (min); LSR means liquid solid ratio (mL/g); SSR means sample solvent ratio (g/mL); P means power (W)/power intensity (W.cm^2)/power density (W/mL)/amplitude (%); EC means ethanol concentration (%); AC means acetone concentration (%); F means frequency (kHz); S means solvent.

from mint leaves, hops, marigold, and fennel utilizing UAE. The findings suggested that the extraction rate (34%) was much more than the traditional approach. When compared to static extraction with a solvent to fruit weight ratio of 40 mL/g, ultrasonic assisted extraction increased Geniposide production by 16.5% (Ji et al., 2006). Hemwimol et al. (2006) used ultrasonic aided ethanol extraction to extract anthraquinones alkaloids from the *Morinda citrifolia* roots. The extraction time was lowered by 75% when compared to a nonsonicated sample. Mantegna et al. (2012) used ultrasonic aided extraction to extract resveratrol from *Polygonum cuspidatum*. The findings indicated a considerable increase in extraction yields. Chemical elements of plant species are mostly found inside cells. As a result, the cell wall should be damaged to allow the active ingredient to be released. However, obtaining the necessary chemicals is problematic owing to a lack of effective extraction methods. Ultrasonic-assisted extraction would be ideal for resolving this issue. The combination of cavitation, mechanical, and thermal processes dramatically increases absorption into the cell. The solvent dissolves an increasing number of intracellular components, which are subsequently collected and purified. As a result of the foregoing studies, UAE technology has the potential to dramatically increase the extraction of bioactive substances such as polyphenols, anthocyanins, and aromatic compounds from medicinal crops. The increased yield and activity of bioactive molecules achieved by the UAE techniques represent significant advancements.

9.9 Conclusions

Ultrasound extraction should be considered as a technologically and economically viable technology for extracting phytoconstituents from various plants. The UAE has demonstrated that it is better suited to extract biologically active compounds from plants and plant byproducts without causing damage to the structure of volatile bioactive chemicals. Ultrasound is a relatively recent innovation in the food and food industry. It provides the benefits of reducing taste loss, enhancing homogeneity, conserving energy, boosting output, increasing efficiency, lowering chemical and physical hazards, and being ecosustainable. These techniques can be used to a variety of underutilized by-products to extract valuable bioactive elements. The process factors have a significant impact on extraction effectiveness and should be carefully investigated in the laboratory about any procedure in the pharmaceutical, cosmetics, or food industries to get bioactive chemicals from medicinal plants and plant extracts.

References

Al-Dhabi, N., Ponmurugan, K., Maran Jeganathan, P., 2017. Development and validation of ultrasound-assisted solid-liquid extraction of phenolic compounds from waste spent coffee grounds. Ultrason. Sonochem. 34, 206–213. https://doi.org/10.1016/j.ultsonch.2016.05.005.

Antolak, H., Kregiel, D., 2017. Food preservatives from plants. Food Addittives. InTech, London, UK, pp. 45–71. http://doi.org/10.5772/intechopen.70090.

Astráin-Redín, L., Raso, J., Condón, S., Cebrián, G., Álvarez, I., 2019. Application of high-power ultrasound in the food industry. Sonochemical Reactions. Intech Open.

Azmir, J., Zaidul, I., Rahman, M., Sharif, K., Mohamed, A., Sahena, F., et al., 2013. Techniques for extraction of bioactive compounds from plant materials: a review. J. Food Eng. 117 (4), 426–436. https://doi.org/10.1016/j.jfoodeng.2013.01.014.

Behzadnia, A., Moosavi-Nasab, M., Tiwari, B.K., 2019. Stimulation of biosurfactant production by *Lactobacillus plantarum* using ultrasound. Ultrason. Sonochem. 59, 104724. https://doi.org/10.1016/j.ultsonch.2019.104724.

Castañeda-Ovando, A., Pacheco-Hernández, M., Páez-Hernández, M., Rodríguez, J., Galán-Vidal, C., 2009. Chemical studies of anthocyanins: a review. Food Chem. 113 (4), 859–871. https://doi.org/10.1016/j.foodchem.2008.09.001.

Charles, D.J., 2013. Antioxidant Properties of Spices, Herbs and Other Sources. Springer, New York, USA http://doi.org/10.1007/978-1-4614-4310-0.

Chemat, F., Zill-e-Huma, Khan, M., 2011. Applications of ultrasound in food technology: processing, preservation and extraction. Ultrason. Sonochem. 18 (4), 813–835. https://doi.org/10.1016/j.ultsonch.2010.11.023.

Chen, Y., Luo, H., Gao, A., Zhu, M., 2011. Ultrasound-assisted extraction of polysaccharides from litchi (Litchi chinensis Sonn.) seed by response surface methodology and their structural characteristics. Innovative Food Sci. Emerg. Technol. 12 (3), 305–309. https://doi.org/10.1016/j.ifset.2011.03.003.

Chen, M., Zhao, Y., Yu, S., 2015. Optimization of ultrasonic-assisted extraction of phenolic compounds, antioxidants, and anthocyanins from sugar beet molasses. Food Chem. 172, 543–550. https://doi.org/10.1016/j.foodchem.2014.09.110.

Chemat, F., Strube, J., 2015. Green Extraction of Natural Products: Theory and Practice. John Wiley & Sons, Weinheim, Germany.

Chemat, F., Rombaut, N., Meullemiestre, A., Turk, M., Perino, S., Fabiano-Tixier, A., Abert-Vian, M., 2017. Review of green food processing techniques. Preservation, transformation, and extraction. Innovative Food Sci. Emerg. Technol. 41, 357–377. https://doi.org/10.1016/j.ifset.2017.04.016.

Cravotto, G., Boffa, L., Mantegna, S., Perego, P., Avogadro, M., Cintas, P., 2008. Improved extraction of vegetable oils under high-intensity ultrasound and/or microwaves. Ultrason. Sonochem. 15 (5), 898–902. https://doi.org/10.1016/j.ultsonch.2007.10.009.

Cravotto, G., & Cintas, P., 2011. Introduction to sonochemistry: a historical and conceptual overview.

De Sousa Falcão, H., Leite, J., Barbosa-Filho, J., De Athayde-Filho, P., De Oliveira Chaves, M., &Moura, M., et al., 2008. Gastric and duodenal antiulcer activity of alkaloids: a review. Molecules 13 (12), 3198–3223. https://doi.org/10.3390/molecules13123198.

Dzah, C.S., Duan, Y., Zhang, H., Wen, C., Zhang, J., Chen, G., Ma, H., 2020. The effects of ultrasound assisted extraction on yield, antioxidant, anticancer and antimicrobial activity of polyphenol extracts: a review. Food Biosci. 35, 100547. https://doi.org/10.1016/j.fbio.2020.100547.

Ercan, S.Ş, Soysal, Ç., 2011. Effect of ultrasound and temperature on tomato peroxidase. Ultrason. Sonochem. 18 (2), 689–695. https://doi.org/10.1016/j.ultsonch.2010.09.014.

Esclapez, M., García-Pérez, J., Mulet, A., Cárcel, J., 2011. Ultrasound-assisted extraction of natural products. Food Eng. Rev. 3 (2), 108–120. https://doi.org/10.1007/s12393-011-9036-6.

Flórez, N., Conde, E., Domínguez, H., 2014. Microwave assisted water extraction of plant compounds. J. Chem. Technol. Biotechnol. 90 (4), 590–607. https://doi.org/10.1002/jctb.4519.

Fu, X., Belwal, T., Cravotto, G., Luo, Z., 2020. Sono-physical and sono-chemical effects of ultrasound: primary applications in extraction and freezing operations and influence on food components. Ultrason. Sonochem. 60, 104726. https://doi.org/10.1016/j.ultsonch.2019.104726.

Gallo, M., Ferrara, L., Naviglio, D., 2018. Application of ultrasound in food science and technology: a perspective. Foods 7 (10), 164. https://doi.org/10.3390/foods7100164.

Galván D'Alessandro, L., Dimitrov, K., Vauchel, P., Nikov, I., 2014. Kinetics of ultrasound assisted extraction of anthocyanins from Aroniamelanocarpa (black chokeberry) wastes. Chem. Eng. Res. Des. 92 (10), 1818–1826. https://doi.org/10.1016/j.cherd.2013.11.020.

Garcia-Castello, E., Rodriguez-Lopez, A., Mayor, L., Ballesteros, R., Conidi, C., Cassano, A., 2015. Optimization of conventional and ultrasound assisted extraction of flavonoids from grapefruit (*Citrus paradisi* L.) solid wastes. LWT - Food Sci. Technol. 64 (2), 1114–1122. https://doi.org/10.1016/j.lwt.2015.07.024.

Garner, W., Hunt, F., 1980. Origins in acoustics: the science of sound from antiquity to the age of Newton. Leonardo 13 (3), 248. https://doi.org/10.2307/1577837.

Ghafoor, K., Choi, Y., Jeon, J., Jo, I., 2009. Optimization of ultrasound-assisted extraction of phenolic compounds, antioxidants, and anthocyanins from grape (*vitis vinifera*) seeds. J. Agric. Food Chem. 57 (11), 4988–4994. https://doi.org/10.1021/jf9001439.

González-Centeno, M., Knoerzer, K., Sabarez, H., Simal, S., Rosselló, C., Femenia, A., 2014. Effect of acoustic frequency and power density on the aqueous ultrasonic-assisted extraction of grape pomace (*Vitis vinifera* L.) – a response surface approach. Ultrason. Sonochem. 21 (6), 2176–2184. https://doi.org/10.1016/j.ultsonch.2014.01.021.

Guandalini, B., Rodrigues, N., Marczak, L., 2019. Sequential extraction of phenolics and pectin from mango peel assisted by ultrasound. Food Res. Int. 119, 455–461. https://doi.org/10.1016/j.foodres.2018.12.011.

Harnedy, P., O'Keeffe, M., FitzGerald, R., 2015. Purification and identification of dipeptidyl peptidase (DPP) IV inhibitory peptides from the macroalga Palmaria palmata. Food Chem. 172, 400–406. https://doi.org/10.1016/j.foodchem.2014.09.083.

Hemwimol, S., Pavasant, P., Shotipruk, A., 2006. Ultrasound-assisted extraction of anthraquinones from roots of *Morinda citrifolia*. Ultrason. Sonochem. 13 (6), 543–548. https://doi.org/10.1016/j.ultsonch.2005.09.009.

Hosseini, S., Khodaiyan, F., Kazemi, M., Najari, Z., 2019. Optimization and characterization of pectin extracted from sour orange peel by ultrasound assisted method. Int. J. Biol. Macromol. 125, 621–629. https://doi.org/10.1016/j.ijbiomac.2018.12.096.

Jacotet-Navarro, M., Rombaut, N., Fabiano-Tixier, A., Danguien, M., Bily, A., Chemat, F., 2015. Ultrasound versus microwave as green processes for extraction of rosmarinic, carnosic and ursolic acids from rosemary. Ultrason. Sonochem. 27, 102–109. https://doi.org/10.1016/j.ultsonch.2015.05.006.

Jakobek, L., Šeruga, M., Medvidović-Kosanović, M., Novak, I., 2007. Antioxidant activity and polyphenols of Aronia in comparison to other berry species. Agriculturae Conspectus Scientificus 72 (4), 301–306.

Ji, J., Lu, X., Cai, M., Xu, Z., 2006. Improvement of leaching process of Geniposide with ultrasound. Ultrason. Sonochem. 13 (5), 455–462. https://doi.org/10.1016/j.ultsonch.2005.08.003.

Joana Gil-Chávez, G., Villa, J., Fernando Ayala-Zavala, J., Basilio Heredia, J., Sepulveda, D., Yahia, E., González-Aguilar, G., 2013. Technologies for extraction and production of bioactive compounds to be used as nutraceuticals and food ingredients: an overview. Compr. Rev. Food Sci. Food Saf. 12 (1), 5–23. https://doi.org/10.1111/1541-4337.12005.

Kazemi, M., Khodaiyan, F., Hosseini, S., 2019. Eggplant peel as a high potential source of high methylated pectin: ultrasonic extraction optimization and characterization. LWT 105, 182–189. https://doi.org/10.1016/j.lwt.2019.01.060.

Khatoon, S., 2016. A novel histological approach for identification of alkaloid bearing plants. Int. J. Botany 13 (1), 28–36. https://doi.org/10.3923/ijb.2017.28.36.

Khezerlou, A., Jafari, S., 2020. Nanoencapsulated bioactive components for active food packaging. In: Handbook of Food Nanotechnology. Elsevier, Amsterdam, Netherlands, pp. 493–532. https://doi.org/10.1016/b978-0-12-815866-1.00013-3.

Kulkarni, V., Rathod, V., 2014. Mapping of an ultrasonic bath for ultrasound assisted extraction of mangiferin from *Mangifera indica* leaves. Ultrason. Sonochem. 21 (2), 606–611. https://doi.org/10.1016/j.ultsonch.2013.08.021.

Kumar, K., Srivastav, S., Sharanagat, V., 2021. Ultrasound assisted extraction (UAE) of bioactive compounds from fruit and vegetable processing by-products: a review. Ultrason. Sonochem. 70, 105325. https://doi.org/10.1016/j.ultsonch.2020.105325.

Lavilla, I., Bendicho, C., 2017. Fundamentals of ultrasound-assisted extraction. In: Water Extraction of Bioactive Compounds. Elsevier, Amsterdam, Netherlands, pp. 291–316.

Li, D., Zhao, H., Muhammad, A.I., Song, L., Guo, M., Liu, D., 2020. The comparison of ultrasound-assisted thawing, air thawing and water immersion thawing on the quality of slow/fast freezing Bighead Carp (Aristichthys nobilis) fillets. Food Chem. 320, 126614. https://doi.org/10.1016/j.foodchem.2020.126614.

Lianfu, Z., Zelong, L., 2008. Optimization and comparison of ultrasound/microwave assisted extraction (UMAE) and ultrasonic assisted extraction (UAE) of lycopene from tomatoes. Ultrason. Sonochem. 15 (5), 731–737. https://doi.org/10.1016/j.ultsonch.2007.12.001.

Liu, K., Wang, H., Zhang, X., 2019. Ductile mode cutting of brittle materials. In: Springer Series in Advanced Manufacturing. Springer, Berlin, Germany, pp. 39–53. https://doi.org/10.1007/978-981-32-9836-1_3.

Lobo, V., Patil, A., Phatak, A., Chandra, N., 2010. Free radicals, antioxidants and functional foods: Impact on human health. Pharmacogn. Rev. 4 (8), 118. https://doi.org/10.4103/0973-7847.70902.

Lu, X., Tang, K., Li, P., 2016. Plant metabolic engineering strategies for the production of pharmaceutical terpenoids. Front. Plant Sci. 7, 1–11. https://doi.org/10.3389/fpls.2016.01647.

Ma, H., Huang, L., Peng, L., Wang, Z., Yang, Q., 2015. Pretreatment of garlic powder using sweep frequency ultrasound and single frequency countercurrent ultrasound: optimization and comparison for ACE inhibitory activities. Ultrason. Sonochem. 23, 109–115. https://doi.org/10.1016/j.ultsonch.2014.10.020.

Mantegna, S. Binello, A., Boffa, L., Giorgis, M., Cena, C., Cravotto, G., 2012. A one-pot ultrasound-assisted water extraction/cyclodextrin encapsulation of resveratrol from Polygonum cuspidatum. Food Chem. 130 (3), 746–750. https://doi.org/10.1016/j.foodchem.2011.07.038.

Maran, J., Priya, B., 2014. Ultrasound-assisted extraction of polysaccharide from Nephelium lappaceum L. fruit peel. Int. J. Biol. Macromol. 70, 530–536. https://doi.org/10.1016/j.ijbiomac.2014.07.032.

Maran, J., Priya, B., Al-Dhabi, N., Ponmurugan, K., Moorthy, I., Sivarajasekar, N., 2017. Ultrasound assisted citric acid mediated pectin extraction from industrial waste of Musa balbisiana. Ultrason. Sonochem. 35, 204–209. https://doi.org/10.1016/j.ultsonch.2016.09.019.

Matsuura, K., Hirotsune, M., Nunokawa, Y., Satoh, M., Honda, K., 1994. Acceleration of cell growth and ester formation by ultrasonic wave irradiation. J. Ferment. Bioeng. 77 (1), 36–40. https://doi.org/10.1016/0922-338x(94)90205-4.

Mohammed, M.E., Alhajhoj, M.R., 2019. Importance and applications of ultrasonic technology to improve food quality. Food Processing. IntechOpen.

Mongenot, N., Charrier, S., Chalier, P., 2000. Effect of ultrasound emulsification on cheese aroma encapsulation by carbohydrates. J. Agric.Food Chem. 48 (3), 861–867. https://doi.org/10.1021/jf990494n.

Minjares-Fuentes, R., Femenia, A., Garau, M., Meza-Velázquez, J., Simal, S., Rosselló, C., 2014. Ultrasound-assisted extraction of pectins from grape pomace using citric acid: A response surface methodology approach. Carbohydr. Polym. 106, 179–189. https://doi.org/10.1016/j.carbpol.2014.02.013.

Moorthy, I., Maran, J., Ilakya, S., Anitha, S., Sabarima, S., Priya, B., 2017. Ultrasound assisted extraction of pectin from waste *Artocarpus heterophyllus* fruit peel. Ultrason. Sonochem. 34, 525–530. https://doi.org/10.1016/j.ultsonch.2016.06.015.

Moorthy, I., Maran, J., Surya, S., Naganyashree, S., Shivamathi, C., 2015. Response surface optimization of ultrasound assisted extraction of pectin from pomegranate peel. Int. J. Biol. Macromol. 72, 1323–1328. https://doi.org/10.1016/j.ijbiomac.2014.10.037.

Naji, O., Al-juboori, R.A., Bowtell, L., Alpatova, A., Ghaffour, N., 2020. Direct contact ultrasound for fouling control and flux enhancement in air-gap membrane distillation. Ultrason. Sonochem. 61, 104816. https://doi.org/10.1016/j.ultsonch.2019.104816.

Nijveldt, R., van Nood, E., van Hoorn, D., Boelens, P., van Norren, K., van Leeuwen, P., 2001. Flavonoids: a review of probable mechanisms of action and potential applications. Am. J. Clin. Nutr. 74 (4), 418–425. https://doi.org/10.1093/ajcn/74.4.418.

Nishad, J., Saha, S., Kaur, C., 2019. Enzyme- and ultrasound-assisted extractions of polyphenols from Citrus sinensis (cv. Malta) peel: a comparative study. J. Food Process. Preserv. 43 (8), e14046. https://doi.org/10.1111/jfpp.14046.

Palma, M., Barroso, C., 2002. Ultrasound-assisted extraction and determination of tartaric and malic acids from grapes and winemaking by-products. Anal. Chim. Acta 458 (1), 119–130. https://doi.org/10.1016/s0003-2670(01)01527-6.

Palma, A., Díaz, M.J., Ruiz-Montoya, M., Morales, E., Giráldez, I., 2021. Ultrasound extraction optimization for bioactive molecules from Eucalyptus globulus leaves through antioxidant activity. Ultrason. Sonochem. 76, 105654.

Pan, Z., Qu, W., Ma, H., Atungulu, G., McHugh, T., 2012a. Continuous and pulsed ultrasound-assisted extractions of antioxidants from pomegranate peel. Ultrason. Sonochem. 19 (2), 365–372. https://doi.org/10.1016/j.ultsonch.2011.05.015.

Pan, G., Yu, G., Zhu, C., Qiao, J., 2012b. Optimization of ultrasound-assisted extraction (UAE) of flavonoids compounds (FC) from hawthorn seed (HS). Ultrason. Sonochem. 19 (3), 486–490. https://doi.org/10.1016/j.ultsonch.2011.11.006.

Ponmurugan, K., Al-Dhabi, N., Maran, J., Karthikeyan, K., Moothy, I., Sivarajasekar, N., Manoj, J., 2017. Ultrasound assisted pectic polysaccharide extraction and its characterization from waste heads of Helianthus annus. Carbohydr. Polym. 173, 707–713. https://doi.org/10.1016/j.carbpol.2017.06.018.

Povey, M.J., Mason, T.J. (Eds.), 1998. Ultrasound in Food Processing. Springer Science & Business Media, Berlin, Germany.

Ramic, M., Vidović, S., Zeković, Z., Vladić, J., Cvejin, A., Pavlić, B., 2015. Modeling and optimization of ultrasound-assisted extraction of polyphenolic compounds from *Aronia melanocarpa* by-products from filter-tea factory. Ultrason. Sonochem. 23, 360–368. https://doi.org/10.1016/j.ultsonch.2014.10.002.

Ranjha, M., Irfan, S., Lorenzo, J., Shafique, B., Kanwal, R., Pateiro, M., et al., 2021. Sonication, a potential technique for extraction of phytoconstituents: a systematic review. Processes 9 (8), 1406. https://doi.org/10.3390/pr9081406.

Raza, A., Li, F., Xu, X., Tang, J., 2017. Optimization of ultrasonic-assisted extraction of antioxidant polysaccharides from the stem of Trapa quadrispinosa using response surface methodology. Int. J. Biol. Macromol. 94, 335–344. https://doi.org/10.1016/j.ijbiomac.2016.10.033.

Richter, B., Jones, B., Ezzell, J., Porter, N., Avdalovic, N., Pohl, C., 1996. Accelerated solvent extraction: a technique for sample preparation. Anal. Chem. 68 (6), 1033–1039.

Rodrigues, S., Fernandes, F., de Brito, E., Sousa, A., Narain, N., 2015. Ultrasound extraction of phenolics and anthocyanins from jabuticaba peel. Ind. Crops Prod. 69, 400–407. https://doi.org/10.1016/j.indcrop.2015.02.059.

Rodrigues, S., Pinto, G., Fernandes, F., 2008. Optimization of ultrasound extraction of phenolic compounds from coconut (Cocos nucifera) shell powder by response surface methodology. Ultrason. Sonochem. 15 (1), 95–100. https://doi.org/10.1016/j.ultsonch.2007.01.006.

Roselló-Soto, E., Galanakis, C., Brnčić, M., Orlien, V., Trujillo, F., Mawson, R., et al., 2015. Clean recovery of antioxidant compounds from plant foods, by-products and algae assisted by ultrasounds processing. Modeling approaches to optimize processing conditions. Trends Food Sci. Technol. 42 (2), 134–149. https://doi.org/10.1016/j.tifs.2015.01.002.

Sabnis, S., Raikar, R., Gogate, P., 2020. Evaluation of different cavitational reactors for size reduction of DADPS. Ultrason. Sonochem. 69, 105276. https://doi.org/10.1016/j.ultsonch.2020.105276.

Safdar, M., Kausar, T., Nadeem, M., 2016. Comparison of ultrasound and maceration techniques for the extraction of polyphenols from the mango peel. J. Food Process. Preserv. 41 (4), e13028. https://doi.org/10.1111/jfpp.13028.

Şahin, S., Şamlı, R., 2013. Optimization of olive leaf extract obtained by ultrasound-assisted extraction with response surface methodology. Ultrason. Sonochem. 20 (1), 595–602. https://doi.org/10.1016/j.ultsonch.2012.07.029.

Samaram, S., Mirhosseini, H., Tan, C., Ghazali, H., Bordbar, S., Serjouie, A., 2015. Optimization of ultrasound-assisted extraction of oil from papaya seed by response surface methodology: oil recovery, radical scavenging antioxidant activity, and oxidation stability. Food Chem. 172, 7–17. https://doi.org/10.1016/j.foodchem.2014.08.068.

Sarheed, O., Shouqair, D., Ramesh, K., Khaleel, T., Amin, M., Boateng, J., Drechsler, M., 2020. Formation of stable nanoemulsions by ultrasound-assisted two-step emulsification process for topical drug delivery: effect of oil phase composition and surfactant concentration and loratadine as ripening inhibitor. Int. J. Pharm. 576, 118952. https://doi.org/10.1016/j.ijpharm.2019.118952.

Sharma, M., Koul, A., Sharma, D., Kaul, S., Swamy, M., Dhar, M., 2019. Metabolic engineering strategies for enhancing the production of bio-active compounds from medicinal plants. Nat. Bio-Active Compd. 287–316. https://doi.org/10.1007/978-981-13-7438-8_12.

ŠicŽlabur, J., Voća, S., Dobričević, N., Brnčić, M., Dujmić, F., RimacBrnčić, S., 2015. Optimization of ultrasound assisted extraction of functional ingredients from Stevia Rebaudiana Bertoni leaves. Int. Agrophys. 29 (2), 231–237. https://doi.org/10.1515/intag-2015-0017.

Singla, M., Sit, N., 2021. Application of ultrasound in combination with other technologies in food processing: a review. Ultrason. Sonochem. 73, 105506. https://doi.org/10.1016/j.ultsonch.2021.105506.

Soria, A., Villamiel, M., 2010. Effect of ultrasound on the technological properties and bioactivity of food: a review. Trends Food Sci. Technol. 21 (7), 323–331. https://doi.org/10.1016/j.tifs.2010.04.003.

Su, Y., Zhang, M., Bhandari, B., Zhang, W., 2018. Enhancement of water removing and the quality of fried purple-fleshed sweet potato in the vacuum frying by combined power ultrasound and microwave technology. Ultrason. Sonochem. 44, 368–379. https://doi.org/10.1016/j.ultsonch.2018.02.049.

Sutariya, S., Sunkesula, V., Kumar, R., Shah, K., 2018. Emerging applications of ultrasonication and cavitation in dairy industry: a review. Cogent Food Agric. 4 (1), 1549187. https://doi.org/10.1080/23311932.2018.1549187.

Tao, Y., Sun, D.W., 2015. Enhancement of food processes by ultrasound: a review. Crit. Rev. Food Sci. Nutr. 55 (4), 570–594. https://doi.org/10.1080/10408398.2012.667849.

Tian, Y., Zeng, H., Xu, Z., Zheng, B., Lin, Y., Gan, C., Lo, Y., 2012. Ultrasonic-assisted extraction and antioxidant activity of polysaccharides recovered from white button mushroom (*Agaricus bisporus*). Carbohydr. Polym. 88 (2), 522–529. https://doi.org/10.1016/j.carbpol.2011.12.042.

Thompson, E., 2000. Sounds of our times: two hundred years of acoustics. Robert T. Beyer. ISIS 91 (4), 763–764. https://doi.org/10.1086/384954.

Upadhyay, S., Dixit, M., 2015. Role of polyphenols and other phytochemicals on molecular signaling. Oxid. Med. Cell. Long. 2015, 504253. http://doi.org/10.1155/2015/504253.

Vardanega, R., Santos, D.T., Meireles, M.A.A., 2014. Intensification of bioactive compounds extraction from medicinal plants using ultrasonic irradiation. Pharmacogn. Rev. 8 (16), 88–95.

Vilkhu, K., Mawson, R., Simons, L., Bates, D., 2008. Applications and opportunities for ultrasound assisted extraction in the food industry — a review. Innovative Food Sci. Emerg. Technol. 9 (2), 161–169. https://doi.org/10.1016/j.ifset.2007.04.014.

Vinatoru, M., 2001. An overview of the ultrasonically assisted extraction of bioactive principles from herbs. Ultrason. Sonochem. *8* (3), 303–313. https://doi.org/10.1016/s1350-4177(01)00071-2.

Wang, W., Ma, X., Jiang, P., Hu, L., Zhi, Z., Chen, J., 2016. Characterization of pectin from grapefruit peel: A comparison of ultrasound-assisted and conventional heating extractions. Food Hydrocoll. 61, 730–739. https://doi.org/10.1016/j.lwt.2018.01.007.

Weissler, A., Cooper, H., Snyder, S., 1950. Chemical effect of ultrasonic waves: oxidation of potassium iodide solution by carbon tetrachloride. J. Am. Chem. Soc. *72* (4), 1769–1775. https://doi.org/10.1021/ja01160a102.

Wen, C., Zhang, J., Zhang, H., Dzah, C., Zandile, M., Duan, Y., et al., 2018. Advances in ultrasound assisted extraction of bioactive compounds from cash crops – a review. Ultrason. Sonochem. 48, 538–549. https://doi.org/10.1016/j.ultsonch.2018.07.018.

Xiong, T., Xiong, W., Ge, M., Xia, J., Li, B., Chen, Y., 2018. Effect of high intensity ultrasound on structure and foaming properties of pea protein isolate. Food Res. Int. 109, 260–267. https://doi.org/10.1016/j.foodres.2018.04.044.

Xu, Y., Zhang, L., Bailina, Y., Ge, Z., Ding, T., Ye, X., Liu, D., 2014. Effects of ultrasound and/or heating on the extraction of pectin from grapefruit peel. J. Food Eng. 126, 72–81. https://doi.org/10.1016/j.jfoodeng.2013.11.004.

Yao, Y., 2016. Enhancement of mass transfer by ultrasound: Application to adsorbent regeneration and food drying/dehydration. Ultrason. Sonochem. *31*, 512–531. https://doi.org/10.1016/j.ultsonch.2016.01.039.

Yan, Y., Yu, C., Chen, J., Li, X., Wang, W., Li, S., 2011. Ultrasonic-assisted extraction optimized by response surface methodology, chemical composition and antioxidant activity of polysaccharides from *Tremella mesenterica*. Carbohydr. Polym. 83 (1), 217–224. https://doi.org/10.1016/j.carbpol.2010.07.045.

Zhao, Y., Yi, J., Bi, J., Chen, Q., Zhou, M., Zhang, B., 2018. Improving of texture and rehydration properties by ultrasound pretreatment for infrared-dried shiitake mushroom slices. Drying Technol. 37 (3), 352–362. https://doi.org/10.1080/07373937.2018.1456449.

Zhang, H., Tsao, R., 2016. Dietary polyphenols, oxidative stress and antioxidant and anti-inflammatory effects. Curr. Opin. Food Sci. 8, 33–42. http://doi.org/10.1016/j.cofs.2016.02.002.

Zhang, Y., Abatzoglou, N., 2020. Review: fundamentals, applications and potentials of ultrasound-assisted drying. Chem. Eng. Res. Des. 154, 21–46. https://doi.org/10.1016/j.cherd.2019.11.025.

Zhang, Z., Wang, L., Li, D., Jiao, S., Chen, X., Mao, Z., 2008. Ultrasound-assisted extraction of oil from flaxseed. Sep. Purif. Technol. 62 (1), 192–198. https://doi.org/10.1016/j.seppur.2008.01.014.

Zhu, Y., Sun, J., Xu, D., Wang, S., Yuan, Y., Cao, Y., 2018. Investigation of (+)-catechin stability under ultrasonic treatment and its degradation kinetic modeling. J. Food Process Eng. 41 (8), e12904. https://doi.org/10.1111/jfpe.12904.

Zhu, Z., Zhang, P., Sun, D., 2020. Effects of multi-frequency ultrasound on freezing rates and quality attributes of potatoes. Ultrason. Sonochem. 60, 104733. https://doi.org/10.1016/j.ultsonch.2019.104733.

CHAPTER TEN

Sonication microwave synergistic extraction of oils from plant sources

Satinder Kaur, Antima Gupta and Arashdeep Singh
Department of Food Science and Technology, Punjab Agricultural University, Ludhiana, Punjab, India

10.1 Introduction

Oilseeds are seeds that are majorly cultivated for the production of edible oils. Oilseeds are major source of oils which are consumed by human beings as well as its by-products by animals. In the broader sense, rapeseed, peanut, soyabean, and cottonseed can be considered as major oilseeds. On the other hand, linseed, flaxseed, coconut, sesame, etc. are some other examples of oilseeds which are used for extraction of oil as edible oil and for other applications as well. Edible oils and oilseeds play vital role in the consumption basket in India. Besides an important contributor of taste; edible oil is a significant source of nutrition in diet. Generally, oilseeds are major source of energy and moderate source of fibers and proteins (30%–45%). Exceptionally, peanuts contain 50%–60% protein whereas sunflower contains 50% of proteins. Fats and oils are used widely for industrial purposes in order to produce processed foods and other toiletries. Meanwhile, oil meals and oilcakes are other applications of oilseeds in animal feed industry. Traditions, level of income, taste preference and relative prices are some important factors influencing production, consumption and other utilization of oilseeds. Some nutritionists claim that minimum quantity of oils and fat in human body annually should be about 6–8 kg. However, culinary traditions are major factor that influence the quantity of consumption. Owing to short time disruption in marketing during prevalence of COVID-19; there is rapid increase in prices of oilseeds. The production of oilseeds is presented in Fig. 10.1. During this period, soyabean production has been increased by 1.1% annually. Expected production of soyabean is about 411 Mt by year 2030 which is double quantity as that of 179 Mt; production of other oilseeds like sunflower, rapeseed, and groundnut.

Ultrasound and Microwave for Food Processing: Synergism for Preservation and Extraction.
DOI: https://doi.org/10.1016/B978-0-323-95991-9.00011-4

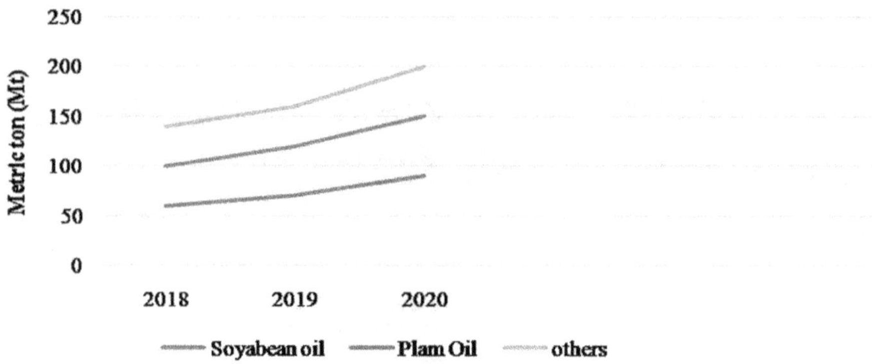

Figure 10.1 Production of Oilseeds (OECD/FAO, 2021).

Besides number one importer of vegetable oil, India is the second largest consumer of vegetable oils in the world. Per capita consumption of vegetable oil is 2.3% p.a. which is reaching 14 kg/capita by 2030. Due to increase in living standards and shifting of population toward urbanization, consumptions of processed food are increased drastically which contains high amount of vegetable oils. In least developed countries; due to their low per capita income, consumption of oils is also low that is 1.3% p.a. Conventional method of oil extraction is pressing of oilseeds and other is solvent extraction method which is a popular and common method at present as well. Solvent residues, longer extraction time, high energy utilization and change in organoleptic properties are some disadvantages of these conventional methods.to replace these; innovative technologies like microwave and ultrasound assisted extraction have been in progress at present. Mentioned technologies have various applications like extraction, transesterification, etc. in oil industries.

Edible plant-based oils can be obtained from various parts of it; fruits, plumule, pulp, and seeds. Besides its application in cooking edible plant oils are also used in health supplements, cosmetics and in other purpose. Processing like cooking leads to diversified flavor and sense of satiety in food. On the other hand, cooking causes change in overall sensory appeal of food; by changing its flavor, texture, color and taste, etc. owing to aggravation of consumption of edible plant-based oils (EPO) and increased environmental pollution; the safety and quality of oils become matter of consideration. For this strict monitoring should be applied while processing transport, cultivation and storage of these plant-based oils. In ancient time, commonly used oils for cooking purpose were olive oil (*Olea europaea*) and sesame oil (*Sesamum indicum*). With the development of new innovative technologies, improved agricultural system, and other processing more plants are involved in oil extraction. High oil content is produced by different herbaceous plants but sesame produce the highest content of oil as compared to other herbal oil crops (Table 10.1). Owing to inefficient processing techniques and low global production; sesame is not a used commonly for cooking. On the other hand, composition, biological activity and oil content vary greatly in different crops (Yi et al., 2017). At present, the largest quantity of Rapeseed (*Brassica napus*) and soyabean oil (*Glycine max*) is produced globally. In addition to this, many other wooden plants are used for oil extraction. The four well-known wooden plant-based sources are coconut, oil palm, olive, and oil-seed camellia which contain considerable amount of oil. Rapeseed and peanut oil are the richest sources of unsaturated fatty oils; about 80%. With advantage of no use of cultivation land, wooden plant sources play vital in development and security of edible plant oils.

Traditionally, most used and common method of extraction of plant-based edible oils was pressed extraction which is common at present as well (Yara-Varon et al., 2017). Pressed extraction has advantage of no solvent pollution as compared to solvent extraction method of oils. On the other hand, pressed extraction retains natural flavor, high quantity and simple operations are other advantages of pressed extraction. Cold pressing extraction and hot-pressing extraction are two types of pressed extraction. In cold pressing method extraction of oil low temperature is used in oil mill in which plant oils undergoes stir frying at low temperatures. This method of extraction ensures high flavor retention (Fawzy, 2013; Yang et al., 2013; Sielicka et al., 2014). On the other hand, in hot pressed extraction oil is extracted by using continuous high temperature environment (Siger et al., 2017). Cold pressed method is more popular as compared to hot pressed as it has major drawback of losing flavors and nutrients like vitamins (Soria and Villamiel, 2010; Teh and Birch, 2014; Koubaa et al., 2016). Recent studies show that hot pressing cause damage to vitamin E as compared to clod pressing extraction in sunflower seed oil which show 45.7 mg/100 g and 13.2 mg/100 g content respectively (Wei and Wang, 2017).

10.2 What are microwave-assisted extraction and ultrasound-assisted extraction?

Microwave-assisted extraction is a process which involves microwave energy for heating of solvents which are in contact with sample matrix in

Table 10.1 Oil-bearing herbaceous plants.

Common name	Botanical name	Production area	Oil content	References
Cotton	Gossypium Spp.	India, Uzbekistan	15%–40%	Shang et al. (2017)
Rape	*Brassica napus L.*	All over the world	37.5%–46.3%	Zhao et al. (2005); Ishaq et al. (2017)
Peanut	*Arachis hypogaea L.*	Africa, America and Asia	46%–57%	Wang et al. (2018)
Soyabean	*Glycine max.*	Brazil, China and USA	18%–24%	Li et al. (2018)
Sunflower	*Helianthus annus*	All over the world	46%–50%	Rauf et al. (2017)
Hemp	*Cannabis sativa L.*	All over the world	25%–35%	Vonapartis et al. (2014)
Sesame	*Sesamum indicum*	Temperate and Tropical region	43%–61%	Latif and Anwar (2011)
Perilla	*Perilla frutescens*	India, Japan, Russia	40%–50%	Liao et al. (2018)
Fiber flax	*Linumusitatissimum L.*	Euro-Asian Temperature zone	35%–45%	Martinchik et al. (2012)
Pecan	*Carya cathayensisSarg.*	China	60%–70%	Huang et al. (2016)
Peony	*Paeonia*	China	27%–33%	Ning et al. (2015); Zhang et al. (2018)
Olive	*Olea europaea L.*	Mediterranean coast	31%–56%	Sun et al. (2017); Olmo-Garcıa et al. ´(2018)
Pine	*Pinus*	Brazil	58%–69%	Bao and Guo (2016)
Cocoa	*Theobroma cacoa L.*	North-South latitude of the equator	45%–60%	Servent et al. (2018)
Shiny-leaved yellow horn	*Xanthacerassorbifolium*	North and Northeast China	50%–60%	Cao (2015)

order to separate analytes from sample matrix into solvents. Microwave-assisted extraction has various advantages over conventional methods like reduction in energy consumption, reduction in time of extraction and operational cost. It is also characterized by low CO_2 emission. UAE is ultrasound or ultrasonication-assisted extraction which involves usage of ultrasound waves for oil extraction. In UAE acoustic waves, passing through solvent causes production of cavitation bubbles; which ranges in kilohertz. Cavitation bubble causes cell wall bursting which leads to increment of mass transfer of phenolic compounds across the cellular membrane inside the solution. Low input power, low solvent residue, etc. are advantages of these methods.

10.3 Advantages and principle of sonication microwave synergism

Microwave-assisted extraction and ultrasound-assisted extraction is recognized as popular and efficient technique of oil extraction as it drastically reduces mechanism timing, enhances quality, and yield of the extract. Beside chief utilization in laboratory, methods are now become popular at industrial level as well especially MAE (G. Cravotto et al., 2007). On the other hand, ultrasound-assisted extraction works on phenomenon of cavitation which leads to enhancement of mass transfer by the release of soluble components from plant cell through disruption. This effect is negligible at 400–800 kHz and show maximum result at 18–40 kHz. Ultrasound causes acceleration in swelling and rehydration in previously dried sample matrix. Along with matrix fragmentation, it accelerates hydration process without causing any effective chemical degradation. Ultrasound-assisted extraction has proved to be a flexible or adaptable technology to step up the industrial applications.

The microwave-assisted extraction works on principle of dielectric heating in which sample matrix absorb microwave energy and convert it into heat energy. As compared to conductive heating, microwave heating leads heating of whole sample volume. By enhancing the rotation of molecular dipoles (effect which is opposed by viscosity of respective medium), it causes disruption of weak hydrogen bond. Moreover, solvent penetration into matrix is increased by the movements of dissolved ions which ultimately facilitate analyte solvation. This effect is efficiently works if, solvent and matrix nature is strong. Microwave often affects mainly matrix and the solvent results into low dielectric constant thus remains cold. This setup has been successfully used in extraction of oil from plant sources owing to its advantages of dealing with thermosensitive compounds. By direct distillation, microwave heating is also convenient way for extraction of essential oils from fresh plant sources. Liquid-phase extraction which is often used for extraction of essential oils from plants based on different dielectric constants for microwave for different absorbance. Solvent having low dielectric constant is mixed with matrix to be extracted which is usually water rich thus the heating effect is mostly concentrated on plant sample. Microwave-assisted extraction has advantages like it take less time, better quality products, and less solvent residue. Biologically active compounds like glycyrrhizic acid from glycyrrhiza glabra, taxanes from Taxus brevifolia and artemisinin from Artemisia annua have been extracted.

10.4 Resource efficient technology for extraction

To reduce the over exploitation of material and energy, to enhance environmental and economic profits; resource efficient technologies are need to be developed. For oil extraction resource efficient technologies can be developed by synergistic technologies associated with process intensification. Development of novel methods or equipment that has ability to produce superior benefits and remarkably higher yields as compared to conventional methodology and equipment. These profits can be perceived by reduction in equipment size, extraction timing, minimum wastage, remarkable increment in quality and quantity of product, energy efficiency, reduction in complexity in production techniques which finally results into safer, cheaper, and sustainable final product (Stankiewicz and Moulijn, 2010). Development of process intensification in equipment and method could focus on development of novel reactor having ability of intense mixing that facilitates heat-mass transfer and on integration of reaction processes like innovation of technology, alternative uses of energy sources respectively; so that processing steps can be minimized. For example, in the context of oil; owing to process intensification heat and mass transfer rate increased in various processes like transesterification, extraction, purification, or separation.

10.5 Process intensification by synergism

Intensification of process by combination of two different technologies leads to synergism that is magnified or increased affect (Huang et al., 2007). Synergism is basically a phenomenon which results from combination of different tools, reagents or technologies that is dominating as compared to their original individual impact (Martinez-Guerra, 2014). To attain synergism, intensification of process should fulfill following:

- Intermolecular and intramolecular should be maximum through creating dynamic conditions in order to promote kinetic control with higher efficiency.
- Verification of uniform heating and gradient-less mixing.
- Maximum specific surface areas and optimization of driving forces in order to improve mass and heat transfer.
- Synergistic effect should be maximum as compared to partial or conventional methods.

The most applicable issues related to intensification of process is structure majorly in catalysis and other molecular reactions, synergy-functional domain in which multifunctional equipment are developed, energy-thermodynamic domain in which energy is transmitted to transport, hydrodynamic and chemical processes and time-temporal domain in which timing of process, events and application of dynamics are controlled (Stankiewicz et al., 2006). Moreover, it is essential to consider, proper organization of two or more conventional methods so that it leads to process intensification ultimately leads to synergy when in combination. Although this intensification leads to synergy but it also has limiting factors with regard to optimization and process control. Combinations of different processes may not be able to intensify the process.

10.6 Synergism of microwave and ultrasound-assisted extraction

Ultrasound- and microwave-mediated organic synthesis is a matter of concern, at present owing to its superiority in providing high quality products in shorter time duration (Gude et al., 2011). Intensification of these two processes is well utilized by synergism in various chemical applications and other engineering purposes. But applications of this synergism in extraction industry have been yet to explore more. Material reactions with the electromagnetic interactions often results into increment of thermal power which produces dominating chemical reaction is due to microwave energy (Robert and England, 2003). Microwave energy is capable of providing immediate heating which is owing to three major reaction environments; dipolar momentum, ionic conduction, and interfacial polarization—a combination of dipolar momentum and ionic conduction (Gude et al., 2013). These reactions provide thermal energy in mechanism. In ultrasound, interaction of the sound waves with reaction materials; generate the acoustic cavitation which leads to intense mixing. Ultimately this process leads to enhancement of heat and mass transfer and thus process efficiency (Gude et al., 2013; Luo et al., 2014). Example is intensification of process by ultrasonication causes increment in mass transfer among liquid and gas molecules, by up to five-fold. On the other hand, solid-liquid mass can be enhanced by 20–25 folds that significantly increase the product yield. The timing of reaction can be dramatically reduced by up to 1250 times due to rapid heating in microwave (Stankiewicz et al., 2006). Owing to higher yield and product conversion, these two-energy irradiations are capable of improving material and energy efficiencies. Owing to higher rate of interaction with reaction material, it results in thermal runaway and hot-spot formation. Mass transfer and a mixing mechanism which ensures uniform heating throughout the reaction materials caused by the microwave energy; is a necessity of this over all mechanism. Similarly, ultrasonication cause intense mixing due to enhanced heat and mass transfer; this is due to generation of acoustic cavitation which leads to formation of microbubbles in air. Accompanied by high thermal and pressure release, the formation-release-collapse of these microbubbles provides heating and cooling effects at microscale levels. This microlevel energy is not sufficient for generating enough heat energy in reaction medium which is dependent on volume of reactor, reaction time, and the exposure time. This is a clear drawback of ultrasound mediated reactions (Martinez-Guerra et al., 2014). In order to increase kinetics of reaction, it needs some external source of reaction. Taking in account above mentioned limitation and prospects for the separate process intensification, there is need to develop a hybrid system that have capability of incorporation of both conventional mixing and heating effects that ultimately cause increment in process outputs. This might lead to efficient chemical (which is considered as green chemistry), material and energy uses (Gude et al., 2015; Cravotto et al., 2007). Superiority of methodology and equipment obtained through integrated process intensification leads to commercial benefits at large scale applications. This hybrid technology is matter of concern as it will be proved as ideal for production of good quality of bioproducts with extraction mechanism. Both for organic and inorganic chemical reactions, ultrasonication, and microwave use low amount of solvents and catalysts and low consumption of energy. Moreover, recovery of the product is greatly enhanced due to increase in kinetics and reaction time decreased drastically (Martinez-Guerra et al., 2015). These facts clearly ensure the principle of green chemistry based on *e-factor* and *atom economy*

(Gude et al., 2015). Efficient use of raw material that is maximizing conversion of desired atoms of employed material in to desired products is referred to as atom economy. The e-factor is associated with waste residue which is generated while delivering the final product. Proportion of green reaction is inversely proportional to e-factor of the reaction. Reducing or eliminating waste emission from process play vital role in sustainable extraction mechanism and resource efficient chemical procedures.

10.7 Microwave and ultrasonication-assisted extraction (MUAE) equipment

With professional multimode oven, all microwave operations are used to carry out at 2.45 GHz-Microsynth, Milestone, Italy. Ultrasonication operations can be operated by various laboratory equipment in collaboration with Danacamerinisas-Torino, Italy; which can work at different frequencies-300, 40, 25, and 19 kHz:

- With immersion titanium horns, a high-power probe system working at 40 and 25 kHz.
- Working at 300 kHz—a titanium cup-horn. Owing to efficient cooling system which is capable of employing an oil circulation system and a chiller; the transducer has ability to work on high power of about 500 W without get being overheated.
- The high-power cavitating tube—an innovative and highly technical cup-horn like system which is consisted of a hollow thin titanium cylinder which if connected to a booster through base.

For measurements of acoustic energy transferring throughout the liquid from its bottom and wall can be determined by using a small hydrophone. Thermostatting fluid circulation is carried out throughout the polytetrafluoroethylene (PTFE) jacket. Two transducers working at different frequencies can be combined by insertion of an immersion horn in the cavitating tube.by inserting a horn made up of polyether ether ketone-PEEK in a microwave; MW/US irradiation can be achieved in a single reaction vessel.

10.8 Microscopic changes and extraction mechanism

The external structure of kernel was intact and no effective destruction was observed before the extraction of oil from plant sources. Owing to damaging of cell structure, external structure of sample is characterized by porosity which is caused by many irregular cavities after microwave ultrasonication-assisted extraction. Morphological changes can be noted in microwave ultrasonication-assisted extraction within 38 minutes as compared to solvent extraction which can be apparent in 6 hours. In solvent extraction heat transfer is conducted from outside to inside whereas in MUAE is inside to outside. In MUAE, mass transfer is also conducted in same way as that of heat transfer but in solvent extraction mass transfer is conducted from inside to outside. With the increase in temperature due to intensified process by combination of microwave and ultrasonication, there is advantage of rapid diffusion of the oil content from the plant structure into the extraction solvent matrix. These outcomes are compatible with the report of Virot et al. (2008). Owing to high-frequency electromagnetic wave of microwave, it is capable of transferring into heat energy which raise the temperature of intracellular structure in shorter time that ultimately cause the oil leaching into the extraction solvent as it generates high pressure to decompose cell wall. Meanwhile, ultrasonic power which is continuously travelling back and forth through the extraction solvent is also destruct cell structure owing to bubble formation and collapsing due to cavitation effect ultimately easy release of oil. According to report, ultrasonication is not operated as single mean but in mixing effects in specific order as; local erosion, shear forces, sonoporation, fragmentation, capillary effect, and detexturation (Chemat et al., 2017a, b; Khadhraoui et al., 2018). The mixing effect caused by ultrasonication is not surface phenomenon as it causes deep destruction throughout cell structure. So, MUAE shows synergism and intensified process of microwave and ultrasonication which provide heat energy and cavitation effect respectively which cause easy disruption of sample structure and facilitates easy release of oil from sample. Similar reports were shown by Wang et al. (2018) and Chen et al. (2016).

10.9 Comparison of energy, cost, scale-up, and environmental impact

As compared to conventional method that is solvent extraction; MUAE is superior in all aspects of cost, energy consumption, high quality and quantity of bioproducts, and environmental beneficial impacts. The process timing taken by 6 hours in solvent extraction is obviously much higher than MUAE which take only 38 minutes. Solvent used in solvent extraction mechanism is also higher as compared to microwave ultrasonication-assisted extraction which causes increase in operation cost in respective method. The energy requirement for conduction of solvent extraction is 0.50 kWh and

1.20 kWh in microwave ultrasonication-assisted extraction. In the context of environment, calculated quantity of carbon emission which is higher in solvent extraction method, that is, 2400 g/g of oil as compared to microwave ultrasonication-assisted extraction which is about 608 g/g of oil. According to literature, results were obtained that 800 g CO_2 emission from 1 kWh is not accepted in environment (Vian et al., 2008). According to this MUAE is considered to be eco-friendly extraction method of oil. Owing to scarcity of industrial unit few years back; operating simultaneously both ultrasonic generator and microwave generator as one, use of microwave ultrasonic-assisted extraction was limited to laboratory scale only. At present due to immense progress of microwave-assisted extraction and ultrasonication-assisted extraction have numerous industrial applications. Recent studies show that use of microwave-assisted extraction is capable of extraction of essential oils from plant sources of quantity 100, 20, and 10 kg operated in single slot (Farhat et al., 2011).

Chemat et al. (2017a, b) study showed that two types industrial equipment are used having high power ultrasonic energy; bath and prob were used. One was developed by Germany's Hielscher Company ultrasonic apparatus which works on range 500–1600 W energy for industrial scale and 50–400 W for laboratory scale. Other than that, was by REUS company of France, ultrasonic equipment for pilot and large-scale production ranges from 30 to 50 L and 500 to 1000 L, respectively. Recent studies show that combination of different equipment ensures safety of synergism of microwave and ultrasonic devices; by simple modifications whole system become capable of scaling-up and automatization (Cravotto and Cintas, 2007). Similarly, combination of ultrasound and microwave is efficient and in future combined reactor will be developed for industrial scale (Leonelli and Mason, 2010). Thus, there is promising fact that single reactor for intensified process would be developed.

10.10 Comparison of extraction efficiencies

In context to the extraction efficiencies, microwave ultrasonication-assisted extraction is compared with several conventional methods like solvent, microwave-assisted and ultrasound/ultrasonication-assisted extraction. Although the yield of oil obtained from MUAE is little bit less as compared to solvent extraction, that is, 34.8 ± 0.31% but solvent to material ratio -8 mL/g and extraction time of MUAE-38 minutes is comparably less than that of SE which is 10 mL/g and 6 hours respectively. In accordance to operation cost and energy consumption, MUAE ensures superiority over solvent extraction method. Meanwhile, UAE and MAE were significantly as compared to MUAE that is 25.13 ± 0.24% and 27.45 ± 28%, respectively. The MUAE is reported high potential mechanism because of synergism of two different mechanisms and their energetic radiations. Oil extraction in MUAE is accelerated due to heating effect of microwave in addition to cavitation effect provided by ultrasound radiation travelling through solvent matrix which increases cell destruction and facilitates oil release (Wang et al., 2018). Consequently, extraction time declined and extraction efficiency have been increased therefore, MUAE is considered as potential method for oil extraction.

10.11 Comparison of compositional properties of oil

10.11.1 Fatty acid composition of oil

The major unsaturated fatty acids are stearic acid (18:0), palmitic acid, linoleic acid (18:2), oleic acid (18:2), palmitic acid (16:0), and linolenic acid (18:3). A small amount of docosadienoic acid (22:2) and cis-Ecosaenoic acid (20:1) is also found. Unsaturated fatty acids content is found around 80% of total fatty acids. All results of fatty acid composition are consistent with previous reports (Wang et al., 2017). Beside these all results were found; in which high content of palmitic acid is obtained with low level of oleic and linoleic acid in SC-CO_2 oil extraction (Rajaei et al., 2005). Meanwhile by subcritical water extraction, oleic acid is found in higher content; linoleic and palmitic acid is in low content (Li et al., 2017). Owing to geographical position, extraction method and cultivated varieties; fatty acids composition of oilseeds can be changed (Ezeh et al., 2016). Attention should be paid to constituents, which are at higher risk of degradation by effect of ultrasound and microwave; besides shorter time and low temperature implementation of MUAE. By acceleration of extraction process, microwave irradiation did not cause any change in volatile composition of oil. Microwave irradiation producing nonionizing radiation leads to rotation of dipoles causes molecular motion of ions, do not destruct over all molecular structure (Chemat et al., 2015). Meanwhile, no considerable difference has been observed in soyabean germ oil extraction by conventional and ultrasonic method (Cravotto et al., 2008). Similarly, no remarkable difference is found in ultrasound and conventional extraction of rapeseed oil and difference is found in oil content degradation in ultrasonic and microwave extraction (Hosseini et al., 2015; Desai et al., 2010). Distinction in fatty acid composition by microwave, ultrasonic

Table 10.2 Comparison of physicochemical and fatty acids composition in tea seed oil extracted by SE and MUAE.
Physicochemical composition (approximate value)

	Solvent extraction (SE)	Microwave ultrasonication-assisted extraction (MUAE)
Relative density (20°C, g/mL)	1.46	1.46
Saponification value (mg KOH/g)	194.38	195.73
Refractive index (25°C)	0.92	0.92
Iodine value (gI_2/100 g)	88.29	87.43
Acid value (mg KOH/g)	4.28	3.54
Unsaponifiable matter	0.58	0.61
Peroxide value (meq O_2/Kg)	16.57	13.18
Fatty acid composition (% approximate)		
Oleic acid	53.95	54.15
Linoleic acid	25.47	25.42
Linolenic acid	0.62	0.63
Stearic acid	16.48	16.41
Palmitic acid	1.60	1.59
Docosadienoic acid	0.83	0.75
Cis-8-eicosenoic acid	1.04	1.05
Total monosaturated fatty acid	53.05	55.18
Total polyunsaturated fatty acid	26.82	26.78
Total saturated fatty acid	18.07	18.02

method is due to different space, solvents and time applied to sample. At the end, no clarified distinction is found on quantity of fatty acid composition in SE and MUAE. Therefore, MUAE do not show negative impact on overall composition of fatty acids in oil content. Table 10.2 depicts the different between the physicochemical and fatty acids composition in Tea seed oil extracted by SE and MUAE

10.11.2 Physicochemical properties of oil

Two different oil samples extracted from different extraction method; do not show remarkable difference in degree of unsaturation, molecular weight and purity. Given results are due to no significant distinction in saponification value, unsaponifiable value, relative density, refractive index, and iodine value. Iodine value can be used to determine the dryness characteristic of oil. If the oil value is below the level that is 100 g I_2/100 g; then oil is classified as nondry category and can be used in other purposes like lubrication and cosmetics (Hernandez et al., 2013). As compared to SE, peroxide value and acid value is significantly lower than MUAE. These lower values of

Table 10.3 Bioactive compositions of oil extracted by SE and MUAE.

Bioactive compounds	SE	MUAE
Total phenols (mg GAE/kg)	87.50 ± 4.97	125.72 ± 5.68
α-tocopherol (mg/kg)	3780.45 ± 10.24	315.33 ± 8.17
β-carotene (mg/kg)	33.4 ± 1.36	24.29 ± 1.06
Phospholipids (%)	0.23 ± 0.01	0.30 ± 0.01
Squalene (mg/100 g)	268.74 ± 11.46	352.87 ± 13.62
Phytosterol (mg/100 g)	356.29 ± 11.13	428.57 ± 12.38

peroxide and acid are indication and proof of good oil quality and stability. Reduced value in microwave-assisted extraction is owing to shorted time duration and reduced temperature range which results into low oxidation; ultimately lowers the acid and peroxide value of oil. Moreover, as compared to conventional extraction method, MUAE consists high bioactive profile including α-tocopherol, β-carotene, and phenolic content which play vital role in protection of oil against oxidative rancidity. Overall sensory appeal and nutritional composition of oil extracted by MUAE is improved because of presence of various nonglyceridic micro components (Jiao et al., 2014). So, results indicate overall physicochemical composition of oil by MUAE is better as compared to that of SE.

10.11.3 Comparison of bioactive profile

Owing to synergism and process intensification in MUAE, overall bioactive profile is higher, process reinforcing energy cause destruction of cytoarchitecture; thereby facilitates easy separation of bioactive contents (Rashed et al., 2016). Mutual reinforcing energy of microwave and ultrasound leads to weakening of complexes made up of bioactive content and polysaccharides or proteins. In context to the SE, lower content of bioactive profile is due to exposure to high temperature for log time duration. Less extraction time and low temperature in MUAE play vital role in preserving activity of thermolabile substances (Table 10.3). Differences were observed in context to difference of cultivar, producing area and postharvesting conditions; in oil obtained from Zhejiang Province of China. Remarkable high content of α-tocopherol with low content of sterols is obtained in SE; with aqueous extraction low content of squalene and high content of α-tocopherol is reported (Wei et al., 2015; Huang et al., 2014). Antioxidants such as α-tocopherol, phenolics and β-carotenes plays vital role in scavenging free radicals or directly termination the chain reaction. MUAE of tea seed oil results into α-tocopherol content around 378.45 ± 10.24 mg/kg, β-carotene content about 33.42 ± 1.36 mg/kg and TP 125.72 ± 5.68 mg GAE/kg approximately which is significantly higher as compared to SE i.e., 315.33 ± 8.17 mg/kg, 24.29 ± 1.06 mg/kg and 87.50 ± 4.97 mg GAE/kg, respectively.

10.12 Applications of MUAE

10.12.1 Biodiesel production

Other than oil extraction MUAE is also applicable in biodiesel production. By esterification and transesterification conversion of many oils like coconut, corn, vegetable, jatropa, rice bran, rapeseed, mustard, sunflower, cottonseed, safflower, castor, and animal fats into biodiesel. Just microwave and microwave ultrasonication-assisted heating energy were evaluated. Moreover, application of MUAE is also reported in Kjeldahl nitrogen and food industry was also reported (Chemat et al., 2004). For SE, microwave and MUAE; yields for esterification in 1 hour are 80%, 91%, and 99%, respectively. Extraction of oils from a marine microalga species and soybean germ is investigated by individual MW, US and MUAE (Cravotto et al., 2008). This study reported that yields enhanced by 50% to 500% and time reduced by up to 10-fold that shows 10 times higher production in 10 time less time. In MUAE, alcohol as solvent requirement is reduced both in transesterification and esterification. For example, in conventional method, oil to methanol molar ratios for esterification1:3 and transesterification is 1:6 and for MUAE mediated reactions esterification is 1:2 and transesterification is 1:4. Meanwhile, reaction time is reduced to 20 and 90 to 15 and 6 minutes, respectively (Gole et al., 2013). Microwave ultrasound-assisted extraction is not only limits to this but also used in stages of other process like feed stock pretreatments. To reinforce combined energy of MUAE, concentrated algal suspension is operated in mechanism which results into release of cell contents mainly lipids. This biological matrix gets converted into in to biocrude for further treatments. Green extraction of valuable bioproducts such as proteins and lipids are promoted due to high temperature and high-pressure reactions under MUAE (Reddy et al., 2014).

10.12.2 Bioethanol production

For bioethanol production, three major feedstock categories are following:

1. Starch-containing feedstock-wheat, corn, rice, milo, cassava, potatoes, and barley
2. Lignocellulosic biomass-wasted crops, straw, wood, grass, and crop residues and,

3. Sucrose-containing feedstock-sugar beet, sugar cane, fruits and sweet sorghum (Balat et al., 2009)

Sustainable feedstock is starch- and sugar-based feedstock; A variety of valuable products can be recovered which includes ethanol, acetone, butanol, acetone, lactic acid, and amino acids. Meanwhile, lignocellulosic material is harder to process, owing to unfavorable economics. The pretreatment methods can be effectively enhanced by this process intensification. Lignocellulosic materials hydrolysis into sugars and their subsequent fermentation to bioethanol can be improved by the process of sonication. Similarly; from maize, corn meal and potato starch degradation can be improved by process of sonication. The effect of sonication was evaluated in many studies; both at low and high frequencies on carbohydrates in sulfuric acid (Czechowska-Biskup et al., 2005). Report shows Microwave application in starch depolymerization to enhance hydrolysis in dilute hydrochloric acid, water and with chloride-based catalyst. Catalytic conversion of starch based industrial waste such as wet potato sludge into sugars evaluated recently by application of MUAE (Hernoux-Villière et al., 2013). Exposure of 2 hours at 60°C results into 46% catalytic conversion.

10.12.3 Biogas production

Application of microwave ultrasound-assisted extraction as process intensification effects may improve in return the energy recovery, the biogas yields and the overall economics of mechanism. Owing to electromagnetic field, which possesses depolarization under effect of microwave energy cause alignment of macromolecules thus improves nonthermal effect apart from heating applications ultimately cause possible hydrogen bond breakdown (Saha et al., 2011). This disruption cause increment in hydrolysis and disintegration of sludge which further enhancement inactivation of fecal coliforms, rate of anaerobic reactions, and dehydration in process of biogas production ultimately lead higher production.

10.13 Prevailing issues with this hybrid technology and future trend

The process of MUAE shows tremendous effect on production of organic and inorganic productions. Various current issues need to be considered for implementation.

10.13.1 Understanding and evaluation of process chemistry

Different materials interact differently with chemicals and solvents. Some materials transmit the microwave energy, some absorb it or some may partially absorb it. In the same way, some chemicals show tremendous effect and some show almost negligible effect under ultrasound (Mason et al., 2011). Since some solvents get easily heated by MWs, this leads to negative impact on thermosensitive materials (like lipids). So, solvents having low dielectric constant should be used for such materials (like hexane).

10.13.2 Improving energetic efficiency in MUAE reactor

Knowledge gap in proper utilization of process intensification of this synergistic process should be covered. For efficient energy use the energy density (W/m^2) and power density (W/m^3) need to be evaluate. The term energy density is related to design of reactor which produces required energy intensity. Meanwhile, the power density provides optimum energy, expressed in small volumes and depends upon volume of reactor. The dimensions of reactor can be determined by the optimum energy intensity (diameter and length).

10.13.3 Increment in the recovery of valuable product

The low ability of microwave energy to penetrate through the material might be limitation of the process. For large scale two matters need to be considered ultrasound horns in contact-type or plug-flow type reactor; microwave irradiation frequency should be 2450 MHz. for ultrasonication long vibrating reactors should be built for continuous flow.

10.14 Conclusion

Superior beneficial outcomes from UAE and MUA when in synergism as MUAE show wide range of challenges. These challenges and limitations with advantages provide new opportunities for further researches on this emerging field. Although mass and heat transfer related limitations have been overcome but limitations of overpower need to be control. Major matter of concern is need at present to make this hybrid technology economically feasible. Efficient utilization of this intensified process is in developing, reactor having microwave energy with longitudinal US plug-type system. With negligible yield difference in oil production by SE and MUAE; this innovative technology leads to reduce in cost, time, and energy consumption with high extraction potential. Fatty acid composition in conventional and MUAE is similar but physicochemical composition is significantly high. Microwave ultrasound-assisted extraction has great extraction potential; thus, this hybrid technology is emerging technology for extraction of oils and phytochemicals from various plant sources.

References

Balat, M., Balat, H., 2009. Recent trends in global production and utilization of bio-ethanol fuel. Appl. Energy 86, 2273–2282.

Bao, Y., and Guo, Y. (2016). Optimization of ultrasonic-assisted aqueous enzymatic extraction of pine nut oil and its oxidative stability. Food Sci. 27, 60–68. doi: 10.7506/spkx1002-6630-201622009

Cao, Y., 2015. Research on quality characteristics of different groups of xanthocerassorbifolia. Food Sci. Technol. 40, 40–44. doi:10.13684/j.cnki.spkj.2015.01.009.

Chemat, S., Lagha, A., Amar, H.A., Chemat, F., 2004. US assisted MW digestion. Ultrason. Sonochem. 11, 5–8.

Chemat, F., Fabiano-Tixier, A.S., Vian, M.A., Allaf, T., Vorobiev, E., 2015. Solvent-free extraction of food and natural products. Trac-Trend Anal. Chem. 71, 157–168.

Chemat, F., Rombaut, N., Meullemiestre, A., Turk, M., Perino, S., Fabiano-Tixier, A.S., AbertVian, M., 2017a. Review of green food processing techniques. Preservation, transformation, and extraction. Innov. Food Sci. Emerg. 41, 357–377.

Chemat, F., Rombaut, N., Sicaire, A.G., Meullemiestre, A., Fabiano-Tixier, A.S., AbertVian, M., 2017b. Ultrasound assisted extraction of food and natural products. Mechanisms, techniques, combinations, protocols and applications. A review. UltrasonSonochem 34, 540–560.

Chen, F.L., Zhang, X.L., Zhang, Q., Du, X.Q., Yang, L., Zu, Y.G., Yang, F.J., 2016. Simultaneous synergistic microwave–ultrasonic extraction and hydrolysis for preparation of trans-resveratrol in tree peony seed oil-extracted residues using imidazolium-based ionic liquid. Ind. Crops Prod. 94, 266–280.

Cravotto, G., Cintas, P., 2007. The combined use of microwaves and ultrasound: improved tools in process chemistry and organic synthesis. Chem. Eur. J. 13, 1902–1909.

Cravotto, G., Cintas, P., 2007. The combined use of MWs and US: improved tools in process chemistry and organic synthesis. Chem. Eur. J. 13, 1902–1909.

Cravotto, G., Cintas, P., 2007. Extraction of flavourings from natural sources. In: Taylor, A., Hort, J. (Eds.), Modifying Flavour in Food. Woodhead Publishing Ltd., Cambridge, pp. 41–63 Chapter 3.

Cravotto, G., Boffa, L., Mantegna, S., Perego, P., Avogadro, M., Cintas, P., 2008. Improved extraction of vegetable oils under high-intensity ultrasound and/or microwaves. Ultrason. Sonochem. 15, 898–902.

Czechowska-Biskup, R., Rokita, B., Lotfy, S., Ulanski, P., Rosiak, J.M., 2005. Degradation of chitosan and starch by 360-kHz US. Carbohydr. Poly. 60, 175–184.

Desai, M., Parikh, J., Parikh, P.A., 2010. Extraction of natural products using microwaves as a heat source. Sep. Purif. Rev. 39, 1–32.

Ezeh, O., Gordon, M.H., Niranjan, K., 2016. Enhancing the recovery of tiger nut (Cyperus esculentus) oil by mechanical pressing: moisture content, particle size, high pressure and enzymatic pre-treatment effects. Food Chem 194, 354–361.

Farhat, A., Fabiano-Tixier, A.S., Maataoui, M.E.I., Maingonnat, J.F., Romdhane, M., Chemat, F., 2011. Microwave steam diffffusion for extraction of essential oil from orange peel: kinetic data, extract's global yield and mechanism. Food Chem 125, 255–261.

Fawzy, M., 2013. Healthy blends of high linoleic sunflower oil with selected cold pressed;oils: Functionality, stability and antioxidative characteristics. Ind. Crops Products 43, 65–72. doi:10.1016/j.indcrop.2012.07.013.

Gole, V.L., Gogate, P.R., 2013. Intensification of synthesis of biodiesel from non-edible oil using sequential combination of MW and US. Fuel Process. Technol. 106, 62–69.

Gude, V.G., Grant, G.E., 2013. Biodiesel from waste cooking oils via direct sonication. Appl. Energy 109, 135–144.

Gude, V.G., Martinez-Guerra, E., 2015. Green chemistry of MW-enhanced biodiesel production. In: Production of Biofuels and Chemicals with MW. Springer, The Netherlands, pp. 225–250.

Gude, V.G., Patil, P.D., Deng, S., Nirmalakhandan, N., 2011. MW enhanced methods for biodiesel production and other environmental applications. In: Green Chemistry for Environmental Remediation. Wiley Interscience, New York, pp. 209–249.

Gude, V.G., Patil, P.D., Martinez-Guerra, E., Deng, S., Nirmalakhandan, N., 2013. MW energy potential for biodiesel production. Sustain. Chem. Process. 1, 1–31.

Gude, V.G., 2015. Synergism of microwaves and ultrasound for advanced biorefineries. Resour.-Efficient Technol. 1 (2), 116–125.

Hernandez, B., Luna, G., Garcia, O., Mendoza, M.R., Azuara, E., Beristain, C.I., Jimenez, M., 2013. Extraction and characterization of Oecopetalum mexicanum seed oil. Ind. Crops Prod. 43, 355–359.

Hernoux-Villière, A., Lassi, U., Hu, T., Paquet, A., Rinaldi, L., Cravotto, G., et al., 2013. Simultaneous MW/US-assisted hydrolysis of starch-based industrial waste into reducing sugars. ACS Sustain. Chem. Eng. 1, 995–1002.

Hosseini, S., Gharachorloo, M., Tarzi, B.G., Ghavami, M., Bakhoda, H., 2015. Effects of ultrasound amplitude on the physicochemical properties of some edible oils. J. Am. Oil Chem. Soc. 92, 1717–1724.

Huang, K., Wang, S.J., Shan, L., Zhu, Q., Qian, J., 2007. Seeking synergistic effect – a key principle in process intensification. Sep. Purif. Technol. 57, 111–120.

Huang, S.S., Wu, S.X., Nie, K.F., 2014. Optimization of aqueous extraction a process for fresh camellia seed oil. Food Mach. 30, 185–190.

Huang, J., Zhang, T., Zhang, Q., Chen, M., Wang, Z., Zheng, B., et al., 2016. The mechanism of high contents of oil and oleic acid revealed by transcriptomic and lipidomic analysis during embryogenesis in Carya cathayensisSarg. BMC Genomics 17, 113. doi:10.1186/s12864-016-2434-7.

Ishaq, M., Razi, R., Khan, S.A., 2017. Exploring genotypic variations for improved oil content and healthy fatty acids composition in rapeseed (Brassica napus L.). J. Sci. Food Agric. 97, 1924–1930. doi:10.1002/jsfa.7997.

Jiao, J., Li, Z.G., Gai, Q.Y., Li, X.J., Wei, F.Y., Fu, Y.J., Ma, W., 2014. Microwave-assisted aqueous enzymatic extraction of oil from pumpkin seeds and evaluation of its physicochemical properties, fatty acid compositions and antioxidant activities. Food Chem 147, 17–24.

Khadhraoui, B., Turk, M., Fabiano-Tixier, A.S., Petitcolas, E., Robinet, P., Imbert, R., Maâtaouic, M.E.I., Chemat, F., 2018. Histo-cytochemistry and scanning electron microscopy for studying spatial and temporal extraction of metabolites induced by ultrasound. Towards chain detexturation mechanism. Ultrason. Sonochem. 42, 482–492.

Koubaa, M., Mhemdi, H., Barba, F.J., Roohinejad, S., Greiner, R., Vorobiev, E., 2016. Oilseed treatment by ultrasounds and microwaves to improve oil yield and quality: an overview. Food Res. Int. 85, 59–66. doi:10.1016/j.foodres.2016.04.007.

Latif, S., Anwar, F., 2011. Aqueous enzymatic sesame oil and protein extraction. Food Chem 125, 679–684. doi:10.1016/j.foodchem.2010.09.064.

Leonelli, C., Mason, T.J., 2010. Microwave and ultrasonic processing: now a realistic option for industry. Chem. Eng. Process. 49, 885–900.

Li, Z.M., Li, J.L., Xiao, Z.H., Wu, H., L, R., k, Z., M.R., LiC.Z., 2017. Optimization of simultaneous extraction of tea seed oil and Saponin with subcritical water applying response surface methodology. J. Chin. Cereals Oils Assoc. 32, 81–87.

Li, Y., Yu, Z., Jin, J., Zhang, Q., Wang, G., Liu, C., et al., 2018. Impact of elevated CO_2 on seed quality of soybean at the fresh edible and mature stages. Front. Plant Sci. 9, 1413. doi:10.3389/fpls.2018.01413.

Liao, B., Hao, Y., Lu, J., Bai, H., Guan, L., Zhang, T., 2018. Transcriptomic analysis of Perilla frutescens seed to insight into the biosynthesis and metabolic of unsaturated fatty acids. BMC Genomics 19, 213. doi:10.1186/s12864-018-4595-z.

Luo, J., Fang, Z., Smith, R.L., 2014. US-enhanced conversion of biomass to biofuels. Prog. Energy Combust. Sci. 41, 56–93.

Martinchik, A.N., Baturin, A.K., Zubtsov, V.V., Viu, M., 2012. [Nutritional value and functional properties of flaxseed]. VoprPitan 81, 4–10.

Martinez-Guerra, E., Gude, V.G., 2014. Synergistic effect of simultaneous MW and US irradiations on transesterification of waste vegetable oil. Fuel 137, 100–108.

Martinez-Guerra, E., Gude, V.G., 2015. Continuous and pulse sonication effects on transesterification of used vegetable oil. Energy Convers. Manag. 96, 268–276.

Mason, T.J., Chemat, F., Vinatoru, M., 2011. The extraction of natural products using US or MWs. Curr. Org. Chem. 15, 237–247.

Ning, C., Jiang, Y., Meng, J., Zhou, C., Tao, J., 2015. Herbaceous peony seed oil: a rich source of unsaturated fatty acids and g-tocopherol. Eur. J. Lipid Sci. Technol. 117, 532–542. doi:10.1002/ejlt.201400212.

Olmo-Garcıa, L., Kessler, N., Neuweger, H., Wendt, K., Olmo-Peinado, J.M., Fernandez-Gutierrez, A., et al., 2018. Unravelling the distribution of secondary metabolites in Olea europaea l.: exhaustive characterization of eight olive-tree derived matrices by complementary platforms (LC-ESI/APCI-MS and GC-APCI-MS). Molecules 23, 2419. doi:10.3390/molecules 23102419.

Rajaei, A., Barzega, M., Yamini, Y., 2005. Supercritical fluid extraction of tea seed oil and its comparison with solvent extraction. Eur. Food Res. Technol. 220, 401–405.

Rashed, M.M.A., Tong, Q.Y., Abdelhai, M.H., Gasmalla, M.A.A., Ndayishimiye, J.B., Chen, L., Ren, F., 2016. Effect of ultrasonic treatment on total phenolic extraction from Lavandula pubescens and its application in palm olein oil industry. Ultrason. Sonochem. 29, 39–47.

Rauf, S., Jamil, N., Tariq, S.A., Khan, M., Kausar, M., Kaya, Y., 2017. Progress in modification of sun flower oil to expand its industrial value. J. Sci. Food Agric. 97, 1997–2006. doi:10.1002/jsfa.8214.

Reddy, H.K., Muppaneni, T., Patil, P.D., Ponnusamy, S., Cooke, P., Schaub, T., et al., 2014. Direct conversion of wet algae to crude biodiesel under supercritical ethanol conditions. Fuel 115, 720–726.

Robert, A., England, M.W. Synthesis: a new wave of synthetic organic chemistry. LabPlus International, April/May 2003.

Saha, M., Eskicioglu, C., Marin, J., 2011. MW, ultrasonic and chemo-mechanical pre-treatments for enhancing methane potential of pulp mill wastewater treatment sludge. Bioresour. Technol. 102, 7815–7826.

Servent, A., Boulanger, R., Davrieux, F., Pinot, M.-N., Tardan, E., Forestier Chiron, N., et al., 2018. Assessment of cocoa (Theobroma cacao L.) butter content and composition throughout fermentations. Food Res. Int. 107, 675–682. doi:10.1016/j.foodres.2018.02.070.

Shang, X., Cheng, C., Ding, J., Guo, W., 2017. Identification of candidate genes from the SAD gene family in cotton for determination of cottonseed oil composition. Mol. Genet. Genomics 292, 173–186. doi:10.1007/s00438-016-1265-1.

Sielicka, M., Małecka, M., Purłan, M., 2014. Comparison of the antioxidant capacity of lipid-soluble compounds in selected cold-pressed oils using photochemiluminescence assay (PCL) and DPPH method. Eur. J. Lipid Sci. Technol. 116, 388–394. doi:10.1002/ejlt.201300356.

Siger, A., Jozefifiak, M., Gornas., P., 2017. Cold-pressed and hot-pressed rapeseed oil: the effects of roasting and seed moisture on the antioxidant activity, canolol, and tocopherol level. Acta Sci. Pol. Technol. Aliment. 16, 69–81. doi:10.17306/J.AFS.2017.0458.

Soria, A.C., Villamiel, M., 2010. Effect of ultrasound on the technological properties and bioactivity of food: a review. Trends Food Sci. Technol. 21, 323–331. doi:10.1016/j.tifs.2010.04.003.

Stankiewicz, A.I., Moulijn, J.A., 2010. Process intensification: transforming chemical engineering. Chem. Eng. Process 96, 22–34.

Stankiewicz, A., 2006. Energy matters: alternative sources and forms of energy for intensification of chemical and biochemical processes. Chem. Eng. Res. Des. 84, 511–521.

Sun, R., Ye, R., Gao, L., Zhang, L., Wang, R., Mao, T., et al., 2017. Characterization and ectopic expression of CoWRI1, an AP2/EREBP domain-containing transcription factor from coconut (Cocos nucifera L.) endosperm, changes the seeds oil content in transgenic Arabidopsis thaliana and rice (Oryza sativa L.). Front. Plant Sci. 8, 63. doi:10.3389/fpls.2017.00063.

Teh, S.S., Birch, E.J., 2014. Effect of ultrasonic treatment on the polyphenol content and antioxidant capacity of extract from defatted hemp, flax and canola seed cakes. Ultrasonics Sonochem 21, 346–353. doi:10.1016/j.ultsonch.2013.08.002.

Vian, M.A., Fernandez, X., Visinoni, F., Chemat, F., 2008. Microwave hydrodiffusion and gravity, a new technique for extraction of essential oils. J. Chromatogr. A 1190, 14–17.

Vonapartis, E., Aubin, M.P., Seguin, P., Mustafa, A.F., Charron, J.B., 2014. Seed composition of ten industrial hemp cultivars approved for production in Canada. J. Food Composition Anal. 39, 8–12. doi:10.1016/j.jfca.2014.11.004.

Wang, X.Q., Zeng, Q.M., Verardo, V., Contreras, M.D.M., 2017a. Fatty acid and sterol composition of tea seed oils: their comparison by the "FancyTiles" approach. Food Chem 233, 302–310.

Wang, X., Xu, P., Yin, L., Ren, Y., Li, S., Shi, Y., et al., 2018. Genomic and transcriptomic analysis identifified gene clusters and candidate genes for oil content in peanut (Arachis hypogaea L.). Plant Mol. Biol. Rep. 36, 518–529. doi:10.1007/s11105-018-1088-9.

Wang, Y., Li, R., Jiang, Z.T., Tan, J., Tang, S.H., Li, T.T., Liang, L.L., He, H.J., Liu, Y.M., Li, J.T., Zhang, X.C., 2018. Green and solvent-free simultaneous ultrasonic-microwave assisted extraction of essential oil from white and black peppers. Ind. Crop Prod. 114, 164–172.

Wei, Z.W., Wang, J.G., 2017. Study on the quality and bioactive substances of hot and cold-pressed sunflower oil. Cereals Oils 30, 28–30 1008-9578(2017)05-0028-03.

Wei, J., Chen, L., Qiu, X.Y., Hu, W.J., Sun, H., Chen, X.l., Bai, Y.Q., Gua, X.Y., Wang, C.l., Chen, H., Hu, R.B., Zhang, H., Shen, G.X., 2015. Optimizing refining temperatures to reduce the loss of essential fatty acids and bioactive compounds in tea seed oil. Food Bioprod. Process 94, 136–146.

Yang, M., Zheng, C., Zhou, Q., Huang, F., Liu, C., Wang, H., 2013. Minor components and oxidative stability of cold-pressed oil from rapeseed cultivars in China. J. Food Composition Anal. 29, 1–9. doi:10.1016/j.jfca.2012.08.009.

Yara-Varon, E., Li, Y., Balcells, M., Canela-Garayoa, R., Fabiano-Tixier, A.S., Chemat, F., 2017. Vegetable oils as alternative solvents for green oleo extraction, purification and formulation of food and natural products. Molecules 22, 1474. doi:10.3390/molecules22091474.

Yi, J.Y., Kim, H.J., Chung, M.S., 2017. Manufacture of low-benzo(a)pyrene sesame seed (Sesamum indicum L. oil using a self-designed apparatus. PLoS One 12, e0173585. doi:10.1371/journal.pone.0173585.

Zhang, Q., Yu, R., Sun, D., Rahman, M.M., Xie, L., Hu, J., et al., 2018. Comparative transcriptome analysis reveals an efficient mechanism of alpha-linolenic acid in tree peony seeds. Int. J. Mol. Sci. 20, 65. doi:10.3390/ijms20010065.

Zhao, J., Becker, H.C., Zhang, D., Zhang, Y., Ecke, W., 2005. Oil content in a european × chinese rapeseed population. Crop Sci 45, 51–59. doi:10.2135/cropsci2005.0051.

CHAPTER ELEVEN

Sonication-microwave synergistic extraction of proteins from plant sources and its effect on protein

Md. Mostafa Kamal[a,f], Md Saifullah[b], Naymul Karim[c], Muhammad Umair[d], Husnain Raza[e] and Mohammad Rezaul Islam Shishir[e]

[a]Department of Food and Agroprocess Engineering, Bangabandhu Sheikh Mujibur Rahman Science and Technology University, Gopalganj 8100, Bangladesh
[b]School of Environmental and Life Sciences, College of Engineering, Science and Environment, University of Newcastle, Ourimbah, NSW, 2258, Australia
[c]Department of Food Science and Nutrition, Zhejiang University, Hangzhou, China
[d]Department of Food Science and Engineering, College of Chemistry and Engineering, Shenzhen University, Shenzhen, China
[e]Shenzhen Key Laboratory of Marine Microbiome Engineering, Institute for Advanced Study, Shenzhen University, Shenzhen, China
[f]Department of Food Processing and Preservation, Hajee Mohammad Danesh Science and Technology University, Dinajpur 5200, Bangladesh

11.1 Introduction

Around 820 million people are associated with hunger problems and the trend of this number has been gradually uplifting with the growing population worldwide (FAO et al., 2019). A significant number of people worldwide are also suffering from malnutrition due to the absence of sufficient amount of high-quality proteins in their daily diets (Semba, 2016). Proteins constitute a vital portion of the human body, maintaining growth, and development. The rising interest in the significance of proteins in the daily diet and health management has stimulated the scientific and research communities to explore the sources of sustainable and high-quality protein (Kumar et al., 2021). In recent decades, plant proteins gaining more popularity/attention over animal proteins due to their easy availability, and different health-beneficial properties, especially to the vegetarian and vegan community. In 2018, the global sale of plant-derived protein products was estimated at $35 billion, which is expected to be greater than 45 billion USD by the year 2023 (Akharume et al., 2021; Cassity, 2019). Moreover, consumption of animal proteins has been associated with several human diseases including cancer, diabetes, heart attack, cardiovascular disorders, obesity,

Ultrasound and Microwave for Food Processing: Synergism for Preservation and Extraction.
DOI: https://doi.org/10.1016/B978-0-323-95991-9.00007-2

and kidney disease, which discourages the consumers, and encourages the researchers to explore the best alternative source of protein (Boukid, 2021; Campbell, 2017; Ko et al., 2020). However, this challenge could be fixed by substituting animal-based proteins with plant-based proteins. Numerous studies demonstrated that developing appropriate technologies could play a significant role in the cost-effective preparation of plant proteins.

Extraction/isolation is a crucial unit operation step for the preparation and processing of macro or micronutrients from natural products. Several factors, such as type of material source, extraction conditions, applied techniques, solvent type, concentration, and initial sample preparation can influence the yield percentage, physicochemical quality, and functional properties of protein (Setyaningsih et al., 2016). Studies on protein science indicate that extraction processing can alter the native structure of protein depending upon the extraction techniques and conditions applied. Conventional methods like alkaline extraction, salt precipitation, solvent precipitation, and heat application are commonly used to extract protein from plant sources (Kain et al., 2009; Ochoa-Rivas et al., 2017). These methods require a long extraction time and have limited extraction efficiency. They are also not adequately efficient to extract quality proteins because of protein aggregation and changes in structural integrity due to heat application to remove the solvents. This further may lead to the loss of protein functionalities and the formation of toxic substances (Tang et al., 2003; Ho et al., 2007; Sereewatthanawut et al., 2008). Therefore, potential nonthermal extraction technologies are highly appreciable for the efficient extraction of proteins with proper structural and functional quality.

In recent years, different nonthermal processing technologies have been emerged that are frequently applied for the extraction of different compounds like polysaccharides, pectin, pigments, bioactive polyphenols, etc. Among the modern technologies, ultrasound, pulsed electric field, high hydrostatic pressure, microwave, ohmic heating, plasma processing, sub- or supercritical fluid, radiofrequency, pressurized liquid, and enzymatic extraction have been reported as the most efficient extraction methods used in the food and nutraceutical industries (Chemat et al., 2017; Panja, 2018; Kamal et al., 2020). Studies recommended that these technologies are highly reliable in respect of time and energy consumption, cost management, and higher extraction efficiency with superior quality of the extracted components. Ultrasound is a nonclassical extraction technology that enhances the extraction process through thermal cavitation, and mechanical force (Li et al., 2016) by the high-frequency sound waves (>20 kHz). This technology has been extensively used due to its excellent extraction efficiency with a moderate impact on the extracted compounds.

Microwave technology is another important physical and eco-friendly extraction method that uses microwave energy to disrupt cell structure to facilitate the extraction of desired components like protein from plant matrices. The microwave-assisted extraction technique is gaining much attention due to its tremendous efficiency, less solvent requirement, and short irradiation time. However, microwave technology has some drawbacks such as it may alter the structural properties of proteins, instant heating accelerates free radical formation, reduce uniform heating, and reduce uniform localization of charged groups in protein molecules (Jiao et al., 2019; Li et al., 2020). These problems can be overcome by combining ultrasonication and microwave technology. Recent research evidenced that synergistic application of sonication and microwave can enhance the extraction process of biomolecules from plant matrices, decrease the extraction time, and modulate the structural and functional characteristics of the extracted molecules (Liu et al., 2017; Xu et al., 2018; Cheng et al., 2021; Liu et al., 2019). Keeping this space, this chapter outlined the recent advancement of sonication-microwave synergistic application for the extraction of plant proteins. This chapter has been divided into four important parts to provide an in-depth conceptualization of the recent progress in the extraction of plant proteins using ultrasonication, microwave, and their synergism approaches: (i) sources of plant proteins and their biological importance, (ii) advanced techniques employed for the extraction of plant proteins, in which major emphasize has been given on the application and working mechanism of ultrasonication, microwave, and their synergism approach, (iii) effect of ultrasonication, microwave, and their synergism on the modification/alteration of major characteristics of the protein, such as hydrophobicity, solubility, foaming, emulsifying, digestibility, and water holding capacity, and (iv) conclusive remarks and future trends.

11.2 Overview of proteins: sources and biological importance

Protein is a vital component of plant and human cells, known as the building block of body. It consists of a long chain of amino acids linked with peptide bonds. Protein can be found in several sources including animals, plants, kinds of seafood, insects, and other sources (i.e., mushrooms, yeast (single-cell protein) (Kamal et al., 2019; Kamal et al., 2021). Fig. 11.1 outlined

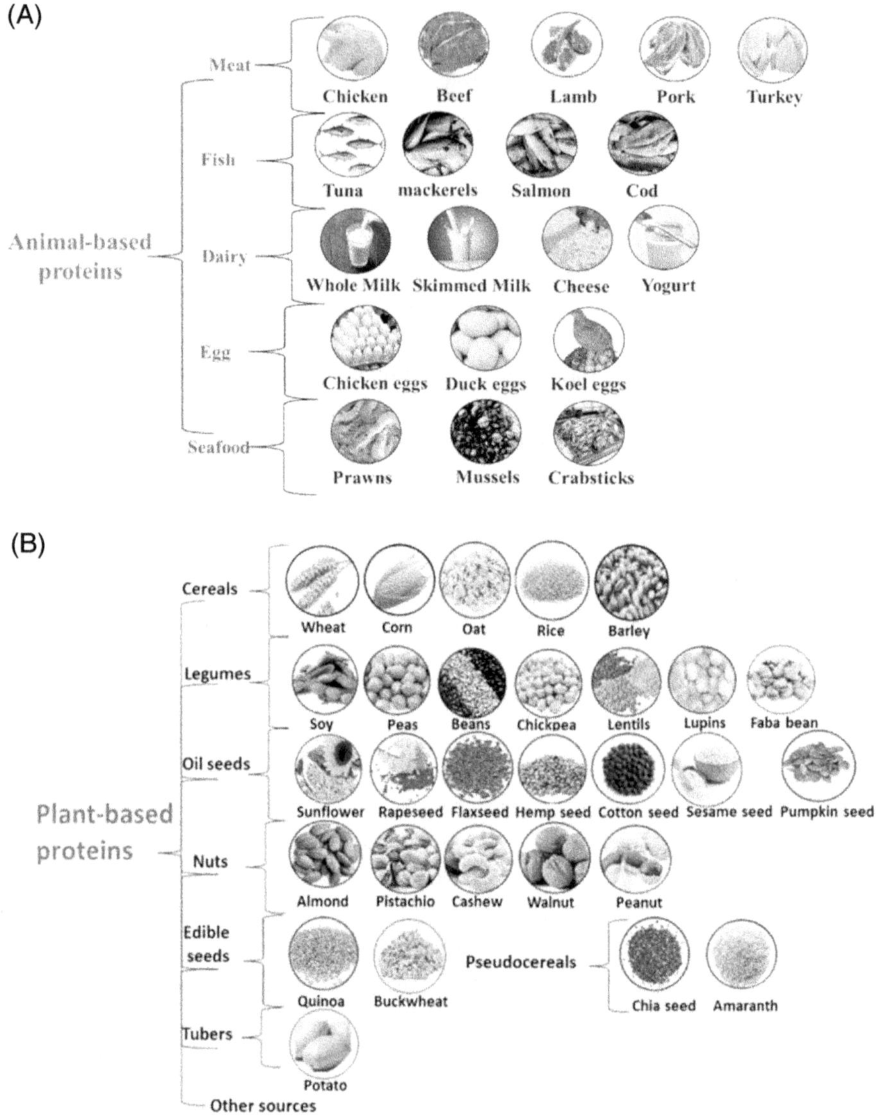

Figure 11.1 Major sources of proteins (A) animal-based and (B) plant-based.

the major sources of animal and plant proteins. Animals are the major source of proteins and essential components of several food products. Animal protein consumption comprises over 60% worldwide. In the human diet, the main contributors of protein are several animal sources including dairy-based products such as whole and skimmed milk, cheese, yogurt, chicken, beef, goat, lamb, pork, and turkey; fishes such as tuna, mackerels, salmon, and cod; eggs such as from chicken, duck, and koel; different seafood such as prawns, mussels, crab sticks; and different edible insect species (Kamal et al., 2021; Nwachukwu and Aluko, 2021). Most of the essential amino acids come from animal-based proteins, however, possess a serious health concern due to higher cholesterol levels and saturated fatty acids (Richter et al., 2015).

In recent years, due to lifestyle diversity and health consciousness, people are trending to show their deep interest in plant-based proteins (Akyüz and Ersus, 2021), because of their production sustainability, low price, as well as nutritional, health-promoting, and functional properties (Tahmasian et al., 2022). Plant protein sources are widely distributed and demonstrate nutrient property, structural diversity, and functional behavior because of various bioactive compounds. The potential sources of plant-based protein include cereals, for example, wheat, maize, rice, oat, barley, and sorghum; pulses, for example, peas, chickpeas, lentils, and kidney beans; pseudocereals such as buckwheat, amaranth, and quinoa; oilseeds, for example, soybean, flaxseed, sunflower, peanuts, sesame, cotton, and hempseed; tubers, for example, potato; edible seeds, for example, chia, basil, and pumpkin; tree nuts, for example, chestnuts, almonds, walnuts, and pecans (Agrawal et al., 2017; Phongthai et al., 2018; Vilcacundo et al., 2018; Nasrabadi et al., 2021; Nwachukwu and Aluko, 2021). Recent research has demonstrated that plant by-products, for example, tomato, cherry, peach, date, and papaya seeds, bran from wheat, rice, buckwheat, and oat are considered important protein sources (Kamal et al., 2021).

Proteins are complex macromolecules actively connected with the physiochemical processes of life including metabolism, movement, defense, cellular communication, molecular recognition, and others. They help to maintain body physiology through building/repairing cells and tissues. It is evidenced from the literature that proteins have multiple health-beneficial properties as summarized in Fig. 11.2. A myriad of studies has demonstrated the health potential of proteins including improved metabolic profile and helped to reduce chronic diseases (Akharume et al., 2021) via exerting antioxidant and anti-inflammatory activity.

Plant-derived proteins can offer a significant contribution to food formulations due to their gelling, thickening, solubility, and foaming properties along with their ability to act as an emulsion stabilizer, water, and fat binder (Boukid, 2021; Kumar et al., 2022). Plant-derived proteins could vary in composition, functionality, and nutritional benefits depending on their several factors, for example, biological source and variety, extraction/processing

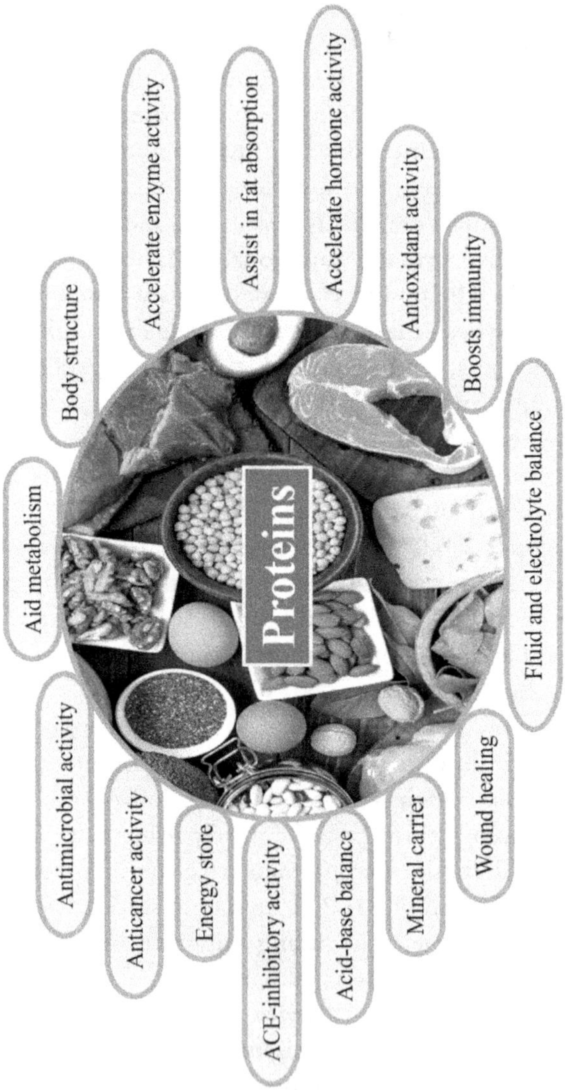

Figure 11.2 Biological functions of proteins in the human body.

conditions, and level of purity (Hoehnel et al., 2019; Boukid, 2021; Nasrabadi et al., 2021; Nwachukwu and Aluko, 2021). For instance, soy protein is considered the most abundantly applied in the food industry because of its balanced amino acid combination along with high functionality (Yuliarti et al., 2021). Wheat protein (gluten) has been used since ancient times for its better solubility, swelling, water holding capacity, rheological, and viscoelastic properties. It also acts as the extenders and binders to reduce the cooking loss during processing (Chiang et al., 2019). Recently pea protein and its isolate have gained popularity because of its good functionality and lower risk for allergenicity. It could be mixed with other sources of proteins to alternate the textural and nutritional properties (Schreuders et al., 2019). Rice, wheat, and oat bran are also used as potential protein sources for their excellent functionality (Kumar et al., 2021). Furthermore, proteins from different beans such as faba and mung beans are more promising for their high nutritional value, high protein content, sustainability, and affordability (Akharume et al., 2021).

In the recent year, plant proteins are used as core/coating materials for the development of nanoparticles and microparticles to encapsulate, protect, and delivery of numerous bioactive compounds including minerals, vitamins, nutraceuticals, probiotics, antimicrobials, and antioxidants (Zink et al., 2016; Shishir et al., 2018, 2021; Akhtar et al., 2021). For example, zein is the major corn protein that has been used to construct colloidal delivery systems for lipophilic bioactive compounds (Zhang et al., 2015). The wheat proteins such as gliadins have been reported to use as an antioxidant emulsifier in emulsion-based delivery systems (Qiu et al., 2015), and as a foaming agent in the food industry (Peng et al., 2018). Furthermore, gliadin-lecithin nanoparticles have been used to encapsulate and protect curcumin (Yang et al., 2018). Furthermore, plant-based proteins have the huge potential to be explored in different areas as an alternative to animal proteins.

11.3 Recent trends in protein extraction from plant sources

Plant proteins are the potential alternative to animal proteins, which contain most of the essential amino acids or which can be combined with other protein sources to compensate for certain amino acids deficiency (Mechmeche et al., 2017). Moreover, proteins have several functionalities, such as solubility, digestibility, and others (Rai et al., 2016; Hou et al., 2017; Contreras et al., 2019) along with their application in different nonfood

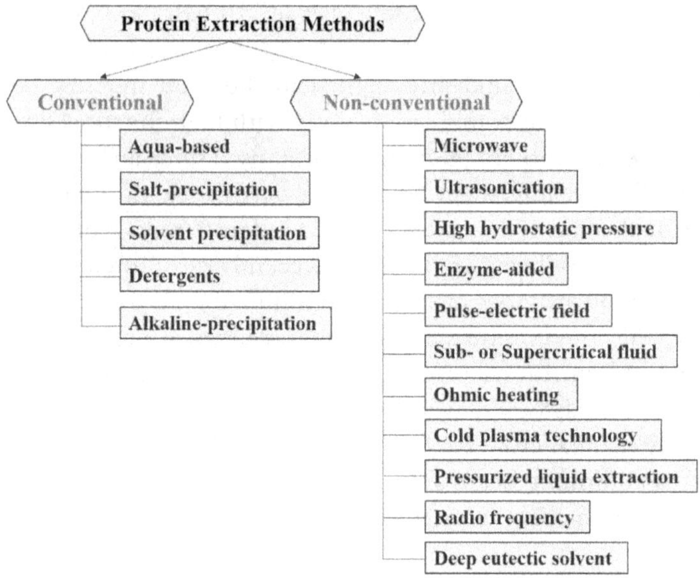

Figure 11.3 Different techniques for protein extraction.

sectors as adhesives, coatings, drug-carrier, and biopolymers (Akhtar et al., 2021; Fetzer et al., 2018; Lin et al., 2021) are mainly depending on the extraction techniques applied. In this perspective, efficient extraction of protein from plant matrices is highly recommended from the techno-functionality point of view. Therefore, this section describes and discusses the most recent advances of different techniques from conventional to modern approaches (Fig. 11.3) proposed for protein extraction with special preference given to the synergistic application of sonication-microwave techniques. The insight could open up overall prospects for protein extraction from plant matrices, which might expand the portfolio of plant resources.

11.3.1 Conventional approaches

The conventional extraction methods are the most frequently used techniques for protein extraction. These techniques are involved with different solvents, including water, organic solvents, alkali, and acids and further, they are integrated with other approaches to enhance the protein yield percentage (Deleu et al., 2019; Bose et al., 2019). The extraction efficiency of proteins depends on the nature of the sources. The extracted protein could be in different forms, such as, concentrate, hydrolysate, and isolate. Protein concentrate is typically achieved by the subtraction of nonprotein substances in presence of hydroalcoholic solutions or acidic water (Rodrigues et al., 2012). Through the solubilization of protein in saline or at a certain pH medium, the protein isolate is produced (Gençdağ et al., 2021). The protein concentrate and protein isolate mainly differ from each other depending upon their extraction method (Yalçin and Çelik, 2007). On the other hand, protein hydrolysates are produced through the hydrolysis of protein by different enzymes, acids, alkaline treatment, or fermentation. They are different from the concentrate and isolate because hydrolysate is a mixture of small peptides and amino acids. According to Kumar et al., three main steps are generally involved in conventional methods, such as defatting of sample, extraction, and precipitation of protein) for extracting protein ingredients (Kumar et al., 2021). Different solvents, for example, n-hexane, petroleum ether, n-pentane, and other organic solvents are frequently used to eliminate those compounds that hinder the extraction process. In the next step, different salts, and ionic- and nondetergents are used in the hot or cold water-based extraction process. Here, different modern extraction techniques are used to improve the extraction yield. Finally, the extracted proteins are concentrated or precipitated using different chemical agents such as citric acid, ethanol, methanol, ammonium sulfate, acetone, TCA (trichloroacetic acid), hydrochloric acid followed by centrifugation to remove impurities (Lee et al., 2017; Bose et al., 2019).

Recently, Chen et al. (2019) recovered protein from *Moringa oleifera* seeds using aqueous extraction techniques, where ethanol and petroleum ether have been used as the extraction solvent. They claimed 33% of proteins were recovered after purification. Embaby et al. (2018) reported 63.70% protein concentrate from *Acacia tortilis* seeds using ethanolic extraction. Several pieces of literature have been found elsewhere that reported alkali-based protein extraction with higher yields. Wang et al. (2014) carried out aqueous extraction of protein from rice bran using NaCl, alcohol, and NaOH. They stated that 13.80% of proteins were recovered with 45–47% purity, while Bernardi et al. (2018) reported that 34.51% of proteins were recovered using 1M NaOH. A study by Pickardt et al. (2015) used the mild acidic extraction condition with NaCl to extract protein from sunflower and claimed that 23–26% of proteins were recovered. For instance, Vaidya et al. (2017) reported a protein yield of 51.58% with a total protein content of 86.29% from custard apple seed flour using 0.6–2 M NaOH solution, while Mechmeche et al. (2017) reported 80.37% protein yield from tomato using 0.5N NaOH and Du et al. (2018) claimed 77.32% protein yield from mung bean using KOH as solvent. Furthermore, Shafiqur et al. found 85.20% protein using 0.2M

NaOH from defatted peanut meal (Shafiqur et al., 2018). In addition, it has been reported that the increase of alcoholic solvents hydration and process temperature can improve the protein fraction from rice bran, however, they also can decrease the protein solubility in the defatted meal (Capellini et al., 2017).

The published works of literature elsewhere argued that traditional methods have plenty of drawbacks that hinder the protein extraction efficiency and cause the huge native structural damage of the extracted protein with their functionalities. For example, protein extraction by conventional methods is often restricted by mass transfer resistances due to the association of several phases in the system (Akharume et al., 2021; Gençdağ et al., 2021; Kamal et al., 2021; Kumar et al., 2021). Furthermore, these methods are time-consuming and may require a fairly long-time depending on the diffusion rates of solvents through the complex routes of materials. Besides, high energy consumption is also one of the notable disadvantages of conventional extraction methods (Lianfu and Zelong, 2008). In addition to these, the native properties of a protein can be reduced due to the heat sensitivity of protein to severe pressure and temperature conditions. The use of chemical solvents may also be detrimental to the humans and environment. The traditional extraction methods are simple, require less initial cost, and extraction efficiency can be improved by optimizing the extraction parameters easily. However, these methods harshly affect the protein quality, for example, digestibility, disintegrate the structure of important amino acids, and lower the commercial and consumer satisfactoriness of extracted protein. This could be overcome by replacing the conventional processes with modern technologies that might enhance the extraction process and retain the desired protein quality without significantly affecting the structural, and functional attributes.

11.3.2 Modern approaches

In recent years, several novel, and innovative approaches have gained much popularity in the food and pharmaceutical research areas. Among these technologies, ultrasound, cold plasma, pulsed electric fields, microwave technology, and high-pressure processing are extensively studied and they are mostly focused on physical pretreatment to improve the extraction process (Pojić et al., 2018). According to Rahman and Lamsal (2021), these novel extraction technologies follow the general principle of disrupting the plant's cell walls to facilitate the targeted component (e.g., protein) diffusion as in liquid. As an emerging technology, ultrasonication with nontoxic and green solvents has gained more attention due to an eco-friendly approach and it is considered useful to boost the green chemistry research (Chemat et al., 2017; Pojić et al., 2018). In Fig. 11.4, a graphical illustration of ultrasonication, microwave oven, and combined ultrasound-microwave assembly for extraction of plant proteins has been presented.

Ultrasonication is a nonthermal and physical processing approach that has a popularity for its wide application to extract numerous bioactive molecules (Işçimen and Hayta, 2018). Besides, ultrasound is an energy and cost-efficient technology that is easy to install and requires less maintenance with less investment, short processing time, less solvent consumption, and higher ability to enhance solvent penetration and improve extraction yield (Chemat et al., 2017). Ultrasound technology is classified as low and high intensity based on the frequency used. Low-intensity ultrasound refers to a high frequency of >100 kHz with low power of <1 W/cm^2, while high-intensity ultrasound includes a low frequency of 20–100 kHz and high power of >1 W/cm^2 (Musielak et al., 2016; Gençdağ et al., 2021). Ultrasound generates acoustic cavitation in a liquid medium using the sound wave ranging from 20 kHz to 1 GHz, which assists the extraction process by the destruction of cellular structure. The acoustic cavitation system is produced by ultrasound consequences in the fast development and implosion of air bubbles in a liquid medium. It causes the generation of high shear stress as well as mechanical energy-induced temperature rise, resulting in the disruption of cell membranes (Bedin et al., 2020; Bernardi et al., 2021; Ghanghas et al., 2022). Moreover, it assists the solvent to penetrate the cellular structure and thereby improving the mass transfer process and greater release of desired molecules (Ly et al., 2018). The principle of sonication-assisted extraction leads to (a) disintegration of the cells to enhance hydration and decrease the cellular size, (b) pitting and erosion of cells surface, (c) enhancing the solvent penetration efficiency into cells, (d) enhancing hydration and swelling, (e) detaching the molecules from the sample surface, and (f) breakdown the physical bonding of molecular aggregates and converting them into smaller molecules (Rahman and Lamsal, 2021).

Ultrasound technology has many potential areas for industrial applications including inactivation of enzymes and microbial growth, crystallization, drying, degassing, extraction, filtration, homogenization, tenderization of meat, oxidation, sterilization, etc. (Jambrak et al., 2014). Over the past decades, ultrasound technology has been extensively studied for the extraction of protein from plant matrices aiming to lessen extraction time, solvent

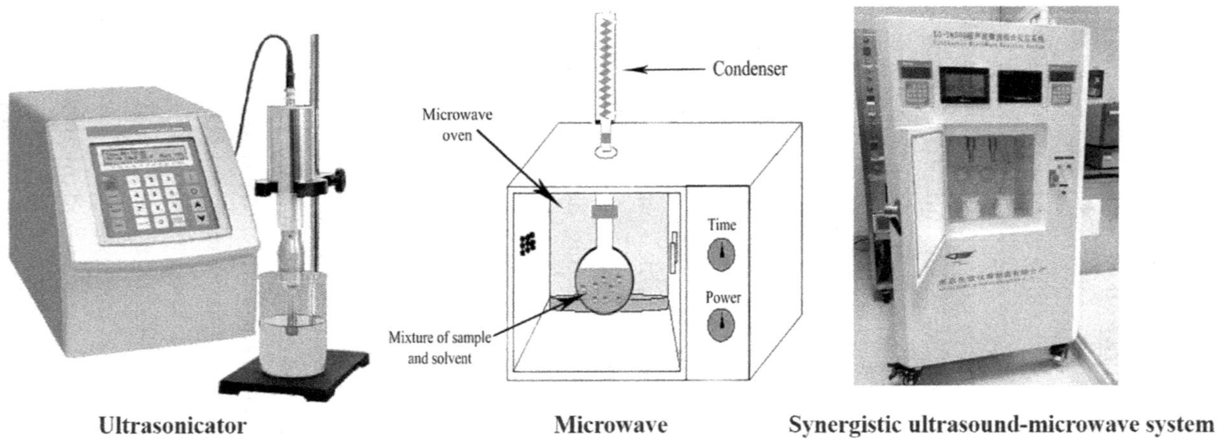

Figure 11.4 Ultrasonication, microwave oven, and combined ultrasound-microwave assembly for extraction of protein from plant sources.

consumption, and energy and obtain greater control over processing conditions (Bernardi et al., 2021; Chemat et al., 2017). Furthermore, researchers have demonstrated the impact of ultrasonication on the physicochemical, microstructural, and functional attributes of extracted proteins. Ultrasound cavitation produces flow turbulent and high shear force that can enhance the air-water interfacial area influencing the physical interactions, such as hydrogen bonds and hydrophobic interactions. These consequently could improve the techno-functional properties, that is, protein solubility, emulsifying capacity, digestibility, etc. (Bernardi et al., 2021). Integration of chemical approach with UAE, alkaline solution (pH $>$8.0) is frequently applied for the extraction of protein. UAE with distilled or deionized water as the solvent can also be used for protein extraction, which allows the green processing, and low-cost technique (Li et al., 2017). The application of sonication technology for protein extraction from plant sources has been tabulated in Table 11.1.

According to Leighton (1995), ultrasonic frequency is one of the main factors that influences the cavitation yield in liquid, because multifrequency sonication generates a greater cavitation yield than that of traditional monofrequency ultrasonic equipment. Furthermore, the ultrasonic technique possesses different working modes that are important for controlling the extraction process and enhancing the process efficiency (Yang et al., 2018). Around 37–57% protein yield was observed for wheat-gram extracted protein using an ultrasound probe (Zhu et al., 2009). Li et al. (2017) obtained 88.4% protein yield from rice dreg flour using ultrasound and alkali extraction methods. Liu et al. (2019) stated around 15% protein yield from wampee seed using an ultrasound bath. Similarly, the application of ultrasound with diluted sulfuric acid-soaking pretreatment on protein extraction from defatted rice bran was reported by Zhang et al. (2019). Results revealed that the ultrasound with 0.9% sulfuric-acid soaking at 90°C pretreatment enhanced the protein content from 53.1% to 72.8%. Dabbour et al. (2020) explored the impact of localized-enzymolysis and sonication treatment on the techno-functional and microstructure properties of sunflower meal protein, and its hydrolysate. The oil absorption capacity and dispersibility of the sonicated samples were improved, whereas the water holding capacity, and molecular weight were reduced.

Morales et al. (2015) noticed that ultrasound treatment can reduce the particle size of protein and thereby improve the foaming capacity of isolated soy protein. According to Jiang et al. (2014), ultrasound treatment can alter the secondary and tertiary protein structures confirmed during the study

Table 11.1 Recent studies on the application of sonication technology for protein extraction from plant sources.

Plant materials	Sonicator configuration	Experimental arrangements	Major Findings	References
Okara solution	20 kHz and power of 400 W	65 W for 0.5, 1, 5, and 15 min using water as solvent	• The treatment enhanced the protein release by around 10% and 30%.	(Preece et al., 2017)
Watermelon seed	Ultrasonic power of 200 W and frequencies of 25 and 40 kHz.	50–200 W at 25 kHz frequency at the ratio from 1:10 to 1:60 (w/v) and pH was adjusted in the varying range from 7 to 11 for 15 min	• The protein extraction yield was >80% with improved functional characteristics and amino acid profile.	(Gadalkar and Rathod, 2020)
Brown tea residue	20 kHz, power output of 150, 300, and 450 W/L	377 W/L for 13 min using various concentrations of NaOH (0.05–0.15 M)	• Protein yield was increased from 56.4% to 63.5%.	(Ayim et al., 2018)
Sunflower (*Helianthus annuus* L.) meals	Frequencies of 20, 28, 35, 40, and 50 kHz and power of 300 W	220 W/L for 15 min and water and alkali solvent at pH 8.0	• Protein yield was enhanced from 28% to 54.26%.	(Dabbour et al., 2018)
Sesame bran	Frequency of 35 kHz	Ultrasound power of 528–836 W, the temperature of 25–55°C, and time of 10–120 min	• The protein yields ranged from 39.8% to 58.5% with improved phenolics and antioxidant activity.	(Görgüç et al., 2019)
Rice dreg flour	Countercurrent ultrasound of 20 kHz, power 320, 416, and 448 W	448 W for 40 min with pulsing, pulse duration/pulse interval ratio 10/6 s using various concentrations of NaOH (0.02–0.10 mol/L)	• Protein extraction was increased from 43.20% to 88.44%.	(Li et al., 2017)

(continued on next page)

Wampee seed	Ultrasonic cleaner with a maximum power output of 240 W and frequency of 40 kHz	Solid-to-solvent ratio (1:20–1:30 g/ml), sonication time (50–70 min), and pH (11–13)	• 15.06% protein was yielded at the optimized condition with improved functional quality.	(Liu et al., 2019)
Defatted peanut meal	20 kHz, Model VC 750, Sonics and Materials Inc., the United States	30 W/g for 15 min and water and alkali solvent at pH 6.8 to 10	• Protein yield was enhanced by 10.6% at pH 6.8.	(Nguyen and Le, 2019)
Rice bran	26 kHz microwave-ultrasonic synthesis extraction processor	For sonication treatment: power (100–600 W) and time: 10–60 min. Sonication combined with 0.9% sulfuric-acid soaking at 90°C.	• The protein yield varied between 53.1% and 72.8%.	(Zhang et al., 2019)
Chickpea and kidney bean flour, defatted soy flakes, and soy flour	20 kHz with a power of 750 W	2.5 and 4.5 W/mL for 5 min using water and alkali at pH 8.5	• Protein yield was increased by flake >22%; by flour 10%; by kidney bean 7%; by chickpea no change.	(Byanju et al., 2020)
Coffee silver skin	Ultrasonic Probe (1.8 cm diameter) with 25 kHz	Temperature: 50°C; sonication time: 10 min; amplitude: 100%; ultrasound intensity: 38.0 W/cm^2; solvent concentration: 0.2 M and 0.6 M	• Protein yield (13.51% and 14.04% for 0.2 M and 0.6 M, respectively) was enhanced by 2.8 and 2.3 times.	(Wen et al., 2020)

(continued on next page)

Table 11.1 Recent studies on the application of sonication technology for protein extraction from plant sources—cont'd

Plant materials	Sonicator configuration	Experimental arrangements	Major Findings	References
Walnut	Ultrasound processor with frequency: 25 kHz and maximum power: 1200 W	The ultrasound power level of 200, 400, or 600 W for 15 and 30 min at a temperature below 45°C	• Observed small change in protein secondary structure after sonication • Appreciable changes in tertiary structure occurred and improved water-solubility, emulsifying activity, and stability.	(Zhu et al., 2018)
Defatted soy flakes (90 PDI) and flour	20 kHz and power of 2.2 kW	360 W/mL for 2 min for flour and 720 W/mL for 4 min for flake using water and alkali at pH 8.5	• Protein extraction yield of flakes and flour was enhanced by 16.58% and 9.76%, accordingly.	(Rahman et al., 2021)
Defatted rice bran	Horn-type ultrasonic probe with a frequency of 20 kHz	Sonication power: 0, 5, 10, 15, 20, and 25 W/g of material dry mass. Sonication time: 0, 1, 2, 3, 4 and 5 min	• The ultrasonic extraction yielded 2.20 times higher than the conventional extraction.	(Ly et al., 2018)
Faba bean	Probe type (8 mm dia.) with a frequency of 20 kHz and output power of. 500/750 W	Ultrasound Amplitude: 45–75% Sonication time: 11.3–33.7 min	• Ultrasound treatment improved the foaming property and foam stability and β conformations. • While decreased the inter-molecular aggregates and relative protein digestibility.	(Martínez-Velasco et al., 2018)

for black bean protein isolates. In the same study, they claimed that protein solubility can be improved owing to the rupture of internal hydrophobic molecular interaction of protein, which increases surface hydrophobicity, and molecular movement. Tian et al. (2020) described that the protein solubility can be increased significantly after sonication. This is because of the conformational change in the secondary and tertiary structures of protein along with the disruption of peptide aggregates, which was also observed in case of soy protein hydrolysates. Oppositely, Flores-Jiménez et al. (2019) stated an increased oil absorption capacity (OAC) was observed for ultrasound-treated canola protein isolates in comparison with the sample having no ultrasound treatment. Furthermore, Li et al. (2018) found OAC improvement occurred by ultrasound-assisted extraction studied for isolated quinoa protein. This study also stated that the greater protein hydrophobicity is associated with the increase of ultrasonic power and resultant higher OAC.

However, Ochoa-Rivas et al. (2017) found an insignificant reduction of peanut protein OAC using both ultrasound and microwave irradiation which has been attributed to the physical interactions between protein molecules and fat droplets. According to Rahman and Lamsal (2021), the efficiency of ultrasound application for protein extraction is dependent on several factors, which are classified as: (a) ultrasound cavitation based, and (b) product-based. Ultrasound cavitation-based factors include intensity, energy density, bath or probe type, temperature, duration of treatment, size, and shape of the sonicator. Besides, product or sample-based factors include, type of substrate or desired protein, sample composition, solutes size, solvent type, and solvent-to-solute ratio. These factors significantly dominate the protein characteristics including their nutritional, structural, and techno-functional attributes.

Microwave-assisted extraction is one of the proven techniques to conform to the purpose of sustainable "green" chemistry and extraction. This technique is well-known in the field of chemical and food industry for its significant impact on the extraction of different bioactive compounds like protein, pectin, and phytochemicals, as well as on the pretreatment method for various operations (Amponsah and Nayak, 2016; Kamal et al., 2020; Zheng et al., 2020; Behere et al., 2021). Microwave technology works based on electromagnetic waves with a frequency ranging from 300 MHz to 300 GHz (Guzik et al., 2021). Microwaves can be reflected, transmitted, and absorbed like infrared, and light waves. In microwave technology, the energy is produced from electrical energy which is transformed into microwave via a microwave power or generator. The standard frequency of microwaves used for commercial food service and domestic kitchens is around 2450 MHz and is used mainly for their ability to heat the products.

The underlying principle of **microwave technology** for the extraction of different biomolecules (e.g., proteins) is the mutual action of dipole rotation and ionic conduction, which leads to the breakdown of H-bonding exist in the plant cell wall (Kumar et al., 2021). This reaction enhances the fluidity or porosity of the cell membrane leading to the better penetration of solvent into the cellular structure as well as greater discharge of intracellular compounds in the solvent medium. This process is mainly accompanied by the transfer of electrical energy from the electromagnetic field in the chamber of the device, and the mechanism takes place at the molecular level. Due to this mechanistic ions acceleration, that is, ionic conductivity, polar particles' movement in the direction of the electric field and colliding with the water molecules, and transmission of kinetic energy (Guzik et al., 2021), many millions of collisions occur that create a frictional force to emit thermal energy within every second (Liu and Ma, 2016; Özer et al., 2018). Moreover, the microwave-tempted dipole rotation of molecules and the movement of ions improve the penetration of solvent into the matrix, cause cell wall disruption and discharge the intracellular product, resulting in the enhanced extraction of desired molecules (Phongthai et al., 2016).

Microwave-assisted extraction has several benefits over traditional extraction methods, such as rapid heating process, short processing time, high reproducibility, less solvent consumption, simplifying manipulation and work-up, higher purity of the extracted compounds, more effective heating, greater energy transfer, decreased thermal gradients, discerning heating, and greater processing yield (Aftari et al., 2015; Kumar et al., 2021; Lee, 2017). The cells wall structure/thickness of plant matrix, type of solvent, volume solid/solvent volume ratio, microwave power, processing time, and temperature are the most influential parameters in the microwave-assisted extraction approach (Moret et al., 2019). However, Martins et al. (2019) also pointed out that irregular and uncontrolled degradation of protein may occur due to the uneven heat distribution generally in a heterogeneous food matrix, resulting in reduced conductivity.

Extraction of protein from plant matrix using microwave technology along with other methods has long been used (Table 11.2). It is evidenced from the previous studies that microwave-assisted extraction yields protein with better biological functionality compared to traditional chemical solvent-based extraction methods. For example, Phongthai et al. (2016) stated that 22.07% protein was recovered from rice bran by applying microwave

Table 11.2 Studies on the application of microwave technology for protein extraction from plant sources.

Plant materials	Microwave configuration	Experimental arrangements	Major findings	References
Soybean	Multimode microwave reactor having frequency 2.45 GHz and power of 1000 W variable in 10 W increments.	Power level (540–810 W), target temperature (60–80°C), and stirring speed (140–180 rpm)	• The extraction yield and protein contents were 24% and 44.4%, respectively. • Increased solubility and digestibility of extracted soymilk.	Varghese and Pare (2019)
Rice bran	Microwave machine with multimode cavities and a frequency of 2450 MHz	Microwave power (600–1000 W), extraction time (60–120 s), solid-to-liquid ratio (0.5–1.5 g/10 mL distilled water), and pH 10.0	• The protein yield of MAE was 1.54-time higher than that of alkaline extraction.	Phongthai et al. (2016)
Defatted rice bran meal	Domestic microwave with a frequency of 2450 MHz and operated at 800 Watt	Microwave irradiation time of 20, 30, 40, 60, and 90 s, homogenized for 10 min and enzyme-treated with 0.1% (w/w) papain and 5.0% (v/w) viscozyme for 1 h at 37°C.	• Protein yield was increased after the treatment.	Bandyopadhyay et al. (2012)
Peanut flour	Commercial microwave (LG, MS1142GW, Mexico)	Peanut flour to distilled water ratio was 1:10 and 1:25; microwave power was 145, 290, 435, 580, or 725, and processing time was 2, 4, 6, 8, and 10 min while maintained pH 9.0	• Maximum protein yield (55% with 100% purity) was obtained at 725 W and 8 min • MAE improved the water absorption index, foam activity index, emulsifying index, and foam stability index.	(Ochoa-Rivas et al., 2017)

(continued on next page)

Table 11.2 Studies on the application of microwave technology for protein extraction from plant sources—cont'd

Plant materials	Microwave configuration	Experimental arrangements	Major findings	References
Soy flour	Microwave reaction system at a standard frequency of 2450 MHz equipped with a carousel of 14 Teflon vessels	Extraction temperatures of 60, 70, and 100°C and times of 0, 5, and 10 min at fixed microwave power 600 W.	• Increased protein content compared to the conventional method.	(Amponsah and Nayak, 2016)
Coffee silver skin	Modified microwave oven equipped with a condenser and a refrigerated circulator	Microwave irradiation for 10 and 20 min at the microwave power of 434.7 W	• Protein yield was increased by 5.8-fold and 7-fold with no significant changes in the molecular weight.	(Wen et al., 2020)
Watermelon seed	Power-controlled microwave system operates at a maximum of 1500 W power, and its p/T sensor controls thermal overshoots.	Solid-to-solvent ratio (1:10–1:40), pH (7–10), microwave power (30 W, 50 W, 70 W), extraction time (30 s–8 min)	• About 90% protein recovery was observed at the optimized condition with excellent functional properties.	(Behere et al., 2021)
Waste sesame bran	Microwave extractor (Sineo, MAS- II Plus, China)	Operation temperature (25–55°C); time (10–120 min) at solid to liquid ratio of 1:10 (w/v) and pH of 9.5 with a fixed microwave power of 750 W (enzyme alcalase (0.12–2.40 AU/100 g) was used for microwave-assisted enzymatic extraction)	• The protein yields ranged from 43.8% to 61.6% and 38.2–94.1% for microwave- and microwave-assisted enzymatic extraction, respectively. • Produced antioxidant-rich protein isolates with excellent amino acid contents.	(Görgüç et al., 2020)

(continued on next page)

Rice bran	The microwave extraction system has a frequency of 50–60 Hz and consists of a metal box, equipped with a carousel of 8 Teflon tubes in a closed system.	Operation temperature (30–55°C); time (60–120 s), and pH (9–11) at a recordable microwave power of 400–450 W	• The extraction yield was 22.04% higher and protein content was 6.19% higher than the alkaline conditions.	(Bedin et al., 2020)
Wheat protein fiber	Not mentioned	The effects of microwave power (100, 200, 300, 400, and 500 W), exposure time (1, 2, 3, 4, and 5 min), and pH (5, 6, 7, and 8)	• The microwaved fiber had increased –S-S- content, α-helices, crystallinity, thermal stability. • Had smooth and dense microstructure with good hygroscopicity, however, decreased the surface hydrophobicity.	(Liu and Ma, 2016)

(continued on next page)

Table 11.2 Studies on the application of microwave technology for protein extraction from plant sources—cont'd

Plant materials	Microwave configuration	Experimental arrangements	Major findings	References
Soybean	Microwave oven (Model: MZG1500S, Huiyan Microwave system engineering limited company, Nanjing, China)	Microwave irradiation for 10 min at the power of 600 W with magnetic stirring at 200 rpm	• Microwaving reduced allergenicity by 24.7%, and influenced molecular interactions, such as free sulfhydryl, hydrophobicity and zeta potential, and secondary structures.	(Li et al., 2016)
Lotus seed	XH-100B variable power 144 microwave system equipped with a time control switch, air cooling system, vacuum pump, and a vacuum pressure balance system; with a frequency of 147 2.45 GHz.	Microwave-vacuum power from 0 to 200 W for 3 min	• MV treatment-induced unfolding of extracted protein isolates, and converted the secondary structures gradually into β-turns and random coils. • While enhancing the flocculation and creaming stability.	(Zheng et al., 2020)

techniques and this yield was 1.54-fold higher than that of the alkaline method with similar digestibility. Ferreira et al. (2018) reported 1.34–7.39% protein yields from broccoli by-products at 500 W for 5 min using microwave-assisted extraction techniques. Wen et al. (2020) claimed a 7-fold higher protein recovery (43.53%) was obtained from coffee silver skin by applying microwave-assisted extraction techniques when compared with conventional technique (6.23% recovery). The application of microwaves at 725 W for 8 min for protein extraction from defatted peanuts yielded 55% protein with 100% purity and was 77% higher protein compared to the control sample (Ochoa-Rivas et al., 2017). They also claimed no significant changes occurred in the primary structure, however, secondary protein structure enhanced several functional attributes, for example, water absorption index, fat absorption index, emulsifying activity, foam activity, and protein digestibility. It is reported that microwaves can change the materials' structure and permanently change their properties. Accordingly, Byaruhanga et al. (2006) found that the kafirin content of β-folding increased after microwave treatment. Görgüç et al. (2020) explored that the recovery of sesame bran protein was obtained ranging from 43.8% to 61.6% and 38.2% to 94.1%, respectively, using the microwave-assisted and combined microwave-assisted enzymatic approach.

Varghese and Pare (2019) demonstrated a significant intensification in the extraction yield (24%) and protein content (44.4%) for microwave-assisted extraction of soy milk in comparison with solvent extraction. Besides, they reported proteins with improved solubility and digestibility of microwave extracted soymilk in comparison with conventionally extracted milk. Li et al. (2016) claimed that microwave treatment can reduce the allergenicity of soy protein isolates by altering their secondary protein structure. Microwave treatment has been associated to mitigate gluten-induced toxicity for celiac patients (Lamacchia et al., 2016). On the contrary, microwave treatment has been found inefficient for detoxifying gluten (Mahroug et al., 2019). Moreover, the digestibility of plant proteins can also be affected by microwave heating. Microwave heated rice protein showed around 10.3% and 16.8% reduction in protein digestibility during gastric digestion and gastrointestinal digestion, respectively, which has been referred to the development of intramolecular disulfide linkages resulting in protease resistance (Liu et al., 2019). Likewise, microwave heating at 1000 W for 5 min allowed crosslinking between amino acids, and thereby reduced in vitro gluten digestibility (Xiang et al., 2020). In contrast, microwave treatment has also been observed to have a positive effect on the soymilk protein (Vanga et al., 2020) and pigeon pea protein digestibility (Sun et al., 2020).

Recently, **enzyme-assisted extraction** gained popularity as an innovative and eco-friendly approach to extracting protein from plant resources. The enzyme-aided protein extraction is based on the disintegration of cells wall by applying one or more enzymes to release protein adhering to the polysaccharide (such as cellulose, hemicellulose, or pectin) matrix of plant materials. Enzymes widely used for protein extraction include cellulose, celluclast, pectinase, viscozyme, amyloglucosidase, protease, alcalase, protamex, α-amylase, nutrase, galactanase, protex, papain, palatase, natuzyme, and kemzyme (Gençdağ et al., 2021; Kamal et al., 2021; Kumar et al., 2021). The enzymatic activity may differ based on different materials used and target compounds. Enzyme-assisted extraction is highly reliant on the operating conditions, for example, substrate-to-enzyme ratio, pH, temperature, and extraction time (Demuez et al., 2015). Furthermore, enzyme-aided protein extraction has some shortcomings, such as lengthy processing, high operating expenditure, greater power consumption, permanent disruption of the carbohydrate-protein matrix, and requires precise adjustments of pH and temperature (Pojić et al., 2018). Nevertheless, it is measured as a mild, noninvasive, and green extraction method with lower environmental effects and superior quality of the extracted protein when compared with the conventional chemical extraction (Sari et al., 2013; Liu and Ma, 2016; Ochoa-Rivas et al., 2017).

Enzymatic and enzymatic-solvent extractions showed higher protein content from flaxseed meals with better emulsifying characteristics than that of alkali extraction (Tirgar et al., 2017). Chirinos et al. (2017) found the greater quality of alcalase enzyme-assisted extraction of proteins from *Plukenetia ubilis* L. kernel meal and it was around 44.7% of proteins with low hydrolysis degree over alkaline extraction (yield of 29.7% proteins). A protein yield of 77% was obtained from rapeseed meal treated with cellulase at operating conditions of 67 U/g enzyme to substrate ratio, at 40°C for 6 h, maintaining the pH of 4.8 (Li and Guo, 2017). Fetzer et al. (2018) found 80.7% and 78.3% protein from cold-pressed rapeseed meal and prepressed rapeseed meal, respectively, using protease (2%, w/v substrate to enzyme ratio, at 50°C, pH of 5.7–5.8). Literature evidenced that enzyme-assisted extraction results in better efficiency under moderate conditions (Gençdağ et al., 2021), special consideration should be given to maintaining the pH (6–8), and 40–60°C temperature to avoid the detrimental effects under extreme conditions.

Another innovative nonthermal approach toward protein extraction is the use of **high hydrostatic pressure (HPP)**, which operates at 100 and 700 MPa (Farkas, 2016; Kumar et al., 2021). HPP works in three ways such as initial pressure increase, pressure balance, and final pressure release. When the reference sample is dissolved with the extractant and placed inside the pressure vessel, high pressure causes cell wall disruption within a short time. As a result, the solvent penetrates easily through the scratched cellular membrane into the cell compartment and enhances the mass transfer process. After a certain time passes, the pressure becomes balanced, however, the solvent continues to penetrate through the cell membrane and dissolve the target component, for example, protein (Huang et al., 2013). Finally, the pressure is released into the atmosphere, causing the cells to swell and distort. This facilitates the increase of samples' surface area and diffusion process causing enhanced extraction efficiency (Chen et al. 2009). Pressure-assisted extraction is governed by several factors, like extraction pressure, operating time, nature, and concentration of extraction solvent, and solid-liquid ratio (Mustafa and Turner, 2011). It is evidenced from the literature that pressure-based protein extraction affects protein denaturation, aggregation, and interactions. However, HHP treatment can enhance the protein hydrophobicity and reduce its solubility, and improve techno-functional properties (Queirós et al., 2018). Zhou et al. (2016) demonstrated an increased surface hydrophobicity of ginkgo seeds protein and stated that HPP altered the secondary structure as well as improved the heat stability and emulsifying properties. On the other hand, HHP enhanced the solubility as well as the water and oil holding capacity of pine nuts protein (Cao et al., 2018). A similar observation was reported for kidney bean protein where high pressure (>600 MPa) was applied (Ahmed et al., 2018). HHP was also found to improve the nutritional importance of soy protein isolate and ginkgo seed proteins, including the reduction of allergenicity (Li et al., 2018; Zhou et al., 2016).

An emerging technology for protein extraction and modification is the application of a **pulsed electric field (PEF)** system. In this technique, a strong electric field with short pulses (10–80 kV/cm) is generated within micro- or milliseconds due to potential difference, thus disruption of the cell wall is caused by the electroporation (Sedaghat Doost et al., 2020). A PEF system comprises a chamber, electrodes, a high-voltage pulse generator, and a computer for monitoring and controlling devices. Zhang et al. (2017) described the enhanced solubility, emulsifying, and foaming properties of PEF-treated canola seed protein. However, it can cause denaturation and aggregation of proteins leading to reduced solubility (Sedaghat Doost et al., 2019). It is reported that PEF can change the secondary and tertiary structure of plant-based proteins. For instance, the secondary structure of canola proteins was changed by altering the proportions of α-helices, β-sheets, and β-turns in the amide-I region using PEF. They also detected the alteration in the tertiary structure of the protein with a higher amount of free sulfhydryl groups and enhanced surface hydrophobicity (Zhang et al., 2017).

Some other innovative and promising protein extraction and modification technologies include pulse light (pulse UV) (Panozzo et al., 2016), cold atmospheric plasma processing (Mehr and Koocheki, 2021; Zhang et al., 2021), high-voltage electrical discharge (Parniakov et al., 2015; Roselló-Soto et al., 2015), radio frequency treatment (Guo et al., 2017; Hassan et al., 2019; Ling et al., 2019), ohmic heating (Li et al., 2018), and subcritical extraction (Lu et al., 2016). Although these techniques are not studied extensively for protein extraction, however, those conducted studies showed excellent performance in the recovery of protein and modified the structural properties that showed improved functional characteristics.

11.3.3 UltrasoundSonication-microwave synergistic extraction (UMSE) of protein

As mentioned in the previous sections, there are several innovative and eco-friendly extraction technologies for the extraction of plant bioactive molecules, especially those applied in the case of protein extraction. This section describes the sonication-microwave synergistic extraction of protein.

11.3.3.1 Basics of synergism

It is well-known that different extraction technologies have some drawbacks when applied individually. However, their combined use has proven to be efficient to overcome these drawbacks. Process synergism refers to the combination of two or more technologies, tools, or reagents applied to a processing system to overcome the negative effects of individual technology on the targeted outputs (Gude, 2015). According to (Gerven and Stankiewicz, 2009), several criteria should be considered synergistic process. These include, (i) maximizing the effectiveness of intra- and inter-molecular interactions by generating dynamic conditions to stimulate kinetic regimes with higher conversion and efficiency; (ii) ensuring uniform and homogeneous mixing and heating; (iii) optimizing driving forces and maximizing specific surface areas to increase the heat and mass transfer, and (iv) maximizing the synergistic effects from conventional or partial processes. In connection with these criteria, Stankiewicz (2006) mentioned that the molecular structure of products, energy sources and their application processes, development and combination of multifunctional tools, and time/process control is the principal contributing factors for the process intensification to achieve synergism (Gude, 2015). Hence, it is highly important to isolate appropriate process configurations when merging the effects of two process technologies to endorse the process strength (Huang et al., 2007). It is noteworthy to mention that all process combinations may not result in a time for process intensification system, even if they offer a synergism. Therefore, several supplementary matters may arise from the novel methods concerning process control and optimization.

11.3.3.2 Principle of ultrasound-microwave synergistic extraction

Ultrasound-microwave synergistic extraction (UMSE) is a novel hybrid technology that integrates the beneficial features of ultrasonic and microwave techniques. It allows maximum usage of the high-energy effect of microwaves and ultrasonic cavitation to overcome the limitations of other techniques, that is, conventional, ultrasonic, and microwave. The ultrasonic-microwave synergistic extraction technique ensures accelerated dielectric heating of matrix and its homogeneous mixture with solvent, as well as acoustic-cavitation induces improved mass transfer of soluble components (Cravotto and Cintas, 2007). According to Wang et al. (2018), a simultaneous UMAE technique has the capability to restrict the uneven heating and weak thermal effect that are critical gaps noticed in MAE and UAE techniques, respectively. It was observed that the effect of mechanical oscillation and stirring action induced by ultrasonic cavitation waves could effectively result in uniform heating by microwave; also the heat energy generated by microwave to the extraction medium could improve the weak thermal effect of ultrasonication (Quaisie et al., 2021). In this way, the higher momentum and energy in UMSE promote the cell rupture and release the targeted compounds, for example, protein to the surrounding solvent (Chan et al., 2011; Xu et al., 2018). In UMSE, the ultrasound facilitates the solvent permeation into the sample matrix and intensifies the contact surface area by disrupting the cell wall and improving mass transfer that stimulates the dissolution of soluble compounds.

Meanwhile, microwaves heat the sample rapidly, snowball the mass transfer rate and solubility of solute, hasten the desorption of the target compound from the sample matrix, and boost the extraction efficiency

(Vinatoru et al., 2017). In actual practice, the rapid sample preparation and fast extraction process are intensified in UMSE, which makes it a time-saving, cost-effective, and highly efficient method (Cravotto and Cintas, 2007; Sun et al., 2019). In UMSE, there are flexibilities in its simultaneous or sequential applications. In simultaneous UMSE, the process is conducted in the same reactor or container built specially to facilitate the ultrasound and microwave energy production to heat the sample in the solvent, while UMSE in sequence can be conducted in two separate machines (i.e., power ultrasound and microwave machines) one irradiation in one machine follows another irradiation in another machine (Gude, 2015; Martina et al., 2016).

11.3.3.3 Application of UMSE in protein extraction

The advancement of ultrasonication and microwave technology for the extraction of high-value components like protein from plants has gained much attention in the academia and scientific community. However, the application of combined sonication and microwave technology, that is, the synergistic use for extraction of protein from plant resources remains in its early stage. The potential application of sonication-microwave synergistic technology for protein extraction from plant resources has been presented in Table 11.3. Recently, Cheng et al. (2021) studied the ultrasound-microwave-assisted extraction of protein from *Moringa oleifera* leaves, and the processing parameters were optimized using the response surface methodology technique. In this study, the processing variable optimized were as follows: solvent-to-solid ratio (80–120), temperature (40–60°C), microwave power (40–160 W), extraction time (125–185 s) in ultrasonic-microwave collaborative extraction equipment with the ultrasonic frequency of 25 kHz and a fixed power of 300 W. The authors reported protein yield of 82.07 mg/g at an optimized condition of solvent/solid ratio 91:1 (v/w), extraction time 148 s, extraction temperature 41°C, microwave power 81 W. They reported that extraction time was the major influential factor on protein yield, followed by the solvent-solid ratio, microwave power, and temperature. They also claimed a higher extraction efficiency by UMSE compared to the conventional solvent extraction method.

Liu et al. (2017) investigated the efficacy of ultrasound-microwave synergistic extraction method for protein extraction from pumpkin seed using a green extraction medium of polyethylene glycol (PEG 200)-based deep eutectic solvent (DES). They optimized the key factors such as responsible PEG 200-based DES concentration (10–40%), liquid to solid ratio (10–40 mL/g), microwave power (50–190 W), temperature (30–60°C) with

Table 11.3 Reports on the synergistic application of sonication-microwave technology for protein extraction from plant sources.

Plant materials	Extraction type	Equipment configuration	Experimental arrangements	Major Findings	References
Moringa oleifera leaves	Simultaneous	Ultrasonic-microwave collaborative extraction equipment consisting ultrasonic frequency of 25 kHz and a fixed power of 300 W	Solvent-to-solid ratio (80–120), temperature (40–60°C), microwave power (40–160 W), and extraction time (125–185 s)	• A protein yield of 82.07 mg/g was found at the optimized condition.	(Cheng et al., 2021)
Peanut flour	Sequential	Probe type sonicator with a 22-mm diameter and 24 kHz; and commercial microwave (LG, MS1142GW, Mexico)	Ultrasound treatment (100% amplitude and 15 min) and then cool down to 23°C, microwave applied (725 W for 8 min)	• A synergic effect was not observed when sonication followed by microwaving was applied sequentially.	(Ochoa-Rivas et al., 2017)
Pumpkin seed	Simultaneous	Ultrasonic-microwave synergistic XO-SM200 system with 1200 W (2450 MHz) maximum microwave power and 800 W (25 kHz) ultrasonic power.	PEG 200-based DES concentration (10–40%), liquid-to-solid ratio (10–40 mL/g), microwave power (50–190 W), temperature (30–60°C) with 240 W ultrasonic power for a fixed reaction time of 4 min	• 93.95% protein yield was observed with an average precipitation rate of 97.97% at optimized conditions.	(Liu et al., 2017)

(continued on next page)

Table 11.3 Reports on the synergistic application of sonication-microwave technology for protein extraction from plant sources—cont'd

Plant materials	Extraction type	Equipment configuration	Experimental arrangements	Major Findings	References
Barley	Simultaneous	Microwave frequency (2450±50) MHz with a maximum output power of 1000 W and the ultrasonic frequency 25 kHz with a maximum output power of 800 W.	Temperature: 45, 55, 65, 75, and 85°C; Microwave power: 185 220 255 290 325 W; Ultrasonic power: 560 600 640 680 720 W; Extraction time: 12 15 18 21 24 min	• Extraction yield of hordein reached 8.84% at optimum conditions.	(Guan et al., 2018)
Soybean dreg	Sequential	Kitchen microwave (MH-250, China) and laboratory ultrasonic cleaner (KQ5200DE, China)	For single-factor study: Ultrasonic power: 200–500 W Ultrasonic time: 10–40 min Microwave power: 200–680 W Microwave process time: 1–7 min Solvent to material ratio: 6:1, 7:1, 8:1, 9:1, 10:1, 11:1, 12:1; pH: 7.0–10.0 For optimization study: pH: 8.5, 9.0, 9.5; ultrasonic time: 25, 30, 35 min; microwave process time: 4, 5, 6 min; solvent-to-material ratio: 9:1, 10:1, 11:1 (%), at a fixed ultrasonic and microwave power of 400 and 440 W, respectively.	• A protein yield of 94.43% was found at optimum conditions. The extracted protein showed homogeneous texture and good sensory quality.	(Yu-lin et al., 2015)

(continued on next page)

Eurycoma longifolia roots	Sequential	ETHOS-Milestone microwave extractor (ATC-FO-300, North America) with a round-bottomed extraction vessel attached to a cooling system and probe type Ultrasonicator	Sequential ultrasound-microwave and sequential microwave-ultrasound-assisted extraction. For microwave-assisted extraction: Temperature (45–85°C), microwave power (168–432 W), and irradiation time (7–33 min). For ultrasound-assisted extraction: Temperature (45–85°C), Ultrasound intensity (0.36–3.05 UI) and sonication time (2–28 min)	• Sequential ultrasound-microwave-assisted extraction and sequential microwave-ultrasound-assisted extraction yielded soluble protein at 27.17% and 25.24%, respectively.	(Elhag et al., 2020)
Sweet potato	Simultaneous	Ultrasonic-microwave extractor (CW-2000; Shanghai XTrust Analytical Instruments Co., Ltd., Shanghai, China).	Ultrasonic frequency of 40 kHz and microwave power 10–800 W by using an	• Improved the functional properties with enhanced antioxidant and sensory characteristics.	(Habinshuti et al., 2020)

(continued on next page)

Table 11.3 Reports on the synergistic application of sonication-microwave technology for protein extraction from plant sources—cont'd

Plant materials	Extraction type	Equipment configuration	Experimental arrangements	Major Findings	References
Walnut	Simultaneous	UWave-1000 Microwave-ultraviolet-ultrasonic trinity synthesis extraction reactor:	For single-factor study: Solid to liquid: 1:10–1:30; pH 8.5–10.5; Ultrasonic power: 300–500 W, Microwave power: 100–300 W; For optimization study: Solid to liquid: 1:15–1:25; pH 8.5–9.5 Ultrasonic power: 350, 400, 450 W; Microwave power: 100, 150, 200 W	• The extraction rate was 93.48% at optimized parameters.	(Yong-ting et al., 2018)
Mung bean	Simultaneous	Ultrasonic-microwave processor (CW2000: Beijing Jinyang Wanda Technology Co., Ltd.) with an ultrasound frequency of 24 kHz	For single-factor study: Microwave power: 100–500 W Temperature: 25–45°C, Time: 10–30 min Solid to liquid: 1:5, 1:10, 1:15, 1:20, 1:25 pH 9.5; For optimization study: Microwave power: 200–400 W, Temperature: 30–40°C Time: 10–20 min, Solid to liquid: 1:10, 1:12.5, 1:15, 1:17.5, 1:20, pH 9.5.	• A protein yield of 19.5% was found at the optimum conditions.	(Chaoyang et al., 2020)

(continued on next page)

Soymilk	Sequential	Microwave (IntelloCook, LG, South Korea) with an output power of 900 W and a bath type ultra- sonicator with a frequency of 28±2 kHz	Microwave irradiation time of 1, 2, 3, 4, and 5 min and sonication time of 30, 60, and 90 min at 40, 50, and 60 °C	• Protein yield increased with the sonication time and temperature increased while decreased with the increase in microwave irradiation time. Produced soymilk had enhanced nutritional quality and low enzymatic activity. Caused conformational deformation of proteins and enhanced the molecular interactions.	(Kumar et al., 2021)

240 W ultrasonic power for 4 min. Results of this study revealed that 93.95% protein yield was observed with an average precipitation rate of 97.97% at optimized conditions (28% PEG 200-based DES concentration, 28 g/mL solid-to-liquid ratio, 140 W microwave power, and 43°C extraction temperature,). The authors also claimed in this study that PEG 200-based DES UMSE offered the highest protein yields, used less solvent, and had a lower extraction time than with PEG 200-based DES ultrasonication, microwave, and water bath assisted extraction methods. Guan et al. (2018) explored the process intensification of ultrasonic-microwave-assisted technology for hordein (the storage protein of barley), in which the yield percentage of hordein was 8.84% extracted at optimum conditions of temperature 78°C, microwave power 298 W, and ultrasonic power 690 W, and time 20 min. A research report by Yu-lin et al. (2015) claimed that the synergistic extraction by ultrasound-microwave technology was quite feasible for the extraction of soybean dreg protein. They obtained a protein yield of 94.43% at an optimum condition of ultrasonic time 30 min, microwave time 5 min, microwave power 440 W, ultrasonic power 400 W, and solvent-to-solid ratio 10.30 mL/g with a pH of 9.5.

Water-soluble protein was extracted from *Eurycoma longifolia* roots using sequential ultrasound-microwave-assisted extraction technology (Elhag et al., 2020). In this study, the protein was extracted by two sequential options: sequential ultrasound-microwave-assisted extraction and sequential microwave-ultrasound-assisted extraction, which yielded 27.17% and 25.24% water-soluble protein, accordingly from *Eurycoma longifolia* roots. Sweet potato protein hydrolysates were produced using ultrasound-microwave (UM)-assisted enzymatic techniques to improve and characterize the antioxidant peptides (Habinshuti et al., 2020). Results showed a synergetic effect of UM and enzymatic treatment. Furthermore, a significant increase in the degree of hydrolysis and production of small and antioxidant-reactive peptides was observed. To enhance the extraction efficiency of mung bean protein, Chaoyang et al. (2020) used the ultrasonic and microwave co-extraction technology and stated the protein yield of 19.5% at the optimum conditions included: microwave power (338 W), temperature (37°C), time (15.6 min), and liquid-solid ratio (17.5 mL/g).

Yong-ting et al. (2018) successfully implemented the ultrasound-microwave technology for protein extraction from walnuts and optimized the extraction parameters for better yielding. It was found that the extraction rate was 93.48% at optimized parameters as follows: the ratio of material to liquid 20 g/mL, pH 9.0, ultrasonic power 400 W, and microwave power

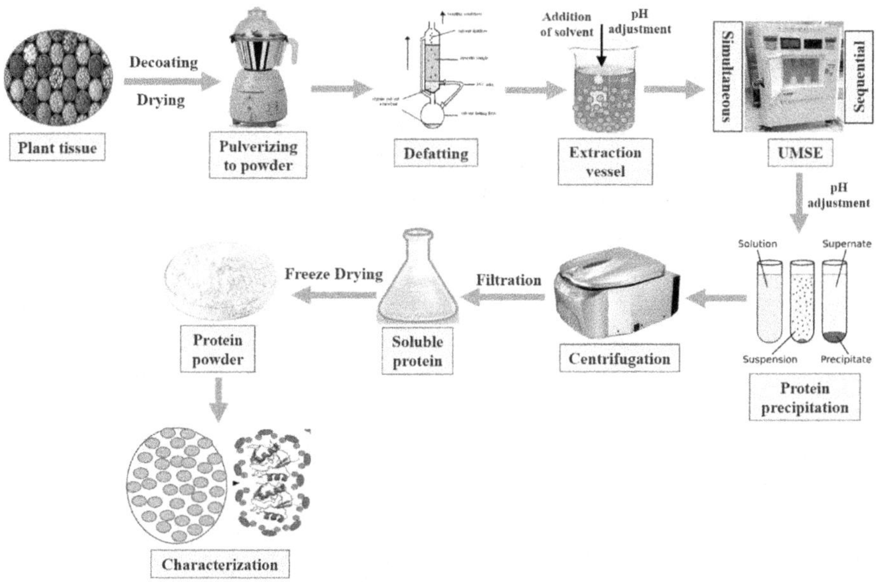

Figure 11.5 Overall processing protocols of protein extraction from plant sources.

200 W with the improvement of protein functionality. However, controversial results were found by Ochoa-Rivas et al. (2017) who evaluated the sequential effects of ultrasonication followed by microwave-assisted alkaline extraction of peanut proteins and reported no synergistic effect. Although the individual application of ultrasound and microwave improved the extraction yield of protein by 136% and 77%, respectively when compared with traditional methods. The overall protein extraction processes are summarized in Fig. 11.5.

11.4 Effect of ultrasound, microwave, and their synergism on protein functionality

The utilization of protein in different foods is preferred due to its functional properties. Kinsella (1976) defined the functional properties as—"functional properties denote those physical and chemical properties that influence the behavior of proteins in food systems during processing, storage, cooking, and consumption." Likewise, Damodaran and Paraf (1997) defined the protein functionality as "the physicochemical properties that influence functional behavior of proteins in food include their size, shape, amino acid composition, and sequence, net charge, charge distribution, hydrophobicity, hydrophilicity, structures (secondary, tertiary, and quaternary), molecular flexibility/rigidity in response to the external environment (pH, temperature, salt concentration) or interaction with other food constituents." Besides, solubility, emulsification capacity, foaming capacity, and rheological behaviors are some of the key functional properties of proteins (Karki et al., 2009). Several factors, such as heating, drying, pH, ionic strength, presence of reducing agents, storage conditions, as well as physical, enzymatic, or chemical modifications determine the functional properties of proteins (Kinsella, 1979). Studies evidenced that native proteins are not readily available for food applications due to a lack of functional properties, therefore, innovative technologies are required to modify the protein for desirable functional properties (Karki et al., 2009). As mentioned in previous sections, the application of high-intensity sonication, microwave irradiation, and their combination is considered as the most effective and applied physical methods of plant protein modifications. A schematic diagram has been shown in Fig. 11.6 to illustrate the effect of sonication, microwave and their synergistic effect on the modification of protein functionality.

11.4.1 Effect on protein structure and hydrophobicity

The formation of protein structure in typical plant-derived protein is displayed in Fig. 11.7, which consists of (Lin et al., 2021) (a) primary structure (an amino acid sequence), (b) secondary structure (actually a structure formed by C–O, and NH hydrogen bonding between different amino acids, such as α-helix, β-turn, and β-sheet), (c) tertiary structure (a stable single protein molecule with a hydrophobic nucleus, disulfide bonds, hydrogen bonds, salt bridges, and other bonds to fold higher protein structure), and (d) quaternary structure (the interaction of different polypeptide chains). The quaternary structure forms the functional protein complexes. Proteins have variable biological properties and functionalities, which are because of the variety of amino acids and their sequences as well as the formation of secondary, tertiary, and quaternary structures (Lin et al., 2021).

Zhang et al. (2018) advocated that the application of high-intensity ultrasound can modify the structural integrity of proteins by interrupting the hydrogen bond and hydrophobic interactions as well as destroying the tertiary structures of proteins, due to the actions of cavitation at higher temperatures resulting in protein refolding. This structural modification can enhance hydrolysis and increase biological activities (Ozuna et al., 2015). Zhu et al. (2018) declared similar effects in the case of walnut protein when treated with high-intensity ultrasound, and reported improvement of

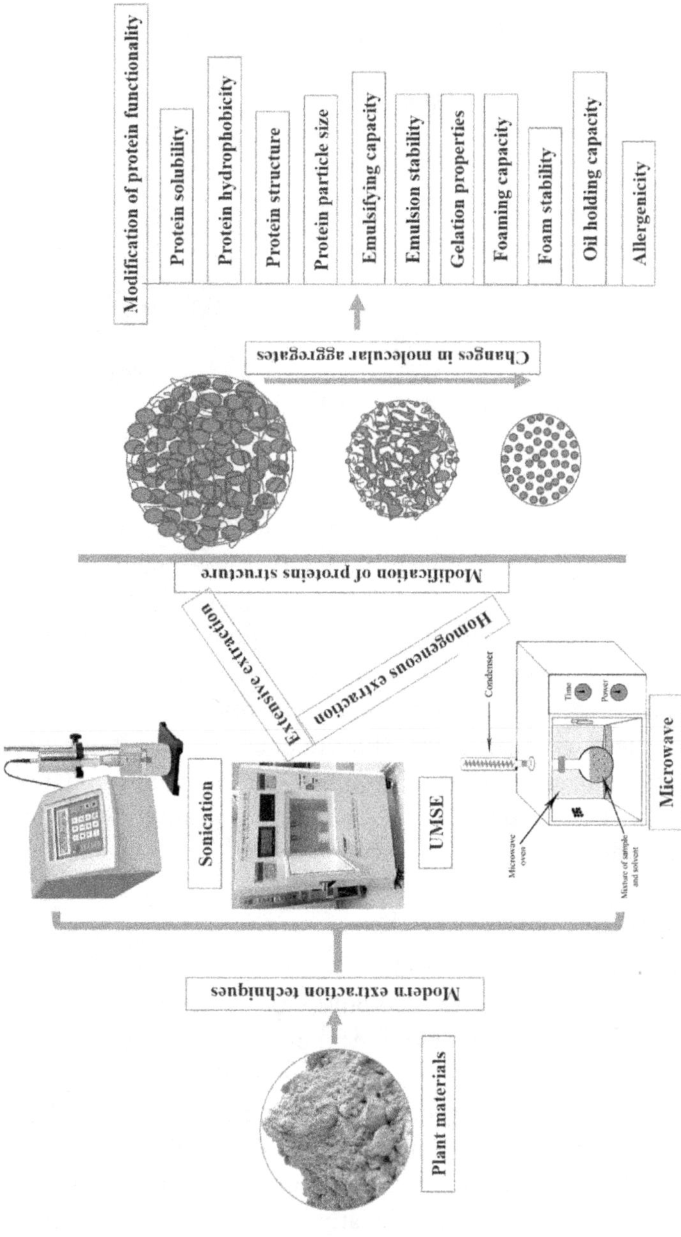

Figure 11.6 Effect of extraction on plant protein functionality using sonication, microwave and their synergism techniques.

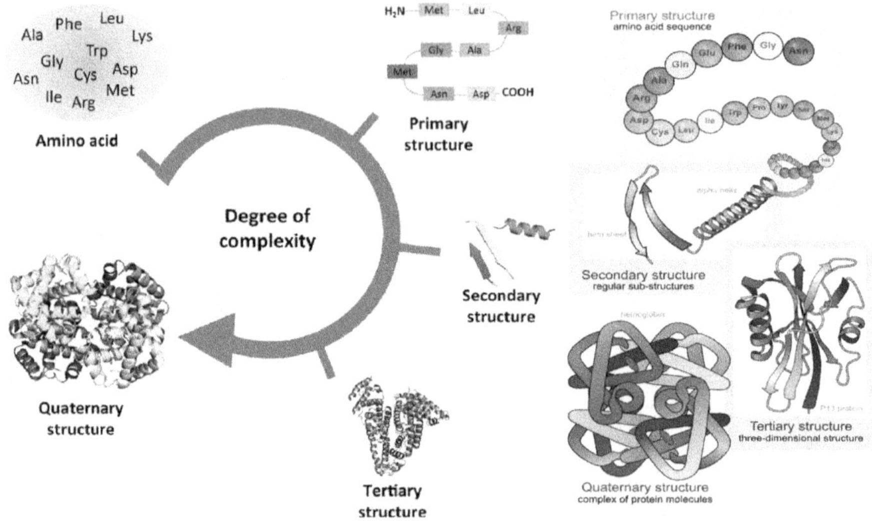

Figure 11.7 Formation of four levels of protein structure.

water solubility, emulsifying activity, and emulsifying stability. In contrast, structural alterations in protein occurred by microwave mainly attributed to the formation of carbon-lead free radicals, which are due to the absorption of microwave energy by the polar groups of protein molecules (Han et al., 2018). These free radicals again interact with the polar residues convoyed in a systematic association of ions under microwave treatment inducing alterations in the protein structures (Ekezie et al., 2017). The heating effect produced under microwave-field modifies the structural and functional properties of proteins. However, the changing magnitude is reliant on the level of protein denaturation (Han et al., 2018; Yao et al., 2018).

11.4.2 Effect on hydrophobicity

The study of Cao et al. (2018, 2021) confirmed that suitable heat-assisted denaturation can cause enhanced emulsifying attributes of a protein by enhancing the surface hydrophobicity. At the same time, an excessive thermal effect due to a fast increase in temperature may induce unwanted protein denaturation, thus negatively affecting the functional properties. These structural arrangements are directly linked with the hydrophobicity and stability of the protein. Exposure of plant tissue to ultrasound and microwave regimes for an extended period partially denatures the proteins, which undergo excessive bonding and reduce the hydrophobicity (Yanjun et al., 2014; Zheng et al., 2020). The hydrophobicity of sunflower protein increased up to a certain time (20 min) during ultrasound-assisted extraction and decreased with a further increase of the exposure time to 30 min (Malik et al., 2017). Previous studies suggested that during sonication, proteins undergo a certain degree of molecular unfolding and are consequently uncovered to more hydrophobic groups, which results in protein accumulation to protect the hydrophobic region of the protein (Chen et al., 2011; Jiang et al., 2014; Malik and Saini, 2017).

11.4.3 Effect on protein solubility

Protein solubility is an important functional property that determines its denaturation and aggregation (Arzeni et al., 2012). It is evidenced that the protein solubility increases after the sonication treatment as a result of structural changes in protein molecules and the formation of insoluble protein aggregates (Hu et al., 2015; Tang et al., 2009). Arzeni et al. (2012) observed that sonication-treated soy protein isolates showed higher solubility at a reduced particle size, and Jiang et al. (2014) stated that sonication-assisted partially unfolded black bean protein isolates had higher solubility. A similar observation was reported by Resendiz-vazquez et al. (2017) for jackfruit seed protein isolates. The ultrasound-cavitation may disrupt the hydrogen bonding and hydrophobic interactions as well as create a molecular unfolding resulting in lower protein aggregates that induce higher solubility of sonicated proteins. However, high power sonication proved to lower the solubility in denatured protein due to the action of more hydrophilic bonds and excessive particle aggregation (Karki et al., 2009). Similar to ultrasonication, microwave-treated proteins also exhibited a higher solubility than their conventional counterpart. For instance, Varghese and Pare (2019) informed a significant improvement of solubility in the case of soy milk protein extracted using microwave technology. The authors claimed microwave power and temperature were the most influential parameters and found a positive correlation with protein solubility. It is evidenced from the published literature that microwave-assisted extraction of protein causes a rapid breakdown of reserve native proteins, which are existent in the intact form in plant tissues (Phongthai et al., 2016). In addition, the lower residence in a microwave cavity and higher temperature resulted in less denatured protein and increased protein solubility (Varghese and Pare, 2019). Moreover, Zheng et al. (2020) investigated an increasing trend of protein solubility at lower to medium power levels in case of microwave-vacuum-assisted extraction of lotus seed protein isolates, and the solubility was gradually decreased with the increased microwave power. The formation of microwave-induced insoluble protein aggregates reduces the protein solubility, which is because of intermolecular disulfide bridges and hydrophobic interactions in thermal and radical environments.

11.4.4 Effect on emulsifying properties

Emulsifying activity is another vital functional property of protein, which was largely connected with its hydrophobicity and solubility. According to Falade et al. (2021), emulsions are thermodynamically unstable systems to enable the fraternization of two immiscible liquids, in which a dispersed phase is distributed into the continuous phase. Hydrolyzed proteins have gained great attention in emulsion formulations for their ability to enhance the peptide diffusion into the emulsion interface to reduce interfacial tension and form a consistent emulsion. Ultrasonication and microwave processing have a profound impact on the emulsifying properties of protein from plant matrices (Zhang et al., 2014; Zheng et al., 2020). It was professed that ultrasound accelerated the structural modification and polydispersibility of sunflower protein isolates, which opened up the molecular flexibility and surface hydrophobicity to permit efficient accessibility of protein molecules at the oil-water interface (Malik et al., 2017; Zhu et al., 2018). Malik et al. (2017) stated that improved emulsifying and emulsion stability of sonication-treated sunflower protein isolates. In case of emulsifying properties of peanut protein isolates, Zhang et al. (2014) advocated rapid temperature increase and high-pressure generation during ultrasound processing as the driving force for changing the emulsifying properties of protein isolates.

Sonication of millet protein at 18.4 W/cm^2 for 5–12.5 min did not alter the emulsifying properties, while 20 min caused a significant improvement. However, beyond 20 min at 73.95 W/cm^2 decreased the emulsification intensity. On the other hand, Zheng et al. (2020) explored the effect of microwave-vacuum power level on the structural and emulsifying properties of lotus seed protein isolates. They reported that mild microwave power intensity had a positive impact on the emulsion stability, and produced smaller-sized oil droplets. The microwave-assisted extraction facilitates the production of a smaller particle-sized protein that provides greater space and molecular flexibility to interact with oil droplets. It boosted the absorbability at the oil-water interface and assisted to decrease the interfacial tension between the water and oil phase during emulsification (Zheng et al., 2020). Studies by Phongthai et al. (2016) and Ochoa-Rivas et al. (2017) also stated the same phenomena, and stated the improved emulsion stability index of microwave-aided extracted rice bran and peanut protein, respectively.

11.4.5 Effect on foaming and foam stability

Ultrasonication and microwave irradiation have significant effects on the foaming capacity and foam stability of proteins. For instance, Zheng et al. (2020) and Zhang et al. (2011) described improved foaming capacity and foam stability of sunflower protein isolates and wheat gluten, respectively after ultrasound treatment. The improvement of foaming properties in soy protein isolates was ascribed to the decreased particle size after ultrasound treatment (Morales et al., 2015). On the other hand, Behere et al. (2021) stated better foaming capacity and foam stability of watermelon seed protein after microwave-assisted in comparison with conventional and ultrasound treatment. According to literature, foam formation depends on the transportation, penetration, and reorganization of molecules on the air-water interface, which are swayed by the hydrophobicity, size, and structural flexibility of the protein molecules (Malik et al., 2017; Malik and Saini, 2017). Research demonstrated that sonication and microwave treatment facilitates the structural modification of native proteins, which become partially denatured and exposed to the hydrophobic regions that enhance the interaction between the air-water interface, which leads to foams with higher stability (Han et al., 2018; Yao et al., 2018; Zheng et al., 2020).

11.4.6 Effect on protein digestibility

Protein digestibility determines the utilization of protein in foods and becomes digested after consumption. Microwave and ultrasound technology significantly affect the plant protein digestibility properties. For example, Varghese and Par (2019) claimed around 4.22% increase in the protein digestibility of soymilk by microwave heating compared to conventional heating. Higher protein digestibility was also reported by Ochoa-Rivas et al. (2017) and Behere et al. (2021) for peanut protein isolates and watermelon seed protein, respectively during extraction using microwave technology. The increasing trends of digestibility of extracted proteins by microwave-cavitation may be accredited to the modifications in tertiary structure as a function of temperature by microwaves. The temperature of microwaves enhances the protein denaturation or unfolding to open the structure permitting proteases to hydrolyze protein chains (Ochoa-Rivas et al., 2017). However, contradictory results were found in case of proteins extracted by ultrasound technology, such as almond milk protein (Vanga et al., 2020), and faba bean protein (Martínez-Velasco et al., 2018). The decreasing tendencies of in vitro protein digestibility of sonication-treated proteins may be caused by the excessive hydrophobic groups and the formation of large aggregates in protein molecules (Martínez-Velasco et al., 2018; Vanga et al., 2020).

11.4.7 Effect on water holding capacity and oil holding capacity

Water holding capacity (WHC) of protein improves juiciness, thickening, and viscosity. These are related to the water content, whereas oil holding capacity (OHC) is associated with the sensory attributes, for example, flavor retention. Sonication and microwave treatment have a significant effect on the water and oil holding capacity of proteins and are negatively correlated with each other (i.e., WHC decrease and OHC increase), which was evidenced by the previous studies of Behere et al. (2021) for watermelon seed protein, Malik and Saini (2017) for protein isolate of dephenolized sunflower meal, and Malik et al. (2017) for protein isolate of sunflower seeds and kernels. Resendiz-vazquez et al. (2017) provided an analogous statement for jackfruit (*Artocarpus heterophyllus*) protein isolate. As described earlier, sonication and microwave treated proteins possess a higher number of hydrophobic groups that allowed the oil to be physically encapsulated, which ultimately declines the WHC (Malik et al., 2017; Resendiz-vazquez et al., 2017).

11.5 Conclusions and future prospective

As discussed above, proteins play a vital role in the human body. Most of the plant proteins come from cereals, legumes, pulses, oilseed crops, and edible seeds. Various innovative and potential technologies have emerged in recent years, which have effectively been used for the extraction and modification of proteins. Previous research findings evidenced that microwave and ultrasound technology are more efficient options for protein extraction, and alteration for specific uses. However, have some drawbacks that hinder their beneficial application. Moreover, with the emergence of combined or hybrid extraction techniques as such, UMSE of proteins from plant tissues gaining much interest for their improved extraction efficiency and functionality, as well as overcoming the limits of their single application. Furthermore, protein utilization depends in many ways, for example, extraction methods, and structural and functional properties. Although there are plenty of reports and reviews on the successful extraction of plant-derived protein using microwave, and ultrasonication technology, still a dearth of literature on the application of UMSE for protein extraction along with the mechanistic explanations of this technology on the structural modification of proteins has not been explored. More precisely, how this technology influences the hydrolysis properties of proteins remains unknown. This novel extraction technique also requires proper design and parameter optimization to establish pilot and further on industrial scale extraction facility. It may lead to a major step in sustainable development and exploitation of proteins from plant sources. Therefore, the research gaps of UMSE on extraction efficiency and protein functionality may create new pathways for improving protein quality.

Conflict of interest

The authors declare that there is no conflict of interest.

References

Aftari, R.V., Rezaei, K., Mortazavi, A., Bandani, A.R., 2015. The optimized concentration and purity of *Spirulina platensis* C-Phycocyanin: A comparative study on microwave-assisted and ultrasound-assisted extraction methods. J. Food Process. Preserv. 39 (6), 3080–3091.

Agrawal, H., Joshi, R., Gupta, M., 2017. Isolation and characterization of enzymatic hydrolyzed peptides with antioxidant activities from green tender sorghum. LWT - Food Sci. and Tech. 84, 608–616.

Ahmed, J., Al-Ruwaih, N., Mulla, M., Rahman, M.H., 2018. Effect of high-pressure treatment on functional, rheological and structural properties of kidney bean protein isolate. LWT - Food Sci. Technol. 91, 191–197.

Akharume, F.U., Aluko, R.E., Adedeji, A.A., 2021. Modification of plant proteins for improved functionality: a review. Comprehens. Rev. Food Sci. Food Saf. 20 (1), 198–224.

Akhtar, A., Aslam, S., Khan, S., McClements, D., Khalid, N., Maqsood, S., 2021. Utilization of diverse protein sources for the development of protein-based nanostructures as bioactive carrier systems: a review of recent research findings (2010–2021). Crit. Rev. Food Sci. Nutr. 1–19.

Akyüz, A., Ersus, S., 2021. Optimization of enzyme assisted extraction of protein from the sugar beet (*Beta vulgaris* L.) leaves for alternative plant protein concentrate production. Food Chem. 335, 127673.

Amponsah, A., Nayak, B., 2016. Effects of microwave and ultrasound assisted extraction on the recovery of soy proteins for soy allergen detection. J. Food Sci. 81 (11), T2876–T2885.

Ayim, I., Ma, H., Alenyorege, E.A., 2018. Optimizing and predicting degree of hydrolysis of ultrasound assisted sodium hydroxide extraction of protein from tea (*Camellia sinensis* L.) residue using response surface methodology. J. Food Sci. Technol. 55 (12), 5166–5174.

Arzeni, C., Martínez, K., Zema, P., Arias, A., Pérez, O.E., Pilosof, A.M.R., 2012. Comparative study of high intensity ultrasound effects on food proteins functionality. J. Food Eng., 108 (3), 463–472.

Bandyopadhyay, K., Chakraborty, C., Barman, A.K., 2012. Effect of microwave and enzymatic treatment on the recovery of protein from Indian defatted rice bran meal. J. Oleo Sci. 61 (10), 525–529.

Bedin, S., Netto, F.M., Bragagnolo, N., Taranto, O.P., 2020. Reduction of the process time in the achieve of rice bran protein through ultrasound-assisted extraction and microwave-assisted extraction. Sep. Sci. Technol. (Philadelphia) 55 (2), 300–312.

Behere, M., Patil, S.S., Rathod, V.K., 2021. Rapid extraction of watermelon seed proteins using microwave and its functional properties. Prep. Biochem. Biotechnol. 51 (3), 252–259.

Bernardi, S., Corso, M.P., Baraldi, I.J., Colla, E., Canan, C., 2018. Obtaining concentrated rice bran protein by alkaline extraction and stirring - optimization of conditions. Int. Food Res. J. 25 (3), 1133–1139.

Bernardi, S., Lupatini-Menegotto, A.L., Kalschne, D.L., Kalschne, D.L., Flores, É.L.M., Bittencourt, P.R.S., Colla, E., Canan, C., 2021. Ultrasound: a suitable technology to improve the extraction and techno-functional properties of vegetable food proteins. Plant Foods Hum. Nutr. 76 (1), 1–11.

Bose, U., Broadbent, J.A., Byrne, K., Hasan, S., Howitt, C.A., Colgrave, M.L., 2019. Optimisation of protein extraction for in-depth profiling of the cereal grain proteome. J. Proteomics 197, 23–33.

Boukid, F., 2021. Plant-based meat analogues: from niche to mainstream. Eur. Food Res. Technol. 247 (2), 297–308.

Byanju, B., Rahman, M.M., Hojilla-Evangelista, M.P., Lamsal, B.P., 2020. Effect of high-power sonication pretreatment on extraction and some physicochemical properties of proteins from chickpea, kidney bean, and soybean. Int. J. Biol. Macromol. 145, 712–721.

Byaruhanga, Y.B., Emmambux, M.N., Belton, P.S., Wellner, N., Ng, K.G., Taylor, J.R.N., 2006. Alteration of kafirin and kafirin film structure by heating with microwave energy and tannin complexation. J. Agric. Food Chem. 54 (12), 4198–4207.

Campbell, T.C., 2017. A plant-based diet and animal protein: questioning dietary fat and considering animal protein as the main cause of heart disease. J. Geriatr. Cardiol. 14 (5), 331–337.

Cao, B., Fang, L., Liu, C., Min, W., Liu, J., 2018. Effects of high hydrostatic pressure on the functional and rheological properties of the protein fraction extracted from pine nuts. Food Sci. Technol. Int. 24 (1), 53–66.

Cao, H., Fan, D., Jiao, X., Huang, J., Zhao, J., Yan, B., Zhou, W., Zhang, W., Zhang, H., 2018. Effects of microwave combined with conduction heating on Surimi quality and morphology. J. Food Eng. 228, 1–11.

Cao, H., Sun, R., Shi, J., Li, M., Guan, X., Liu, J., Huang, K., Zhang, Y., 2021. Effect of ultrasonic on the structure and quality characteristics of quinoa protein oxidation aggregates. Ultrason. Sonochem. 77, 105685.

Capellini, M.C., Giacomini, V., Cuevas, M.S., Rodrigues, C.E.C., 2017. Rice bran oil extraction using alcoholic solvents: physicochemical characterization of oil and protein fraction functionality. Ind. Crops Prod. 104, 133–143.

Cassity, J. (2019). The State of the Global Plant-based Protein Market. 1–13. Available at: https://www.kerry.com/global-en/insights/kerrydigest/2019/the-state-of-the-global-plant-based-protein-market

Chan, C.H., Yusoff, R., Ngoh, G.C., Kung, F.W.L., 2011. Microwave-assisted extractions of active ingredients from plants. J. Chromatogr. A 1218 (37), 6213–6225.

Chaoyang, L., Jingjing, D., Liangyu, L., 2020. Research of total protein extraction from mung bean by ultrasound-microwave synergistic. J. Heilongjiang Bayi Agric. Univ. 32 (5), 49–55.

Chemat, F., Rombaut, N., Sicaire, A.G., Meullemiestre, A., Fabiano-Tixier, A.S., Abert-Vian, M., 2017. Ultrasound assisted extraction of food and natural products. Mechanisms, techniques, combinations, protocols and applications. A review. Ultrason. Sonochem. 34, 540–560.

Chen, L., Chen, J., Ren, J., Zhao, M., 2011. Effects of ultrasound pretreatment on the enzymatic hydrolysis of soy protein isolates and on the emulsifying properties of hydrolysates. J. Agric. Food Chem. 59 (6), 2600–2609.

Chen, R., Wang, X.J., Zhang, Y.Y., Xing, Y., Yang, L., Ni, H., Li, H.H., 2019. Simultaneous extraction and separation of oil, proteins, and glucosinolates from *Moringa oleifera* seeds. Food Chem. 300, 125162.

Chen, R., Meng, F., Zhang, S., Liu, Z., 2009. Effects of ultrahigh pressure extraction conditions on yields and antioxidant activity of ginsenoside from ginseng. Sep. Purif. Technol. 66 (2), 340–346.

Cheng, F., Shu, G., Chen, L., Dai, C., Wan, H., Chen, H., Dong, X., 2021. Ultrasound-microwave assisted extraction of proteins from *Moringa oleifera* leaves: comparative optimization study and LC-MS analysis of the protein concentrate. J. Food Process. Preserv. 45 (6), 1–13.

Chiang, J.H., Loveday, S.M., Hardacre, A.K., Parker, M.E., 2019. Effects of soy protein to wheat gluten ratio on the physicochemical properties of extruded meat analogues. Food Struct. 19, 100102.

Chirinos, R., Aquino, M., Pedreschi, R., Campos, D., 2017. Optimized methodology for alkaline and enzyme-assisted extraction of protein from sacha inchi (*Plukenetia volubilis*) kernel cake. J. Food Process Eng. 40 (2), e12412.

Contreras, M., Lama-muñoz, A., Gutiérrez-pérez, J.M., Espínola, F., Moya, M., Castro, E., 2019. Protein extraction from agri-food residues for integration in biorefinery: potential techniques and current status. Bioresour. Technol. 280, 459–477.

Cravotto, G., Cintas, P., 2007. The combined use of microwaves and ultrasound: improved tools in process chemistry and organic synthesis. Chem. Eur. J. 13 (7), 1902–1909.

Dabbour, M., Xiang, J., Mintah, B., He, R., Jiang, H., Ma, H., 2020. Localized enzymolysis and sonochemically modified sunflower protein: physical, functional and structure attributes. Ultrason. Sonochem. 63, 104957.

Dabbour, M., He, R., Ma, H., Musa, A., 2018. Optimization of ultrasound assisted extraction of protein from sunflower meal and its physicochemical and functional properties. J. Food Process Eng. 41 (5), 1–11.

Damodaran, S., Paraf, A., 1997. Food Proteins and Their Applications, First Ed. Mercel Dekker, New York.

Deleu, L.J., Lambrecht, M.A., Van de Vondel, J., Delcour, J.A., 2019. The impact of alkaline conditions on storage proteins of cereals and pseudo-cereals. Curr. Opin. Food Sci. 25, 98–103.

Demuez, M., Mahdy, A., Tomás-Pejó, E., González-Fernández, C., Ballesteros, M., 2015. Enzymatic cell disruption of microalgae biomass in biorefinery processes. Biotechnol. Bioeng. 112 (10), 1955–1966.

Du, M., Xie, J., Gong, B., Xu, X., Tang, W., Li, X., Li, C., Xie, M., 2018. Extraction, physicochemical characteristics and functional properties of mung bean protein. Food Hydrocoll. 76, 131–140.

Ekezie, F.G.C., Sun, D.W., Cheng, J.H., 2017. Acceleration of microwave-assisted extraction processes of food components by integrating technologies and applying emerging solvents: a review of latest developments. Trends Food Sci. Technol. 67, 160–172.

Elhag, H.E.A., Sulaiman, A.Z., Ajit, A., 2020. Sequential ultrasound-microwave assisted extraction of water soluble proteins from *Eurycoma longifolia* roots. IOP Conf. Ser.: Mater. Sci. Eng. 736, 022115. doi:10.1088/1757-899X/736/2/022115.

Embaby, H.E., Swailam, H.M., Rayan, A.M., 2018. Preparation and physicochemical properties of protein concentrate and isolate produced from *Acacia tortilis* (Forssk.) Hayne ssp. raddiana. J. Food Sci. Technol. 55 (2), 489–495.

Falade, E.O., Mu, T.H., Zhang, M., 2021. Improvement of ultrasound microwave-assisted enzymatic production and high hydrostatic pressure on emulsifying, rheological and interfacial characteristics of sweet potato protein hydrolysates. Food Hydrocoll. 117, 106684.

FAO, IFAD, UNICEF, WFP, & WHO. (2019). The Food and Agriculture Organization of the United Nations: The state of food security in the world. In *Food and Agriculture Organization of the United Nations*. Available at: http://www.fao.org/publications.

Farkas, D.F., 2016. A short history of research and development efforts leading to the commercialization of high-pressure processing of food. In: Balasubramaniam, V.M., Gustavo, V., Barbosa-Cánovas, Huub L.M., Lelieveld (Eds.), High Pressure Processing of Food: Principles, Technology and Applications, Food Engineering Series. Springer Nature, Springer New York, NY, pp. 19–36.

Ferreira, S.S., Passos, C.P., Cardoso, S.M., Wessel, D.F., Coimbra, M.A., 2018. Microwave assisted dehydration of broccoli by-products and simultaneous extraction of bioactive compounds. Food Chem. 246, 386–393.

Fetzer, A., Herfellner, T., Stäbler, A., Menner, M., Eisner, P., 2018. Influence of process conditions during aqueous protein extraction upon yield from pre-pressed and cold-pressed rapeseed press cake. Ind. Crops Prod. 112, 236–246.

Flores-Jiménez, N.T., Ulloa, J.A., Silvas, J.E.U., Ramírez, J.C.R., Ulloa, P.R., Rosales, P.U.B., Carrillo, Y.S., Leyva, R.G., 2019. Effect of high-intensity ultrasound on the compositional, physicochemical, biochemical, functional and structural properties of canola (*Brassica napus* L.) protein isolate. Food Res. Int. 121, 947–956.

Gadalkar, S.M., Rathod, V.K., 2020. Extraction of watermelon seed proteins with enhanced functional properties using ultrasound. Prep. Biochem. Biotechnol. 50 (2), 133–140.

Gençdağ, E., Görgüç, A., Yılmaz, F.M., 2021. Recent advances in the recovery techniques of plant-based proteins from agro-industrial by-products. Food Rev. Internat. 37 (4), 447–468.

Gerven, T.V, Stankiewicz, A., 2009. Structure, energy, synergy, times: the fundamentals of process intensification. Industr. Eng. Chem. Res. 48, 2465–2474.

Ghanghas, N., M.T, M., Sharma, S., Prabhakar, P.K, 2022. Classification, composition, extraction, functional modification and application of rice (*Oryza sativa*) seed protein: a comprehensive review. Food Rev. Int. 38 (4), 364–383. doi:https://doi.org/10.1080/87559129.2020.1733596.

Görgüç, A., Bircan, C., Yılmaz, F.M., 2019. Sesame bran as an unexploited by-product: effect of enzyme and ultrasound-assisted extraction on the recovery of protein and antioxidant compounds. Food Chem. 283, 637–645.

Görgüç, A., Özer, P., Yılmaz, F.M., 2020. Microwave-assisted enzymatic extraction of plant protein with antioxidant compounds from the food waste sesame bran: comparative optimization study and identification of metabolomics using LC/Q-TOF/MS. J. Food Process. Preserv. 44 (1), 1–11.

Guan, X., Li, L., Liu, J., Li, S., 2018. Effects of ultrasonic-microwave-assisted technology on Hordein extraction from barley and optimization of process parameters using response surface methodology. J. Food Qual. 1–8. doi:https://doi.org/10.1155/2018/9280241.

Gude, V.G., 2015. Synergism of microwaves and ultrasound for advanced biorefineries. Res.-Effic. Technol. 1 (2), 116–125.

Guo, C., Zhang, Z., Chen, J., Fu, H., Subbiah, J., Chen, X., Wang, Y., 2017. Effects of radio frequency heating treatment on structure changes of soy protein isolate for protein modification. Food Bioprocess Technol. 10 (8), 1574–1583.

Guzik, P., Kulawik, P., Zając, M., Migdał, W., 2021. Microwave applications in the food industry: an overview of recent developments. Crit. Rev. Food Sci. Nutr. 1–20. doi:https://doi.org/10.1080/10408398.2021.1922871.

Habinshuti, I., Mu, T.H., Zhang, M., 2020. Ultrasound microwave-assisted enzymatic production and characterisation of antioxidant peptides from sweet potato protein. Ultrason. Sonochem. 69, 105262.

Han, Z., Cai, M.J., Cheng, J.H., Sun, D.W., 2018. Effects of electric fields and electromagnetic wave on food protein structure and functionality: a review. Trends Food Sci. Technol. 75, 1–9.

Hassan, A.B., von Hoersten, D., Ahmed, M., I., A., 2019. Effect of radio frequency heat treatment on protein profile and functional properties of maize grain. Food Chem. 271, 142–147.

Ho, C.H.L., Cacace, J.E., Mazza, G., 2007. Extraction of lignans, proteins and carbohydrates from flaxseed meal with pressurized low polarity water. LWT - Food Sci. Technol. 40 (9), 1637–1647.

Hoehnel, A., Axel, C., Bez, J., Arendt, E.K., Zannini, E., 2019. Comparative analysis of plant-based high-protein ingredients and their impact on quality of high-protein bread. J. Cereal Sci. 89, 102816.

Hou, F., Ding, W., Qu, W., Oladejo, A.O., Xiong, F., Zhang, W., He, R., Ma, H., 2017. Alkali solution extraction of rice residue protein isolates: influence of alkali concentration on protein functional, structural properties and lysinoalanine formation. Food Chem. 218, 207–215.

Hu, H., Cheung, I.W.Y., Pan, S., Li-Chan, E.C.Y., 2015. Effect of high intensity ultrasound on physicochemical and functional properties of aggregated soybean β-conglycinin and glycinin. Food Hydrocoll. 45, 102–110.

Huang, H.W., Hsu, C.P., Yang, B.B., Wang, C.Y., 2013. Advances in the extraction of natural ingredients by high pressure extraction technology. Trends Food Sci. Technol. 33 (1), 54–62.

Huang, K., Wang, S.J., Shan, L., Zhu, Q., Qian, J., 2007. Seeking synergistic effect – a key principle in process intensification. Sep. Purif. Technol. 57 (1), 111–120.

Işçimen, E.M., Hayta, M., 2018. Optimisation of ultrasound assisted extraction of rice bran proteins: effects on antioxidant and antiproliferative properties. Qual. Ass. Saf. Crops Foods 10 (2), 165–174.

Jambrak, A.R., Mason, T.J., Lelas, V., Paniwnyk, L., Herceg, Z., 2014. Effect of ultrasound treatment on particle size and molecular weight of whey proteins. J. Food Eng. 121 (1), 15–23.

Jiang, L., Wang, J., Li, Y., Wang, Z., Liang, J., Wang, R., Chen, Y., Ma, W., Qi, B., Zhang, M., 2014. Effects of ultrasound on the structure and physical properties of black bean protein isolates. Food Res. Int. 62, 595–601.

Jiao, X., Cao, H., Fan, D., Huang, J., Zhao, J., Yan, B., Zhou, W., Zhang, W., Ye, W., Zhang, H., 2019. Effects of fish oil incorporation on the gelling properties of silver carp surimi gel subjected to microwave heating combined with conduction heating treatment. Food Hydrocoll. 94, 164–173.

Kain, R.J., Chen, Z., Sonda, T.S., Abu-Kpawoh, J.C., 2009. Study on the effect of control variable on the extraction of eanut rotein isolates from peanut meal (*Arachis hypohanea* L.). Am. J. Food Technol. 4 (1), 47–55.

Kamal, H., Le, C.F., Salter, A.M., Ali, A., 2021. Extraction of protein from food waste: an overview of current status and opportunities. Comprehens. Rev. Food Sci. Food Saf. 20 (3), 2455–2475.

Kamal, M.M., Kumar, J., Mamun, M.A.H., Ahmed, M.N.U., Shishir, M.R.I., Mondal, S.C., 2021. Extraction and characterization of pectin from *Citrus sinensis* peel. J. Biosyst. Eng. 46 (1), 16–25.

Kamal, M.M., Ali, M.R., Shishir, M.R.I., Saifullah, M., Haque, M.R., Mondal, S.C., 2019. Optimization of process parameters for improved production of biomass protein from *Aspergillus niger* using banana peel as a substrate. Food Sci. Biotechnol. 28 (6), 1693–1702.

Kamal, M.M., Ali, M.R., Hossain, A., Shishir, M.R.I., 2020. Optimization of microwave-assisted extraction of pectin from *Dillenia indica* fruit and its preliminary characterization. J. Food Process. Preserv. 44 (6), e14466.

Karki, B., Lamsal, B.P., Grewell, D., Pometto, A.L., Van Leeuwen, J., Khanal, S.K., Jung, S., 2009. Functional properties of soy protein isolates produced from ultrasonicated defatted soy flakes. JAOCS, J. Am. Oil Chemists' Soc. 86 (10), 1021–1028.

Kinsella, J.E., 1976. Functional properties of proteins in foods: A survey. C R C Crit. Rev. Food Sci. Nutr. 7 (3), 219–280.

Kinsella, J.E., 1979. Functional properties of soy proteins. J. Am. Oil Chemists' Soc. 56 (3), 242–258.

Ko, G.J., Rhee, C.M., Kalantar-Zadeh, K., Joshi, S., 2020. The effects of high-protein diets on kidney health and longevity. J. Am. Soc. Nephrol. 31 (8), 1667–1679.

Kumar, M., Tomar, M., Potkule, J., Reetu, Punia, S., Dhakane-Lad, J., Singh, S., Dhumal, S., Chandra Pradhan, P., Bhushan, B., Anitha, T., Alajil, O., Alhariri, A., Amarowicz, R., Kennedy, J.F., 2022. Functional characterization of plant-based protein to determine its quality for food applications. Food Hydrocoll. 123, 106986.

Kumar, R., Kumar, A., Jayachandran, L.E., Rao, P.S., 2021. Sequential microwave – ultrasound assisted extraction of soymilk and optimization of extraction process. LWT - Food Sci. Technol. 151, 112220.

Kumar, M., Tomar, M., Potkule, J., Verma, R., Punia, S., Mahapatra, A., Belwal, T., Dahuja, A., Joshi, S., Berwal, M.K., Satankar, V., Bhoite, A.G., Amarowicz, R., Kaur, C., Kennedy, J.F., 2021. Advances in the plant protein extraction: mechanism and recommendations. Food Hydrocoll. 115, 106595.

Lamacchia, C., Landriscina, L., D'Agnello, P., 2016. Changes in wheat kernel proteins induced by microwave treatment. Food Chem. 197, 634–640.

Lee, C.S., 2017. Extraction of Bio-Flocculant From Okra Using Hydrothermal and Microwave Extraction Methods Combined With a Techno-Economic Assessment A Thesis Submitted in the Department of Chemical and Environmental Engineering, Faculty of Engineering, University of Nottingham in Fulfillment of the Requirements for the Degree of Doctor of Philosophy.

Lee, H.X., Ahmad, F., Saad, B., Ismail, M.N., 2017. Evaluation of extraction methods for the identification of proteins from date palm (*Phoenix dactylifera* L.) seed and flesh. Prep. Biochem. Biotechnol. 47 (10), 998–1007.

Leighton, T.G., 1995. Bubble population phenomena in acoustic cavitation. Ultrason. – Sonochem. 2 (2), 123–136.

Li, H., Jia, Y., Peng, W., Zhu, K., Zhou, H., Guo, X., 2018. High hydrostatic pressure reducing allergenicity of soy protein isolate for infant formula evaluated by ELISA and proteomics via Chinese soy-allergic children's sera. Food Chem. 269, 311–317.

Li, H., Zhu, K., Zhou, H., Peng, W., Guo, X., 2016. Comparative study of four physical approaches about allergenicity of soybean protein isolate for infant formula. Food Agric. Immunol. 27 (5), 604–623.

Li, J., Guo, Z., 2017. Complete utilization of rapeseed meal to produce lipophilic antioxidants, protein, and monosugars in a concordant manner. ACS Sustain. Chem. Eng. 5 (7), 6218–6226.

Li, K., Ma, H., Li, S., Zhang, C., Dai, C., 2017. Effect of ultrasound on alkali extraction protein from rice dreg flour. J. Food Process Eng. 40 (2), e12377.

Li, S., Yang, X., Zhang, Y., Ma, H., Liang, Q., Qu, W., He, R., Zhou, C., Mahunu, G.K., 2016. Effects of ultrasound and ultrasound assisted alkaline pretreatments on the enzymolysis and structural characteristics of rice protein. Ultrason. Sonochem. 31, 20–28.

Li, X., Da, S., Li, C., Xue, F., Zang, T., 2018. Effects of high-intensity ultrasound pretreatment with different levels of power output on the antioxidant properties of alcalase hydrolyzates from Quinoa (*Chenopodium quinoa* Willd.) protein isolate. Cereal Chem. 95 (4), 518–526.

Li, X., Ye, C., Tian, Y., Pan, S., Wang, L., 2018. Effect of ohmic heating on fundamental properties of protein in soybean milk. J. Food Process Eng. 41 (3), 1–10.

Li, Z., Wang, J., Zheng, B., Guo, Z., 2020. Impact of combined ultrasound-microwave treatment on structural and functional properties of golden threadfin bream (*Nemipterus virgatus*) myofibrillar proteins and hydrolysates. Ultrason. Sonochem. 65, 105063.

Lianfu, Z., Zelong, L., 2008. Optimization and comparison of ultrasound/microwave assisted extraction (UMAE) and ultrasonic assisted extraction (UAE) of lycopene from tomatoes. Ultrason. Sonochem. 15 (5), 731–737.

Lin, D., Zhang, Q., Xiao, L., Huang, Y., Yang, Z., Wu, Z., Tu, Z., Qin, W., Chen, H., Wu, D., Zhang, Q., Li, S., 2021. Effects of ultrasound on functional properties, structure and glycation properties of proteins: a review. Crit. Rev. Food Sci. Nutr. 61 (15), 2471–2481.

Lin, Q., Ge, S., McClements, D., Li, X., Jin, Z., Jiao, A., Wang, J., Long, J., Xu, X., Qiu, C., 2021. Advances in preparation, interaction and stimulus responsiveness of protein-based nanodelivery systems. Crit. Rev. Food Sci. Nutr. 1–14.

Ling, B., Ouyang, S., Wang, S., 2019. Effect of radio frequency treatment on functional, structural and thermal behaviors of protein isolates in rice bran. Food Chem. 289, 537–544.

Liu, C., Ma, X., 2016. Study on the mechanism of microwave modified wheat protein fiber to improve its mechanical properties. J. Cereal Sci. 70, 99–107.

Liu, J., Mu, T., Sun, H., Fauconnier, M.L., 2019. Optimization of ultrasonic–microwave synergistic extraction of flavonoids from sweet potato leaves by response surface methodology. J. Food Process. Preserv. 43 (5), 1–10.

Liu, K., Zheng, J., Chen, F., 2019. Effect of domestic cooking on rice protein digestibility. Food Sci. Nutr. 7 (2), 608–616.

Liu, R.L., Yu, P., Ge, X.L., Bai, X.F., Li, X.Q., Fu, Q., 2017. Establishment of an aqueous peg 200-based deep eutectic solvent extraction and enrichment method for pumpkin (*Cucurbita moschata*) seed protein. Food Anal. Methods 10 (6), 1669–1680.

Liu, Y., Ma, X.Y., Liu, L.N., Xie, Y.P., Ke, Y.J., Cai, Z.J., Wu, G.J., 2019. Ultrasonic-assisted extraction and functional properties of Wampee seed protein. Food Sci. Technol. 39, 324–331.

Liu, Z., Qiao, L., Yang, F., Gu, H., Yang, L., 2017. Brönsted acidic ionic liquid based ultrasound-microwave synergistic extraction of pectin from pomelo peels. Int. J. Biol. Macromol. 94, 309–318.

Lu, W., Chen, X.W., Wang, J.M., Yang, X.Q., Qi, J.R., 2016. Enzyme-assisted subcritical water extraction and characterization of soy protein from heat-denatured meal. J. Food Eng. 169, 250–258.

Ly, H.L., Tran, T.M.C., Tran, T.T.T., Ton, N.M.N., Le, V.V.M., 2018. Application of ultrasound to protein extraction from defatted rice bran. Int. Food Res. J. 25 (2), 695–701.

Mahroug, H., Ribeiro, M., Rhazi, L., Bentallah, L., Zidoune, M.N., Nunes, F.M., Igrejas, G., 2019. How microwave treatment of gluten affects its toxicity for celiac patients? A study on the effect of microwaves on the structure, conformation, functionality and immunogenicity of gluten. Food Chem. 297, 124986.

Malik, M.A., Saini, C.S., 2017. Polyphenol removal from sunflower seed and kernel: effect on functional and rheological properties of protein isolates. Food Hydrocolloids 63, 705–715.

Malik, M.A., Sharma, H.K., Saini, C.S., 2017. High intensity ultrasound treatment of protein isolate extracted from dephenolized sun flower meal: effect on physicochemical and functional properties. Ultrason. – Sonochem. 39, 511–519.

Martina, K., Tagliapietra, S., Barge, A., Cravotto, G., 2016. Combined microwaves/ultrasound, a hybrid technology. Top. Curr. Chem. 374 (6), 175–201.

Martínez-Velasco, A., Lobato-Calleros, C., Hernández-Rodríguez, B.E., Román-Guerrero, A., Alvarez-Ramirez, J., Vernon-Carter, E.J., 2018. High intensity ultrasound treatment of faba bean (*Vicia faba* L.) protein: effect on surface properties, foaming ability and structural changes. Ultrason. Sonochem. 44, 97–105.

Martins, C.P.C., Cavalcanti, R.N., Couto, S.M., Moraes, J., Esmerino, E.A., Silva, M.C., Raices, R.S.L., Gut, J.A.W., Ramaswamy, H.S., Tadini, C.C., Cruz, A.G., 2019. Microwave processing: current background and effects on the physicochemical and microbiological aspects of dairy products. Comprehens. Rev. Food Sci. Food Saf. 18 (1), 67–83.

Mechmeche, M., Kachouri, F., Chouabi, M., Ksontini, H., Setti, K., Hamdi, M., 2017. Optimization of extraction parameters of protein isolate from tomato seed using response surface methodology. Food Anal. Methods 10 (3), 809–819.

Mehr, H.M., Koocheki, A., 2021. Physicochemical properties of Grass pea (*Lathyrus sativus* L.) protein nanoparticles fabricated by cold atmospheric-pressure plasma. Food Hydrocoll. 112, 106328.

Morales, R., Martínez, K.D., Pizones Ruiz-Henestrosa, V.M., Pilosof, A.M.R, 2015. Modification of foaming properties of soy protein isolate by high ultrasound intensity: particle size effect. Ultrason. Sonochem. 26, 48–55.

Moret, S., Conchione, C., Srbinovska, A., Lucci, P., 2019. Microwave-based technique for fast and reliable extraction of organic contaminants from food, with a special focus on hydrocarbon contaminants. Foods 8 (10), 503.

Musielak, G., Mierzwa, D., Kroehnke, J., 2016. Food drying enhancement by ultrasound – a review. Trends Food Sci. Technol. 56, 126–141.

Mustafa, A., Turner, C., 2011. Pressurized liquid extraction as a green approach in food and herbal plants extraction: a review. Anal. Chim. Acta 703 (1), 8–18.

Nasrabadi, M.N., Doost, A.S., Mezzenga, R., 2021. Modification approaches of plant-based proteins to improve their techno-functionality and use in food products. Food Hydrocoll. 118, 106789.

Nguyen, T.H., Le, V.V.M., 2019. Effects of technological parameters of ultrasonic treatment on the protein extraction yield from defatted peanut meal. Int. Food Res. J. 26 (3), 1079–1085.

Nwachukwu, I.D., Aluko, R.E., 2021. Chapter 1: Food protein structures, functionality and product development. In: Udenigwe, Chibuike C (Ed.), Food Proteins and Peptides: Emerging Biofunctions, Food and Biomaterial Applications, Food Chemistry, Function and Analysis, 27. The Royal Society of Chemistry, Burlington House, Piccadilly, London, W1J 0BA, UK, pp. 1–33.

Ochoa-Rivas, A., Nava-Valdez, Y., Serna-Saldívar, S.O., Chuck-Hernández, C., 2017. Microwave and ultrasound to enhance protein extraction from peanut flour under alkaline conditions: Effects in yield and functional properties of protein isolates. Food Bioprocess Technol. 10 (3), 543–555.

Özer, P., Görgüç, A., Yılmaz, F.M., 2018. The use of microwave technology on the extraction of macro and micro components from plant tissues. Gida / J. Food 43 (5), 765–775.

Ozuna, C., Paniagua-Martínez, I., Castaño-Tostado, E., Ozimek, L., Amaya-Llano, S.L., 2015. Innovative applications of high-intensity ultrasound in the development of functional food ingredients: production of protein hydrolysates and bioactive peptides. Food Res. Int. 77, 685–696.

Panja, P., 2018. Green extraction methods of food polyphenols from vegetable materials. Curr. Opin. Food Sci. 23, 173–182.

Panozzo, A., Manzocco, L., Lippe, G., Nicoli, M.C., 2016. Effect of pulsed light on structure and immunoreactivity of gluten. Food Chem. 194, 366–372.

Parniakov, O., Roselló-Soto, E., Barba, F.J., Grimi, N., Lebovka, N., Vorobiev, E., 2015. New approaches for the effective valorization of papaya seeds: extraction of proteins, phenolic compounds, carbohydrates, and isothiocyanates assisted by pulsed electric energy. Food Res. Int. 77, 711–717.

Peng, D., Jin, W., Tang, C., Lu, Y., Wang, W., Li, J., Li, B., 2018. Foaming and surface properties of gliadin nanoparticles: influence of pH and heating temperature. Food Hydrocoll. 77, 107–116.

Phongthai, S., D'Amico, S., Schoenlechner, R., Homthawornchoo, W., Rawdkuen, S., 2018. Fractionation and antioxidant properties of rice bran protein hydrolysates stimulated by in vitro gastrointestinal digestion. Food Chem. 240, 156–164.

Phongthai, S., Lim, S.T., Rawdkuen, S., 2016. Optimization of microwave-assisted extraction of rice bran protein and its hydrolysates properties. J. Cereal Sci. 70, 146–154.

Pickardt, C., Eisner, P., Kammerer, D.R., Carle, R., 2015. Pilot plant preparation of light-coloured protein isolates from de-oiled sunflower (*Helianthus annuus* L.) press cake by mild-acidic protein extraction and polyphenol adsorption. Food Hydrocoll. 44, 208–219.

Pojić, M., Mišan, A., Tiwari, B., 2018. Eco-innovative technologies for extraction of proteins for human consumption from renewable protein sources of plant origin. Trends Food Sci. Technol. 75, 93–104.

Preece, K.E., Hooshyar, N., Krijgsman, A., Fryer, P.J., Zuidam, N.J., 2017. Intensified soy protein extraction by ultrasound. Chem. Eng. Process. - Process Intensif. 113, 94–101.

Qiu, C., Zhao, M., Decker, E.A., McClements, D.J., 2015. Influence of protein type on oxidation and digestibility of fish oil-in-water emulsions: Gliadin, caseinate, and whey protein. Food Chem. 175, 249–257.

Quaisie, J., Ma, H., Golly, M.K., Tuly, J.A., Amaglo, N.K., Jiaqi, Z., 2021. Effect of ultrasound-microwave irradiation hybrid technique on extraction, physicochemical, antioxidative, and structural properties of stearic acid-rich *Allanblackia parviflora* seed oil. Chem. Pap. 75 (9), 4527–4541.

Queirós, R.P., Saraiva, J.A., da Silva, J.A.L., 2018. Tailoring structure and technological properties of plant proteins using high hydrostatic pressure. Crit. Rev. Food Sci. Nutr. 58 (9), 1538–1556.

Rahman, M.M., Lamsal, B.P., 2021. Ultrasound-assisted extraction and modification of plant-based proteins: Impact on physicochemical, functional, and nutritional properties. Comprehens. Rev. Food Sci. Food Saf. 20 (2), 1457–1480.

Rahman, M.M., Dutta, S., Lamsal, B.P., 2021. High-power sonication-assisted extraction of soy protein from defatted soy meals: Influence of important process parameters. J. Food Process Eng. 44 (7), 1–11.

Rai, A.K., Kumari, R., Sanjukta, S., Sahoo, D., 2016. Production of bioactive protein hydrolysate using the yeasts isolated from soft chhurpi. Bioresour. Technol. 219, 239–245.

Resendiz-vazquez, J.A., Ulloa, J.A., Urías-silvas, J.E., Bautista-rosales, P.U., Ramírez-ramírez, J.C., 2017. Effect of high-intensity ultrasound on the technofunctional properties and structure of jackfruit (*Artocarpus heterophyllus*) seed protein isolate. Ultrason. – Sonochem. 37, 436–444.

Richter, C.K., Skulas-Ray, A.C., Champagne, C.M., Kris-Etherton, P.M., 2015. Plant protein and animal proteins: do they differentially affect cardiovascular disease risk? Adv. Nutr. 6, 712–728.

Rodrigues, I.M., Coelho, J.F.J., Carvalho, M.G.V.S., 2012. Isolation and valorisation of vegetable proteins from oilseed plants: methods, limitations and potential. J. Food Eng. 109 (3), 337–346.

Roselló-Soto, E., Barba, F.J., Parniakov, O., Galanakis, C.M., Lebovka, N., Grimi, N., Vorobiev, E., 2015. High voltage electrical discharges, pulsed electric field, and ultrasound assisted extraction of protein and phenolic compounds from olive kernel. Food Bioprocess Technol. 8 (4), 885–894.

Sari, Y.W., Bruins, M.E., Sanders, J.P.M., 2013. Enzyme assisted protein extraction from rapeseed, soybean, and microalgae meals. Ind. Crops Prod. 43 (1), 78–83.

Schreuders, F.K.G., Dekkers, B.L., Bodnár, I., Erni, P., Boom, R.M., van der Goot, A.J., 2019. Comparing structuring potential of pea and soy protein with gluten for meat analogue preparation. J. Food Eng. 261, 32–39.

Sedaghat Doost, A., Nikbakht Nasrabadi, M., Kassozi, V., Dewettinck, K., Stevens, C.V., Van der Meeren, P., 2019. Pickering stabilization of thymol through green emulsification using soluble fraction of almond gum – whey protein isolate nano-complexes. Food Hydrocoll. 88, 218–227.

Sedaghat Doost, A., Nikbakht Nasrabadi, M., Sadžak, A., Van der Meeren, P., 2020. Production of food nanomaterials by specialized equipment. In: Jafari, Seid Mahdi (Ed.), Handbook of Food Nanotechnology. Academic Press (Elsevier), 125 London Wall, London EC2Y 5AS, United Kingdom, pp. 161–204.

Semba, R.D., 2016. The rise and fall of protein malnutrition in global health. Ann. Nutr. Metab. 69 (2), 79–88.

Sereewatthanawut, I., Prapintip, S., Watchiraruji, K., Goto, M., Sasaki, M., Shotipruk, A., 2008. Extraction of protein and amino acids from deoiled rice bran by subcritical water hydrolysis. Bioresour. Technol. 99 (3), 555–561.

Setyaningsih, W., Duros, E., Palma, M., Barroso, C.G., 2016. Optimization of the ultrasound-assisted extraction of melatonin from red rice (*Oryza sativa*) grains through a response surface methodology. Appl. Acoust. 103, 129–135.

Shafiqur, R., Islam, M., Rahman, M.M., Uddin, M.B., Mazumder, M.A.R., 2018. Isolation of protein from defatted peanut meal and characterize their nutritional profile. Chem. Res. J. 3 (2), 187–196.

Shishir, M.R.I., Gowd, V., Suo, H., Wang, M., Wang, Q., Chen, F., Cheng, K.W., 2021. Advances in smart delivery of food bioactive compounds using stimuli-responsive carriers: Responsive mechanism, contemporary challenges, and prospects. Comprehens. Rev. Food Sci. Food Saf. 20 (6), 5449–5488.

Shishir, M.R.I., Xie, L., Sun, C., Zheng, X., Chen, W., 2018. Advances in micro and nano-encapsulation of bioactive compounds using biopolymer and lipid-based transporters. Trends Food Sci. Technol. 78, 34–60.

Stankiewicz, A., 2006. Energy matters: Alternative sources and forms of energy for intensification of chemical and biochemical processes. Chem. Eng. Res. Des. 84 (7 A), 511–521.

Sun, H., Li, C., Ni, Y., Yao, L., Jiang, H., Ren, X., Fu, Y., Zhao, C., 2019. Ultrasonic/microwave-assisted extraction of polysaccharides from *Camptotheca acuminata* fruits and its antitumor activity. Carbohydr. Polym. 206, 557–564.

Sun, X., Ohanenye, I.C., Ahmed, T., Udenigwe, C.C., 2020. Microwave treatment increased protein digestibility of pigeon pea (*Cajanus cajan*) flour: elucidation of underlying mechanisms. Food Chem. 329, 127196.

Tahmasian, A., Broadbent, J.A., Juhász, A., Nye-Wood, M., Le, T.T., Bose, U., Colgrave, M.L., 2022. Evaluation of protein extraction methods for in-depth proteome analysis of narrow-leafed lupin (*Lupinus angustifolius*) seeds. Food Chem. 367, 130722.

Tang, C.H., Wang, X.Y., Yang, X.Q., Li, L., 2009. Formation of soluble aggregates from insoluble commercial soy protein isolate by means of ultrasonic treatment and their gelling properties. J. Food Eng. 92 (4), 432–437.

Tang, S., Hettiarachchy, N.S., Eswaranandam, S., Crandall, P., 2003. Protein extraction from heat-stabilized defatted rice bran: II. The role of amylase, celluclast, and viscozyme. J. Food Sci. 68 (2), 471–475.

Tian, R., Feng, J., Huang, G., Tian, B., Zhang, Y., Jiang, L., Sui, X., 2020. Ultrasound driven conformational and physicochemical changes of soy protein hydrolysates. Ultrason. Sonochem. 68, 105202.

Tirgar, M., Silcock, P., Carne, A., Birch, E.J., 2017. Effect of extraction method on functional properties of flaxseed protein concentrates. Food Chem. 215, 417–424.

Vaidya, A.M., Saxena, D.C., Somani, V.B., 2017. Response surface methodology for optimization of protein isolate preparation from custard apple seed flour (*Annona squmosa*). Int. J. Curr. Microbiol. Appl. Sci. 6 (8), 3079–3089.

Vanga, S.K., Wang, J., Orsat, V., Raghavan, V., 2020. Effect of pulsed ultrasound, a green food processing technique, on the secondary structure and in-vitro digestibility of almond milk protein. Food Res. Int. 137, 109523.

Vanga, S.K., Wang, J., Raghavan, V., 2020. Effect of ultrasound and microwave processing on the structure, in-vitro digestibility and trypsin inhibitor activity of soymilk proteins. LWT - Food Sci. and Technol. 131, 109708.

Varghese, T., Pare, A., 2019. Effect of microwave assisted extraction on yield and protein characteristics of soymilk. J. Food Eng. 262, 92–99.

Vilcacundo, R., Miralles, B., Carrillo, W., Hernández-Ledesma, B., 2018. In vitro chemopreventive properties of peptides released from quinoa (*Chenopodium quinoa* Willd.) protein under simulated gastrointestinal digestion. Food Res. Int. 105, 403–411.

Vinatoru, M., Mason, T.J., Calinescu, I., 2017. Ultrasonically assisted extraction (UAE) and microwave assisted extraction (MAE) of functional compounds from plant materials. TrAC - Trends Anal. Chem. 97, 159–178.

Wang, C., Li, D., Xu, F., Hao, T., Zhang, M., 2014. Comparison of two methods for the extraction of fractionated rice bran protein. J. Chem. 1–10.

Wang, Y., Li, R., Jiang, Z.T., Tan, J., Tang, S.H., Li, T.T., Liang, L.L., He, H.J., Liu, Y.M., Li, J.T., Zhang, X.C., 2018. Green and solvent-free simultaneous ultrasonic-microwave assisted extraction of essential oil from white and black peppers. Ind. Crops Prod. 114, 164–172.

Wen, L., Álvarez, C., Zhang, Z., Poojary, M.M., Lund, M.N., Sun, D.W., Tiwari, B.K., 2020. Optimisation and characterisation of protein extraction from coffee silverskin assisted by ultrasound or microwave techniques. Biomass Convers. Biorefin. 11, 1575–1585.

Xiang, S., Zou, H., Liu, Y., Ruan, R., 2020. Effects of microwave heating on the protein structure, digestion properties and Maillard products of gluten. J. Food Sci. Technol. 57 (6), 2139–2149.

Xu, N., Sun, Y.H., Guo, X.L., Liu, C., Mao, Q., Hou, J.M., 2018. Optimization of ultrasonic-microwave synergistic extraction of polysaccharides from *Morchella conica*. J. Food Process. Preserv. 42 (2), e13423.

Yalçin, E., Çelik, S., 2007. Solubility properties of barley flour, protein isolates and hydrolysates. Food Chem. 104 (4), 1641–1647.

Yang, S., Dai, L., Sun, C., Gao, Y., 2018. Characterization of curcumin loaded gliadin-lecithin composite nanoparticles fabricated by antisolvent precipitation in different blending sequences. Food Hydrocoll. 85 (17), 185–194.

Yang, X., Li, Y., Li, S., Oladejo, A.O., Wang, Y., Huang, S., Zhou, C., Ye, X., Ma, H., Duan, Y., 2018. Effects of ultrasound-assisted α-amylase degradation treatment with multiple modes on the extraction of rice protein. Ultrason. Sonochem. 40, 890–899.

Yanjun, S., Jianhang, C., Shuwen, Z., Hongjuan, L., Jing, L., Lu, L., Uluko, H., Yanling, S., Wenming, C., Wupeng, G., Jiaping, L., 2014. Effect of power ultrasound pre-treatment on the physical and functional properties of reconstituted milk protein concentrate. J. Food Eng. 124, 11–18.

Yao, J., Zhou, Y., Chen, X., Ma, F., Li, P., Chen, C., 2018. Effect of sodium alginate with three molecular weight forms on the water holding capacity of chicken breast myosin gel. Food Chem. 239, 1134–1142.

Yong-ting, Y., Lei, F., Wan-li, S., Yan, G., Qiong, X., 2018. Study on the characterizations and optimization of walnut protein extraction by ultrasonic-microwave synergistic assisted technique. Food Res. Dev. 39 (22), 35–40.

Yu-lin, X., Yu-xiu, X., Li-peng, W., Yu-rong, J., 2015. Study on ultrasonic/microwave synergistic extraction of protein from soybean dreg. Sci. Technol. Food Indust. 36 (14), 266–270.

Yuliarti, O., Kiat Kovis, T.J., Yi, N.J., 2021. Structuring the meat analogue by using plant-based derived composites. J. Food Eng. 288, 110138.

Zhang, H., Claver, I.P., Zhu, K.X., Zhou, H., 2011. The effect of ultrasound on the functional properties of wheat gluten. Molecules 16 (5), 4231–4240.

Zhang, L., Wang, L.J., Jiang, W., Qian, J.Y, 2017. Effect of pulsed electric field on functional and structural properties of canola protein by pretreating seeds to elevate oil yield. LWT - Food Sci. and Technol. 84, 73–81.

Zhang, L., Pan, Z., Shen, K., Cai, X., Zheng, B., Miao, S., 2018. Influence of ultrasound-assisted alkali treatment on the structural properties and functionalities of rice protein. J. Cereal Sci. 79, 204–209.

Zhang, Q.T., Tu, Z.C., Xiao, H., Wang, H., Huang, X.Q., Liu, G.X., Liu, C.M., Shi, Y., Fan, L.L., Lin, D.R., 2014. Influence of ultrasonic treatment on the structure and emulsifying properties of peanut protein isolate. Food Bioprod. Process. 92 (1), 30–37.

Zhang, S., Huang, W., Feizollahi, E., Roopesh, M.S., Chen, L., 2021. Improvement of pea protein gelation at reduced temperature by atmospheric cold plasma and the gelling mechanism study. Innov. Food Sci. Emerg. Technol. 67, 102567.

Zhang, Y., Wang, B., Zhang, W., Xu, W., Hu, Z., 2019. Effects and mechanism of dilute acid soaking with ultrasound pretreatment on rice bran protein extraction. J. Cereal Sci. 87, 318–324.

Zhang, Y., Cui, L., Che, X., Zhang, H., Shi, N., Li, C., Chen, Y., Kong, W., 2015. Zein-based films and their usage for controlled delivery: origin, classes and current landscape. J. Control Release 206 (2699), 206–219.

Zheng, Y., Li, Z., Zhang, C., Zheng, B., Tian, Y., 2020. Effects of microwave-vacuum pre-treatment with different power levels on the structural and emulsifying properties of lotus seed protein isolates. Food Chem. 311, 125932.

Zhou, H., Wang, C., Ye, J., Chen, H., Tao, R., Cao, F., 2016. Effects of high hydrostatic pressure treatment on structural, allergenicity, and functional properties of proteins from ginkgo seeds. Innovative Food Sci. Emerg. Technol. 34, 187–195.

Zhu, K.X., Sun, X.H., Zhou, H.M., 2009. Optimization of ultrasound-assisted extraction of defatted wheat germ proteins by reverse micelles. J. Cereal Sci. 50 (2), 266–271.

Zhu, Z., Zhu, W., Yi, J., Liu, N., Cao, Y., Lu, J., Decker, E.A., McClements, D.J., 2018. Effects of sonication on the physicochemical and functional properties of walnut protein isolate. Food Res. Int. 106, 853–861.

Zink, J., Wyrobnik, T., Prinz, T., Schmid, M., 2016. Physical, chemical and biochemical modifications of protein-based films and coatings: An extensive review. Int. J. Mol. Sci. 17 (9), 1376.

CHAPTER TWELVE

Sonication microwave synergistic extraction of bioactive compounds from plants, animals and others agro-industrial waste sources

Muhammad Sajid Arshad[a], Waseem Khalid[a], Muhammad Zubair Khalid[a], Zahra Maqbool[a], Anwar Ali[b], Safura Kousar[a], Muhammad Faizan Afzal[a], Shanza Mukhtar[c], Muhammad Abdul Rahim[a], Afifa Aziz[a] and Mohammad Javed Ansari[d]

[a]Department of Food Science, Government College University, Faisalabad, Pakistan
[b]Department of Epidemiology and Health Statistics, Xiangya School of Public Health, Central South University, Changsha, China
[c]Department of Nutrition and Dietetics, The University of Faisalabad, Faisalabad, Pakistan
[d]Department of Botany, Hindu College Moradabad (Mahatma Jyotiba Phule Rohilkhand University, Bareilly, UP), India

12.1 Introduction

Food waste is composed of substances that are valuable for human use but these are considered deteriorated, polluted, and discarded. Food waste is becoming more of a problem that influences all aspects of waste management, including developing long-term solutions with the participation of all in the food supply chain counting industry, agriculture, merchants, collection to disposal, and consumers (Girotto et al., 2015). Globally, with the passage of every day, industrial wastes are rising in volume that having a significant impact on ecosystem health and ultimately the human population. As a result, every agro-industrial sector has a pressing requirement for the safe consumption of agricultural products through waste recycling. Agro-industries are generating a large amount of waste that may be used in a variety of industries, including energy generation and composting (Yusuf, 2017). In recent years, increased energy use, and economic pressure on the industry have facilitated the need for resource sustainability, and optimal production. Agro-industrial wastes are a viable option for meeting the needs of the present generation without harming coming generations, hence there

Ultrasound and Microwave for Food Processing: Synergism for Preservation and Extraction.
DOI: https://doi.org/10.1016/B978-0-323-95991-9.00008-4

is a pressing need for a greater understanding of the complexity of agro-industrial waste usage and recycling procedures (Yusuf, 2017).

Agro-industrial waste involves a tremendous amount of bioactive compound loss; thus it is a mandatory task to recover these bioactive compounds through industries to maintain the economy as well as to meet human demand (Campos et al., 2020). Compounds with bioactivity are being recovered from a variety of industrial food wastes counting leaves, barriers, peels, seeds, culls, wood, rinds, pulp, pits, press cakes, malts, marc, hops, algae, husks, hulls, wasted grain, shrimp carapaces, crab, and other fish byproducts (Kallel et al., 2014). Generally, complex polysaccharides, proteins, and polyphenolic components are the major portion of agro-industrial wastes (Yadav et al., 2009; Marinari et al., 2000). The current food waste recovery trend is based on a rising understanding of the advantages derived from potentially marketable components included in food wastes and coproducts (Laufenberg et al., 2003). The waste is not only biodegradable but also high in nutritious contents such as carbohydrates, proteins, fibers, minerals, vitamins, etc. (Panesar et al., 2015).

Bioactive compounds loss in agro-industrial waste are recovered by various extraction technologies by the use of organic solvent that is the need for sustainability of alternatives. The widely used green extraction technologies for bioactive extraction are supercritical fluid extraction (SFE), ultrasound-assisted extractions (UAE), and microwave-assisted extraction (Panzella et al., 2020). Microwave-assisted extraction (MAE) has been recognized as a promising and effective method for extracting bioactive chemicals from food and industrial wastes. MAE is a technique used to extract soluble compounds from a variety of materials into a fluid using microwave radiation (Dagui and Qing, 2009). Microwave-assisted extraction is better for the environment since it uses less energy and can utilize a wider choice of nontoxic solvents. Microwave radiation has been demonstrated to improve bioactive chemical extraction by speeding up the extraction process (Baiano et al., 2014). In synergistic extractions, the role of the solvent has been widely explored by many authors. When two extracting agents are present, the impact of extraction from an aqueous to an organic phase of certain metal species is larger than the total of the extraction by each of the extractants alone (Mazille et al., 2009).

12.2 Types of agro-industrial waste and management

The demand for agro-industrial products is increasing extensively with growing population. Food products that are manufactured in large quantities

Table 12.1 Bioactive compounds in agro-industrial wastes.

Types of agro-industrial waste	By-products	Bioactive compounds	References
Plants	Green tea residue	Arabinose, xylose and catechins	(Tsubaki et al., 2008)
	Wheat bran	Cellulose	(Grover et al 2012)
	Rice bran	Vitamin E and anthocyanidins	(Huang et al., 2016)
	Corn steep liquor	Ferulic acid and sinapic acid	(Vecino et al., 2016)
	Coconut oil cake and groundnut oil cake	Rifamycin	(Vastrad, 2012)
	Potato peel	Xanthan	(Vidhyalakshmi et al., 2012)
Animals	Sheep wool	Antioxidant and peptides	(Fakhfakh et al, 2013)
	Chicken feathers	Antioxidants	(Fontoura et al., 2014)
	Ergosterol	Dairy waste	(Németh et al., 2015)
Nonfood	Jute retting units	Lignocellulose residues	(Ansi et al., 2017)

to fulfill this immense demand for agro-industrial products which is the major cause of agro-industrial waste. The agro-industry especially the food industry imparts a pivotal role in waste production which causes climate change, environmental pollution and adverse economic loss (FAO, 2019). The highly growing food processing industries in the European Union produced a high quantity of inedible food waste approximately 30 Mt each year. The reduction in agro-industrial waste by about 30–50% can improve the available food supply to 15% (Wang et al., 2016). Agro-industrial waste is recognized in several means along with the development of various valuable products and by-products like bio-products, bio-fuel, bio-degradable plastics as well as organic fertilizers in the form of manure. The agro-based industries are categorized into three different sections including animal-based food, plant-based, as well as nonfood-based agro-industrial wastes (Table 12.1).

12.2.1 Plant-based agro-industrial wastes

Plant-based wastes are classified into process residues and field residues. The waste material obtained from the field resides involved leaves, seedpods, stalks, and stems that are the remains of postharvest crops in the field. A small amount of field waste is used for various purposes like on-farm used as animal feed as well as for horticulture. On the other side, the processed residues are produced from the processing and modification of the raw materials. The plant-based agro wastes are commonly produced from the sugar industries, fermentation-based industries, by-products of grains from flour mills as well as food manufacturing and processing industries.

Plant-based industrial wastes are rich in various nutrients including fibers, lignin, cellulose, pectin, sugar, proteins, vitamins, polyphenols, and lipids (Barcelos et al., 2020). Lignocellulose residues are considered as a plentiful renewable resource on the earth and a large amount of agro-industrial wastes enriched with lignocellulose are not appropriately valorized and port for degradation (Ramírez et al, 2019). Lignocellulose residues are mainly converted into biofuels, biofertilizers, valuable products as well as used to produce mushrooms, enzymes, and other important products (Ansi, 2017; Kumla et al., 2020). A large amount of agro-industrial waste was obtained from the vegetables, fruits, tubers, roots, and cereals. The frequently cultivated fruit crops include banana, apples, grapes, citrus fruits, and mangoes as well as in vegetable onions, tomatoes, cabbages carrots, potatoes, and cucumber are widely cultivated. Moreover, the most important isolated bioactive compounds involve fatty acids, tannins, flavanols, essential minerals, anthocyanins, and pigments (Ben-Othman et al., 2020).

The residue acquired from the processing of cereals, pulses, and grains is the auspicious source of nutrients as well as bioactive compounds. The enormous number of phytochemicals used in cosmetics, food and medicine are mainly obtained from shells, husks, barks, shells, leaves, and roots of several vegetables, crops, and crops (Shirahigue and Antonini, 2020; Usmani et al., 2020). Flavonoids, tannins, anthocyanins, and alkaloids are polyphenols. And phenolic compounds present in plants and wood smoke-like hydroquinone, gallic acid and cresol are used as food preservatives (Quinto et al., 2019). The polyphenols including oleuropein, cyanidin chloride, and apigenin extracted from olive mill water waste are employed to describe its effectiveness in cystic fibrosis cells by preventing DNA complexes (Lampronti et al., 2013). NF kappa B activity is accountable for inflammation in rheumatoid arthritis, osteoporosis, cancer, and cystic fibrosis. It is viable to draw out phenolic compounds using modern technologies like ultrasound as well as microwave-supported extraction (Panzella, 2020). Biostimulants are another category of compounds obtained from wastes such as sewage sludge, chitin, chitosan, organic waste streams. The waste produced from the brewery industry involves organic waste such as hops and spent malt which is useful in soil improvement as well as animal feed (Thomas and Rahman, 2006).

12.2.2 Animal-based agro-industrial wastes

Meat, as well as meat products, are excellent sources of bioactive compounds as well as nutrients that are not extensively taken from plant-based foods like fruits and vegetables, therefore, a major portion of the human diet consists of these products (Pogorzelska-Nowicka et al., 2018). The meat and poultry industries play a pivotal role to meet the increasing demand as a result of population growth. Besides that, these industries produced a huge quantity of waste which has high concentration of valuable substances like, phosphorus, grease, and nitrogenous compounds. In poultry industries, animal blood is a by-product but it is rich in protein as well as iron (Wan et al., 2002). Blood is a by-product due to its numerous properties including the source of bioactive compounds, used in blood cake, blood pudding, blood sausages in Asia and Europe as well as used as food stabilizer, emulsifier, color additive, and clarifier (Silva and Silvestre, 2003). The immense amount of organic remains obtained from the poultry, fish, and meat processing areas are amassing, which are rich in protein caused the public annoyance through profane smell.

Dairy industries usually encompass the processing of raw milk into different dairy products such as condensed butter, milk, consumer milk, ice cream, yogurt, dried milk, and cheese by applying various techniques including pasteurization, chilling, and drying. On the other hand, the waste produced by these industries causes severe environmental problems. This water-based waste contains a large number of valuable nutrients like organic milk products, dissolved sugars, fats, proteins as well as mineral (Watkins and Nash, 2010). Quinoline is mostly attained from plants and a small amount of quinoline alkaloids including kokusagine, quinidine, Quinine as well as dictamnine are obtained from fungi, insects, algae, and other terrestrial animals. Quinoline present in natural compounds has a variety of pharmacology, and biological effects. It is mostly used in making dyes, hydroxyquinoline sulfate, a solvent for resins synthesis and also possesses antifungal, antibacterial, anti-inflammatory, antimalarial, anthelmintic activities. The bioactive protein and peptides are mainly acquired from the animals-based agro wastes Products. The fish and meat like tendons, bones, skin, internal organs, parts of the gastrointestinal tract, and blood are the source of derived products. The dairy industry is also considered one of the best origins of bioactive peptides and proteins especially procured from cheese production (Corrêa et al., 2014).

12.2.3 Nonfood agro-industrial wastes

As food processing industries like dairy, meat, marine, brewery, fruit, and vegetables as well as grain processing industries are involved to produce a large amount of wastes as by product. Similarly, the other industries take account in the agro-industrial waste that can be useable for the welfare of human. The nonfood based agro-industries produced biodegradable waste effluents. Except tannery and textile industries all effluents are nontoxic. Most of these wastes are pristine imparts threat to the human, environment, and animal health.

12.3 Bioactive compounds in agro-industrial wastes

Bioactive compounds are those compounds that are high in food and found in minute amounts in waste foods. To see their effects on people's health, these compounds have been investigated thoroughly. Numerous bioactive chemicals have been discovered that differ greatly in functional and structural properties (Teodoro, 2019). Plant-derived bioactive chemicals are natural chemical substances found in minute concentrations in plants (Uwineza and Waśkiewicz, 2020). These compounds have the capability to interact with one or more living tissue components by providing a wide range of possible consequences. Bioactive chemicals are naturally occurring essential and nonessential molecules that are part of the food chain and have been found to affect human health (González, 2020). Vitamins and minerals are considered essential as they regulate a specific biochemical process in the body, whereas phenolics, phytosterols, carotenoids, flavonoids, glucosinolates, essential oils, alkaloids, and saponins are considered nonessential because they maintain cellular health and increase the effect of other essential nutrients (Rodríguez García & Raghavan, 2021).

Out of a total waste of 1.3 billion tons, vegetables and fruits account for 0.5 billion tons of food waste each year (FAO, 2019). Agriculture is one of the world's most productive industrial sectors that produces the most waste. Food losses during postharvest might be as high as 50%. This waste is a source of contamination for the water, air, and soil (Leyva-López et al., 2020). This remaining material is not utilized and disposed of properly, that causes major environmental issues. The agro-industrial waste consists of a diverse range of items, including stems, husks, roots, and seeds. Because these residues have high mineral, carbohydrate, and protein content, therefore they could be exploited to develop new human and animal feeds. There is a lot of concern in the development and recovery of natural-source chemicals with a lot of biological or bioactive potential. These chemicals have applications in a variety of fields, including food, cosmetics, and pharmaceuticals. It also aids in the maintenance of good health and the prevention of degenerative diseases. Nowadays, incorporating bioactive substances that promote the above-mentioned benefits in daily life that is very important activity (Londoño-Hernandez et al., 2020). Seeds can be used as a source of macronutrients or as a raw material for extracting vegetable oils as agro-industrial waste because they contain a large number of bioactive components (Varzakas et al., 2016).

The constantly expanding farming industry remains a crucial component of the global economy, while being mostly overtaken by technological progress. Year after year, the number of poultry, cattle, and ovine livestock units grows (Girotto and Cossu, 2017). As a cost-effective solution to dispose of waste from one of today's most polluting industries, the utilization of animal wastes as bioactive compounds has been a lot of research in recent years. The key byproducts of slaughterhouses and the meat business, collagen, and blood, have been tested the most in this regard. Despite the fact that peptides with various bioactivities, such as angiotensin-converting enzyme (ACE) inhibitory, antimicrobial, and opioid have been identified. Some studies suggest that ACE-inhibitory sequences are the most common. In invertebrates, collagen is the most abundant protein and it is readily available and inexpensive in the meat industry. It is the primary component of skin, cartilage, bones, and hide. However, it has a low nutritional value due to a lack of necessary amino acids (Pogorzelska-Nowicka et al., 2018).

Polysaccharides, pigments, sterols, antioxidants, proteins, and polyunsaturated fatty acids are abundant in the marine which is a rich natural resource of biologically active compounds like polyunsaturated fatty acids (PUFAs). It can also be derived from a wide range of sources, including sponges, microorganisms, and marine plants. Marine bioactive components and other marine-derived nutrients have great promise of being functional food ingredients, as they have beneficial health benefits in addition to physiological effects (anti-inflammatory and anticancer activity) (Varzakas et al., 2016).

Extraction is achieved by methods such as supercritical extraction, mechanical expelling, solvent extraction, Ultrasound-assisted extraction, and microwave extraction. These are key technology during bioactive components recovery (Kumar et al., 2021). For extraction of bioactive compounds

UAE is one of the potential technologies using mixing effects, vibration, and acoustic cavitation in liquid medium. To extract useful chemicals from plant materials effectively, frequency runs from 20 kHz to 100 kHz. Another approach that can be used in conjunction with traditional solvent extraction is MAE. This method outperforms traditional extraction methods because of the shorter extraction time, the use of fewer solvents, and the high extraction rate. Microwaves have an electromagnetic field that ranges from 300 MHz to 300 GHz (Pattnaik et al., 2021). Antiaging effects, prevention and treatment of cancer, control and prevention of metabolic diseases, protection against cardiovascular diseases, and protection against neurodegenerative diseases are all benefits of direct supplementation of plants containing bioactive compounds, and their frequent consumption (Ortega and Campos, 2019).

Bioactive chemicals are a great source of molecules for manufacturing of functional foods, food additives, and nutraceuticals. Few bioactive chemicals possess anticancer effects, anti-inflammatory, and antioxidant, as well as the ability to influence metabolic processes. Fruits having high carotenoids and polyphenols are demonstrated to have antioxidant properties and reduce the incidence of some cancers. The by-products and biomolecules retrieved can be employed to the production of functional foods, as well as medical and pharmaceutical treatments. Bioactive phytochemicals isolated from agricultural by-products, such as sterols, terpenes, carotenes, tocopherols, and polyphenols, possess substantial antioxidant activity (Kumar et al., 2017).

12.4 Extraction of bioactive compound from agro-industrials waste

Extraction is the method of separating the desired substance from mixed substances. The mixture is exposed to a solvent that is soluble in the target material but insoluble in the other compounds present. Plant-based component extraction has been around for a long time and is still popular. Thanks to technological advancements, extracts with higher yields are now possible. While Global yield is an important statistic since it evaluates extraction performance. It may be beneficial to concentrate on the extraction of certain chemicals (specific metabolites) to enrich the sample and avoid the extraction of undesirable ones such as primary metabolites (triacylglycerols and carbohydrates) (Lefebvre et al., 2021).

The usage of the bioactive components in different variety of commercial sectors which include the food, chemical, and pharmaceutical industries, those necessitates the development of the most appropriate and consistent technology for extracting these active components from plant materials. Numerous novel approaches have been developed in addition to traditional methods but no single method is now considered as the gold standard in the extraction of bioactive chemicals from plants. The efficiency of traditional and nonconventional extraction procedures is largely determined by crucial input parameters such as scientific expertise, the chemistry of bioactive compounds, and gaining a better understanding of the plant matrix (Azmir et al., 2013).

The uses of food extraction methods that are cutting-edge have been widely investigated. As a result of rising consumer expectations for greener options that do not involve harmful chemicals, as well as corporate concerns about nontoxic and sustainable extraction techniques (Wen et al., 2020). Bioactive components have been extracted from numerous natural sources using both traditional and recent extraction strategies (Mehmood et al., 2019). Traditional extraction procedures have been used in the past such as maceration, solid-liquid extraction (Soxhlet), and hydro-distillation with organic solvents (Giacometti et al. 2018). However, they require a lot of high-purity solvents that are expensive with low extraction selectivity and take a lot of time (Aires, 2017). Because of the limitations of traditional extraction, environmentally friendly, sustainable, and more efficient extraction technologies are needed.

In the extraction sector of the food industry, an appropriate extraction method is needed that balances production costs, process efficiency, product quality, and should be environmentally acceptable. Novel extraction strategies have been created to address some of the shortcomings of traditional extraction methods (Putnik et al., 2018). Novel extraction techniques, such as enzyme assisted extraction, are of interest to the food industry (Marathe et al., 2019), ultrasound-assisted (Bindes et al., 2019), pulsed electric field-assisted (Redondo et al., 2018), supercritical fluid (Pimentel-Moral et al., 2019), pressurized liquid (Pereira et al., 2019), and microwave-assisted (Rodsamran and Sothornvit, 2019).

Combining innovative extraction strategies has been shown in numerous studies to be effective for quick and efficient extraction (Tiwari, 2015; Aires, 2017; Chemat et al., 2017). UMAE, microwave-assisted enzymatic extraction (MAEE), and ultrasound-assisted enzymatic extraction (UAEE) are three methods that can be used together to increase the potential extraction ability (Lefebvre et al., 2021).

12.4.1 Sonication microwave synergistic extraction of bioactive compounds

The ultrasonic microwave-assisted extraction, which is the grouping MAE and of UAE, is the most promising extraction technology. In UMAE, the extraction process is accelerated and samples may be prepared quickly. UMAE is a short extraction time, cost effective method with high extraction yield, and high efficiency (Cravotto et al., 2008). By increasing mass transfer and disrupting cell walls, ultrasound can improve the solvent penetration into the sample. It increases the contact surface area and promotes the solvation of soluble compounds. At the same time, microwave heating boosts the solubility of the solute and mass transfer rate by rapidly heating the sample. It accelerates the removal of important compounds from the sample matrix by desorption and thereby improving extraction efficiency (Vinatoru et al., 2017).

Irradiation with microwaves and ultrasound can be done continuously or occasionally. The ultrasonic or microwave irradiation ceases during the extraction process, for a while in the intermittent method, and this on/off cycle is repeated numerous times. Microwave and ultrasound irradiation on/off cycles can be coordinated or independent. When, In the extraction system, there is no refluxing mechanism fitted, the excessive energy input can cause high evaporation of the solvent especially methanol and ethanol, for example, have a low boiling point. The method of intermittent use might assist in controlling the temperature or reducing the evaporation of the solvent (Mari et al., 2018).

Two machines microwave and power ultrasound can be used individually to finish the irradiation in succession with one irradiation in one machine followed by another irradiation in the other. Irradiations that occur at the same time would bring additional energy and heat to the extraction solvent in the same amount of time but to combine both irradiations in the same reactor/container there is a need for a specifically constructed apparatus. According to Cravotto and Cintas (2007), a UMAE system was created by inducing ultrasound into a reactor in a microwave oven using an ultrasonic probe. The UMAE system should be designed with a safety and technical instrument setup. For example, when a metallic probe or an ultrasonic probe composed of nonmetallic materials is put within the microwave chamber such as pyrex, polytetrafluoroethylene, polyether ether ketone, or quartz is used then vessel rupture or even an explosion could occur. The inherent structure of these materials' probes might be irrevocably destroyed, and these probes operate at maximum safe operating power of 90 W. However, they claimed that the probe material's power upper limit did not prevent UMAE from being used since the UMAE system's simultaneous irradiation used significantly less power than either microwave or ultrasonic irradiation alone. When deciding whether or not to utilize the metallic probe (Cravotto and Cintas, 2007).

The previous studies revealed that the best conditions for UMAE of lotus seed were microwave power 250.00 W, liquid-solid ratio 10.00 mL/g (L/S ratio), extraction time 325.00 s, and ultrasonic power 300.46 W. In which yields of total tetrasaccharides, oligosaccharides, and trisaccharides increased by 27.21 percent, 76.59 percent, and 17.47 percent as per the sequence. While having a comparison between microwave-assisted extraction, UAE, and hot water extraction, the extraction time was reduced by 1.16, 8.92, and 12.18 times, respectively MAE. These findings showed that UMAE alone has more efficiency for oligosaccharide extraction than microwave aided, ultrasonic, or hot water extraction (Lu et al., 2017). In comparison to UAE, Lianfu and Zelong (2008) investigated the extraction of lycopene from tomatoes utilizing UMAE. The proportion of lycopene output was 97.4 percent for UMAE and 89.4 percent for UAE. UMAE has a better efficiency because of the acoustic cavitation of ultrasound and the heating action of microwaves. UAE, solvent-free UMAE, and MAE, were used to extract the essential oils from black and white peppers (Piper nigrum L.). The superoxide radical scavenging activities and oil yield of the white pepper extracted by UMAE were much greater than those obtained by UAE and MAE. Results indicated that UMAE was a much more efficient extraction method than UAE and MAE (Wang et al., 2018). UAE, SE, heat reflux extraction (HRE), MAE, and UMAE extractions were explored and compared in extracting flavonoids from Spatholobus suberectus. The results revealed that HRE and UMAE had greater flavonoids extraction yields than the other extraction methods. However, HRE required 4 h of extraction time to reach the greatest extraction yields, whereas UMAE required just 450 s. The use and extraction of pectin from jackfruit (Artocarpus heterophyllus) peels could add value to jackfruit waste products. The ultrasonic-microwave-assisted extraction (UMAE) of pectin with citric acid was optimized using a response surface technique. The antioxidant and physicochemical characteristics of pectin produced through UMAE (UM-P), and conventional heating (CH-P) were examined. The best UMAE conditions were the solid-liquid ratio of 1:48 (w/v), 29 min extraction time, and 86°C extraction temperature. Pectin yield was 21.5 percent under these circumstances (Cheng et al., 2011) (Table 12.2).

Table 12.2 Extraction of bioactive compounds by sonication and microwave synergism.

Sources	Extraction yield	Conditions	Bioactive compound	Authors
Pomelo peels	UMAE give high yield of pectin as compared to MUAE, MAE and UAE	1. Sonication 27.52 min MW 643.44 W for 6.4 min	Pectin	(Liew et al., 2016)
Burdock leaves	UMAE more yield than HRE	1. MW 400 W	Phenolic compounds	(Lou et al., 2012)
Lotus seed	UMAE has a greater yield of oligosaccharides than traditional hot water, MAE and UAE	1. Extraction time 325.00 s 2. US 300.46 W 3. MW 250.00 W	Oligosaccharides	(Lu et al., 2017)
Jackfruit peel	UMAE is considered potential alternative conventional extraction	1. Temperature 86°C 2. Extraction time 29 min	Pectin	(Xu et al., 2018)
Spatholobus Suberectus	UMAE has high yield than UAE, MAE, SE, SE and HRE	1. Temperature 80°C 2. Extraction time 450 s 3. US power 50 W 4. MW power 300 W	Flavonoids	(Cheng et al., 2011)
A. nodosum	UMAE generated higher yields of compounds compared to UAE and MAE	1. MW 250, 600, 1000W 2. Sonication probe 500W	Fucose-sulphated polysaccharides and antioxidants	(Garcia-Vaquero et al., 2020)
Potato pulp	Extraction pectin yield was 22.86%	1. Temperature 93°C 2. extraction time 50 min 3. US power 50 W	Pectin	Yang, Mu, and Ma (2019)
Cabbage outer leaves	UAE + MAE higher contents of extractable bioactive compounds	1. MW 850 W 2. Time 2 min	Glucosinolates, sulforaphane, vitamin C, phenolics, and antioxidants	Pongmalai et al., 2015

12.5 Conclusion

It is concluded that synergism of different extraction technologies such as MAE, (UAE) may give high yield, better quality with shorter time period. With the increase in the consumption of agro-foods, the production of waste has been reported to be high that directly enhance the environmental pollution. The industries are trying to manage the waste by using different procedures. The agro-industrial waste contains different bioactive compounds that can extract by using novel technologies. Currently, novel technologies are being used to extract bioactive chemicals. These technologies may give high yield and efficiency. Nowadays, synergism of different technologies is being used to get clean, nontoxic and sustainable production.

References

Aires, A., 2017. Phenolics in foods: extraction, analysis and measurements. Phenolic Comp 61–88.

Ansi, L., 2017. Comparitive studies of the effect of different microorganisms on coir pith composting. Indian J. Life Sci. 6 (1), 55.

Azmir, J., Zaidul, I.S.M., Rahman, M.M., Sharif, K.M., Mohamed, A., Sahena, F., Omar, A.K.M., 2013. Techniques for extraction of bioactive compounds from plant materials: A review. J Food Eng 4, 426–436.

Baiano, A., Bevilacqua, L., Terracone, C., Contò, F., Del Nobile, M.A., 2014. Single and interactive effects of process variables on microwave-assisted and conventional extractions of antioxidants from vegetable solid wastes. J. Food Eng. 120, 135–145.

Barcelos, M.C., Ramos, C.L., Kuddus, M., Rodriguez-Couto, S., Srivastava, N., Ramteke, P.W., Molina, G., 2020. Enzymatic potential for the valorization of agro-industrial by-products. Biotechnol Lett 1, 29.

Ben-Othman, S., Jõudu, I., Bhat, R., 2020. Bioactives from agri-food wastes: present insights and future challenges. Molecules 253, 510.

Bindes, M.M.M., Reis, M.H.M., Cardoso, V.L., Boffito, D.C., 2019. Ultrasound-assisted extraction of bioactive compounds from green tea leaves and clarification with natural coagulants (chitosan and Moringa oleífera seeds). Ultrasonics sonochem. 51, 111–119.

Campos, D.A., Gómez-García, R., Vilas-Boas, A.A., Madureira, A.R., Pintado, M.M., 2020. Management of fruit industrial by-products—a case study on circular economy approach. Molecules 25 (2), 320.

Chemat, F., Rombaut, N., Sicaire, A.G., Meullemiestre, A., Fabiano-Tixier, A.S., Abert-Vian, M., 2017. Ultrasound assisted extraction of food and natural products. Mechanisms, techniques, combinations, protocols and applications. A review. Ultras Sonochem 34, 540–560.

Cheng, X.L., Wan, J.Y., Li, P., Qi, L.W., 2011. Ultrasonic/microwave assisted extraction and diagnostic ion filtering strategy by liquid chromatography–quadrupole time-of-flight mass spectrometry for rapid characterization of flavonoids in *Spatholobus suberectus*. J. Chromatogr. A 1218 (34), 5774–5786.

Corrêa, A.P.F., Daroit, D.J., Fontoura, R., Meira, S.M.M., Segalin, J., Brandelli, A., 2014. Hydrolysates of sheep cheese whey as a source of bioactive peptides with antioxidant and angiotensin-converting enzyme inhibitory activities. Peptides 61, 48–55.

Cravotto, G., Boffa, L., Mantegna, S., Perego, P., Avogadro, M., Cintas, P., 2008. Improved extraction of vegetable oils under high-intensity ultrasound and/or microwaves. Ultrason. Sonochem. 15 (5), 898–902.

Cravotto, G., Cintas, P., 2007. The combined use of microwaves and ultrasound: improved tools in process chemistry and organic synthesis. Chem.–Eur. J., 13 (7), 1902–1909.

Fakhfakh, N., Ktari, N., Siala, R., Nasri, M., 2013. Wool-waste valorization: production of protein hydrolysate with high antioxidative potential by fermentation with a new keratinolytic bacterium, *Bacillus pumilus* A1. J. Appl. Microbiol. 115 (2), 424–433.

FAO, 2019. The State of Food and Agriculture 2019. Moving Forward on Food Loss and Waste Reduction. FAO, Rome, pp. 2–13.

Fontoura, R., Daroit, D.J., Correa, A.P., Meira, S.M., Mosquera, M., Brandelli, A., 2014. Production of feather hydrolysates with antioxidant, angiotensin-I converting enzyme-and dipeptidyl peptidase-IV-inhibitory activities. New Biotechnol. 31 (5), 506–513.

Garcia-Vaquero, M., Ummat, V., Tiwari, B., Rajauria, G., 2020. Exploring ultrasound, microwave and ultrasound–microwave assisted extraction technologies to increase the extraction of bioactive compounds and antioxidants from brown macroalgae. Mar. Drugs 18 (3), 172.

Giacometti, J., Kovačević, D.B., Putnik, P., Gabrić, D., Bilušić, T., Krešić, G., Jambrak, A.R., 2018. Extraction of bioactive compounds and essential oils from mediterranean herbs by conventional and green innovative techniques: A review. Food Res Inter 113, 245–262.

Girotto, F., Alibardi, L., &Cossu, R., 2015. Food waste generation and industrial uses: a review. Waste Manage. (Oxford) 45, 32–41.

Girotto, F., Cossu, R., 2017. Animal waste: opportunities and challenges. Sust Agri Rev 22, 1–13.

González, S., 2020. Dietary bioactive compounds and human health and disease. Nutrients 2, 348.

Grover, S., Kathuria, R.S., Kaur, M., 2012. Energy values and technologies for non woody biomass: as a clean source of energy. IOSR J. Electr. Electron. Eng. 1 (2), 10–14.

Huang, Y.P., Lai, H.M., 2016. Bioactive compounds and antioxidative activity of colored rice bran. J. Food Drug Anal. 24 (3), 564–574.

Kallel, F., Driss, D., Chaari, F., Belghith, L., Bouaziz, F., Ghorbel, R., Chaabounia, S.E., 2014. Garlic (*Allium sativum* L.) husk waste as a potential source of phenolic compounds: influence of extracting solvents on its antimicrobial and antioxidant properties. Ind. Crop Prod. 62, 34–41.

Kumar, K., Srivastav, S., Sharanagat, V.S., 2021. Ultrasound assisted extraction (UAE) of bioactive compounds from fruit and vegetable processing by-products: a review. Ultrason. Sonochem. 70, 105325.

Kumar, K., Yadav, A.N., Kumar, V., Vyas, P., Dhaliwal, H.S., 2017. Food waste: a potential bioresource for extraction of nutraceuticals and bioactive compounds. Bioresour. Bioprocess. 4 (1), 1–14.

Kumla, L., Suwannarach, N., Sujarit, K., Penkhrue, W., Kakumyan, P., Jatuwong, K., Lumyong, S., 2020. Cultivation of mushrooms and their lignocellulolytic enzyme production through the utilization of agro-industrial waste. Molecules 25, 2811.

Lampronti, I., Borgatti, M., Vertuani, S., Manfredini, S., Gambari, R., 2013. Modulation of the expression of the proinflammatory IL-8 gene in cystic fibrosis cells by extracts deriving from olive mill waste water. Evid.-Based Complement. Altern. Med. 2013, 1–11.

Laufenberg, G., Kunz, B., Nystroem, M., 2003. Transformation of vegetable waste into value added products: (A) the upgrading concept; (B) practical implementation. Bioresour. Technol. 87, 167–198.

Lefebvre, T., Destandau, E., Lesellier, E., 2021. Selective extraction of bioactive compounds from plants using recent extraction techniques: A review. J. Chro A 1635, 461770.

Leyva-López, N., Lizárraga-Velázquez, C.E., Hernández, C., Sánchez-Gutiérrez, E.Y., 2020. Exploitation of agro-industrial waste as potential source of bioactive compounds for aquaculture. Foods, 9 (7), 843.

Lianfu, Z., Zelong, L., 2008. Optimization and comparison of ultrasound/microwave assisted extraction (UMAE) and ultrasonic assisted extraction (UAE) of lycopene from tomatoes. Ultrason. Sonochem. 15 (5), 731–737.

Liew, S.Q., Ngoh, G.C., Yusoff, R., Teoh, W.H., 2016. Sequential ultrasound-microwave assisted acid extraction (UMAE) of pectin from pomelo peels. Int. J. Biol. Macromol. 93, 426–435.

Londoño-Hernandez, L., Chávez-González, M.L., Ascacio-Valdés, J.A., Ruiz, H.A., Toro, C.R., Aguilar, C.N., 2020. Fermentative bioprocesses for detoxification of agri-food wastes for production of bioactive compounds. In: Food Microbiology and Biotechnology. Apple Academic Press, New York; Ortega and Campos, 2019, Mexico, pp. 287–318.

Lou, Z., Wang, H., Zhu, S., Chen, S., Zhang, M., Wang, Z., 2012. Ionic liquids based simultaneous ultrasonic and microwave assisted extraction of phenolic compounds from burdock leaves. Anal. Chim. Acta 716, 28–33.

Lu, X., Zheng, Z., Li, H., Cao, R., Zheng, Y., Yu, H., Zheng, B., 2017. Optimization of ultrasonic-microwave assisted extraction of oligosaccharides from lotus (Nelumbo nucifera Gaertn.) seedsInd. Crops Prod 107, 546557.

Marathe, S.J., Jadhav, S.B., Bankar, S.B., Dubey, K.K., Singhal, R.S., 2019. Improvements in the extraction of bioactive compounds by enzymes. Current Opin Food Sci 25, 62–72.

Marić, M., Grassino, A.N., Zhu, Z., Barba, F.J., Brnčić, M., Brnčić, S.R., 2018. An overview of the traditional and innovative approaches for pectin extraction from plant food wastes and by-products: ultrasound-, microwaves-, and enzyme-assisted extraction. Trends Food Sci. Technol. 76, 28–37.

Marinari, S., Masciandaro, G., Ceccanti, B., &Grego, S., 2000. Influence of organic and mineral fertilisers on soil biological and physical properties. Bioresour. Technol. 72 (1), 9–17.

Mazille, F., Schoettl, T., Pulgarin, C., 2009. Synergistic effect of TiO_2 and iron oxide supported on fluorocarbon films. Part 1: effect of preparation parameters on photocatalytic degradation of organic pollutant at neutral pH. Appl. Catal. B 89 (3-4), 635–644.

Mehmood, A., Ishaq, M., Zhao, L., Yaqoob, S., Safdar, B., Nadeem, M., Wang, C., 2019. Impact of ultrasound and conventional extraction techniques on bioactive compounds and biological activities of blue butterfly pea flower (Clitoria ternatea L.). Ultras Sonochem 51, 12–19.

Németh, Á.R.O.N., Kaleta, Z.O.L.T.Á.N., 2015. Complex utilization of dairy waste (whey) in biorefinery. WSEAS Trans. Environ. Dev. 11, 80–88.

Ortega, A.M.M., Campos, M.R.S., 2019. Bioactive compounds as therapeutic alternatives. In: Maira Rubi Segura Campos (Ed.), Bioactive Compounds. Woodhead Publishing, pp. 247–264.

Panesar, R., Kaur, S., Panesar, P.S, 2015. Production of microbial pigments utilizing agro-industrial waste: a review. Curr. Opin. Food Sci. 1, 70–76.

Panzella, L., Moccia, F., Nasti, R., Marzorati, S., Verotta, L., Napolitano, A., 2020. Bioactive phenolic compounds from agri-food wastes: an update on green and sustainable extraction methodologies. Front. Nutr. 7, 60.

Pattnaik, M., Pandey, P., Martin, G.J., Mishra, H.N., Ashokkumar, M., 2021. Innovative technologies for extraction and microencapsulation of bioactives from plant-based food waste and their applications in functional food development. Foods 10 (2), 279.

Pereira, D.T.V., Tarone, A.G., Cazarin, C.B.B., Barbero, G.F., Martinez, J., 2019. Pressurized liquid extraction of bioactive compounds from grape marc. J. Food Eng. 240, 105–113.

Pimentel-Moral, S., Borrás-Linares, I., Lozano-Sánchez, J., Arráez-Román, D., Martínez-Férez, A., Segura-Carretero, A., 2019. Supercritical CO_2 extraction of bioactive compounds from Hibiscus sabdariffa. J. Supercrit. Fluids. 147, 213–221.

Pogorzelska-Nowicka, E., Atanasov, A.G., Horbańczuk, J., Wierzbicka, A., 2018. Bioactive compounds in functional meat products. Molecules 23 (2), 307.
Pongmalai, P., Devahastin, S., Chiewchan, N., Soponronnarit, S., 2015. Enhancement of microwave-assisted extraction of bioactive compounds from cabbage outer leaves via the application of ultrasonic pretreatment. Sep. Purif. Technol. 144, 37–45.
Putnik, P., Lorenzo, J.M., Barba, F.J., Roohinejad, S., Režek Jambrak, A., Granato, D., Bursać Kovačević, D., 2018. Novel food processing and extraction technologies of high-added value compounds from plant materials. Foods 7, 106.
Qing, Y., Dagui, Z., 2009. Rapid analysis of the essential oil components of dried *Peril-lafrutescens* (L.) by magnetic nanoparticle-assisted microwave distillation and simultaneous headspace solid-phase microextraction followed by gas chromatography-mass spectrometry. Anal. Methods 1, 39.
Quinto, E.J., Caro, I., Villalobos-Delgado, L.H., Mateo, J., De-Mateo-Silleras, B., Redondo-Del-Río, M.P., 2019. Food safety through natural antimicrobials. Antibiotics 8 (4), 208.
Redondo, D., Venturini, M.E., Luengo, E., Raso, J., Arias, E., 2018. Pulsed electric fields as a green technology for the extraction of bioactive compounds from thinned peach by-products. Inno Food Sci Eme Tech 335–343.
Rodríguez García, S.L., Raghavan, V., 2021. Green extraction techniques from fruit and vegetable waste to obtain bioactive compounds: a review. Crit Rev Food Sci Nutr 1–21.
Rodsamran, P., Sothornvit, R., 2019. Extraction of phenolic compounds from lime peel waste using ultrasonic-assisted and microwave-assisted extractions. Food Biosci. 28, 66–73.
Shirahigue, L.D., Ceccato-Antonini, S.R., 2020. Agro-industrial wastes as sources of bioactive compounds for food and fermentation industries. Ciência Rural 50, 1–17.
Silva, V.D., Silvestre, M.P., 2003. Functional properties of bovine blood plasma intended for use as a functional ingredient in human food. LWT-Food Sci. Technol. 36 (7), 709–718.
Teodoro, A.J., 2019. Bioactive compounds of food: their role in the prevention and treatment of diseases. Oxid. Med. Cell. Long. Available at: https://doi.org/10.1155/2019/3765986.
Thomas, K.R., Rahman, P., 2006. Brewery wastes. Strategies for sustainability. A review. Aspects Appl. Biol. 80, 1–11.
Tiwari, B.K., 2015. Ultrasound: A clean, green extraction technology. TrAC Tren Anal Chem 71, 100–109.
Tsubaki, S., Iida, H., Sakamoto, M., Azuma, J.I., 2008. Microwave heating of tea residue yields polysaccharides, polyphenols, and plant biopolyester. J. Agric. Food Chem. 56 (23), 11293–11299.
Usmani, Z., Sharma, M., Sudheer, S., Gupta, V.K., Bhat, R., 2020. Engineered microbes for pigment production using waste biomass. Curr. Genomics 21 (2), 80–95.
Uwineza, P.A., Waśkiewicz, A., 2020. Recent advances in supercritical fluid extraction of natural bioactive compounds from natural plant materials. Molecules 25 (17), 3847.
Varzakas, T., Zakynthinos, G., Verpoort, F., 2016. Plant food residues as a source of nutraceuticals and functional foods. Foods 5 (4), 88.
Vastrad, B.M., Neelagund, S.E., 2012. Optimization of process parameters for rifamycin b production under solid state fermentation from Amycolatopsis mediterranean MTCC14. Int. J. Curr. Pharm. Res. 4 (2), 101–108.
Vecino, X., Barbosa-Pereira, L., Devesa-Rey, R., Cruz, J.M., Moldes, A.B., 2015. Optimization of liquid-liquid extraction of biosurfactants from corn steep liquor. Bioprocess Biosys Eng. 38, 1629–1637.
Vidhyalakshmi, R., Vallinachiyar, C., Radhika, R., 2012. Production of xanthan from agro-industrial waste. J. Adv. Sci. Res. 3 (2), 56–59.
Vinatoru, M., Mason, T.J., Calinescu, I., 2017. Ultrasonically assisted extraction (UAE) and microwave assisted extraction (MAE) of functional compounds from plant materials. TrAC Trends Anal. Chem. 97, 159–178.

Wan, Y., Ghosh, R., Cui, Z., 2002. High-resolution plasma protein fractionation using ultrafiltration. Desalination 144 (1-3), 301–306.

Wang, B., Dong, F., Chen, M., Zhu, J., Tan, J., Fu, X., … Chen, S., 2016. Advances in recycling and utilization of agricultural wastes in China: based on environmental risk, crucial pathways, influencing factors, policy mechanism. Proc. Environ. Sci. 31, 12–17.

Wang, Y., Li, R., Jiang, Z.T., Tan, J., Tang, S.H., Li, T.T., … Zhang, X.C., 2018. Green and solvent-free simultaneous ultrasonic-microwave assisted extraction of essential oil from white and black peppers. Ind. Crops Prod. 114, 164–172.

Watkins, M., Nash, D., 2010. Dairy factory wastewaters, their use on land and possible environmental impacts—a mini review. Open Agric. J. 4 (1), 1–9.

Wen, L., Zhang, Z., Sun, D.W., Sivagnanam, S.P., Tiwari, B.K., 2020. Combination of emerging technologies for the extraction of bioactive compounds. Crit Rev Food Sci Nutr 11, 1826–1841.

Xu, S.Y., Liu, J.P., Huang, X., Du, L.P., Shi, F.L., Dong, R., … Cheong, K.L., 2018. Ultrasonic-microwave assisted extraction, characterization and biological activity of pectin from jackfruit peel. LWT 90, 577–582.

Yadav, S., Yadav, P.K., Yadav, D., Yadav, K.D.S., 2009. Purification and characterization of pectin lyase produced by *Aspergillus terricola* and its application in retting of natural fibers. Appl. Biochem. Biotechnol. 159 (1), 270–283.

Yang, J.S., Mu, T.H., Ma, M.M., 2019. Optimization of ultrasound-microwave assisted acid extraction of pectin from potato pulp by response surface methodology and its characterization. Food Chem. 289, 351–359.

Yusuf, M., 2017. Agro-industrial waste materials and their recycled value-added applications. In: Handbook of Ecomaterials. Springer, pp. 1–11.

CHAPTER THIRTEEN

Synergistic effects of sonication and microwave on safety and quality of foods

Manju Nehra[a], Jogender Duhan[b], Mohsin Khan[c], Kawaljit Sandhu[d], Amanjyoti[a,b,c,d] and Mohammad Javed Ansari[e]

[a]Department of Food Science and Technology, Chaudhary Devilal University, Sirsa, Haryana, India
[b]Department of Biotechnology, Chaudhary Devilal University, Sirsa, Haryana, India
[c]Department of Biosciences, Jamia Milia University, New Delhi, India
[d]Department of Food Science and Technology, Maharaja Ranjit Singh University, Bathinda, Punjab, India
[e]Department of Botany, Hindu College Moradabad (Mahatma Jyotiba Phule Rohilkhand University Bareilly), India

13.1 Introduction

In high-intensity method of Sonication, broadly applicable in food sector like fermentation (Ojha M. et al., 2017), crystallization (Povey, 2017), cutting (Liu J. et al., 2015), food preservation (Demirdöven, 2008), and drying (Musielak M. et al., 2016). Sonication, according to Musielak M. et al. (2016), is used to modify the matrix through which waves pass and enhance the productivity of food processing. Sonication is done by ultrasonication path, mechanical homogenizer, ultrasonic probe (sonicator), and ultrasonic vibrator following the magnetic stirring of mixture in magnetic stirrer. Ultrasonic frequencies are reported to be 20–40 kHz, so the process is also known as ultrasonication. The heat and mass transfer method also improved sonication process in food industries.

A snack food, Meng-sha Huang et al. studied that the pumpkin chips were prepared by ultrasound and microwave-assisted vacuum frying. The results showed that applying ultrasound to the pumpkin chip can increase their quality. In this experiment, ultrasound, and microwave vacuum frying chips had significantly lower moisture contents than microwave vacuum frying, and vacuum frying. The moisture content differences can be attributed to both the higher temperature and the ultrasound energy of ultrasound, and microwave vacuum frying samples. In compared to MVF and VF samples, the combined use of ultrasound and microwave technologies also aids in fat reduction while improving food texture. The ultrasound method also favored the microstructure of pumpkin chips and preserved their shape.

Ultrasound and Microwave for Food Processing: Synergism for Preservation and Extraction.
DOI: https://doi.org/10.1016/B978-0-323-95991-9.00006-0

Color is the only color parameter that was not significantly influenced by ultrasound. The a* value was not affected by the ultrasound treatment, and thus it did not have any significant synergistic effect with microwaves on pumpkin chips. Moreover, the ultrasonic wavelength had no discernible effect on the quality metrics of pumpkin chips.

Various researchers showed that ultrasonication, microwaving, or a combination of both methods is safe ways to process sugarcane juice (SCJ). This process can prevent the loss of nutrients and reduce the number of microbes in SCJ.

Sonication of orange juice was tested by different quality parameters of Brix (sugar), pH, and TA (total acidity). No significant differences were found. However, browning and color values were observed to have changed significantly. Also, cloud value enhancement was found to have occurred. The Browning index, L*, a*, and b*, as well as the cloud value, were utilized to assess the efficacy of ultrasonication on TCD in this study. The results showed that these indexes followed zero-order kinetics or first-order kinetics. Ultrasonic intensity was found to be in correlation with the rate constants. This is evidence that ultrasound on TCD was either distinct or very distinct (Tiwari et al., 2008).

The power of the ultrasound and microwave ovens were set to different levels in order to determine which would be more effective in removing moisture and oil from the chicken nuggets. In terms of evaporation of water and oil absorption, the higher the power level, the better the outcomes. For good results at higher power levels, less frying time was required, resulting in reduced oil uptake.

The texture and color outcomes of the experiment showed that higher power levels produced better results. Furthermore, greater ultrasound and microwave powers revealed more damaged and porosity microstructures. The more powerful the ultrasound as well as microwave, the more probable the items would be of higher quality (Chitrakar et al., 2019).

13.2 Effect on texture

Sonication and microwave may affect the textural properties also. These properties are among the first parameters to be noticed and studied. Different studies are there claiming the effect of sonication and microwave on various foods in terms of textural changes. The textural changes of Chinese yam slices during frying were measured by hardness value. The toughness of a sample of brass was examined using ultrasonic and microwave radiation at different ranges. The result showed that increasing ultrasound power or increasing microwave power caused a significant decrease in hardness value. In a study by Su et al. (2016), the hardness value of fried potato chips was reported to decrease as the microwave power increased. The microwave power was found to have the greatest impact on moisture content, and therefore, the hardness value of the chips.

Similarly, Su et al. (2018) studied that hardness of purple sweet potato decreased in the microwave and ultrasound vacuum frying processes, with an increase in power. Microwave and ultrasound are both used for pretreatment of food instead of using high-temperature heating methods. Microwave heating can be done quickly with low energy consumption and is convenient to carry out. However, it only heats the surface layer of the product evenly, leaving the interior cold. Ultrasound increases the temperature in a short time but not evenly. In the study of texture of fried chips, water evaporation is an important factor. During frying, the water evaporates, and leaves behind pores. The oil occupies these pores when the chips are fried because it is a less viscous substance than water.

13.3 Effect on color

Likewise, the color of fried Chinese yams were recorded by a spectrophotometer. It was found that the lightness value (L*) increased while the total color difference value (ΔE) decreased with increasing ultrasound and microwave powers. When the samples were exposed to 600 W ultrasound power, a color difference of 1.62 was found between the yellow and white samples. When the samples were exposed to 3000 W microwave power, a color difference of 2.18 was found between the yellow and white samples. The redness value (a*) for both types of exposure showed almost no change. The effect of different ultrasound and microwave powers on frying time and color development in yam chips. The both ultrasound and microwave powers decreased the required frying time by about 10%. However, the heat treatment was reduced by 75% for samples processed with ultrasound and 50% for those treated with microwave. These results indicate that less heat exposure time. When ultrasound was introduced to electromagnetic vacuum frying of sliced potatoes, Su et al. (2018) found that color parameters improved. The authors explained that the inclusion of ultrasound in a microwave heating system led to improved coloration because the sonication process increased the temperature of cooking media and accelerated the diffusion of water molecules inside the potato tissue.

Fan et al. (2005) reported that the change of the color of fried products was majorly due caramelization, Maillard reaction, and chemical oxidation all these consider in nonenzymatic browning.

Ultrasound treatment affects the moisture, oil absorption capacity, texture, color, and structure of chips that prepared from sweet potato. The results found that there were significant differences in all variables except for texture after frying. It was determined that the optimal stage of sweet potato chips is dehydration at 20°C for 30 min before deep-frying. The use of ultrasound pretreatment to make fried sweet potato chips crisp can speed up the evaporation of water and reduce drying time. This method can also help to minimize oil absorption capacity and improve crispness in fried product of sweet potato chips. A technology was invented to solve the above problem. It was a process of using microwave power in high frequency electromagnetic wave to heat food, which could shorten the heating time, and reduce thermal damage to food color. The amount of time ultrasound pretreatment power is increased, the less drying time needed, and higher water evaporation rate. This results in a crisp structure and less oil absorption. The structure and integrality of the cell wall and samples were not disrupted by ultrasound, according to SEM micrographs of the different sample groups. Currently, the majority of fried potato chips are produced using conventional frying methods. Ultrasound pretreatment (proper ultrasound pretreatment level and time) is capable of producing healthier (lower oil content), crisper (lower breaking force), and more attractive (low ΔE value) fried sweet potato chips (Qiu L. et al., 2018).

In order to preserve food and drinks, they are usually processed. Pasteurization is one of the most common heat processing methods which widely use in preservation of beverages for preventing microbial growth.

The use of high temperatures is a disadvantage of this treatment, since it may cause unwanted nutritional, and biochemical changes, which may compromise the end product's quality attributes. Most of the people prefer fresh juices that are extracted from the whole fruit to prevent nutrient loss. Maintaining the quality of food would be beneficial to consumers and producers. In order for a product to be successful in the marketplace, it must satisfy customers' expectations. One key element in meeting customer's expectations is ensuring that products are safe, wholesome, and correctly labeled. Ultrasonic food processing has received a lot of interest as a feasible alternative to standard thermal procedures. Ultrasound can be used for the disaggregation, disruption, emulsification, and dispersion of food ingredients.

Ultrasound is not only used to process foods but also to remove micro-organisms and their spores from foods.

In recent years, the ability of ultrasonic processing to maintain juice quality while increasing bioactive ingredients and shelf life has gained interest from scientists. Ultrasound can help prevent spoilage micro-organisms by destroying them.

13.4 Improvement the quality of fruit juices

In food industry, quality control is a crucial factor in the success of any business. Quality control involves monitoring of various parameter like acidity, pH, electrical conductivity, Brix index (percentage), cloudiness, and color. The DPPH (2,2-diphenyl-1-picrylhydrazyl) is used to measure the total antioxidant activity which is initiated by phenolic, flavonoid, and flavonol compound of fruits. Ultrasound treatment of fruit juices at low temperatures, even as low as room temperature or below, has the potential to save food processors, and producers a lot of money. Ultrasound is a nonthermal processing technology that can be applied to juice or whole fruit products. It is based on cavitation mechanisms that enable disruption of cell membranes and fluids without the need.

Antioxidant capacity, DPPH scavenging activity, ascorbic acid, total phenolics, flavonoids, and flavonols all enhance significantly when grapefruit is sonicated for varied durations of time (30, 60, and 90 min) at a specific temperature (20°C) and frequency (28 kHz). The increase in these compounds may be due to the specific effects of ultrasound on the cellular structure of grapefruit (Aadil et al., 2013). Researchers have found that some differences in color values of grapefruit juice exist. After making changes to the manufacturing process, the quality of the juice was improved with no changes to acidity levels, pH, or Brix. No changes to pH, Brix, or acidity levels generally improved the quality of the juice.

M. Abid et al. investigated the effect of ultrasound on apple juice (2014). To kill enzymes (polyphenolase, peroxidase, and pectin methyl esterase), and bacteria, ultrasonic was mixed with thermal treatment (total plate count, yeast, and mold). The highest inactivation of enzymes was achieved using an ultrasonic probe at 60°C for 10 min. The use of thermosonication in sour orange juice significantly improved other juice parameters such as Brix and acidity. This process also deactivated the pectin methyl esterase. The thermosonication process used to inactivate pectin methyl esterase was effective at lower temperatures. The optimal times were 21.8 min at 75°C and 9.8 min at 63°C, respectively (Koshani et al., 2015). Sonication is a process that breaks up the bonds of different molecules in food (M. Abid, S. Jabbar et al., 2014). This process examines how it affected sugars, carotenoids, polyphenolic compounds, and minerals in apple juice. The sonicator is attached to a power source and its frequency, and amplitude are increased by turning dials. Samples are placed in the sonicator with different concentrations of sodium chloride and distilled water. The sonication of apple juice for 30 min and 60 min showed significant changes in the concentration of polyphenolic compounds, minerals such as Na, K, and Ca, viscosity, carotenoids, and sugars. In Japan, the researchers looked at how ultrasound affected apples. They found that a 30 min ultrasonic treatment did not significantly alter the color, value, antioxidant capacity, or scavenging activity of apple juices (M. Abid, 2013). However, the study observed a significant reduction of microbial population in the sonicated juice. This shows that ultrasonic treatment of juice increases both its quality and its safety. Pulsed ultrasound was already utilized in the food industry to improve product quality. Consumer panels preferred the taste of sonicated food samples during taste tests (M. Abid, S. Jabbar et al., 2015).

The effects of sonication and microwave treatments on the physical and chemical properties of phalsa drink after four months of storage were examined. Sonication and microwave treatments were combined to create a phalsa drink with valuable stability of physical and chemical parameters. The two treatments did not have a significant effect on the pH of the phalsa drink, but microwave treatment caused a lower increase in acidity. Sonication increased TSS more than microwave treatment. Sonication also significantly increased ascorbic acid content, while microwave treatment had no significant effect on it. The antioxidant activity of phalsa drink was drastically reduced by microwave cooking, but phenols, flavonoids, and reducing power steadily increased. The researchers compared two different methods of preparing the extract: sonication (using sound waves) and microwaves. They found that both methods reduced microbial load, but that microwaves were more effective at reducing phytochemical content. Commercial-scale sonication and microwave processing of phalsa beverages without chemical preservation are used to improve quality and shelf stability (Nadeem et al., 2021). Whether compared to control and single MW and US treatments, SCJ processed with microwave-ultrasonication (MW-US) used to have a longer shelf life. This technology may be useful in retaining bioactive compounds while reducing microorganisms. A new process has been developed called the microwave-ultrasonication (MW-US) treatment, which could be used for the processing of sugarcane juice at an industrial level. This method can also be used to produce a packaged drink that is safe for all consumers (Zia et al., 2019).

13.5 Enhancement of bioactive ingredients

With the recent developments in ultrasound technology, it is now possible to obtain a higher concentration of bioactive ingredients while at the same time obtaining significantly higher yields and greater product quality. Ultrasound is a nonthermal method of processing liquid foods. Unlike other methods of food processing, it does not involve high temperatures. Many nutrients are degraded while others are oxidized when processed at higher temperatures. Ultrasound has been shown to retain several bioactive chemicals, both immediately after processing, and after storage. This makes it a great option for those looking to keep their products fresh and effective. The grape juice was made by cleaning grapes using sonication and then incubating them for 6 h in the dark at 25 degrees. As a result, the amount of resveratrol in the finished juice increased by factors of 1.15, 1.24, and 1.53, respectively (Hasan et al., 2014). Jiang et al. (2015) evaluated the effect of different sterilization methods on different fruits like black mulberry juice to evaluate bioactive and antioxidant compounds. The thermal method heated the juice to 12.81 kJ/mol, the microwave method heated it to 13.07 kJ/mol, and the ultrasonic processing heated it to 15.99 kJ/mol.

After juicing, the samples were kept for 8 days at 5, 15, and 25°C. They discovered that the antioxidant activity (AOA) of heating method-treated juices dropped with the storage period, but that the antioxidant activity (AOA) of microwave and ultrasound-treated juices increased during the first two days before declining with storage period. Researchers examined the effectiveness of various fruit and vegetable juices treated with ultrasonic (15 Hz, 100 W at 15 min) and Ultraviolet (2 Uv-c lamps, 254 nm, 8 W) to those treated with traditional heat pasteurization (P. Khandpur and P.R. Gogate, 2016). The goal of this research was to see how different types of juice affected vitamin C levels, carbohydrate content, and antioxidant activity in a sample. After 3 weeks of refrigeration, orange juice treated with ultrasound retained the majority of its nutritious properties. When compared to thermally processed juices, the same method on spinach juice was successful in maintaining nutrients for up to 18 days, resulting in a longer storage period. In a recent study, researchers found that ultrasounds did not affect the juice's content of ascorbic acid (vitamin C), which is an antioxidant. The panel preferred the ultrasound treated juice over untreated because it was sweeter, although the dose of sugar was higher in the untreated samples (P. Khandpur and P.R. Gogate, 2015). Other studies have also indicated that juices retain their quality when they are processed ultrasonically. For example, carambola fruit juice (M.L. Rojas et al., 2016), pear juice (M. Saeeduddin, 2015), and five different fruit juices (S. Saikia, et al., 2015) are all ultrasonically processed without losing the qualities of their juices. Cactus pear juice (N. Del Socorro Cruz-Cansino et al., 2015) and Chokanan Mango fruit juice (V. Santhirasegaram, et al., 2015; V. Santhirasegaram et al., 2013), apple juice (L.C.A. Silva et al., 2015); apple and nectar juices (M. Šimunek et al., 2013; M. Šimunek et al., 2014); strawberry juice (B. Tomadoni et al., 2016), and juice of kiwifruit (B. Tomadoni et al., 2015).

All researchers mentioned above stated that antioxidants and microflora reduction preserves juice quality parameters. A negative finding is reported by one author (Y. Sun, L. Zhong et al., 2015). At the levels of sonication required to keep fresh apple juice from browning, total phenolics, flavonoids, chlorogenic acid, and antioxidant activity were all reduced. The presence of a specific enzyme, whose activity is determined by the amount of oxygen in the juice and whether or not phenolic compounds are present, is required for enzyme browning of fresh apple juice, according to the scientists. Despite an increase in PPO activity, the authors believe that the reduction in enzymatic browning they observed was due to degassing of the apple juice during sonication, which resulted in oxygen removal. During ultrasonic processing, the juice was exposed to ultrasonic waves that created cavitations in the liquid. This caused the generation of free radicals that were found to have a major effect in accelerating the degradation of phenolic and antioxidant compounds. In addition, these free radicals also degraded other nutrients in the juice. All of these factors combined to produce an undesirable final product.

Probiotic drinks made without dairy

Consumers are increasingly more interested in nondairy probiotic juices. People are seeking these items in greater numbers because they want to minimize or eliminate dairy from their diet. Product development that contains helpful bacteria has the potential to grow in the market. The use of ultrasound to promote the growth of probiotic bacteria in fruit juices, as well as a variety of drinks that have been infected and treated with sonication, has proven to be useful. The survivability of Lactobacillus casei in pineapple juice was investigated in a study. After ultrasonically treating the samples, it was discovered that microbial viability was highest at 31°C and pH 5.8. In a study by Weida et al. (M.G.M. Costa et al., 2013), the microbial stability of lactoferrin was determined for 42 days at 4 *f*C, and found that there was no significant difference between the sweetened and nonsweetened samples ($P > 0.05$).

The impact of high-intensity ultrasound on polyphenoloxidase activity, juice viscosity, and orange juice color was investigated. Juice was divided into two groups: one sonicated, and the other nonsonicated. The sonicator was set to a power of 376 Wcm^2 for 10 min. The activity of polyphenoloxidase was reduced by 20%. Lactobacillus casei was tested on cantaloupe melon juice as a nondairy probiotic juice substrate (Fonteles et al., 2013).

The ultrasonically treated juice was kept at 4°C for 42 days. Various criteria were assessed, including color, acidity, and viable cell counts. The caloric value was lowered due to sugar consumption by the remaining bacteria during storage, but all other parameters remained unchanged. Sonication, or sound energy, has been discovered to be beneficial in the development of nondairy probiotic drinks. If there is enough consumer interest, this technology could be adopted in commercial production. An example of consumer interest in nondairy probiotic juices is the increasing demand for products such as Yakult, a Japanese drink that has been consumed since the 1920s. In fact, its popularity has grown to the point where it is now available in 30 countries.

13.6 Dairy products and milk

In recent years, the use of ultrasound in the food processing industry was already investigated. Ultrasound's effect on the physical and functional features of dairy products, as well as its safety, shelf life, and potential cost savings to farmers haves been studied. Emulsification of milk products is important for the consistency and taste of the final product, and several studies have examined how ultrasound affects the overall homogenization and degree of emulsification (A. Shanmugam et al., 2014; J. Beevers et al., 2015). Other researchers employ ultrasonic to fractionate milk liquid and remove fat layers, resulting in milk with a lower fat content (T. Leong et al., 2014a, T. Leong et al., 2014b). They've also employed it to break down casein micelles and create milk with smaller particles. This milk shows the protein-digesting enzyme renneting ability that could increase efficiency in cheese production.

13.7 Emulsification degree

Milk product stability is frequently assessed in terms of milk homogeneity, fat globule sizes, and degree of emulsification. Flavor components are also monitored to determine how much the product has deteriorated. (P. Juliano et al., 2014). A number of researches have shown that ultrasonically treated milk can emulsify fat globules and reduce their size. It has been observed that ultrasonic treatment can produce a more stable end product (A. Shanmugam et al., 2014). When treated with ultrasound, the production of cavitation bubbles in liquids aids in the creation of more stable emulsions.

In the past, a number of different techniques have been used in order to reduce fat from milk. Recently, modern technology has allowed for the creation of a new method called microfiltration. In this process, milk is forced through membranes that are very small in size. This allows for the separation of fat particles which are too large to pass through these pores. Sonication time is directly related to droplet size reduction and emulsion stability (S. Abbas et al., 2013; T.S.H. Leong et al., 2009; S. Kentish et al., 2008). In the lab, stable nanoemulsions were created by combining sonication with high pressure homogenization. The nanoemulsion required less energy to emulsify and the fat globules were smaller in size than those in traditional emulsions (S. Calligaris et al., 2016). Two 7% flaxseed oil/milk emulsions were subjected to 20 kHz ultrasound for 1–8 min. The sonication procedure increased the emulsion's gelation properties, gel strength, and elastic nature, according to the findings.

It is thought that thermosonication has the potential to reduce fat globule size due to the existence of smaller oil globules stabilized by partially denatured whey proteins. A 22 mm probe was used to evaluate the effect of sonication on fat globules after thermosonication in another investigation (D. Bermdez-Aguirre et al., 2008).

The milk samples used in the experiment were smaller than untreated milk samples and had a more homogeneous particle structure. The fat globules in thermally treated milk had been disrupted, and the fat droplets became much smaller, but they became even more uniform after homogenization. Three dairy proteins (sodium caseinate, whey protein isolate (WPI), and milk protein isolate (MPI)) were subjected to ultrasound treatment. J. O'Sullivan et al. (2014) sonicated the samples for 2 min at a power intensity of 34 Wcm^2. According to a study conducted by C. N. Fisk et al., ultrasound treatment reduced the size of both milk and pea protein isolates to the nanoscale (150 nm) from micron sized aggregates under similar ultrasonic settings (20 lm). In the food industry, emulsions are used to mix ingredients that would otherwise not mix. We found that using ultrasonically treated milk proteins compared to untreated or unadsorbed ones produces smaller emulsion droplet sizes. This shows that interfacial proteins rearrange, allowing for the generation of smaller emulsion droplets with a thicker interfacial layer and stronger electrostatic repulsive force. Ultrasound was also used to lower the particle sizes of fat globules in goat's milk (Karlovic et al., 2014), and this was shown in skimmed milk as well (Vijayakumar et al., 2015).

Emulsification makes extensive use of ultrasound. The most common frequency is 20 kHz, but research is being done for the same procedure up to 2 MHz. Currently, the power ultrasound area continues to be the most effective for producing stable emulsions. Extreme temperatures and shear conditions, on the other hand, can generate hydroxyl and hydrogen radicals at this frequency. These radicals can cause lipid oxidation and milk product degradation (Makino et al., 1983). In this study, ultrasonic treatment was used to examine the stability of milk and the lipid oxidation that occurs during storage. Frequencies (20, 400, 1000, 1600, and 2000 kHz) and specific energies (up to 409 kJ/kg) were used to treat milk samples (P. Juliano et al., 2014). The goal of the experiment was to see how sonication and temperature affected the fat content of milk. The findings revealed that lowering the sonication time and temperature can reduce lipid oxidation in milk. To some extent, this is determined by the fat content of the processed milk. The most common degradation products were found at 1000 kHz and energies greater than 271 kJ/kg for raw milk and 102 kJ/kg for pasteurized skim milk in the investigation. This indicates that this energy level is suitable for emulsification. This knowledge could be useful for developing larger scale up reactors for the dairy industry. These reactors may be used to maintain milk sensory quality. Ultrasound appears to be a viable alternative processing technique in food emulsion making. Stable emulsions could be prepared without the use of emulsification enhancers, resulting in more stable additive-free goods.

13.8 Fractionation

Ultrasound can be used to separate milk into cream and skim. In order to achieve this, a high-power ultrasound is used to create tiny bubbles in the milk and then these bubbles push the fat molecules up and out of the liquid. Fractionation is a crude oil processing technology that works on the principle of ultrasonic standing waves, which are created in a liquid when it is sonicated, resulting in the development of low-pressure nodes and high-pressure anti-nodes. Instead of the very disruptive force present at low frequencies, ultrasound is employed in fat separation/fractionation to induce fat mobility. Fat naturally migrates to high-pressure locations (anti nodes). This causes the fat to cluster together. Ultrasound increases this process, causing the fat to separate more quickly (T. Leong et al., 2014a, b). Ultrasound causes fat to coalesce in milk. Properties such as temperature, frequency and other factors affect the rate of this process. However, a constant ultrasonic field is required for full coalescence (P. Juliano et al., 2011). The work was then scaled up to 6 L of volume. Various frequencies and transducer configurations were used to enhance creaming. The most efficient results were obtained with 0.4 MHz, using single or double vertical transducers. (P. Juliano et al., 2013).

A process in which ultrasound was used to enhance gravity has been examined for its ability to separate milk (T. Leong et al., 2015). By sonicating milk for various amounts of time. After each stage, the fat content and size distribution of the fat droplets were determined. Following that, an aliquot was collected for particle size characterization after sonicating for five minutes using single and dual transducer configurations at 1 MHz and 2 MHz. Ultrasound-enhanced fractionation produced more globules in semi-skim milk and fat-enriched fractions than traditional procedures. In vitro studies have shown that higher frequency ultrasound can help remove fat cells when compared to lower frequency ultrasound. Ultrasound-assisted fractionation of lipoaspirate was performed on fat samples, and the effects on fat removal were compared between 1 MHz and 2 MHz. The results showed that 2 MHz removed 59±2% of the fat in the initial sample, compared to only 47±2% removed with 1 MHz after 3 ultrasound-assisted fractionation stages.

A technique that uses ultrasonic waves to separate fat from milk was developed. The process was scaled up by using ultrasonic waves at a higher frequency and temperature (T. Leong, 2016). A method to separate milk fat into two streams, a high-fat and a low-fat stream was developed. The higher frequencies of ultrasound were used to produce oxidative free radicals that separated the milk fat.

Temperature impacts on fat coalescence, the effect of cavitation and noncavitation on separation, and subsequent scale-up design factors are all part of fractionation investigations. Ultrasound has been found in studies to improve the efficiency of fractionation procedures.

13.9 Proteins in milk are affected

The casein micelles in milk are hypothesized to be destabilized by ultrasound, resulting in a better coagulation process. This can be used for various milk sources, such as reconstituted milk or goat's milk. Goat's milk is known to coagulate less effectively than cow's milk.

The coagulant produced in this study had smaller and more uniform particle sizes thanks to the application of ultrasonic at a 20 Hz and 800 watts. As a result, the final product is firmer and coagulation occurs more quickly.

The process of making reconstituted milk was sonicated at 30°C with frequencies of 20, 400, or 1600 kHz with a 286 kJ/kg energy input (Z. Liu et al., 2014a, b). This is thought to be the reason that reconstituted milk has a smaller aggregate size. A source of milk was then treated with rennet, which is an enzyme used to curdle milk. In terms of rennet gelation rate, curding rate, and curd hardness, the procedure was shown to have improved qualities. This technique of production could result in a product with different qualities than thermally treated milk, and a more effective manufacturing procedure. The coagulum strength, gel firmness, cohesiveness, and water retention capacity of milk, as well as the final storage modulus, were found to be improved by ultrasonic treatment. Some factors that led to the development of the supermarket industry in the US were stated. An increase in insoluble high molecular weight coaggregates generated as a result of protein denaturation during sonication was one of the causes. (P. Sfakianakis, C. Tzia, 2013; P. Sfakianakis et al., 2014; O. Gursoy et al., 2016). Scientists have learned that the frequency of 20 kHz is best for sonicating proteins. The study discovered that this frequency increased protein solubilization while delaying serum separation and creating minimal change in the end product's quality or sensory qualities.

The processes, Sonication and microwave, may positively affect shelf life and preservation of various food items if done correctly and carefully. The standardization and duration of application methods are two key factors to be kept in mind while adopting these methods.

13.10 Conclusion

Sonication and microwave may affect the physical, chemical and textural properties also. These properties are among the first parameters to be noticed and studied. Different studies are there claiming the effect of

Sonication and microwave on various foods in terms of behavioral changes. The processes, Sonication and microwave, may positively affect shelf life and preservation of various food items if done correctly and carefully. The standardization and duration of application methods are two key factors to be kept in mind while adopting these methods.

References

Aadil, R.M., et al., 2013. Effects of ultrasound treatments on quality of grapefruit juice. Food Chem 141, 3201–3206.

Abbas, R., et al., 2013. An overview of ultrasound-assisted foodgrade nanoemulsions. Food Eng. Rev. 5, 139–157.

Abid, M., et al., 2014. Sonication enhances polyphenolic compounds, sugars, carotenoids and mineral elements of apple juice. Ultrason. Sonochem. 21, 93–97.

Abid, M., et al., 2013. Effect of ultrasound on different quality parameters of apple juice. Ultrason. Sonochem. 20, 1182–1187.

Abid, M., et al., 2014. Thermosonication as a potential quality enhancement technique of apple juice. Ultrason. Sonochem. 21, 984–990.

Abid, M., et al., 2015. Qualitative assessment of sonicated apple juice during storage. J. Food Process. Preserv. 39, 1299–1308.

Beevers, J., et al., 2015. Comparative assessment of the effect of ultrasound treatment on protein functionality pre and post emulsification, Colloids Surf. Physicochem. Eng. Aspects 484, 89–98.

Bermúdez-Aguirre, D., et al., 2008. Microstructure of fat globules in whole milk after thermosonication treatment. J. Food Sci. 73, E325–E332.

Calligaris, S., et al., 2016. Nanoemulsion preparation by combining high pressure homogenization and high power ultrasound at low energy densities. Food Res. Int. 83, 25–30.

Chitrakar, A. et al. (2019). The synergistic effect of ultrasound and microwave on the physical, chemical, textural, and microstructural properties of vacuum fried Chinese yam (Dioscorea polystachya).

Costa, M.G.M., et al., 2013. High-intensity ultrasound processing of pineapple juice. Food Bioprocess Technol. 6, 997–1006.

Del Socorro Cruz-Cansino, M., et al., 2015. Shelf life, physicochemical, microbiological and antioxidant properties of purple cactus pear (Opuntia ficus indica) juice after thermoultrasound treatment. Ultrason. Sonochem. 27, 277–286.

Demirdöven, A., Baysal, T., 2008. The use of ultrasound and combined technologies in food preservation. Food Rev. Internat. 25 (1), 1–11.

Fan, L.P., et al., 2005. Vacuum frying of carrot chips. Drying Technol. 23 (3), 645–656.

Fonteles, T.V., et al., 2013. Rodrigues, Stability and quality parameters of probiotic Cantaloupe melon juice produced with sonicated juice. Food Bioprocess Technol 6, 2860.

Gursoy, M., et al., 2016. Effect of ultrasound power on physicochemical and rheological properties of yoghurt drink produced with thermosonicated milk. Emirates J. Food Agric. 28, 235–241.

Hasan, M.L., et al., 2014. Preparation of resveratrolenriched grape juice from ultrasonication treated grape fruits. Ultrason. Sonochem. 21, 729–734.

Huanga, M., et al., 2022. Synergistic effects of ultrasound and microwave on the pumpkin slices qualities during ultrasound-assisted-microwave vacuum frying. J. of Food Process Engineer.

Jiang, B., et al., 2015. Evaluation of bioactive compounds of black mulberry juice after thermal, microwave, ultrasonic processing, and storage at different temperatures. Food Sci. Technol. Int. 21, 392–399.

Juliano, P., et al., 2011. Enhanced creaming of milk fat globules in milk emulsions by the application of ultrasound and detection by means of optical methods. Ultrason. Sonochem. 18, 963–973.

Juliano, P., et al., 2013. Creaming nhancement in a liter scale ultrasonic reactor at selected transducer configurations and frequencies. Ultrason. Sonochem. 20, 52–62.

Juliano, P., et al., 2014. Lipid oxidation volatiles absent in milk after selected ultrasound processing. Ultrason. Sonochem. 21, 2165–2175.

Karlovic, S., et al., 2014. Reducing fat globules particle-size in goat milk: ultrasound and high hydrostatic pressures approach. Chem. Biochem. Eng. Q. 28, 499–507.

Kentish, S., et al., 2008. The use of ultrasonics for nanoemulsion preparation. Innov. Food Sci. Emerg. Technol. 9, 170–175.

Khandpur, P., Gogate, R., 2015. Understanding the effect of novel approaches based on ultrasound on sensory profile of orange juice. Ultrason. Sonochem. 27, 87–95.

Khandpur, P., Gogate, R., 2016. Evaluation of ultrasound-based sterilization approaches in terms of shelf life and quality parameters of fruit and vegetable juices. Ultrason. Sonochem. 29, 337–353.

Koshani, R., et al., 2015. Optimization of thermal and thermosonication treatments on pectin methyl esterase inactivation of sour orange juice (Citrus aurantium). J. Food Process. Preserv. 39, 567–573.

Leong, S., et al., 2014b. Design parameters for the separation of fat from natural whole milk in an ultrasonic litre-scale vessel. Ultrason. Sonochem. 21, 1289–1298.

Leong, T., et al., 2014a. Temperature effects on the ultrasonic separation of fat from natural whole milk. Ultrason. Sonochem. 21, 2092–2098.

Leong, T., et al., 2015. Continuous flow ultrasonic skimming of whole milk in a liter-scale vessel. Ind. Eng. Chem. Res. 54, 12671–12681.

Leong, T., et al., 2016. Ultrasonically enhanced fractionation of milk fat in a litre-scale prototype vessel. Ultrason. Sonochem. 28, 118–129.

Leong, T.S.H., et al., 2009. Minimizing oil droplet size using ultrasonic emulsification. Ultrason. Sonochem. 16, 721–727.

Liu, L., et al., 2015. Applications of Ultrasonic Cutting in Food Processing. J. Food Processing and Preserv. 39 (6), 1762–1769.

Liu, Y., et al., 2014a. Ultrasound effects on the assembly of casein micelles in reconstituted skim milk. J. Dairy Res. 81, 146–155.

Liu, Z., et al., 2014b. Ultrasound improves the renneting properties of milk. Ultrason. Sonochem. 21, 2131–2137.

Makino, J., et al., 1983. Chemical effects of ultrasound on aqueous solutions. Formation of hydroxyl radicals and hydrogen atoms. J. Phys. Chem. 87, 1369–1377.

Musielak, G., et al., 2016. Food drying enhancement by ultrasound – A review. Trends in Food Sci. & Technol. 56, 126–141.

Nadeem, M. et al. (2021). Sonication and microwave processing of phalsa drink: A synergistic approach., pp. 993–1007.

O'Sullivan, J., et al., 2014. The effect of ultrasound treatment on the structural, physical and emulsifying properties of dairy proteins. Food Hydrocoll. 42, 386–396.

Ojha, K.S., et al., 2017. Ultrasound technology for food fermentation applications. Ultrason Sonochem 34, 410–417.

Povey, M.J., 2017. Applications of ultrasonics in food science - novel control of fat crystallization and structuring. Curr. Opin. in Colloid & Interface Sci. 28, 1–6.

Qiu, L., et al., 2018. Effects of ultrasound pretreatments on the quality of fried sweet potato (Ipomea batatas) chips during microwave-assisted vacuum frying. J. of Food Process Engineer. 12879. doi:10.1111/jfpe.12879.

Rojas, M.L., et al., 2016. Peach juice processed by the ultrasound technology: changes in its microstructure improve its physical properties and stability. Food Res. Int. 82, 22–33.

Saeeduddin, M., et al., 2015. Quality assessment of pear juice under ultrasound and commercial pasteurization processing conditions. LWT Food Sci. Technol. 64 (1), 452–458.

Saikia, S., et al., 2015. A comparative study on the effect of conventional thermal pasteurization, microwave and ultrasound treatments on the antioxidant activity of five fruit juices. Food Sci. Technol. Int. 22, 288–301.

Santhirasegaram, V., et al., 2013. Effects of thermal treatment and sonication on quality attributes of Chokanan mango (Mangifera indica L.) juice. Ultrason. Sonochem. 20, 1276–1282.

Santhirasegaram, V., et al., 2015. Effects of thermal and non-thermal processing on phenolic compounds, antioxidant activity and sensory attributes of Chokanan mango (Mangifera indica L.) juice. Food Bioprocess Technol. 8, 2256–2267.

Sfakianakis, P., Tzia, C., 2013. Flavor and sensory characteristics of yogurt derived from milk treated by high intensity ultrasound. In: Ho, C.T., Mussinan, C., Shahidi, F., Contis, E.T. (Eds.), Nutrition, Functional and Sensory Properties of Foods. RSC publishing, pp. 92–97.

Sfakianakis, P., et al., 2014. Comparative study on high-intensity ultrasound and pressure milk homogenization: effect on the kinetics of yogurt fermentation process. Food Bioprocess Technol. 8, 548–557.

Shanmugam, A., Ashokkumar, M., 2014. Functional properties of ultrasonically generated flaxseed oil dairy emulsions. Ultrason. Sonochem. 21, 1649–1657.

Silva, L.C.A., et al., 2015. Inactivation of polyphenoloxidase and peroxidase in apple cubes and in apple juice subjected to high intensity Power ultrasound processing. J. Food Process. Preserv. 39, 2081–2087.

Šimunek, M., et al., 2013. Aroma profile and sensory properties of ultrasound-treated apple juice and nectar. Food Technol. Biotechnol. 51, 101–111.

Šimunek, M., et al., 2014. Rheological properties of ultrasound treated apple, cranberry and blueberry juice and nectar. J. Food Sci. Technol. 51, 3577–3593.

Su, Y., et al., 2016. Application of novel microwaveassisted vacuum frying to reduce the oil uptake and improve the quality of potato chips. LWT - Food Sci. and Technol. 73, 490–497.

Su, Y., et al., 2018. Ultrasonic microwave-assisted vacuum frying technique as a novel frying method for potato chips at low frying temperature. Food and Bioprod. Process. 108, 95–104. doi:10.1016/j.fbp.2018.02.001.

Sun, Y., et al., 2015. Sonication inhibited browning but decreased polyphenols contents and antioxidant activity of fresh apple (Malus pumila mill, cv. Red Fuji) juice. J. Food Sci. Technol. 52, 8336.

Tiwari, B.K., et al., 2008. Effects of sonication on the kinetics of orange juice quality parameters. J. Agric. Food Chem. 56, 2423–2428.

Tomadoni, B., et al., 2016. Optimization of ultrasound, vanillin and pomegranate extract treatment for shelf-stable unpasteurized strawberry juice. LWT Food Sci. Technol. 72, 475–484.

Tomadoni, B., et al., 2015. Individual and combined effects of pomegranate extract and ultrasonic treatments on kiwifruit juice quality parameters. J. Food Process Eng. doi:10.1111/JFPE.12339, Wiley Online.

Vijayakumar, S., et al., 2015. Quality characteristics and plasmin activity of thermosonicated skim milk and cream. J. Dairy Sci. 98, 6678–6691.

Zia, S., et al., 2019. Combined effect of microwave and ultrasonication treatments on the quality and stability of sugarcane juice during cold storage. Internat. J. of Food Sci. and Technol. 54, 2563–2569.

CHAPTER FOURTEEN

Synergistic effect of sonication and microwave on shelf-life of foods

Balarabe B. Ismail[a,b], Umar Garba[b] and Ahmad Gambo[b]
[a]Department of Food Science and Nutrition, Zhejiang Key Laboratory for Agro-Food Processing, Zhejiang University, Hangzhou, Zhejiang, China
[b]Department of Food Science and Technology, Faculty of Agriculture, Bayero University, Kano, Nigeria

14.1 Introduction

The definition of shelf-life has yet to be exhausted. However, one of the widely accepted definitions is "shelf-life is the definite length of time, following manufacture and packaging, during which the food product retains the required level of quality acceptable for consumption." Shelf-life is, therefore, a result of the relationship between food quality and its acceptability at consumption (Nicoli, 2012). Food spoilage is highly probable following harvest or slaughter and happens at different stages of food handling from farm to fork, leading to the deterioration of quality, and ending food shelf-life. That is, food spoilage is likely from raw materials acquisition to the level of consumption (Gould, 1996). Apart from spoilage, with rising foodborne-related diseases coupled with the complexity of the food chain, food safety and quality are made crucial at every stage, beginning from production, and processing to transportation and consumption. Foods to be presented to consumers are expected to be of optimal quality and safety, and at a reasonable price (D'Souza et al., 2017).

Food spoilage can be due to a number of causes which could be physical, chemical, or microbiological. Factors affecting microbial growth in foods are categorized as intrinsic (i.e., food) and extrinsic (i.e., environment) factors. While intrinsic factors include certain food properties, including nutrients, pH and acidity, water activity (aw), and redox potential, the main extrinsic factors include the temperature, and composition of where the food is handled. Besides, the growth and survival of micro-organisms in foods depend on these factors relative to the microbial needs (Augusto et al., 2018). Some of the preservation methods developed to prevent or inhibit microbial growth include using preservatives, curing, acidification, fermentation, or

Ultrasound and Microwave for Food Processing: Synergism for Preservation and Extraction.
DOI: https://doi.org/10.1016/B978-0-323-95991-9.00005-9

physical techniques, including chilling, freezing, drying, vacuum packing, modified atmosphere packing. Some other techniques are designed for targeted inactivation of both pathogenic and spoilage micro-organisms, including pasteurization, sterilization, and irradiation, or to restrict their access to foods via aseptic processing and packaging (Gould, 1996). Some of these techniques involve the use of heat, which has been shown to be detrimental to some important food components and quality, particularly sensorial, and nutritional properties (Augusto et al., 2018).

Currently, the consumer demand for high-quality foods that maintain their natural properties and nutrition is higher than ever, necessitating the development of innovative technologies. Among the various innovative techniques for food preservation, ultrasound and microwave technologies are increasingly utilized by the food industry due to increased production output within the shortest period besides extending shelf-life and improving organoleptic quality without negative environmental consequences (Li et al., 2021). Furthermore, the capacity to combine these two techniques has been shown to shorten processing time, improve food quality, cost-effectiveness, and overcome each method's shortcomings.

14.2 Food shelf-life as a food quality index

Food's shelf-life is the period in which foods maintain its acceptability from safety and organoleptic standpoints. Four major factors that form the basis of food shelf-life include formulations, processing, packaging, and storage conditions (Saha, 2020). Nicoli (2012) provided several illustrations to depict the concept of shelf-life and shelf-life strategy across new product and process development stages, taking into account the four shelf-life factors (Fig. 14.1). Shelf-life assessment is, therefore, an essential component in new product and process development that is geared towards maintaining acceptable food quality.

Though shelf-life assessment does not represent an exciting venture for food scientists, it has become critical for food businesses to preserve their emblem popularity within the marketplace (Calligaris et al., 2016). The retention of food quality is essential to ensure that foods maintain their nutritional value and do not constitute any health risk to consumers. In this regard, preservation techniques are the most appropriate ways to keep food quality and prevent deterioration (Sharif et al., 2017). The shelf-life of foods is limited by the continuous deterioration of quality and safety,

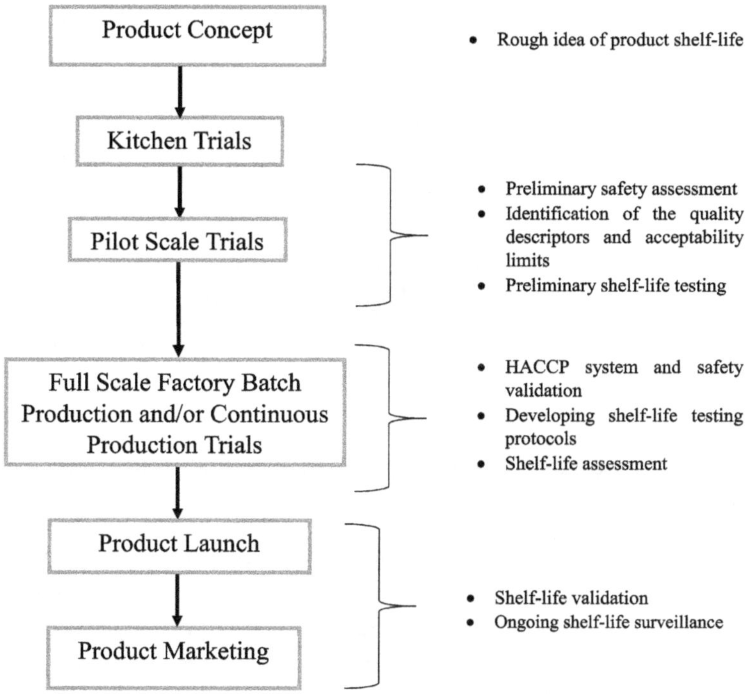

Figure 14.1 Shelf-life strategy at different product and process development stages (Nicoli, 2012).

and the situation becomes exacerbated due to unfavorable temperature and relative humidity conditions during distribution or storage (Corradini, 2018). Furthermore, one of the topmost criteria for rejecting foods, particularly those produced for intermediate or prolonged shelf-life, is the loss of nutritional value. This is because the degradation of labile vitamins and bioactive compounds may result in changes in their concentration which is usually found below the levels declared on the food label, and this may happen much earlier than the onset of sensory characteristics defects (Corradini, 2018). Any food, even as harvested, starts to evolve spoilage responses which often results in food wastage. Hence, food preservation presents a sustainable answer to countering food wastage issues. Our ancestors developed food preservation techniques during the earliest times, and these were possible as they searched for strategies to keep the foods fresh and healthy for a little longer than the period of harvest or slaughter. Concepts like solar drying, salting, and pasteurization have been developed based on climatic and seasonal factors.

Humans used food preservation to shape communities, stop killing animals, and catalyze a leisure mindset by retaining foods for additional time (Sridhar et al., 2021).

With the advent of novel analytical equipment for determining food quality attributes in situ, coupled with present-day reporting technology and befitting predictive tools and computational strategies, real-time and accurate predictions of food shelf-life can be made. In addition, with the potential for integration and optimization of these innovative methods, food safety related issues could be reduced to an acceptable level thanks to the improved rapid alert systems. This in turn, leads to lower food waste and decreased dependence on additives to ensure shelf-life (Corradini, 2018).

14.3 Conventional methods of ensuring shelf-life

Food preservation dates back to the evolution of humans. Hence strategies for food preservation developed hand in hand with human evolution. During this period, numerous techniques were developed, including drying, heating, cooling, salting or pickling, pasteurization or sterilization, edible coating, modified atmosphere packaging (MAP), controlled atmosphere packaging (CAP), and electromagnetic radiation. Rapid industrialization and the advent of new strategies gave rise to new processes particularly, thermal processing, which includes canning and freezing. These processes allowed for a better shelf-life extension and greater control of food pathogens (Sridhar et al., 2021). Besides, a number of conventional heat or thermal treatments are regarded as novel strategies for food preservation. For decades, this approach has been nicely examined in various food sectors: from bakery and dairy to fruits and vegetables (Sridhar et al., 2021). Despite the widespread application of conventional thermal treatments, the use of these treatments may also result in undesirable nutritional and sensorial quality modifications, eventual loss of essential nutrients, protein denaturation, change in color due to Maillard reaction, and off-flavor development (Gholamhosseinpour et al., 2020). Of the several food preservation techniques humankind has advanced over the centuries, canning holds a special status. This technology has stood the test of time, supported humanity in times of peace and wars, and is still substantially used in the food industry (Misra et al., 2019).

In MAP, the perishable product is maintained within the changed headspace that its composition is different from the regular air. In this approach, the crop respiration rate, packaging pouch specifications, storage temperature, and gas ratio are crucial. The most crucial advantages of MAP encompass decreased respiration rate, lowering the speed of the fruit ripening process, and retaining the quality of nutrients (Alden et al., 2019). The CAP storage on the other hand can increase seasonal and perishable produce's storage life as refrigeration alone is not enough. This technique can be used for a variety of fruits and vegetables. To efficiently keep fruit for lengthy intervals, the product's natural ripening process must be delayed without adversely affecting quality. First, fruits' temperature is reduced to the lowest degree feasible without causing damage due to freezing or low-temperature breakdown. Then the environment inside the storage room is modified to decrease oxygen and permit carbon dioxide to increase (Bishop, 1997).

14.4 Novel/innovative techniques to maintain the shelf-life of foods

Several new innovative technologies have been developed over the years. It is worth noting that food producers' acceptance of novel technologies depends mainly on the resources available, the level of awareness regarding operational parameters, and their effect on product quality and safety (Charoux et al., 2017).

The use of edible coatings is an emerging food preservation technique based on soluble formulations that can be applied on food surfaces in such a way that the edible film's thin layer forms an additional layer on the surface of the food, thereby preventing the moisture, oxygen, and solute migration into the food. Edible coatings have found applications in many food products, including meat and fish, where it is applied by dipping, spraying, casting, rolling, brushing, and foaming (Lacroix and Vu, 2014). Biopolymers are biodegradable and eco-friendly compounds, and they may be able to lessen the amount of chemicals required for the preservation and reduce domestic wastes. Various polysaccharides were used to make edible films and coatings, including starch, tapioca, corn, cellulose, and cellulose derivatives (Salehi, 2020). Chitosan, one of the abundant natural polymers, can also form edible coatings commonly used to preserve different foods (Ismail, Funtua, Garba & Sanusi, 2021; Zhang et al., 2021). Microwave heating has been determined to offer an exceedingly powerful means of low-temperature treatment, wherein selective dissipation because of the water inside the material removes overheating in dried zones while accelerating the mass transfer and influencing the molecular structure. Microwave heating hastens chemical reactions and increases enzyme functions considerably, which can be utilized in cell activation or inactivation, for instance, in accelerating yeast activity (Krasnikov, 1989).

The advent of active packaging enabled the substitution of traditional food processing methods (excessive thermal application, brining, acidification, dehydration, and preservation with additives) (Majid et al., 2018). Active packaging is a cutting-edge idea designed to meet the continuous consumer needs and market trends (Qian et al., 2021). This approach emphasizes the use of active materials, such as antioxidants and antimicrobials, and includes those that can adsorb/absorb CO2, O2, flavor/odors, moisture, and ethylene. Some of the active materials are positioned inside the packaging device to maintain a desirable packaging environment correctly, including O2, CO2, ethylene scavengers, and antimicrobial agents. Moreover, spices and plants are utilized to produce essential oils with antifungal and antimicrobial characteristics and could be used in active packaging without producing off-odor or other undesirable product attributes (Mohapatra et al., 2013). These types of packaging could be used in specialized food applications, including the packaging of perishable foods (Qian et al., 2021). Several advancements in active food packaging have been recorded, including developing informative and corrective response packaging, which can effectively be utilized to achieve farm to fork food safety and quality (Zhang et al., 2021).

Pulse electric field is a short-time food processing operation commonly employed for predrying treatment. The foods to be treated are usually placed between electrodes, and then a pulse is supplied with high voltage (50 kV/cm) briefly. The electric field causes the breakdown of foods' cell membrane matrix, resulting in enhanced nutritional value, increased safety, and shelf-life (Sridhar, Ponnuchamy, Kumar, & Kapoor, 2021).

Ultrasound is an emerging technique developed to lessen processing time and cost of processing, maximize quality, and ensure that food products' safety is not compromised (Ismail et al., 2020). The positive effects of ultrasound in food processing include improved mass transfer, food preservation, etc. The high-power ultrasound operates at a frequency range of 20 and 500 kHz and intensities of more than $1W \cdot cm^{-2}$. This type of ultrasound can be disruptive and could detrimentally affect foods' mechanical, physical, or chemical (biochemical) characteristics. Ultrasound has shown great promise in food industry preservation, extraction, and processing areas and is at the forefront of food preservation innovations (Ismail, Guo, et al., 2019; Singla & Sit, 2021). Another food processing approach involves the use of microwaves in food processing. Microwaves are utilized mainly in food industries for drying, pasteurization, sterilization, thawing, tempering, baking, etc.

High-pressure processing (HPP) is one of the nonthermal food preservation methods used for inactivating food pathogens and vegetative spoilage micro-organisms through pressure rather than heat. HPP employs extreme pressure (about four hundred –600 MPa) at refrigeration or mild temperatures (<45°C). The technique is useful in preserving different kinds of foods with minimal effect on taste, texture, appearance, or dietary value. Pressure processing may be used to process high or low moisture foods, either liquid, or solid. Although deadly to micro-organisms, pressure treatment does not break covalent bonds and has little effect on food chemistry (Muntean et al., 2016). HPP applies uniform or even notably-excessive pressures to the product from all sides in a couple of minutes. This approach causes no harm or distorts the food if the food product is not hollow or has a space within (Elamin et al., 2015). The need for HPP technology as a promising method of increasing food shelf-life and delivering healthy and quality food is developing very rapidly because the technology draws a large number of consumers (Elamin et al., 2015).

Cold plasma is a unique nonthermal technology that has established an enormous potential for use in food industries. Earlier, cold plasma became broadly explicitly used within the polymer and virtual industry for surface modification and functionalization of different polymers. However, recently, cold plasma applications have rapidly grown into the treatment of biomedical gadgets and organic materials, food inclusive (Pankaj and Keener, 2017). Cold plasma processing has been tested to affect the quality attributes of food products during processing and storage (Pankaj et al., 2018). A salient feature of this technique is establishing a plasma inside a sealed package. This technique is known as the "in-package plasma" technology and works in parallel with canning, and HPP because it can be applied to sealed packaged food. However, that does not make it a sterilization technology for foods, as it is mainly a surface treatment method. However, it shows the opportunity to raise the bar for surface decontamination techniques for shelf-life extension of minimally processed and raw foods (Misra et al., 2019).

14.5 Principles of ultrasound and microwave in food processing

Ultrasound is a mechanical wave similar to sound waves that are propagated in a medium with a frequency in the range of 20 kHz and 10 MHz above the human ear's audible range. When ultrasonic waves pass through a medium, longitudinal displacement of particles happens, resulting

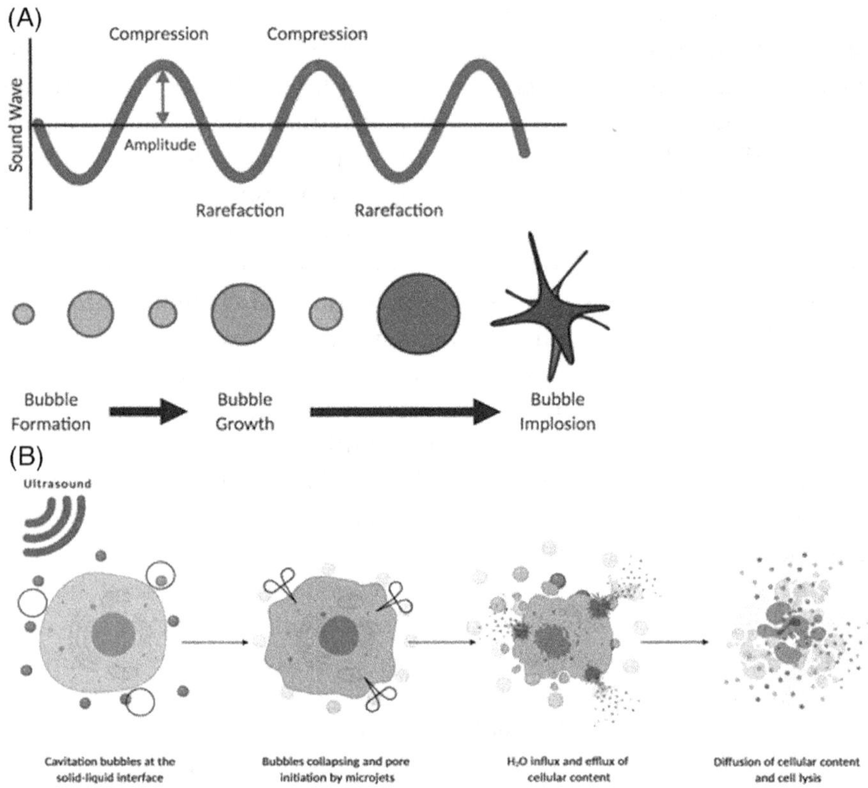

Figure 14.2 Mechanism of US cavitation and formation of bubbles (Perera and Alzahrani, 2021).

in a series of compression and rarefaction in the medium and consequent breakdown of cavitation bubbles (Fig. 14.2A) (Chemat and Khan, 2011; Perera and Alzahrani, 2021). The cavitation phenomenon yields high shear forces in the medium, and this is thought to be the main mechanism for the efficiency of ultrasound in food processing (Fig. 14.2B) (Perera and Alzahrani, 2021).

These mechanisms are responsible for the myriad application of ultrasound in emerging fields such as purification (Belwal et al., 2019; Wang et al., 2019; Yu et al., 2020) and processing (Balthazar et al., 2019; Monteiro et al., 2018; Ordóñez-Santos et al., 2017; Scudino et al., 2020; Wang et al., 2020) of bioactive compounds in foods. Moreover, ultrasound has been reported to improve the bioactive profile and biological potentials of different foods and bioresources and has been used to produce enriched, and purified bioactive compounds (Ismail et al., 2021; Ismail, Guo, et al., 2019;

Ismail, Pu, et al., 2019). Its potential application alone or in combination with other treatments in ensuring the safety of foods through microbial inactivation has also been reported severally (Fan, Hou, et al., 2020; Fan, Ismail, et al., 2020; Fan, Ismail, Hou, Muhammad, Ding, et al., 2019; Fan, Ismail, Hou, Muhammad, Zou, et al., 2019).

There are two widely known methods of applying ultrasound cavitation in a liquid medium: a sonotrode or ultrasonic water bath, both having various applications in food processing (Fig. 14.3a–b). In a typical ultrasound setup, low-frequency alternating current is transformed into high-frequency sound waves through the piezoelectric transducer attached to the bottom of the processing vessel (Fig. 14.3A), or in the form of a sonotrode that can be submerged into the cleaning liquid (Fig. 14.3B). The main function of the transducer is to produce high-intensity waves into the solution, which creates compression. Eventually, the compression leads to the breakdown of liquid into macroscopic cavities or bubbles (Bhargava et al., 2021).

Despite the numerous applications of ultrasound in food processing, some of its major drawbacks include over-dependence on sample matrix and the presence of a disperse phase which may reduce its effectiveness.

Microwaves are electromagnetic waves with frequencies ranging from 300 MHz to 300 GHz. Domestic microwave appliances generally operate at a frequency of 2.45 GHz, while frequencies in the range of 915 MHz and 2.45 GHz are typically found in industrial microwave systems (Chandrasekaran et al., 2013). The microwave heating mechanism is mainly due to materials' microwave energy absorption capacity and their ability to convert this energy into heat. Microwave heating of food materials depends on dipolar and ionic mechanisms (Chandrasekaran et al., 2013). Microwave technology has found applications in different food processing aspects, including drying, tempering, sterilization, baking, etc. Besides, some of the advantages of using microwaves in these processes include less energy consumption due to volumetric heating, reduced processing time and operational cost, and the preservation of food components (Ekezie et al., 2017). Despite its success as a widely regarded processing method with advantages that superseded its counterparts, microwave heating could result in nonuniform temperature distribution, which causes the under-processing of materials. Furthermore, uneven moisture distribution is possible during drying operations which pose significant challenges to process control and difficulty in achieving a complete process and the desirable moisture quality of the final products (Ekezie et al., 2017).

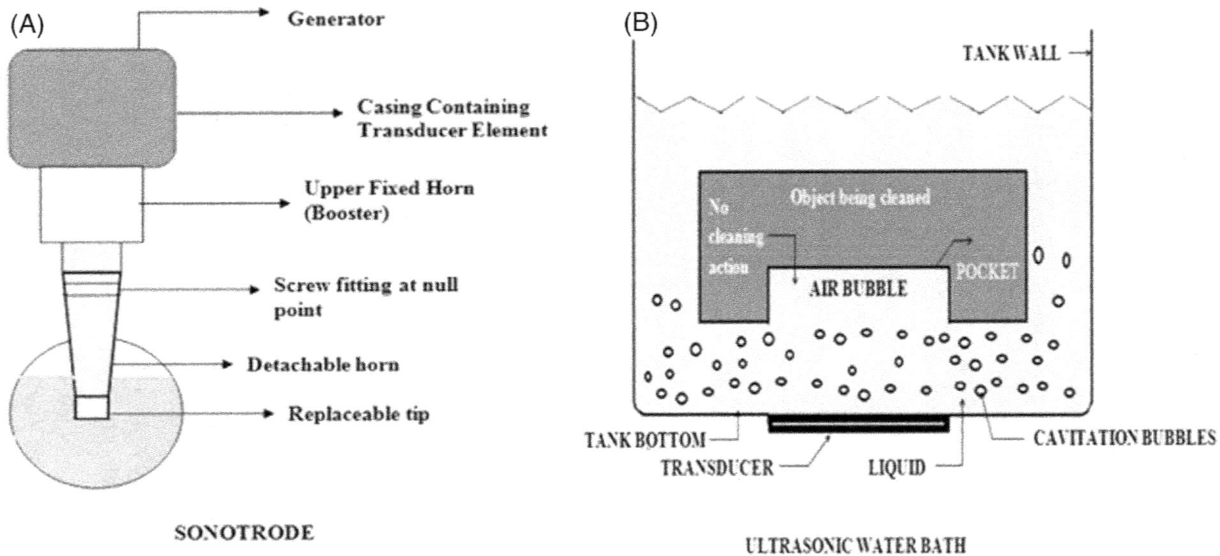

Figure 14.3 Typical methods for ultrasound application (Bhargava et al., 2021).

In summary, the understanding of the complexities and limitations of ultrasound and microwave technologies has led to the emerging area of food preservation where the two techniques could be combined to achieve the desirable shelf-life and quality of food products.

14.6 Applications of ultrasound and microwave synergistically in the processing of different food products

In contrast to the conventional pasteurization methods involving high temperatures for the inactivation of micro-organisms, many researchers have demonstrated the synergistic effect of combined ultrasonic and microwave treatment in achieving minimal or no changes in sensory properties (color, taste, and aroma), and nutrition value of foods (Hosseinzadeh Samani et al., 2016; Jyoti et al., 2020; Samani et al., 2015). Additionally, improved foods' physical, chemical, and biochemical properties due to applying the combined treatments had been extensively researched. As supported by many studies, ultrasound brings about the cavitation effect, which induces the physical, mechanical, and chemical/biochemical changes in food. This, together with the effect of microwave, could undoubtedly improve many processes such as extraction, drying, emulsification, freeze-drying, and inactivation of pathogenic bacteria in foods (Ambadgatti et al., 2020; Awad et al., 2012; Curi et al., 2019). Also, Bhargava et al. (2021) excellently and extensively reviewed the advances in the application of ultrasound technology in food processing and revealed its capacity to enhance process efficiency, yield better quality products, and improve shelf-life besides being a rapid process; thus, eliminating several laborious process steps. Various application of ultrasound combined with microwave for food processing and preservation has been presented in Table 14.1.

14.6.1 Synergism of ultrasound and microwave in pasteurization

Pasteurization of orange juice was conducted using microwave heating at 70°C for 1 min. Among the carotenoid compounds quantified, it was reported that zeaxanthin and β-carotene content decreased by about 26%. However, a slight decrease in vitamin C content was reported, and no significant differences ($P < 0.05$) were found for β-cryptoxanthin content

Table 14.1 Application of combined ultrasound-microwave treatment in foods/food processing.

Processing	Optimum condition	Food item/ product	Desirable properties	References
Pretreatment	Microwaves (2450 MHz, 539 W for 4 min), ultrasonic (35 kHz, for 10 min) and drying at 60°C	Pear	Improved color, antioxidant, and phenolic compound retention	(Önal et al., 2021)
Pasteurization	Microwave power (350 W), temperature (35°C), ultrasonic power (778.2 W), and exposure time (11 min)	Orange Juice	Reduction in Saccharomyces cerevisiae	(Samani et al., 2015)
Vacuum frying	Microwave powers (600 W, 800 W, and 1,000 W) and ultrasound powers (0 W, 300 W, and 600 W)	Pumpkin slices	Low-oil absorption, lower moisture content, better crispness, and good color	(Huang & Zhang, 2018)
Convective drying	Combination of ultrasound powers (75, 125 and 200 W) and microwave power of 100 W	Carrot	Reduced drying time in the range of 9–81, improved color, carotenoids, polyphenols, and antioxidant activity	(Kroehnke et al., 2018)
Pectin extraction	Sonication time (21.35 min), microwave power of 580.9 W and 24.66 mL/g liquid-to-solid ratio	Fig skin	Low-methoxyl pectin (yield of 13.97%) with high lightness and a higher number of natural monosaccharides, uronic acids and	(Mohammad et al., 2019)
Pectin extraction	Temperature (86°C), extraction time (29 min), and solid-liquid ratio (1:48 w/v).	Jack fruit	pectin yield (21.5%) was higher than conventional (17.2%). Rapid process, efficient and safe energy.	(Xu et al., 2018)

(continued on next page)

Oligosaccharides Extraction	Extraction time 325.00 s, liquid-solid ratio (10mL/g), ultrasonic power (300.46 W), and microwave power (250 W)	Lotus seeds (Nelumbo nucifera Gaertn.)	Higher yield of oligosaccharides and shorter extraction time than conventional hot water extraction	(Lu et al., 2017)
Extraction of polysaccharides	The ratio of liquid to raw Material (30 mL/g), microwave time (20 min), microwave power (570 W), and a fixed ultrasonic power (50 W)	Camptotheca acuminata fruit	Higher yields in a shorter time than conventional hot water extraction (HWE): 20 vs. 120 min	(Sun et al., 2018)
Extraction of single cell oil	Microwave power (400 W), time (3.7 min); ultrasonic power (500 W), time 40.5, and material to solvent ratio (1:12.3 w/v) at room temperature of 25°C ±5	Mortierella isabelline	Higher oil yield up to 92.24%	(Zhou et al., 2018)
Enzyme assisted-protein hydrolysis	Ultrasound power (50 W), microwave (10–800 W) at optimum time and temperature of 8 min and 55°C.	Chinese sturgeon protein hydrolysate	Improved antioxidant and protein hydrolysis	(Noman et al., 2020)

(Cinquanta et al., 2010). Also, microwave treatment alone was reported to have caused some losses in blackberry jelly's nutritional quality. Although this effect occurred at a low-power microwave, high-power presented high antioxidant activity (Curi et al., 2019). However, the work of Kernou et al. (2021) demonstrated the effectiveness of the combined sonication and microwave treatments. The authors observed that ultrasonic treatment at 42 kHz for 60 min has not shown any significant effect on the survival rate of *Escherichia coli* ATCC 25922 in an orange juice beverage as only 1.3 log reduction was achieved. When the combined ultrasound and microwave treatment were used, it presented a better result than using a single ultrasonic treatment. An 8.0 log reduction was achieved by an ultrasonic treatment for 20 min followed by microwave treatment at 900 W for 30 s. Although microwave radiation was reported to inactivate many micro-organisms such as *Escherichia coli*, *Staphylococcus aureus*, *Streptococcus faecalis*, *Clostridium perfringens*, *Salmonella* spp, *Listeria* spp., spores of bacteria and mold, bacteriophage PL-1 specific to *Lactobacillus casei* were reported to be sensitive to microwave radiation (Woo et al., 2000). Thus, a more severe treatment is required to inactivate such a bacteriophage, which brings about the need for the combined treatment. Samani et al. (2015) have observed the effect of the combined ultrasound and microwave treatment in the inactivation of yeast, particularly *Saccharomyces cerevisiae*, in orange juice. However, during individual treatments, it was observed that the influence of ultrasonic waves in reducing this micro-organism is much stronger than that of microwave, but this was attributed to the longer duration of exposure. Similarly, Jyoti et al. (2020) demonstrated the effectiveness of combined treatment of ultrasound and microwave in the pasteurization of bottled gourd juice. The treatment efficiently reduced heat-resistant *Listeria monocytogenes*, and the process had helped retain the color and other desirable properties such as total phenolic content, antioxidant, terpenoids, vitamins, total protein, and amino acids. For example, the percentage of 2,2-diphenyl-1-picrylhydrazyl (DPPH) radical scavenging activity in ultrasound-microwave treated juice was 43.67%, whereas for raw and conventionally treated juice were 36.54% and 34.32%, respectively. Hosseinzadeh Samani et al. (2016) developed an automatic control device comprising ultrasonic and microwave systems used in the pasteurization of sour cherry juice. Their synergistic effect had shown better retention in the quality properties of the juice. For example, vitamin C content (14%), total phenolic content 1%, and total anthocyanin contents were better maintained than conventional methods (Hosseinzadeh Samani et al., 2016).

14.6.2 Synergistic effect on pretreatment of foods

Combining ultrasound and microwave as an innovative pretreatment of Rocha pears before drying revealed low color changes and better preservation of total phenolic content and antioxidant activities compared to the pear pretreated with either ultrasound or microwave separately and dried under the same conditions (Önal et al., 2021). The similar effect of the combined ultrasound and microwave as a pretreatment method on extraction kinetics, chemical composition, and antioxidant activity of kumquat fruit peel essential oil obtained through hydro-distillation methods was investigated. The process increased yield and the DPPH radical scavenging activity besides decreased extraction time. This was achieved under combined ultrasound (210 W for 30 min) and microwave (300 W and 6 min) conditions. Notably, the pretreatment did not affect the chemical composition of the essential oils (Yu et al., 2021).

14.6.3 Synergistic effect on drying and frying processes

Kowalski et al. (2016) have tested the influence of combined ultrasound and microwave in a hybrid dryer on the drying kinetic, total energy consumption, and quality attribute of the dried product. Their synergistic effect showed significant improvement in the heat and mass transfer efficiency and improved drying rate. The application of the combined methods in another convective drying of carrots using a hybrid dryer reduced the drying time by up to 81% (Kroehnke et al., 2018). Another hybrid convective drying process for strawberries and green paper was enhanced with ultrasound and microwave separately, and in combination. The synergistic effect had resulted in an accelerated drying rate, shortened drying time, reduced energy consumption, and improved quality parameters. During the process, ultrasound-induced the vibration effect while microwave generated the heating effect that synergistically improved the process efficiency in terms of heat and mass transfer (Kowalski et al., 2016; Łechtan et al., 2017). In most documented research investigating the synergistic effect of the ultrasound-microwave application, individual treatments are carried out to ascertain the specific treatments responsible for the improvement or otherwise in foods. For example, Kowalski et al. (2016) noted that convective drying assisted with ultrasound alone at lower power positively affects the carotenoids, phenolic compounds, and color of the strawberries.

Ultrasound-assisted microwave frying of pumpkin slices was investigated to assess the effect of the treatments on moisture and oil contents, texture, color, and microstructure. The synergistic effect of the process was found to promote the quality parameters of the pumpkin chips. Comparing the effect of the individual treatment methods, moisture content, and oil absorption were significantly lowered due to the effect of ultrasound than microwave and vacuum frying (Huang & Zhang, 2018).

14.6.4 Synergistic effect on hydrolysis of foods

Ultrasound and microwave-assisted enzymatic hydrolysis treatment has demonstrated a great potential for producing antioxidant peptides in sweet potato protein hydrolysates. The ultrasound and microwave-enzyme-assisted hydrolysate significantly increased the degree of hydrolysis and production of a peptide with higher antioxidant activity than the untreated samples (Habinshuti et al., 2020). A similar attempt made by Yang et al. (2020) to pretreat grass carp (*Ctenopharyngodon idella*) protein hydrolysate with only ultrasound had also promoted the evolution of Maillard reaction with an improved flavor of the reaction products. Similarly, ultrasound treatment demonstrated an increase in emulsion stability and texture of myosin in myofibrilla protein (Chen et al., 2020).

14.7 Changes in food quality upon application of ultrasound and microwave

Wang et al. (2020) examined the synergistic effect of ultrasound and microwave treatment in improving the physicochemical, and functional properties of chestnut starch. Their findings revealed more severe surface damage, weaker birefringence, and lower relative crystallinity and gelatinization enthalpy of the treated starch than native, and individual treatments. The synergism of the ultrasound and microwave treatment could also affect protein structures, and hydrolysis properties. The application of ultrasound at 300 W couple with 100 W microwave has resulted in a partially denatured structure of the myofibrillar protein of golden threadfin bream (*Nemipterus virgatus*) with increased susceptibility to exogenous enzymes that accelerate the hydrolytic process, thus, yielding high peptide concentration. The peptides produced could inhibit free radical and angiotensin converting enzymes activity, which also decreases the production of proinflammatory cytokines in RAW 264.7 cells stimulated by hydrogen peroxide (H_2O_2). An optimum combination of 300 W ultrasound and 100 W microwave treatment could produce bioactive myofibrilla protein with the most potent effect against cell damage induced by H_2O_2 (Li et al., 2020). The combined treatment was also promising in increasing amino acid content and protein solubility of Chinese sturgeon protein hydrolysate. This is in addition to the high antioxidant of the hydrolysate and reducing power assay. However, the highest degree of protein hydrolysis was observed in samples processed using ultrasound only (19.41, 14.18, and 16.91% for ultrasound-assisted enzymatic treatment (EN), microwave-EN, and ultrasound-microwave-EN, respectively) (Noman et al., 2020). Uluko et al. (2014) also observed the highest radical scavenging activity in enzyme pretreated milk protein concentrate hydrolysate assisted with ultrasound than single microwave and the combined one. This was previously highlighted by Noman et al. Despite the significant improvement in the free radical activity caused by treatment with microwave alone and the combined method, ultrasound alone, in many cases, presents the highest degree of peptide hydrolysis and thereby releasing more peptides, and amino acids with more potent antioxidant activity.

Most of the high antioxidant activity and improved bioactive compound reported could be attributed to the lower temperatures used in the ultrasound and microwave treatments and higher stability of the compounds. This has been observed in the extraction of phenolic compounds from *Satureja macrostema* (Alonso-carrillo et al., 2017). Similarly, the antitumor activity of polysaccharides extracted from *Inonotus obliquus* using the ultrasound-microwave assisted methods was found to be higher than that of conventional hot water extraction, and this, again, could be attributed to the lower temperature associated with the process (Chen et al., 2010).

The effect of the combined treatment on the chemical and microbial composition of rose aromatic water was also investigated. There was an acceptable decrease in phenethyl alcohol, a main component of the rose water and the *E. coli* counts. But a complete loss of the main compound and some other compounds like linalool and cryptone was observed in the conventional pasteurization method. The optimal treatment conditions in this study were 326.24 W, 100 W, 43.32°C, and 4 min for microwave, ultrasound, temperature, and time combinations, respectively (Rostami et al., 2018).

The synergistic effect of ultrasound and microwave improved the quality parameters (color and texture) of vacuum fried Chinese yam (*Dioscorea polystachya*). It was noted that the higher power of both ultrasound and microwave resulted in a higher rate of moisture removal and shorter frying time which also resulted in lowering oil uptake (Chitrakar et al., 2019). In another research, vacuum frying led to an increase in the drying kinetic, effective moisture diffusivity, lower oil and moisture content, improved texture (crispiness), and color parameters relative to the conventional vacuum frying (Faruq et al., 2018; Islam, Zhang, & Fan, 2019; Islam, Zhang, & Mujumdar, 2019). Zhang et al. (2020) have further confirmed the ability of ultrasound to improve the overall flavor of meatballs by increasing the contents of some free fatty acids (such as Lys, Glu, Ala, Gly, Ser, Try, and Cys) that further promote the generation of volatiles.

14.8 Effects on consumer perception/sensory changes

To demonstrate the synergism between the ultrasound and microwave treatment on sensory properties of foods, Sasikumar, Vivek, and Deka (2019) subjected a fruit (*Haematocarpus validus*) juice to these combined processes and then conducted a sensory evaluation using twenty trained panelists. The juice treated with the combined methods was ranked the most acceptable (very good) by the panelists compared to other juice samples subjected to a single treatment of either ultrasound, microwave, or conventional thermal treatment. The process also has numerous other advantages, such as shorter time, less temperature, and high nutrition. Similarly, a soymilk beverage treated with the combined ultrasound and microwave methods and then assessed by 26 panelists became the superior soymilk-based beverage more acceptable by the judges. The quality changes were believed to have resulted from the destruction of lipoxygenase enzymes by the ultrasound and microwave treatments' synergistic effect; hence, no oxidation and off-flavor (beany flavor) were observed (Kumar et al., 2021). In addition to this, acceptability in terms of color and flavor of orange juice pasteurized using the combined methods was found to be significant, according to the findings of Samani et al. (2015). It is worth mentioning that the appearance of the orange juice treated with the combined method was better compared to those of the conventional method (57% vs. 43%). These combined treatments had also shown a positive impact on the nutritional properties of phalsa drink in terms of increased antioxidant activity, total phenolic content, and flavonoids while reducing the microbial load of the juice. This was, however, after the storage period of 120 days (Nadeem et al., 2021).

14.9 Conclusions

Consumers' consistent demands for foods with high quality and safety and devoid of synthetic additives have led to the emergence of several novel techniques which aim to satisfy consumer needs and achieve the myriad objectives of modern food processing methods in terms of shortening process time, reduced costs, and energy consumption as well as environmental friendliness. The combined ultrasound and microwave treatments have shown great promise in achieving the above objectives and were reported severally to be superior to the conventional methods, and individual treatments. Additionally, due to the growing consumer demand for ready-to-eat products processed using greener technologies without much changes to their organoleptic characteristics, these technologies have been widely employed, and currently experimented with in many processes. For example, the technology has been used to modify starches to replace the conventional chemical and enzymatic modifications, and the starches produced can be used for industrial applications. Similarly, its application in protein hydrolysis to generate bioactive peptides useful in fighting or reducing the risk of several diseases has been extensively explored. Also, this combined technique has demonstrated a promising result in improving bioactive compounds during pretreatment, improving extraction yield, and efficiency, achieving complete pasteurization etc. From an economic standpoint, these technologies are also promising and effective due to lower energy consumption, especially when the raw material involved is costly. Although not many challenges were reported in various literature, most of these combined techniques were employed at a pilot or laboratory scale. Thus, proper design and process optimization, especially in hybrid drying, extraction and frying, and other processes, are required to scale up the process.

References

Alden, K.M., Omid, M., Rajabipour, A., Tajeddin, B., 2019. Quality and shelf-life prediction of cauliflower under modified atmosphere packaging by using artificial neural networks and image processing. Comput. Electron. Agric. 163, 104861. https://doi.org/10.1016/j.compag.2019.104861.

Alonso-carrillo, N., Aguilar-santamaría, M.D.L.Á., Vernon-carter, E.J., Jiménez-alvarado, R., Cruz-sosa, F., Román-guerrero, A., et al., 2017. Extraction of phenolic compounds from Satureja macrostema using microwave-ultrasound assisted and reflux methods and evaluation of their antioxidant activity and cytotoxicity. Indust. Crops Prod. 103, 213–221. https://doi.org/10.1016/j.indcrop.2017.04.002.

Ambadgatti, S., Patil, S., Dabade, A., SS, A., Bhushette, P., Sonawane, S.K., 2020. A review on recent trends of ultrasound assisted processing in food segment. J. Microbiol., Biotechnol. Food Sci. 10 (1), 1–4. doi:10.15414/jmbfs.2020.10.1.1-4.

Augusto, P.E.D., Soares, B.M.C., Castanha, N., 2018. Conventional technologies of food preservation. In: Innov. Technol. Food Preserv. Elsevier Inc, pp. 1–24. doi:10.1016/B978-0-12-811031-7.00001-7.

Awad, T.S., Moharram, H.A., Shaltout, O.E., Asker, D., Youssef, M.M., 2012. Applications of ultrasound in analysis, processing and quality control of food : A review. FRIN 48 (2), 410–427. doi:10.1016/j.foodres.2012.05.004.

Balthazar, C.F., Santillo, A., Guimarães, J.T., Bevilacqua, A., Corbo, M.R., Caroprese, M., Marino, R., Esmerino, E.A., Silva, M.C., Raices, R.S.L., Freitas, M.Q., Cruz, A.G., Albenzio, M., 2019. Ultrasound processing of fresh and frozen semi-skimmed sheep milk and its effects on microbiological and physical-chemical quality. Ultrason. Sonochem. 51 (October 2018), 241–248. doi:10.1016/j.ultsonch.2018.10.017.

Belwal, T., Li, L., Yanqun, X., Cravotto, G., Luo, Z., 2019. Ultrasonic-assisted modifications of macroporous resin to improve anthocyanin purification from a pyrus communisvar starkrimson extract. Ultrason. Sonochem. doi:10.1016/j.ultsonch.2019.104853, September.

Bhargava, N., Mor, R.S., Kumar, K., Sharanagat, V.S., 2021. Advances in application of ultrasound in food processing: A review. Ultrason. Sonochem. 70 (July 2020), 105293. doi:10.1016/j.ultsonch.2020.105293.

Bishop, D., 1997. Controlled atmosphere storage. In: Dellino, C.V.J. (Ed.), Cold and Chilled Storage Technology. Chapman & Hall, pp. 53–92.

Calligaris, S., Manzocco, L., Anese, M., Nicoli, M.C., 2016. Shelf-life assessment of food undergoing oxidation–A review. Crit. Rev. Food Sci. Nutr. 56 (11), 1903–1912. doi:10.1080/10408398.2013.807222.

Chandrasekaran, S., Ramanathan, S., Basak, T., 2013. Microwave food processing–A review. FRIN 52 (1), 243–261. doi:10.1016/j.foodres.2013.02.033.

Charoux, C.M.G., O'Donnell, C.P., Tiwari, B.K., 2017. Ultrasound processing and food quality. In: Bermudez-Aguirre, D. (Ed.), Ultrasound: Adv. Food Proc. Preserv. Elsevier Inc., pp. 215–233.

Chemat, F., Khan, M.K., 2011. Applications of ultrasound in food technology: Processing, preservation, and extraction. Ultrason. Sonochem. 18 (4), 813–835. doi:10.1016/j.ultsonch.2010.11.023.

Chen, J., Zhang, X., Chen, Y., Zhao, X., Anthony, B., Xu, X., 2020. Effects of different ultrasound frequencies on the structure, rheological, and functional properties of myosin : Significance of quorum sensing. Ultrason. Sonochem. 69 (June). doi:10.1016/j.ultsonch.2020.105268.

Chen, Y., Gu, X., Huang, S., Li, J., Wang, X., Tang, J., 2010. Optimization of ultrasonic/microwave assisted extraction (UMAE) of polysaccharides from inonotus obliquus and evaluation of its antitumor activities. Int. J. Biol. Macromol. 46 (4), 429–435. doi:10.1016/j.ijbiomac.2010.02.003.

Chitrakar, B., Zhang, M., Fan, D., 2019. The synergistic effect of ultrasound and microwave on the physical, chemical, textural, and microstructural properties of vacuum fried Chinese yam (*Dioscorea polystachya*). J. Food Process. Preserv. 1–12. doi:10.1111/jfpp.14073, Maydoi:.

Cinquanta, L., Albanese, D., Cuccurullo, G., Matteo, D., 2010. Effect on orange juice of batch pasteurization in an improved pilot-scale microwave oven. Food Engineer. Phys. Propert. Eff. 75 (1), 46–50. doi:10.1111/j.1750-3841.2009.01412.x.

Corradini, M.G., 2018. Shelf life of food products: From open labeling to real-time measurements. Annual Rev. Food Sci. Technol. 9, 251–269. doi:10.1146/annurev–food-030117-012433.

Curi, P.N., Salgado, D.L., Mendonça, K., Pio, R., Luiz, J., Ferreira, G., Souza, V.R.De, 2019. Influence of microwave processing on the bioactive compounds, antioxidant activity and sensory acceptance of blackberry jelly. Food Sci. Technol. 2061, 386–391.

D'Souza, A.A., Kumari, D., Banerjee, R., 2017. Nanocomposite biosensors for point-of-care-evaluation of food quality and safety. In: Nanobiosensors. Elsevier Inc, pp. 629–676. doi:10.1016/b978-0-12-804301-1.00015-1.

Ekezie, F.C., Sun, D., Han, Z., Cheng, J., 2017. Microwave-assisted food processing technologies for enhancing product quality and process efficiency : A review of recent developments. Trends Food Sci. Technol. 67, 58–69. doi:10.1016/j.tifs.2017.05.014.

Elamin, W.M., Endan, J.B., Yosuf, Y.A., Shamsudin, R., Ahmedov, A., et al., 2015. High pressure processing technology and equipment evolution: A review. J. Eng. Sci. Technol. Rev. 8 (5).

Fan, L., Hou, F., Muhammad, A.I., Ismail, B.B., Ding, T., Liu, D., 2020. Proteomic responses of spores of bacillus subtilis to thermosonication involve large-scale alterations in metabolic pathways. Ultrason. Sonochem. 104992. doi:10.1016/j.ultsonch.2020.104992.

Fan, L., Ismail, B.B., Hou, F., Guo, M., Ding, T., Liu, D., 2020. Thermosonication pretreatment enhances the killing of germinated bacillus spores adhered to stainless steel surface. LWT 136 (P1), 110248. doi:10.1016/j.lwt.2020.110248.

Fan, L., Ismail, B.B., Hou, F., Muhammad, A.I., Ding, T., Liu, D., 2019. Ultrasound pre-treatment enhances the inhibitory effects of nisin/carvacrol against germination, outgrowth and vegetative growth of spores of bacillus subtilis ATCC6633 in laboratory medium and milk: Population and single-cell analysis. Int. J. Food Microbiol. 108329. doi:10.1016/j.ijfoodmicro.2019.108329.

Fan, L., Ismail, B.B., Hou, F., Muhammad, A.I., Zou, M., Ding, T., Liu, D., 2019. Thermosonication damages the inner membrane of bacillus subtilis spores and impels their inactivation. Food Res. Int. 125 (February), 108514. doi:10.1016/j.foodres.2019.108514.

Faruq, A.Al, Zhang, M., Adhikari, B., 2018. A novel vacuum frying technology of apple slices combined with ultrasound and microwave state key laboratory of food science and technology, Jiangnan University, 214122 wuxi, Jiangsu province key laboratory of advanced food manufacturing equipment and. Ultrason. Sonochem. Elsevier B.V doi:10.1016/j.ultsonch.2018.12.033.

Gholamhosseinpour, A., Hashemi, S.M.B., Raoufi Jahromi, L., Sourki, A.H, 2020. Conventional heating, ultrasound and microwave treatments of milk: Fermentation efficiency and biological activities. Int. Dairy J. 110. doi:10.1016/j.idairyj.2020.104809.

Gould, G.W., 1996. Methods for preservation and extension of shelf life. Int. J. Food Microbiol. 33 (1), 51–64. doi:10.1016/0168-1605(96)01133-6.

Habinshuti, I., Mu, T., Zhang, M., 2020. Ultrasound microwave-assisted enzymatic production and characterization of antioxidant peptides from sweet potato protein. Ultrason. Sonochem. 69 (May), 105262. doi:10.1016/j.ultsonch.2020.105262.

Hosseinzadeh Samani, B., Khoshtaghaza, M.H., Minaei, S., Zareifourosh, H., Eshtiaghi, M.N., Rostami, S., 2016. Design, development and evaluation of an automatic fruit-juice pasteurization system using microwave–Ultrasonic waves. J. Food Sci. Technol. 53 (1), 88–103. doi:10.1007/s13197-015-2026-6.

Huang, M., Zhang, M., 2018. Synergistic effects of ultrasound and microwave on the pumpkin slices qualities during ultrasound-assisted microwave vacuum frying. J. Food Process Eng. 1–8. doi:10.1111/jfpe.12835, May.

Islam, M., Zhang, M., Fan, D., 2019. Ultrasonically enhanced low-temperature microwave-assisted vacuum frying of edamame : Effects on dehydration kinetics and improved quality attributes. Drying Technol. 0 (0), 1–18. doi:10.1080/07373937.2018.1558234.

Islam, M., Zhang, M., Mujumdar, A.S., 2019. Low temperature vacuum frying of edamame assisted by ultrasound and microwave : Effects on the kinetics of oil and product storage properties. Drying Technol. 0 (0), 1–12. doi:10.1080/07373937.2019.1700272.

Ismail, B.B., Guo, M., Pu, Y., Wang, W., Ye, X., Liu, D., 2019. Valorisation of baobab (*Adansonia digitata*) seeds by ultrasound assisted extraction of polyphenolics. optimization and comparison with conventional methods. Ultrason. Sonochem. 52 (November 2018), 257–267. doi:10.1016/j.ultsonch.2018.11.023.

Ismail, B.B., Liu, D., Pu, Y., He, Q., Guo, M., 2021. High-intensity ultrasound processing of baobab fruit pulp: Effect on quality, bioactive compounds, and inhibitory potential on the activity of α-amylase and α-glucosidase. Food Chem. 361 (January), 130144. doi:10.1016/j.foodchem.2021.130144.

Ismail, B.B., Pu, Y., Guo, M., Ma, X., Liu, D., 2019. LC-MS/QTOF identification of phytochemicals and the effects of solvents on phenolic constituents and antioxidant activity of baobab (*Adansonia digitata*) fruit pulp. Food Chem. 277 (March 2019), 279–288. doi:10.1016/j.foodchem.2018.10.056.

Ismail, BB., Funtua, S.F., Garba, U., Sanusi, S.N, 2021. Edible films and coatings based on chitosan with essential oils for antimicrobial food packaging application. La Rivista Italiana Sostanze Grasse 981, XCVIII.

Ismail, B.B., Yusuf, L.Y., Pu, Y., Zhao, H., Guo, M., Liu, D., et al., 2020. Ultrasound-assisted adsorption/desorption for the enrichment and purification of flavonoids from baobab (*Adansonia digitata*) fruit pulp. Ultrasonics - Sonochemistry 65 (104980). https://doi.org/10.1016/j.ultsonch.2020.104980.

Jyoti, M., Arup, D., Das, J., Chakraborty, S., Baishya, P., Ramteke, A., Chandra, S., 2020. Effects of microwave combined with ultrasound treatment on the pasteurization and nutritional properties of bottle gourd (*Lagenaria siceraria*) juice. J. Food Process. Preserv. (June) 1–19. doi:10.1111/jfpp.14904.

Kernou, O.-.N., Belbahi, A., Amir, A., Bedjaoui, K., Kerdouche, K., Dairi, S., Aoun, O., Madani, K., 2021. Effect of sonication on microwave inactivation of escherichia coli in an orange juice beverage. J. Food Process Eng. 44 (January), 1–9. doi:10.1111/jfpe.13664.

Kowalski, S.J., Stasiak, M., Szadzin, J., 2016. Microwave and ultrasound enhancement of convective drying of strawberries : Experimental and modeling efficiency. Int. J. Heat Mass Transfer 103, 1065–1067.

Krasnikov, V.V., 1989. Electromagnetic Radiation in Food-Processing Technology. NBER Working Paper Series 58, 99–104.

Kroehnke, J., Szadzi, J., Stasiak, M., 2018. Ultrasound- and microwave-assisted convective drying of carrots–Process kinetics and product's quality analysis. Ultrason. Sonochem. 48 (May), 249–258. doi:10.1016/j.ultsonch.2018.05.040.

Kumar, R., Ghosh, P., Rao, P.S., Singh, S., Vashishth, R., Vivek, K., 2021. Sensory evaluation of microwave assisted ultrasound treated soymilk beverage using fuzzy logic. J. Saudi Soc. Agri. Sci. 20 (4), 257–264. doi:10.1016/j.jssas.2021.02.005.

Lacroix, M., Vu, K.D., 2014. Edible Coating and Film Materials: Proteins. In: Han, J.H. (Ed.), Innovations in Food Packaging. Elsevier Ltd, pp. 277–304.

Łechtan, J., Kowalski, S.J., Stasiak, M., 2017. The effect of high-power airborne ultrasound and microwaves on convective drying effectiveness and quality of green pepper. Ultrason. Sonochem. 34 (2017), 531–539. doi:10.1016/j.ultsonch.2016.06.030.

Li, S., Zhang, R., Lei, D., Huang, Y., Cheng, S., Zhu, Z., Wu, Z., Cravotto, G., 2021. Impact of ultrasound, microwaves and high-pressure processing on food components and their interactions. Trends Food Sci. Technol. 109 (December 2020), 1–15. doi:10.1016/j.tifs.2021.01.017.

Li, Z., Wang, J., Zheng, B., Guo, Z., 2020. Impact of combined ultrasound-microwave treatment on structural and functional properties of golden threadfin bream (*Nemipterus virgatus*) myofibrillar proteins and hydrolysates. Ultrason. Sonochem. 65 (July 2019), 105063. doi:10.1016/j.ultsonch.2020.105063.

Lu, X., Zheng, Z., Li, H., Cao, R., Zheng, Y., Yu, H., Xiao, J., 2017. Optimization of ultrasonic-microwave assisted extraction of oligosaccharides from lotus (*Nelumbo nucifera gaertn*) seeds. Indust. Crops Prod. (1–12) doi:10.1016/j.indcrop.2017.05.060, May.

Majid, I., Nayik, G.A., Dar, S.M., Nanda, V., 2018. Novel food packaging technologies: Innovations and future prospective. J. Saudi Soc. Agric. Sci. 17 (4), 454–462. https://doi.org/10.1016/j.jssas.2016.11.003.

Misra, N.N., Yepez, X., Keener, K., 2019. In-package cold plasma technologies. J. Food Eng. 244, 21–31. https://doi.org/10.1016/j.jfoodeng.2018.09.019.

Mohammad, S., Gharibzahedi, T., Smith, B., Guo, Y., 2019. Ultrasound-microwave assisted extraction of pectin from fig (*Ficus carica l* .) skin : Optimization, characterization, and bioactivity. Carbohydr. Polym. 222 (June), 114992. doi:10.1016/j.carbpol.2019.114992.

Mohapatra, D., Mishra, S., Giri, S., Kar, A., 2013. Application of hurdles for extending the shelf life of fresh fruits. Trends Post-Harvest Technol. 1 (1), 37–54.

Monteiro, S.H.M.C., Silva, E.K., Alvarenga, V.O., Moraes, J., Freitas, M.Q., Silva, M.C., Raices, R.S.L., Sant'Ana, A.S., Meireles, M.A.A., Cruz, A.G., 2018. Effects of ultrasound energy density on the non-thermal pasteurization of chocolate milk beverage. Ultrason. Sonochem. 42 (November 2017), 1–10. doi:10.1016/j.ultsonch.2017.11.015.

Muntean, M.V., Marian, O., Barbieru, V., Cătunescu, G.M., Ranta, O., Drocas, I., Terhes, S., et al., 2016. High pressure processing in food industry–characteristics and applications. Agric. Agric. Sci. Proc. 10, 377–383.

Nadeem, M., Ghaffar, A., Hashim, M.M., Murtaza, A., Modassar, M., Nawaz, A., Mehmood, A., Riaz, M.N., 2021. Sonication and microwave processing of phalsa drink : A synergistic approach sonication and microwave processing of phalsa drink : A synergistic. Int. J. Fruit Sci. 21 (1), 993–1007. doi:10.1080/15538362.2021.1965942.

Nicoli, M.C., 2012. An introduction to food shelflife: Definitions, basic concepts, and regulatory aspects. In: Nicoli, M.C. (Ed.), Shelf Life Assessment of Food, vol. 1. CRC Press, Boca Raton, FL, pp. 1–8.

Noman, A., Qixing, J., Xu, Y., Abed, S.M., Obadi, M., Ali, A.H., Wenshui, W.Q.A., 2020. Effects of ultrasonic, microwave, and combined ultrasonic- microwave pretreatments on the enzymatic hydrolysis process and protein hydrolysate properties obtained from Chinese sturgeon (*Acipenser sinensis*). J. Food Biochem. (November 2019) 1–13. doi:10.1111/jfbc.13292.

Önal, B., Adiletta, G., Matteo, M.Di, Russo, P., Ramos, N., Silva, C.L.M, 2021. "Rocha " pear : Impact on phytochemical parameters. Foods 10 (853), 1–18. doi:10.3390/foods10040853.

Ordóñez-Santos, L.E., Martínez-Girón, J., Arias-Jaramillo, M.E., 2017. Effect of ultrasound treatment on visual color, vitamin C, total phenols, and carotenoids content in cape gooseberry juice. Food Chem. 233, 96–100. doi:10.1016/j.foodchem.2017.04.114.

Pankaj, S.K., Keener, K.M., 2017. Cold plasma: Background, applications and current trends. Curr. Opin. Food Sci. 16, 49–52. https://doi.org/10.1016/j.cofs.2017.07.008.

Pankaj, S.K., Wan, Z., Keener, K.M., 2018. Effects of cold plasma on food quality: A review. Foods 7 (4). https://doi.org/10.3390/foods7010004.

Perera, C.O., Alzahrani, M.A.J., 2021. Ultrasound as a pre-treatment for extraction of bioactive compounds and food safety: A review. LWT 142, 111114. doi:10.1016/j.lwt.2021.111114.

Qian, M., Liu, D., Zhang, X., Yin, Z., Ismail, B.B., Ye, X., Guo, M., 2021. A review of active packaging in bakery products: Applications and future trends. Trends Food Sci. Technol. 114 (May), 459–471. doi:10.1016/j.tifs.2021.06.009.

Rostami, S., Behruzian, M., Samani, B.H., Lorigooini, Z., Hosseinabadi, T., Zareiforoush, H., Behruzian, A., 2018. Study of combined ultrasound-microwave effect on chemical compositions and E. colicount of rose aromatic water. Iranian J. Pharmaceut. Res. 17 (Special Issue 2), 146–160. doi:10.22037/ijpr.2018.2349.

Saha, S., 2020. IMF: To enhance the shelf-life of food. Int. J. Mod. Trends Sci. Technol. 6 (10), 103–108. doi:10.46501/ijmtst061018.

Salehi, F., 2020. Edible coating of fruits and vegetables using natural gums: A review. Int. J. Fruit Sci. ISSN: 20 (S2), S570–S589. https://doi.org/10.1080/15538362.2020.1746730.

Samani, B.H., Khoshtaghaza, M.H., Lorigooini, Z., Minaei, S., Zareiforoush, H., 2015. Analysis of the combinative effect of ultrasound and microwave power on saccharomyces cerevisiae in orange juice processing. Innov. Food Sci. Emerg. Technol. 32, 110–115. doi:10.1016/j.ifset.2015.09.015.

Sasikumar, R., Vivek, K., Deka, S.C., 2019. Sensory evaluation of ultrasound assisted microwave treated (*Haematocarpus validus*) juice through fuzzy logic approach fruit. Int. Food Res. J. 26 (4), 1229–1236.

Scudino, H., Silva, E.K., Gomes, A., Guimarães, J.T., Cunha, R.L., Sant'Ana, A.S., Meireles, M.A.A., Cruz, A.G., 2020. Ultrasound stabilization of raw milk: Microbial and enzymatic inactivation, physicochemical properties, and kinetic stability. Ultrason. Sonochem. 67 (September 2019), 105185. doi:10.1016/j.ultsonch.2020.105185.

Sharif, Z., Mustapha, F., Jai, J., Mohd Yusof, N., Zaki, N., 2017. Review on methods for preservation and natural preservatives for extending the food longevity. Chem. Eng. Res. Bull. 19 (September), 145. doi:10.3329/cerb.v19i0.33809.

Singla, M., Sit, N., 2021. Application of ultrasound in combination with other technologies in food processing: A review. Ultrason. Sonochem. 73, 105506. doi:10.1016/j.ultsonch.2021.105506.

Sridhar, A., Ponnuchamy, M., Kumar, P.S., Kapoor, A., 2021. Food preservation techniques and nanotechnology for increased shelf life of fruits, vegetables, beverages and spices: A review. Environ. Chem. Lett. 19 (2), 1715–1735. doi:10.1007/s10311-020-01126-2.

Sun, H., Li, C., Ni, Y., Yao, L., Jiang, H., Ren, X., Fu, Y., Zhao, C., 2018. Ultrasonic/microwave-assisted extraction of polysaccharides from camptotheca acuminata fruits and its antitumor activity. Carbohydr. Polym. doi:10.1016/j.carbpol.2018.11.010.

Uluko, H., Zhang, S., Liu, L., Tsakama, M., Lu, J., Lv, J., 2014. Effects of thermal, microwave, and ultrasound pretreatments on antioxidative capacity of enzymatic milk protein concentrate hydrolysates. J. Funct. Foods 2. doi:10.1016/j.jff.2014.11.024.

Wang, J., Wang, J., Vanga, S.K., Raghavan, V., 2020. High-intensity ultrasound processing of kiwifruit juice: Effects on the microstructure, pectin, carbohydrates, and rheological properties. Food Chem. 313 (December 2019), 126121. doi:10.1016/j.foodchem.2019.126121.

Wang, L., Boussetta, N., Lebovka, N., Vorobiev, E., 2019. Ultrasound assisted purification of polyphenols of apple skins by adsorption/desorption procedure. Ultrason. Sonochem. 55 (January), 18–24. doi:10.1016/j.ultsonch.2019.03.002.

Wang, M., Wu, Y., Liu, Y., Ouyang, J., 2020. Effect of ultrasonic and microwave dual-treatment on the physicochemical properties of chestnut starch. Polymers. 12 (8), 1–12. doi:10.3390/POLYM12081718.

Woo, I.S., Rhee, I.K., Park, H.D., 2000. Differential damage in bacterial cells by microwave radiation on the basis of cell wall structure. Appl. Environ. Microbiol. 66 (5), 2243–2247. doi:10.1128/AEM.66.5.2243-2247.2000.

Xu, S., Liu, J., Huang, X., Du, L., Shi, F., Dong, R., Huang, X., Zheng, K., Yang, L., Cheong, K., 2018. Ultrasonic-microwave assisted extraction, characterization and biological activity of pectin from jackfruit peel. LWT - Food Sci. Technol. doi:10.1016/j.lwt.2018.01.007.

Yang, X., Li, Y., Li, S., Ren, X., Olayemi, A., Lu, F., 2020. Effects and mechanism of ultrasound pretreatment of protein on the maillard reaction of protein-hydrolysate from grass carp (*Ctenopharyngodon idella*). Ultrason. Sonochem. 64 (December 2019), 104964. doi:10.1016/j.ultsonch.2020.104964.

Yu, F., Wan, N., Zheng, Q., Li, Y., Yang, M., Wu, Z., 2021. Effects of ultrasound and microwave pretreatments on hydrodistillation extraction of essential oils from kumquat peel. Food Sci. Nutrit. (November 2020) 2372–2380. doi:10.1002/fsn3.2073.

Yu, Q., Fan, L., Li, J., 2020. A novel process for asparagus polyphenols utilization by ultrasound assisted adsorption and desorption using resins. Ultrason. Sonochem. 63 (December 2019). doi:10.1016/j.ultsonch.2019.104920.

Zhang, J., Zhang, Y., Wang, Y., Xing, L., Zhang, W., 2020. Influences of ultrasonic-assisted frying on the fl avor characteristics of fried meatballs. Innovat. Food Sci. Emerg. Technol. 62 (March), 102365. doi:10.1016/j.ifset.2020.102365.

Zhang, X., Ismail, B.B., Cheng, H., Jin, T.Z., Qian, M., Ali, S., Liu, D., Guo, M., 2021. Emerging chitosan-essential oil films and coatings for food preservation-A review of advances and applications. Carbohydr. Polym. 273 (June), 118616. doi:10.1016/j.carbpol.2021.118616.

Zhang, X., Guo, M., Ismail, B.B., Qiao, H., Jin, T.Z., Liu, D., et al., 2021. Informative and corrective responsive packaging: Advances in farm-to-fork monitoring and remediation of food quality and safety. Comprehen. Rev. Food Sci. Food Saf. (June) 1–25. doi:10.1111/1541-4337.12807.

Zhou, C., Sun, D., Sun, X., Zhu, C., Wang, Q., 2018. Combining ultrasound and microwave to improve the yield and quality of single-cell oil from mortierella isabellina NTG1−121. J. Am. Oil Chem. Soc. 1–13. doi:10.1002/aocs.12134.

CHAPTER FIFTEEN

Synergistic effect of sonication and microwave on physicochemical and textural properties of foods

Ali Ikram[a], Farhan Saeed[b], Chasheen Fizza[c], Shahid Bashir[a], Muhammad Afzaal[b] and Mohammad Javed Ansari[d]
[a]University Institute of Food Science and Technology, The University of Lahore, Lahore, Pakistan
[b]Department of Food Sciences, Government College University Faisalabad, Faisalabad, Pakistan
[c]School of Food Science and Engineering, South China University of Technology, Guangzhou, China
[d]Department of Botany, Hindu College Moradabad (Mahatma Jyotiba Phule Rohilkhand University Bareilly), India

15.1 Introduction

Processing of food is the purposeful transformation of agricultural goods into more sustainable, stable, convenient, usable, value-added products that are benign for the consumption of humans. Drying, salting, soaking, smoking, pickling, frying, and other traditional processing and preservation processes are still widely and efficiently employed to prepare raw food products. Most conventional food processing procedures rely on high temperatures to reduce germ development and suppress food-borne pathogens, rendering the food secure for eating. But the conventional thermal procedures waste much energy, and time which is less efficient in terms of production rate. Several food products risk bacterial, or virus contamination, making thermal treatment unsuitable. Thermally sensitive food items may experience physical, chemical, and microbiological changes due to heat treatment, including changes in flavor, color, and texture. This has necessitated research and development into the most efficient use of existing technologies and the birth of creative and successful alternative technologies (Bhargava et al., 2021). Under difficult or nonclassical situations, today, food technology is extensively used in the applied study and industrial subjects. The efficiency of production can be improved by exchanging with better techniques of extraction, processing, and preservation; additionally, it can contribute to environmental safeguarding via lesser use of solvent or water, wastewater minimization, and contaminants removal from processing

Ultrasound and Microwave for Food Processing: Synergism for Preservation and Extraction.
DOI: https://doi.org/10.1016/B978-0-323-95991-9.00019-9

(Singla & Sit, 2021). Various innovative technologies in food processing have now been studied as well as established to alter or substitute existing processing procedures to provide higher-quality, more consumer-preferred meals. Much emphasis has been placed on improving process efficiencies, productivity, quality, and safety of food items in a health admiring manner (Priyadarshini et al., 2019).

Sonication is a cutting-edge technology with several uses in food processing. It might be characterized as a louder sound wave beyond what human ears can detect. Nevertheless, frequencies of US higher than 20 kHz cannot be heard. The energy intensity in the United States varies from 10 to 1000 W/cm^2 at a frequency that ranges from 20 to 100 kHz and significantly influences the chemical and physical characteristics of fluids (Al-Hilphy et al., 2020). Ultrasound (US) is also used in several unit processes: defoaming, extraction, dehydration, freezing, tenderization, emulsification, thawing, and concentration. Furthermore, the US has increased polysaccharide and protein functionalities and bioactivity. All the same, the US has no power against the microbial proliferation and inactivation of enzymes. It may have deleterious effects linked with hydroxyl radical formation, such as off-flavors, physical characteristics, and component deterioration (Morales-soriano et al., 2021).

Microwaves are waves that are nonionizing, electromagnetic, capable of creating micro-motion plus molecular friction in meals due to the effect known as the dielectric loss effect. Microwave treatment can disrupt glycosidic linkages, cause starch molecules to reorganize, and modify the starch granules shape, relative crystallinity, causing differences in swelling capacity, emulsifying stability, as well as emulsifying activity, and pasting viscosity (Wang et al., 2020). In the most common industry wherein microwaves are utilized in food technology, the benefits of microwave energy frequently present a feasible alternative to traditional, more cumbersome thermal treatment techniques. Food technology is the most well-known industry in which microwaves are utilized. Microwaves are used to pasteurize, sterilize, dry, or defrost food goods, among other things. Many businesses employ microwaves in conjunction with traditional thermal technologies. Microwave assistance in the food industry decreases process costs and time while maintaining high sensory properties and generating minor alterations in the food nutrients compared to traditional techniques (Guzik et al., 2021). Though each technique has its benefits, it also has many drawbacks that can cause serious quality problems in food. To overcome this issue, science is putting more effort into combining different technologies to surpass the shortcomings of other technologies and enhance the quality parameters of the product by exceeding each other's efficiency. As a result, hybrid technology emerges (Hii et al., 2021). In This paper, we will discuss the hybrid actions of sonication and microwave and their influence on the textural properties of food, their mode of mechanism, and applications of this compound technology in the industry of food.

15.2 Effect of sonication on physicochemical properties of food

Because of health issues and food safety challenges, consumer demand for greater nutritional properties, higher quality, and modestly refined food has increased rapidly in recent years. The US is an intriguing option for improving nutritional and physical and properties. Different researchers have investigated the utilization of US as a substitute in the processing of food, benefiting from the given physicochemical effects that result from this technology. The killing of micro-organisms by US has been attached primarily to physical changes in particle size by producing the cavitation effect or chemical: formation of free radical precepts. Many internal and external regulating factors influence sonication efficacy. Alternation within and among the molecule would cause either hype or decline in texture, emulsifying capacity, and antioxidant potential. In conjunction with heat processing techniques, ultrasonication can successfully achieve the desired level of enzyme inactivation by lowering the thermal resistance (Salehi, 2020).

According to one study, ultrasound improved the extraction efficiency, color parameter, acidity, peroxide values, and total phenolic contents. The use of US pretreatment has no noticeable effects on the refractive index of the extracted oils. Using ultrasound pretreatments on black cumin seeds before the cold press extraction process may improve the extracted oil's quantitative and qualitative properties. However, given the benefits of ultrasound pretreatment and its applications and effects on extraction efficiency and phenolic compounds, it can be demonstrated that this method can be used in medicinal and valuable plant applications on a large scale (Moghimi et al., 2018). Ultrasound's physiochemical effects may harm food due to developing hydroxyl radicals by the "cavitation" impact causing ascorbic acid degradation and lowering total content of phenol. Ultrasound-induced degradation of lipids, particularly in oily foods, produces bad odors. The secondary reaction by-products decrease the nutritive value and food safety (Sattar et al., 2019). Since sonication treatment cannot achieve desired sterilization alone, it must be coupled with other technologies to confirm food safety. One of the significant reasons that prevent ultrasound in the food industry is the high energy input required for industrial applications. As a result, research into US technology must be expanded to maintain the circumstances, or by combining it with other technologies, we can avoid the deleterious effects of sonication. The study found that ultrasonication has improved the color of juices in parameters of hue, lightness, and chroma. Compared to pasteurization, the samples treated with sonication retained higher antioxidant potential of 5–30%, particularly in the ripe fruit juice, but noticeably lessened enzyme activities such as polygalacturonase and polyphenol oxidase peroxidase and pectin methyl esterase up to 2–10% (Charan & Bavisetty, 2021).

15.3 Effect of microwave on physicochemical properties of food

According to a study by (Li et al., 2019), microwave processing reduced meat color quality and enhanced meat cooking loss in yak meat processing. When wheat starch was microwave pretreated, the results showed that microwave pretreatment might significantly alter and customize the functionality of the starch by affecting the surface and partly destroying the internal morphology of the starch granules. Compared to untreated samples, samples pretreated with microwave had enhanced acetylation, water-solubility, gelatinization, peak viscosity, and pasting temperatures (Zhao et al., 2018). Microwave-heated camu camu juices had slight color variation, which may appeal to consumers because there were no noticeable differences compared to juice that was not treated. The proven use of low energy (350 W/15 s) can boost the physicochemical properties of camu camu juice but accelerates the degradation of juice.

Compared to conventional heating, the treatment of higher energy, that is, 940 W/45 s demonstrated better reduction of bacteria, yeasts, and molds while retaining bioactive components (pasteurization) (de Souza Comapa et al., 2019; Martins et al., 2019). Excessive heat is still a significant issue in the heat for milk and other dairy products, reducing nutritional, and sensory qualities. The microwave technique has been attributed to producing higher-quality food with a longer shelf life, making it a viable substitute for primitively used heat treatment. The major disadvantage of MW is the nonuniform distribution of temperature, which results in cold plus hot spots on food products; nevertheless, the microwave is appropriate for liquid food, particularly in a system of continuous fluids (Martins et al., 2019).

15.4 Effect of sonication on textural properties of food

The impacts of US waves on the texture of food materials are mainly determined by the type of food type, operating conditions, and the process variables. Enough energy by microbubble cavitation to disrupt tissue structure can be produced by using low intensity and higher power waves of sonication, resulting in turgor pressure loss and a softness in the texture of targeted product. Ultrasonication is used to substitute traditional thermal treatments of pasteurization and sterilization of various foods. The exact principle for microbe inactivation all through US irradiation is "cavitation," resulting in cellular membrane thinning, the production of free radicals and localized high-heating (Al-Hilphy et al., 2020). Ultrasound treatment at higher intensities can cause microstructure deformations, resulting in tissue softening. In one study, fresh-cut coconut was pasteurized using high-power ultrasound, and the textural parameter of hardness was studied beyond a time of refrigerated storage. Cavitation mechanisms disrupt the microstructure, causing a lowering of the hardness, and variations in protein structure. However, These effects can be advantageously utilized by carefully maintaining the process variables viz, exposure time, ultrasound intensity, and frequency (Aslam et al., 2021; Al-Hilphy et al., 2020).

In the study (Nowak et al., 2019), the ultrasound treatment was demonstrated on blueberries, and the ultrasound treatment caused just unwanted color changes in the blueberries, with no significant effect on the texture (Nowacka & Dadan, 2021). Sonication of *Ginkgo biloba L.* seeds (GBS) revealed the same point as the study results, which showed that the GBS was pretreated by ultrasonication-soaked water. Sonication resulted in the best texture preservation because the cellular structure had become turgid and softer. It was because of ultrasonic cavitation that destroyed the cell structure of GBS and formed a microchannel. Appropriate sonication treatment can also lessen the chewing force and the shear force in meat food and have a prominent tenderizing effect (Al-Hilphy et al., 2020).

A yogurt study revealed that increasing ultrasound frequency caused a spike in all textural properties. The hardness of samples was increased by sonication. The ultrasound accelerates curd hardening and increases the final product's firmness. Homogenization; cavitation-induced destruction and reduction of size in the fat globules; increased mobility of particles due to high turbulence; denaturation of the globular type whey proteins, and membranes of protein surrounding the fat globule; promotion of the aggregate formation; protein-protein interactions; restriction of protein chains opening (Kenari & Razavi, 2021). This technology impacts the water, fat retention capacity, texture, as well as yield of meat emulsions. (Cichoski et al., 2019). Ultrasonication and its impact on texture reported the existence of microporous surfaces as well as the cracks and fissures formation in cereals. Sonication augmented the percentage of pores, allowing easy water penetration while cooking and promoting soft texture. Furthermore, when sonicated cereals were soaked in a mineral solution, sonication resulted in a lessening of inner iron and phosphorus percentage while increasing its capacity for uptake of iron by fortification (Bonto et al., 2020).

15.4.1 Mechanism of ultrasound

US is an acoustic wave produced by moving objects (Fig. 15.1). Its vibration frequency is typically greater than 20 kHz, exceeding the upper limit of a living person's hearing capacity of sound waves. Ultrasound is classified into three categories based on frequency: diagnostic ultrasound ranging from 1 to 500 MHz, high-frequency ultrasound ranging: 100 kHz–1 MHz, and power ultrasound from 20 to 100 kHz. Fig. 15.2 depicts an ultrasonic technique in food analysis and processing (Gharibzahedi et al., 2018). Power ultrasound owing to a lower frequency but higher energy, can cause large cavitation bubbles with high temperature, and pressure (Gharibzahedi et al., 2018).

The cavitation is the fundamental technique that drives most of the sonication-induced chemical retorts in liquids. Interactions between bubbles and waves within the medium are the primary cause. When US waves from a medium, they generate compression and rarefaction forces that change regularly. US waves compress the medium and press the particles together during a compression cycle. During a rarefaction cycle, the molecules in the medium are exposed to negative force and become loose (Jiang et al., 2020) The successful implementation of ultrasonic technology in industrial decontamination requires an accurate ultrasonic sensor or processor. It has recently been noticed that extensive research is being carried out concerning characteristics, geometry, operating conditions, and probe design to meet the requirements of desired applications in various food items or give optimum ultimatums for each application. (Yuan et al., 2021) The sponging effect that consists of channeling the medium for better transfer of mass due to the alternative compressions and refractions owing to ultrasound waves is used

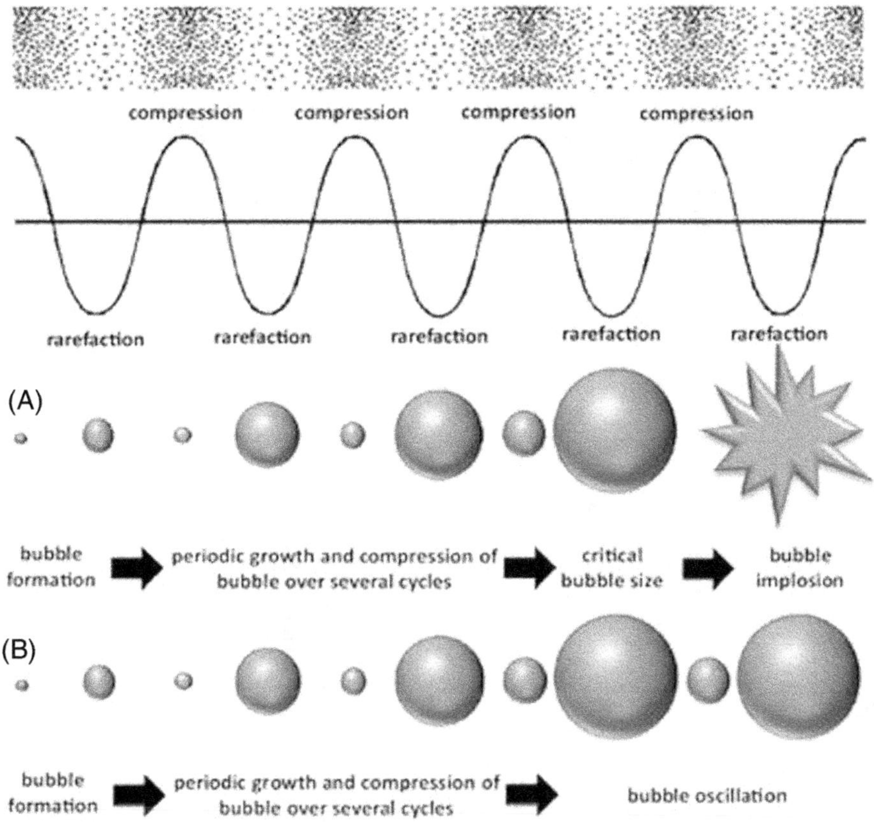

Figure 15.1 Ultrasonication processing in Food Industry (Gharibzahedi et al., 2018).

in operations of extraction, and drying. Due to the above effects, applying high pressure and temperature causes mechanical erosion, rupture, chemical changes, and dispersion in processed foods (Téllez-Morales et al., 2020).

15.5 Effect of microwave on the texture of food

Microwaves (MWs) have many advantages, including time savings, improved final product quality, and faster generation of heat (Table 15.1). Even though MW treatment used in the processing food with advancing technology has a beneficial impact in terms of nutrient value, energy, or time, it is also critical to know how they will affect the textural qualities of the food to which it is applied (Sungsinchai et al., 2019). The effect of microwave treatment on different food textures is given in the Table 15.1. In some cases,

Table 15.1 Effect of microwave on textural attributes of food product.

Material	Condition	Texture results	References
Grape tomato	Microwave heating -Power: Medium (50%) and high (100%). Heating time: 10, 20, 30, 40, 50 s	At central power for 40s, the texture quality is optimal.	(Lu et al., 2011)
Soy protein isolate	Microwave heating –210 W power Time of heating 30, 40, 80, and 90 s	Better viscoelastic properties when compared to traditional heating	(Phuhongsung et al., 2020)
Wheat starch-papaya	Microwave heating – 60, 70, 80 W for 4 min	At 80W, the viscoelasticity and hardness are improved.	(Xu et al., 2020)
thigh and breast of Chicken	Microwave cooking– 180 W for 10 min	Shear force is greater. The increased hardness of thigh and breast meats	(Taşkıran et al., 2020)
Clitoria ternatea, jasmine white rice	Microwave cooking– 800 W for 11 min	Chewiness, stickiness, and hardness values that are higher	(Chusak et al., 2019)
Broccoli, Carrot, Brazilian zucchini	Microwave cooking– 320 W for: 6–16 min	It has been discovered that using a low power microwave for dry heat cooking of vegetables is not appropriate.	(de Castro et al., 2021)
Mozzarella cheese, Cheddar cheeses	Microwave cooking– 900 W for 2 min	Stretchability scores are higher. Worse firmness and stringiness ratings	(Gulzar et al., 2020)
Potato strips	Microwave thawing 270 W Time: 3 min	The values of max cutting force had not noticeable difference.	(Abdul Hamid et al., 2019)
Chicken breast meat	Microwave thawing For 6 min	Maximum drip loss and shear force	(Augustyńska-Prejsnar et al., 2018)
Raspberry	Microwave thawing Optimum thawing conditions	The few drips lost. Sensory evaluation showed the best results of hardness in microwave thawed samples were comparable to fresh fruits of raspberry.	(Liu et al., 2020)

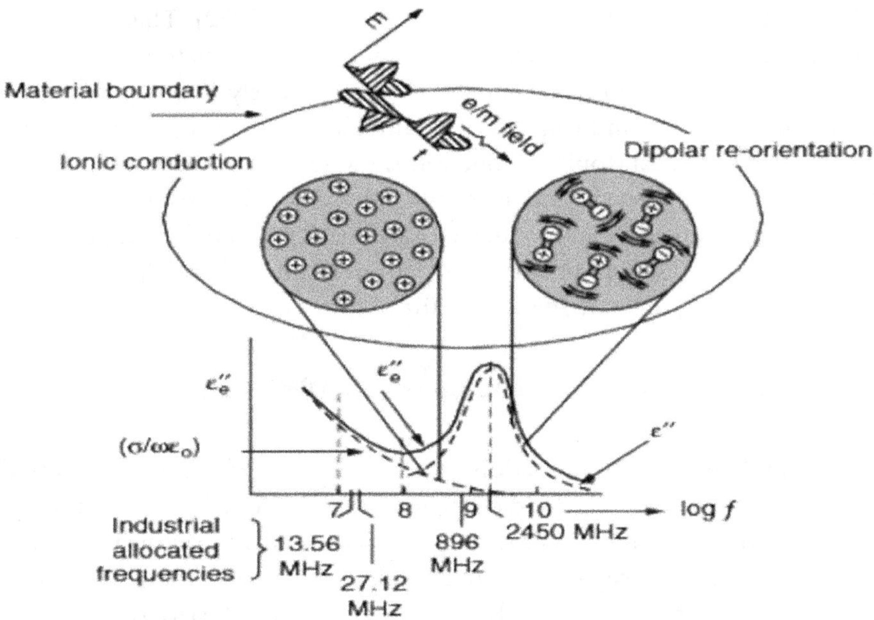

Figure 15.2 The ionic and dipolar loss mechanisms.

the truth is that the shorter time for microwave time can negatively affect the structure of the food. As for fixing it, microwave-assisted processes, including microwave-vacuum, microwave-hot air, microwave-radio frequency, and microwave-freeze, are recommended. At the time of resolving issues with the industrial scale-up microwave (capacity of production and initial investment cost, etc.), this technique will be top choice for better processing of food and enhanced texture (Kutlu et al., 2021).

15.5.1 Mechanism of microwave

Microwave refers to electromagnetic radiation with a wavelength of 1 mm–1 m and a frequency range of 300 MHz–300 GHz. Electromagnetic energy travels through space due to changing magnetic fields plus electric fields (Kumar & Karim, 2019). Microwave invades the specimen unless it finds the moisture and heats it in volumetric order, permitting a more significant pressure gradient and diffusion rate to drive the water out of the product. MW heating is caused by two main processes: 1-dipolar reorientation, 2-ionic conduction. Molecules of water are dipolar and attempt to pursue a highly alternating electric field. The electric field shifts directions with a frequency of 2.45 billion turns per second at a standard MW frequency of

2450 MHz, causing the dipoles to move (Viji et al., 2022). This molecular rotation creates heat as well as friction within the material of food (Atuonwu & Tassou, 2018). Food matrix consists of approximately 80% dipolar water mainly in vegetables and fruits. As a result, microwaves can produce significant volumetric heat. Ionic conduction is the second primary mechanism. Ions, particularly those found in salt-based foods, relocate because of an electric field influence and produce heat. Fig. 15.2 presented the dipolar and ionic loss mechanisms and their contributions to the dielectric properties as a function of frequency (Jiang et al., 2018).

15.5.2 Ultrasound and microwave (US-MW) assisted extraction

UAE and MW-assisted extraction (UMAE) is one of the most admired hybrid combinations. These techniques can be carried out in either a centrifugation of a single reactor or in two independent reactors that could function contemporarily or sequentially. UMAE has successfully been used to extract bioactive compounds from various biomaterials (Guo et al., 2019). The cavitation phenomenon, which is associated with the ultrasound, encourages the emission of desired compounds by causing physical damage to membranes, and walls of the cell. As a result, mass transfer increases, and solvent penetration is accelerated. It is assumed that MW radiation will rapidly heat the biomaterial, causing dissolved molecules to migrate. Ultimately, synchronous or successive US-MW processing provides greater efficiency of extraction (Khadhraoui et al., 2021). UMAE saved time and energy in the oil extraction from tea (*Camelia sinensis*) seed with its high extraction efficiency. While oil by UMAE had better physical and chemical properties and a high proportion of bioactive components, its fatty acid composition was lower. Furthermore, study results revealed that the primary key driver for fast extraction was raw material structure destruction caused by merging ultrasonic and microwave. UMAE was a successful and promising method to extract tea seed oil, and it can be utilized to extract the phytochemicals or oil from many other plant sources (Hu et al., 2019). The schematic diagram of UMAE is given below (Fig. 15.3) (Zeng et al., 2015).

15.5.3 US-MW assisted frying

The effect of vacuum frying coupled with microwave and ultrasonication (UMVF) on the value of fried mushrooms was studied. The US was found to weaken the structure of mushroom, allowing moisture to move from the tissues into the oil while the process of frying. The synergistic effect

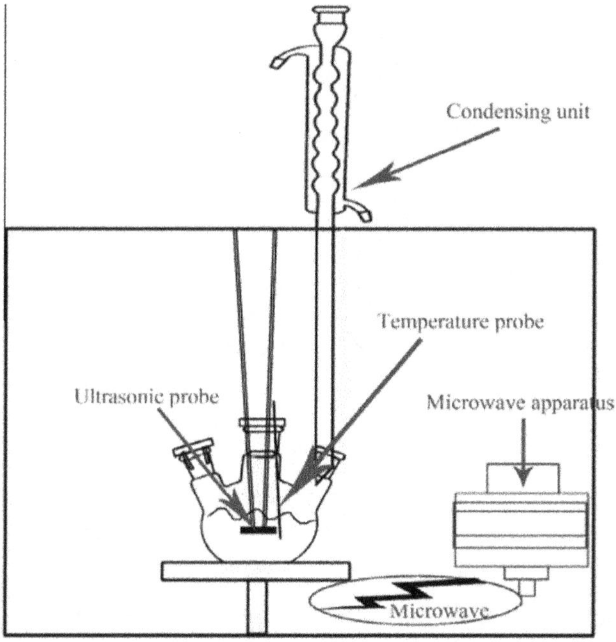

Figure 15.3 Schematic Diagram of UMAE apparatus.

in UMVF, was observed, where microwave caused volumetric heating and the vibration was caused by the ultrasonication, which increased the oil temperature and significantly enhanced the drying rate (Khadhraoui et al., 2021). A more highly permeable microstructure in USMVF samples helped explain why fried samples had less oil uptake and a crisper texture. In terms of product quality and energy efficiency, the frying by vacuum technology that integrates microwave, and sonication was even more impactful than single vacuum frying technology. Whenever the ultrasonic power was increased, the water content and quality attributes improved noticeably. Like the oil level (height) raised, the ultrasonic power density decreased, also, it was divided nonuniformly in the frying chamber (Zhang & Zhang, 2020; Su et al., 2018). Another study on pumpkin chips found that the application of sonication raised the quality attributes of the pumpkin chips. The use of ultrasound cut short the frying time. In addition, compared to traditional frying techniques, the merged implication of microwave and ultrasound lessen oil content, and enhances textural properties. Furthermore, this hybrid-assisted technique advocated the microstructure in chips while maintaining the shape. The only parameter that has not been influenced was color (Huang & Zhang, 2018). The Fig. 15.4 depicts the schematic representation of the USMVF instrument. The microwave source, located

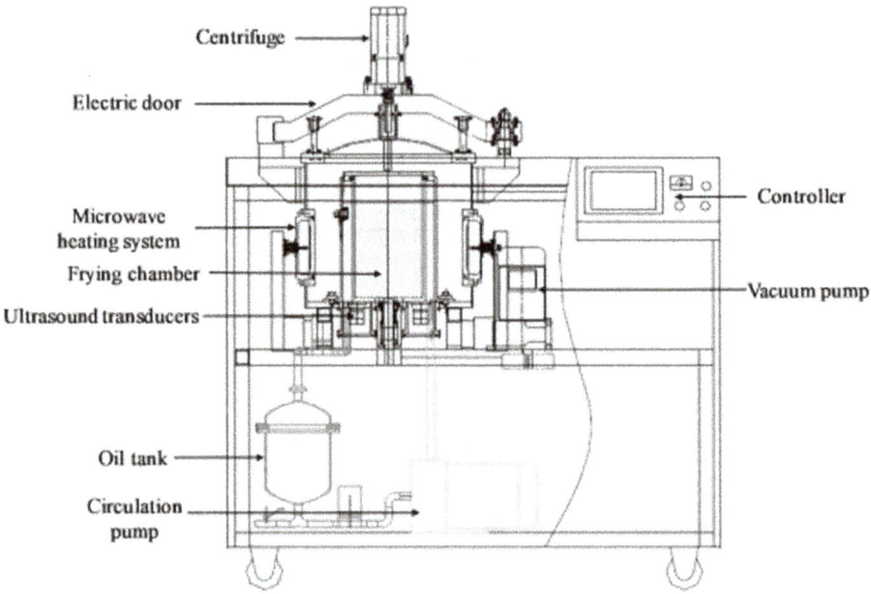

Figure 15.4 Schematic diagram of the ultrasonic microwave-assisted vacuum frying instrument (Su et al., 2021).

near the frying chamber, serves as a rapid heating source. The ultrasonic energy generated by the ultrasonic transducer on the bottom of the fryer could cause a series of ultrasonic impacts on the food material as it passes through the oil medium, including the micro-channels generation in the cellular structure of the specimen, increased heat as well as mass transfer rate, and more uniform heating (Su et al., 2021). The schematic diagram of the ultrasonic microwave-assisted vacuum frying instrument is presented in Fig. 15.4.

15.5.4 US-MS assisted drying

US-MS assisted drying has had various effects on food product quality, including changes in color, density, rigidity, shrinkage, capacity of rehydration, and chemical properties such as phenolic compounds, flavonoids as well as antioxidants (Khadhraoui et al., 2021). This technology in the drying can speed up the drying of fish, fruits, meat, and vegetable. It shortens the time of drying while increasing the heat transfer rate to preserve product quality. US-MW assisted technology is a viable substitute to the conventional drying process (Walayat et al., 2022). Removal of water is sponge effect aided which improves diffusion of water from inner surface to outer surface. Water cavitation creates new micro-channels both inside and outside the cell. In addition, air turbulence is caused by ultrasound at the air-product interface, which helps in moisture removal from the upper surface (Bhargava et al., 2021). When MW and US intensified convective drying was applied, variations in carrot drying kinetics were noticed, and comparable drying rates were observed in this in situ US-MW accounted for the second to MW alone, with the former ingesting more energy. These findings were most likely attributed to the substrate's low permeability and relatively high airflow rate. In materials with restricted porosity and high rigidity, such as carrot, the US can have lesser effectiveness (Nowacka et al., 2019) MW (100 W) and US (200 W) were used to improve the convective dehydration of strawberries. The energy consumed elevated by only 0.5 kWh when ultrasonication and microwave intensified convective drying. It significantly improved the efficiency of heat transfer rate in these fruits because the former produces a heating effect," while the latter produces a vibration effect, and the two together speed up the drying rate through a synergistic effect (Morales-soriano et al., 2021)."

15.6 Conclusion

Over the last few years, ultrasonication and microwave applications in food processing has progressively increased, leading to an enduring influence on food products in liquid form via cavitation. The ultrasound processes, particularly when combined with high temperatures and high techniques, activate microbes to sustain or decontaminate foods. A growing number of manufacturing processes use the sonication to help in the mixing of material, the formation of agglomerates and foams, the dust precipitation, the improvement of filtration efficiency, the drying of products, and the extraction of the bioactive components from vegetables and food. Likewise, microwaves have been used successfully in various food processes, including cooking, drying, and pasteurization of food materials. However, there are several issues with using these technologies separately, but when used together, these two technologies of microwaves and ultrasounds provide improved finished product characteristics, both in physicochemical and textural properties.

References

Abdul Hamid, N., Omar, S., Sanny, M, 2019. c. J Saudi Soc. Agri. Sci. 18 (4), 396–400. doi:10.1016/j.jssas.2018.01.002.

Al-Hilphy, A.R., Al-Temimi, A.B., Al Rubaiy, H.H.M., Anand, U., Delgado-Pando, G., Lakhssassi, N., 2020. Ultrasound applications in poultry meat processing: A systematic review. J. Food Sci. 85 (5), 1386–1396. doi:10.1111/1750-3841.15135.

Aslam, R., Alam, M.S., Kaur, J., Panayampadan, A.S., Dar, O.I., Kothakota, A., Pandiselvam, R., 2021. Understanding the effects of ultrasound processing on texture and rheological properties of food. J. Texture Stud. November, 1–25. doi:10.1111/jtxs.12644.

Atuonwu, J.C., Tassou, S.A., 2018. Quality assurance in microwave food processing and the enabling potentials of solid-state power generators: A review. J. Food Eng. 234, 1–15. doi:10.1016/j.jfoodeng.2018.04.009.

Augustyńska-Prejsnar, A., Ormian, M., Sokołowicz, Z., 2018. Physicochemical and sensory properties of broiler chicken breast meat stored frozen and thawed using various methods. J. Food Qual. 2018. doi:10.1155/2018/6754070.

Bhargava, N., Mor, R.S., Kumar, K., Sharanagat, V.S., 2021. Advances in application of ultrasound in food processing: A review. Ultrason. Sonochem. 70 (July 2020), 105293. doi:10.1016/j.ultsonch.2020.105293.

Bonto, A.P., Tiozon, R.N., Rojviriya, C., Sreenivasulu, N., 2020. Ultrasonics Sonochemistry sonication increases the porosity of uncooked rice kernels a ff ording softer textural properties, loss of intrinsic nutrients and increased uptake capacity during forti fi cation. Ultrason. Sonochem. 68 (June), 105234. doi:10.1016/j.ultsonch.2020.105234.

Charan, S., Bavisetty, B. (2021). Physicochemical qualities and antioxidant properties of juice extracted from ripe and overripe wax apple as affected by pasteurization and sonication. *January*, 1–12. doi:10.1111/jfpp.15524.

Chusak, C., Ying, J.A.Y., Zhien, J.L., Pasukamonset, P., Henry, C.J., Ngamukote, S., Adisakwattana, S., 2019. Impact of clitoria ternatea (butterfly pea) flower on in vitro starch digestibility, texture and sensory attributes of cooked rice using domestic cooking methods. Food Chem. 295 (January), 646–652. doi:10.1016/j.foodchem.2019.05.157.

Cichoski, A.J., Silva, M.S., Leães, Y.S.V., Brasil, C.C.B., de Menezes, C.R., Barin, J.S., Wagner, R., Campagnol, P.C.B., 2019. Ultrasound: A promising technology to improve the technological quality of meat emulsions. Meat Sci. 148 (July 2018), 150–155. doi:10.1016/j.meatsci.2018.10.009.

de Castro, N.T., de Alencar, E.R., Zandonadi, R.P., Han, H., Raposo, A., Ariza-Montes, A., Araya-Castillo, L., Botelho, R.B.A., 2021. Influence of cooking method on the nutritional quality of organic and conventional brazilian vegetables: A study on sodium, potassium, and carotenoids. Foods 10 (8), 1–12. doi:10.3390/foods10081782.

de Souza Comapa, S., Carvalho, L.M.S., Lamarão, C.V., Souza, F., das C. do, A., Aguiar, J.P.L., da Silva, L.S., Mar, J.M., Sanches, E.A., dos Santos, F.F., de Araújo Bezerra, J., Campelo, P.H., 2019. Microwave processing of camu-camu juices: Physicochemical and microbiological parameters. J. Food Process. Preserv. 43 (7), 1–11. doi:10.1111/jfpp.13989.

Gharibzahedi, S.M.T., Roohinejad, S., George, S., Barba, F.J., Greiner, R., Barbosa-Cánovas, G.V., Mallikarjunan, K., 2018. Innovative food processing technologies on the transglutaminase functionality in protein-based food products: Trends, opportunities and drawbacks. Trends Food Sci. Technol. 75, 194–205. doi:10.1016/j.tifs.2018.03.014.

Gulzar, N., Sameen, A., Aadil, R.M., Sahar, A., Rafiq, S., Huma, N., Nadeem, M., Arshad, R., Saleem, I.M., 2020. Descriptive sensory analysis of pizza cheese made from mozzarella and semi-ripened cheddar cheese under microwave and conventional cooking. Foods 9 (2), 1–16. doi:10.3390/foods9020214.

Guo, Z., Zhao, B., Li, H., Miao, S., Zheng, B., 2019. Optimization of ultrasound-microwave synergistic extraction of prebiotic oligosaccharides from sweet potatoes (Ipomoea batatas L.). Innov. Food Sci. and Emerg. Technolog. 54 (August 2018), 51–63. doi:10.1016/j.ifset.2019.03.009.

Guzik, P., Kulawik, P., Zając, M., Migdał, W., 2021. Microwave applications in the food industry: An overview of recent developments. Crit. Rev. Food Sci. Nutr. 0 (0), 1–20. doi:10.1080/10408398.2021.1922871.

Hii, C.L., Ong, S.P., Yap, J.Y., Putranto, A., 2021. Hybrid drying of food and bioproducts : A review. Drying Technol. 39 (11), 1554–1576. doi:10.1080/07373937.2021.1914078.

Hu, B., Li, C., Qin, W., Zhang, Z., Liu, Y., Zhang, Q., Liu, A., Jia, R., Yin, Z., Han, X., Zhu, Y., Luo, Q., Liu, S., 2019. Industrial crops & products a method for extracting oil from tea (*Camelia sinensis*) seed by microwave in combination with ultrasonic and evaluation of its quality. Indust. Crops & Prod. 131 (July 2018), 234–242. doi:10.1016/j.indcrop.2019.01.068.

Huang, M., & Zhang, M. (2018). Synergistic effects of ultrasound and microwave on the pumpkin slices qualities during ultrasound-assisted microwave vacuum frying. *May*, 1–8. doi: 10.1111/jfpe.12835

Jiang, H., Liu, Z., Wang, S., 2018. Microwave processing: Effects and impacts on food components. Crit. Rev. Food Sci. Nutr. 58 (14), 2476–2489. doi:10.1080/10408398.2017.1319322.

Jiang, Q., Zhang, M., Xu, B., 2020. Application of ultrasonic technology in postharvested fruits and vegetables storage: A review. Ultrason. Sonochem. 69 (July), 105261. doi:10.1016/j.ultsonch.2020.105261.

Kenari, R.E., Razavi, R., 2021. Effect of sonication conditions: Time, temperature and amplitude on physicochemical, textural and sensory properties of yoghurt. Int. J. Dairy Technol. 74 (2), 332–343. doi:10.1111/1471-0307.12761.

Khadhraoui, B., Ummat, V., Tiwari, B.K., Fabiano-tixier, A.S., Chemat, F., 2021. Ultrasonics sonochemistry review of ultrasound combinations with hybrid and innovative techniques for extraction and processing of food and natural products. Ultrason. Sonochem. 76, 105625. doi:10.1016/j.ultsonch.2021.105625.

Kumar, C., Karim, M.A., 2019. Microwave-convective drying of food materials: A critical review. Crit. Rev. Food Sci. Nutr. 59 (3), 379–394. doi:10.1080/10408398.2017.1373269.

Kutlu, N., Pandiselvam, R., Saka, I., Kamiloglu, A., Sahni, P., Kothakota, A., 2021. Impact of different microwave treatments on food texture. J. Texture Stud. 1–28. doi:10.1111/jtxs.12635, *September*.

Li, S., Tang, S., Yan, L., Li, R., 2019. Effects of microwave heating on physicochemical properties, microstructure and volatile profiles of yak meat. J. Applied Animal Res. 47 (1), 262–272. doi:10.1080/09712119.2019.1624553.

Liu, L., Lv, C., Meng, X., Xin, G., Li, B., 2020. Effects of different thawing methods on flavor compounds and sensory characteristics of raspberry. Flav. and Fragrance J. 35 (5), 478–491. doi:10.1002/ffj.3580.

Lu, Y., Turley, A., Dong, X., Wu, C., 2011. Reduction of salmonella enterica on grape tomatoes using microwave heating. Int. J. Food Microbiol. 145 (1), 349–352. doi:10.1016/j.ijfoodmicro.2010.12.009.

Martins, C.P.C., Cavalcanti, R.N., Couto, S.M., Moraes, J., Esmerino, E.A., Silva, M.C., Raices, R.S.L., Gut, J.A.W., Ramaswamy, H.S., Tadini, C.C., Cruz, A.G., 2019. Microwave processing: Current background and effects on the physicochemical and microbiological aspects of dairy products. Comprehen. Rev. in Food Sci. and Food Saf. 18 (1), 67–83. doi:10.1111/1541-4337.12409.

Moghimi, M., Farzaneh, V., Bakhshabadi, H., 2018. The effect of ultrasound pretreatment on some selected physicochemical properties of black cumin (*Nigella Sativa*). Nutrire 43 (1), 1–8. doi:10.1186/s41110-018-0077-y.

Morales-soriano, E., Villamiel, M., & Condezo-hoyos, L. (2021). *Ultrasonics sonochemistry hybrid high-intensity ultrasound and microwave treatment : A review on its effect on quality and bioactivity of foods. 80*. doi: 10.1016/j.ultsonch.2021.105835

Nowacka, Małgorzata, & Dadan, M. (2021). *applied sciences current applications of ultrasound in fruit and vegetables osmotic dehydration processes*.

Nowacka, Malgorzata, Wiktor, A., Anuszewska, A., Dadan, M., Rybak, K., Witrowa-Rajchert, D., 2019. The application of unconventional technologies as pulsed electric field, ultrasound and microwave-vacuum drying in the production of dried cranberry snacks. Ultrason. Sonochem. 56 (February), 1–13. doi:10.1016/j.ultsonch.2019.03.023.

Nowak, K.W., Zielinska, M., Waszkielis, K.M., 2019. The effect of ultrasound and freezing/thawing treatment on the physical properties of blueberries. Food Sci. Biotechnol. 28 (3), 741–749. doi:10.1007/s10068-018-0528-5.

Phuhongsung, P., Zhang, M., Bhandari, B., 2020. 4D printing of products based on soy protein isolate via microwave heating for flavor development. Food Res. Int. 137 (May), 109605. doi:10.1016/j.foodres.2020.109605.

Priyadarshini, A., Rajauria, G., Donnell, C.P.O., Brijesh, K., 2019. Emerging food processing technologies and factors impacting their industrial adoption. Crit. Rev. Food Sci. Nutr. 59 (19), 3082–3101. doi:10.1080/10408398.2018.1483890.

Salehi, F., 2020. Physico-chemical properties of fruit and vegetable juices as affected by ultrasound: a review. Int. J. Food Prop. 23 (1), 1748–1765. doi:10.1080/10942912.2020.1825486.

Sattar, S., Imran, M., Mushtaq, Z., Haseeb, M., Holmes, M., Maycock, J., Imran, M., Adeela, K., Khan, M. K., & Muhammad, N. (2019). Functional quality of optimized peach - based beverage developed by application of ultrasonic processing. *August*, 3692–3699. doi: 10.1002/fsn3.1227

Singla, M., Sit, N., 2021. Application of ultrasound in combination with other technologies in food processing: A review. Ultrason. Sonochem. 73, 105506. doi:10.1016/j.ultsonch.2021.105506.

Su, Y., Gao, J., Tang, S., Feng, L., Azam, S.M.R., Zheng, T., 2021. Recent advances in physical fields-based frying techniques for enhanced efficiency and quality attributes. Crit. Rev. Food Sci. Nutr. doi:10.1080/10408398.2021.1882933.

Su, Y., Zhang, M., Adhikari, B., Mujumdar, A. S., & Zhang, W. (2018). Improving the energy efficiency and the quality of fried products using a novel vacuum frying assisted by combined ultrasound and microwave technology. *50*(May 2017), 148–159. doi: 10.1016/j.ifset.2018.10.011

Sungsinchai, S., Niamnuy, C., Wattanapan, P., Charoenchaitrakool, M., Devahastin, S., 2019. Texture modification technologies and their opportunities for the production of dysphagia foods: A review. Compreh. Rev. in Food Sci. and Food Saf. 18 (6), 1898–1912. doi:10.1111/1541-4337.12495.

Téllez-Morales, J.A., Hernández-Santo, B., Rodríguez-Miranda, J., 2020. Effect of ultrasound on the techno-functional properties of food components/ingredients: a review. Ultrason. Sonochem. 61 (August 2019). doi:10.1016/j.ultsonch.2019.104787.

Taşkıran, M., Olum, E., Candoğan, K., 2020. Changes in chicken meat proteins during microwave and electric oven cooking. J. Food Process. Preserv. 44 (2), 1–8. doi:10.1111/jfpp.14324.

Viji, P., Madhusudana Rao, B., Debbarma, J., Ravishankar, C., 2022. Research developments in the applications of microwave energy in fish processing: A review. Trends Food Sci. Technol. 123 (December 2021), 222–232. doi:10.1016/j.tifs.2022.03.010.

Walayat, N., Liu, J., Nawaz, A., Aadil, R.M., López-Pedrouso, M., Lorenzo, J.M., 2022. Role of food hydrocolloids as antioxidants along with modern processing techniques on the surimi protein gel textural properties, developments, limitation and future perspectives. Antioxidants 11 (3), 1–21. doi:10.3390/antiox11030486.

Wang, M., Wu, Y., Liu, Y., Ouyang, J., 2020. Effect of ultrasonic and microwave dual-treatment on the physicochemical properties of chestnut starch. Polymers 12 (8), 1–12. doi:10.3390/POLYM12081718.

Xu, K., Zhang, M., Bhandari, B., 2020. Effect of novel Ultrasonic-Microwave Combined pretreatment on the quality of 3D printed wheat starch-papaya system. Food Biophys. 15 (2), 249–260. doi:10.1007/s11483-019-09615-w.

Yuan, S., Li, C., Zhang, Y., Yu, H., Xie, Y., Guo, Y., Yao, W., 2021. Ultrasound as an emerging technology for the elimination of chemical contaminants in food: A review. Trends Food Sci. Technol. 109 (May 2020), 374–385. doi:10.1016/j.tifs.2021.01.048.

Zeng, H., Zhang, Y., Lin, S., Jian, Y., Miao, S., Zheng, B., 2015. Ultrasonic-microwave synergistic extraction (UMSE) and molecular weight distribution of polysaccharides from fortunella margarita (Lour.) swingle. Sep. Purif. Technol. 144, 97–106. doi:10.1016/j.seppur.2015.02.015.
Zhang, X., Zhang, M., 2020. Trends in food science & technology recent developments in frying technologies applied to fresh foods. Trends Food Sci. Technol. 98 (February), 68–81. doi:10.1016/j.tifs.2020.02.007.
Zhao, K., Saleh, A.S.M., Li, B., Wu, H., liu, Y., Zhang, G., Li, W., 2018. Effects of conventional and microwave pretreatment acetylation on structural and physicochemical properties of wheat starch. Int. J. Food Sci. Technol. 53 (11), 2515–2524. doi:10.1111/ijfs.13845.

CHAPTER SIXTEEN

Synergistic effect of sonication and microwave on the extraction of bioactive peptides

Farhan Saeed[a], Aiza Qamar[b], Muhammad Afzaal[a], Bushra Niaz[a] and Muzzamal Hussain[a]
[a]Department of Food Sciences, Government College University Faisalabad, Faisalabad, Pakistan
[b]Department of Nutrition and Health Promotion, University of Home Economics, Lahore, Pakistan

16.1 Introduction

Bioactive components, particularly bioactive-peptides in food, have sparked attention in recent studies on functional foods. Bioactive peptides (BP) are protein constituents that can affect and change metabolic processes in the human body (Wang. N., 2018). They can be found in both natural and synthetic foods, and if their safety for human consumption has been demonstrated, they have unique metabolic or physiological effects. Bioactive component levels in fruits are affected by causes such as, growing circumstances, cultivar, storing, and shipping (Bennett et al., 2011). Because of their antioxidant characteristics, bioactive substances can be employed as food additives. By oxidizing cell enzymes, proteins, DNA, and cell membrane lipids, reactive oxygen species and free radicals can trigger apoptosis natural antioxidants in food can hunt these oxidative radicals, reducing the evolution of many prolonged diseases caused by oxidation of cellular components (Szymanowska & Baraniak, 2019).

The focus on foods has evolved away from giving vital nutrients for growth and life, towards avoiding, and even preserving many ailments. Also, new scientific advancements, demographic lifestyle changes, and worldwide socio-economic developments all point to a greater need for healthier foods. (Betoret et al., 2011). These are the primary determinants and powerful factors for the global market's new progress and construction of useful foods. Bioactive compounds including vitamins, phytochemicals, and peptides, which are naturally present in functional foods, created during processing, or separated from additional sources, provide health advantages (Fernandes, 2019).

Ultrasound and Microwave for Food Processing: Synergism for Preservation and Extraction.
DOI: https://doi.org/10.1016/B978-0-323-95991-9.00010-2

Antioxidant peptides are one of these bioactive molecules that has recently attracted a lot of interest in the food sector. Antioxidant peptides are proteins with antioxidant action that can be used for mankind well-being, food security and quality by lowering oxidative stress, and lipid peroxidation caused by free radicals (Hema et al., 2017). Scavenging free radicals and reduction of hydro peroxides are the main methods by which antioxidant peptides suppress oxidation (Elias et al., 2008).

Antioxidants originating from natural places like and vegetables fruits are becoming increasingly popular as a result of their numerous health benefits. Fruits and vegetables, on the other hand, must be treated due to their perishability, and short shelf life. Bioactive chemicals are extracted utilizing solvents such as water, methanol, and ethanol in standard processing procedures (Dziki, 2019). Ultrasound-assisted and agitation extractions have recently been thought to be more efficient for extracting important bioactive chemicals from various sources (Solomon & Shimelis, 2020).

BP generated from low-value proteins are becoming increasingly popular. They are usually encoded inside dietary protein sequences and might be released during proteolysis or food preparation with their effects on blood pressure, oxidative stress, inflammation, type-2 diabetes, and these specific dietary protein fragments have a lot of promise for avoiding and curing a variety of long-lasting diseases. Immune responses are regulated by antioxidant defense and the renin-angiotensin system. As a result, antioxidant and antihypertensive peptides frequently demonstrate dual antiobesity and anti-inflammation effects, which has aided the improvement of useful foods, biopharmaceuticals and nutraceuticals (Mora et al., 2012).

BP are unique protein fragments that have favorable effect on physiological functioning of health (Walther & Sieber, 2011; Kitts & Weiler, 2003). There are more than 1500 distinct BPs documented in the "Biopep" database so far (Singh et al., 2014). Proteins are polypeptides that have a higher molecular weight, whereas BPs are organic molecules made by amino acids connected by covalent bonds. Proteins and blood pressure play critical characters in the metabolic operations of active organisms, like humans. They have hormone-like properties and are classed as antithrombotic, antihypertensive, antimicrobial, immunomodulatory, opioid, and mineral binding depending on their method of action.

Bioactive food peptides (BFPs) are encoded in the structure of food proteins. BFPs are thus "locked" in their original form, but can be free during food preparation (through enzymatic hydrolysis), and in the digestive system (López-Barrios et al., 2014). BFPs have been discovered in a many animal and plant dietary proteins. BFPs have physiological roles that are important for human health maintenance, influencing all of the body's major systems (Udenigwe & Aluko, 2012). Angiotensin converting enzyme produces strong blood vessel contractions and restricts their relaxation, resulting in high blood pressure.

There are three mechanisms for releasing BP: enzymatic breakdown by digestive enzymes such as trypsin, chymotrypsin, and pepsin; proteolysis by enzymes produced by microorganisms or plants; and fermentation of milk with proteolysis starter cultures (Korhonen, 2009a). Calcium-binding phosphopeptides, antihypertensive, antibacterial, antioxidative, immunomodulatory and BP have been liberated from casein and whey protein fractions using enzymes like pepsin, alcalase, thermolysin, subtilisin, trypsin, and chymotrypsin (Korhonen, 2009).

The activity of peptides is firm by their amino acid sequence and content once they are free from the ancestor protein where they were encoded. The communication of unique amino acid arrangements that make up proteins modulates natural processes within the body (Kenakin & Miller, 2010). Proteins are classed as endogenous if they are synthesized by an organism from amino acids, or exogenous if they are received from the diet or from an outside source, and they are one of the fundamental parts of food. Proteins derived from plants and animals could include a variety of BP that is encoded in their arrangement (Bhat et al., 2015). Though the relationship among structure and function is not well understood, many BP have structural characteristics such as peptide residue lengths of two to twenty amino acids and the inclusion of hydrophobic amino acids in addition to lysine, proline, or arginine groups (Moller et al., 2008). BP has also been demonstrated to be resistant to digesting peptidases (Kitts & Weiler, 2003).

BPs are a new group of physiologically energetic controllers that can avoid microbial deterioration oxidation and oxidation in foods, for example. They can be used to treat a variety of medical ailments, hence improving one's life (Basilicata et al. 2018). Functional foods and nutraceuticals (Saeed et al., 2021) have recently fascinated much attention, owing to their potential impact on human health and usage in the avoidance of many diseases.

16.2 Sources

Peptides and proteins are the most important macronutrients in diets for the reason that they deliver the essential constituents for protein assembly and function as a source of energy (Dziuba & Dziuba, 2014). They're also a part of a sophisticated chain of organic alterations that take place during the manufacturing and storage of meals, all of which contribute to their sensory qualities. Food proteins and peptides have diverse biological actions and improved nutritional status (Moller et al., 2008).

16.2.1 Peptides from animal sources

Different health consequences have been linked to peptides generated from animal proteins (Apone et al., 2019). Blood is a prospective source of BP and a valuable source of proteins (about 20%). Serum albumin, the major blood protein, has established slight thoughtfulness, despite the fact that blood dumping is a serious challenge for meat producers. In a fresh study, serum albumin was hydrolyzed with various amounts of trypsin, and the peptide arrangements in the hydrolysates were found to have the following activities: glucose regulation, angiotensin converting enzyme inhibition that is, antihypertensive action and Antioxidation (Arrutia et al., 2016a).

16.2.2 Peptides from vegetal sources

The amount of BP generated through in vitro GI digestion of soy milk and soybean seeds was studied. The tests were carried out on protein models derived from soybean milk and seeds, as well as raw soy milk. Research findings showed that soybean proteins degraded during GI digestion, resulting in huge amount of BP, several through known antibacterial action and others by potential antimicrobial activity. The existence of peptides was also investigated using endogenous proteases (Singh et al., 2014). It's possible that the peptides discovered in soy milk models were produced during food handling (Capriotti et al., 2015).

16.3 Extraction of bioactive compounds

Various latest technologies including microwave (MW) assist, high pressure treatment, temperature at different level and moderate pressures, ultrasonication, and super-high frequency electromagnetic field, have been extensively used in the food business and related fields, with the prospective to improve quality, upsurge extraction yield, or protect this environment, among others. The mechanism is related to heat, mechanical efficacies and, cavitation, which can improve bulk transference, enhance substrate-enzyme interaction rate of recurrence, or change substrate configuration (Alarcon-Rojo et al., 2019). Ultrasound-assisted enzymolysis might boost the effectiveness of enzymatic protein hydrolysis on behalf of the formation of BP by causing protein unfolding and enhancing proteolysis by increasing the exposure of vulnerable peptide linkages (Boateng et al., 2019).

The extraction of bioactive substances is subjective to many parameters, comprising the extraction process, raw materials, and extraction solvent (Tiwari, 2015). There are two kinds of procedures: conventional and non-conventional. Organic solvents, agitation and heating are all essential in traditional procedures. Maceration, soxhlet, and hydro-distillation are cases of this kind of procedure. Current methods, too well-known as nonconventional techniques, are green as they need a lesser amount of energy and employ organic solvents, together of which are respectable for the environment (Rodriguez Perez et al., 2015). Innovative skills are being researched to substitute or expand traditional technologies and improve the purposeful qualities of food as customers expect safe and healthy food with extraordinary nutritional and sensual characteristics.

16.3.1 Techniques used in extraction and purification

16.3.1.1 Solid–liquid conventional extraction

Aqueous buffers including chaotropic and reducing reagents (dithiothreitol, mercaptoethanol, urea, etc.) surfactants, and other ingredients are used in traditional protein extraction procedures. Furthermore, protein removal is frequently trailed by a finishing purification stage that consist of using an organic solvent or acidifying the material. Because maximum food proteins have low isoelectric points, protein removal using NaOH solutions and stirring at pHs ranging from 7.5 to 12 is particularly popular, at pH ranging from 3.8 to 5.3. To promote protein solubilization, cultivation temperatures of 40–50°C were used. Fan et al. (2014); Popovi et al. (2013) employed this technique to extract proteins from jujube, watermelon tomato, Chinese cherry, pumpkin seeds, African breadfruit and pumpkin oil cake (Fan et al., 2014; Popovi et al., 2013). Despite the procedure's acceptance and industrialized applicability, its discernment and removal yield are generally low.

16.3.1.2 Protein extraction assisted by ultrasound

High-intensity focused ultrasound extraction was initially created about 1950 and is now commonly used to speed up these processes. High-intensity focused ultrasound generates automated energy and removal is based on the cavitation phenomena. Ultrasonic waves produce rapid pressure variations in a solution, causing tiny gas bubbles to rupture and release large amount of energy. Energy aids the interruption of tissues and cell walls, as well as the extraction of proteins. (Ozuna et al., 2015). Although the fact that this is an ancient approach, up scaling to experimental scale has so far failed (Belwal et al., 2018), and the solvent and flavors utilized now are not suited for food manufacturing.

16.3.1.3 Liquid extraction under pressure

To extract desired substances, pressurized liquid extraction (PLE) applied temperatures and pressures at 50–200°C and 35–200 bar, respectively. The high temperature improves the mass transfer rate and solubility of solvents while lowering viscosity and surface tension. High pressure permits the solvent to reach temperatures over its typical boiling point while remaining liquid. Solvent penetration in the sample is favored under these conditions. PLE uses a fraction of the solvents used in traditional solid–liquid extraction, and it also produces a greater yield and shorter extraction time. Moreover, PLE is commonly performed with ecologically welcoming solvents like ethanol or water (Grosso et al., 2015). Ethanol can be made from agricultural waste and residues as well as from fresh agricultural commodities (cereals and sugar beet) (straws). Though PLE has primarily been used to extract tiny compounds, it can also be used to extract proteins, as evidenced by prior studies on the extraction of algal proteins (Mustafa & Turner, 2011).

16.4 Technology for green extraction

The pharmaceutical and dietary supplement industries rely heavily on phytochemicals produced from plant components. The extraction of medicinal plant ingredients is the first step in both qualitative and quantitative investigation. Even with the greatest chromatographic technology, a partial withdrawal process resulting in a poorly produced extract is enough to provide the most inaccurate findings (Chandrapala et al., 2012). An extraction technique should, in theory, be exhaustive in terms of the contents to be studied, quick, simple, inexpensive, and automated to a high degree. Traditional extraction procedures such as maceration, reflux soxhlet, and hydro distillation, which have been used for decades, are usually the first choice for phytochemical extraction. These approaches, however, have significant limitations, for example, low effectiveness and a long extraction time, especially while only small amounts of chemicals are present. Furthermore, several natural compounds are thermally unstable and may degrade as the temperature rises during extraction. The use of huge amounts of organic solvent in traditional procedures is harmful to the environment, and their disposal is also a worry. In this regard, a technology that can get the majority of the active ingredients in the quickest processing time with the lowest production cost and the least amount of organic solvent will be optimal.

16.4.1 Sonication

The process of using sound energy to agitate particles or discontinuous fibers in a liquid is known as sonication. The method is also known as ultrasonication since ultrasonic frequencies (>20 kHz) are commonly used. An ultrasonic bath or an ultrasonic probe can be used to perform sonication (sonicator). Ultrasonication is a technique that can help you get more yield out of your extraction. Frequency sound waves are sent through the sample during ultrasonication. High and low energy sonication are two types of sonication. For nondestructive quality assurance, low energy sonication (low power, low frequency, >100 kHz, with intensities less than 1 Wcm^{-2}) is commonly utilized (Awad et al., 2012). High-energy sonication (high power, high frequency, 20–500 kHz, with intensities more than 1 Wcm^{-2}) is used to change the physical state of food products, such as emulsification, and defoaming. The sonication treatments are thought to disrupt the covalent connections and links between protein, polysaccharide, and fiber molecules, enhancing protein extractability. Sorghum flours have been demonstrated to generate more protein when sonicated (Beazley, 2017). Tang et al. studied protein extraction from rice bran after sonication treatments and discovered that sonication boosted overall protein yield considerably (Tang et al., 2002). Similar sonication effects on soy protein extraction have been observed, but there is no data on sonication effects on oat protein extraction.

Sonication treatments do not usually modify the fundamental structure of protein extracts (Quansah et al., 2013), but they can change the tertiary and secondary structures of proteins, resulting in a change in functioning. Long-term high-power sonication treatments can cause "cavitation," which is the creation and bursting of air bubbles, which releases energy and heat (O'Sullivan et al., 2016). Cavitation can alter a protein's physical and chemical characteristics (Bean et al., 2006). Protein extraction must be optimized as more individuals turn to food-derived nutraceutical to treat ailments like diabetes.

16.4.2 Extraction assisted by ultrasonication

Ultrasonic treatment declared a viable "green" approach for extraction in the food, chemistry, and medicinal industries over the last three decades. Green extraction attempts to decrease solvent, waste, energy, and pollution while achieving the best product yield (Wei et al., 2015). Ultrasonic treatment is generally applied in the different food sectors for a variety of purposes, including tenderization, extraction, and preservation. Ultrasonication can produce complete withdrawal with improved transparency of the end product in a limited action with good reproducibility, reducing the use of solvent and eliminating waste water treatment post processing procedures (Yu et al., 2016). Ultrasonic treatment has ability to attain the FDA's 5 log decline of germs in drinks criteria (Aneja et al., 2014). Though, ultrasonic therapy has specific negative impacts on bioactive molecules. Tiwari et al. (2008), for example, found a 3.2 percent drop in anthocyanin levels in strawberry juice. Allergens are a fascinating collection of chemicals that have been examined with regards to ultrasonic methods. Food allergies are caused by allergens, which are generally proteinaceous substances that stimulate an excessive immune response. Because ultrasonic treatment affects protein native confirmation (Nazari et al., 2018). Ultrasonication may change the allergenic potential of drugs, creating them comparatively safe for ingesting by select consumer groups. Food products' physico-chemical properties are also influenced by ultrasonication.

The application of ultrasonic technology in meat dealing out has sparked interest due to a variety of generated phenomena and effects. These things are employed aimed at diagnostic reasons, course monitoring, and manipulation of nutritional, biological, and physicochemical properties of diverse media by selecting the suitable interaction parameters. Materials are subjected to enhanced breakdown and depolymerization in the ultrasonic field, reactions are catalyzed, and new complex chemical compounds are generated (Al-Hilphy et al., 2020). The development of ultrasonic phenomena is influenced by a variety of factors. They are impacted by the ultrasonic field's characteristics as well as the investigated materials' basic physicochemical qualities. chemical composition of the environment, and, in particular, the efficacy of acoustic coupling, all have an impact on the events and consequences that result from their interaction (Awad et al., 2012).

The sources of influence of ultrasound are complex biological, physical and chemical processes based on mechanical stresses, energy dissipation, and cavitation, as well as a wide spectrum of secondary events. In particular, mass transfer mechanisms are accelerated, and cell tissue assemblies and microorganisms are destroyed after reaching the ultrasonic cavitation threshold (Téllez-Morales et al., 2020).

16.4.3 Microwave

Abu-Samra et al. were the first to mention the usage of MW energy in 1975. In the laboratory, they employed household ovens to treat biological samples for metal trace analyses (Abu-Samra, 1975). Ganzler et al. (1986) explored that the extraction of organic compounds using MW irradiation became a reality. Pare filed the major patent for MW withdrawal of a natural substance in 1995 (Paré et al., 1995). MW extraction, which was first utilized as a digestion method for a variety of different samples, resulting in advancements in this procedure. MW extraction appears to be one of the most promising ways for extracting valuable chemicals from BP, especially as interest in green technology grows. Furthermore, it is flexible on a small or big size, allowing for technology transfer to advanced sectors.

MW and ultrasound (US) are two essential viable and proficient food treating technologies that are attaining popularity. The consequences of using them as a pretreatment are widely known. By breaking/weakening/cleaving hydrophobic, and electrostatic connections, cavitation and cut off powers generated by sonication cause structural variations in milk proteins (Fang et al., 2012). MWs can cause a diversity of thermal and nonthermal alterations in the assembly of proteins. MW irradiation produces denaturation of proteins, which results in aggregation formation. Protein folding and unfolding are caused by molecule vibrations and collisions, resulting in casein-whey and casein-casein communications. During milk proteolysis, these structural variations affect substrate-enzyme interactions and the production of BP.

MWs are electromagnetic fields having two vertical oscillating fields that is, electric field and magnetic field ranges from 300 MHz to 300 GHz. MWs reducing temperature gradients and speeding up energy transmission. This approach may effectively separate a variety of chemicals, including antioxidants, pigments, essential oils, flavorings, and other organic compounds. In related to conventional heating methods, MW apparatus is a thriving technology as it allows easy, safe, and repeatable access to higher temperatures; the response time can be shortened; the profit can be enhanced; and the cleanliness can be enhanced. This procedure can be carried out with or without the use of a solvent (Surati et al., 2012).

16.4.4 Microwave assisted extraction

Microwave assisted extraction (MAE), a modern technology for extracting phytochemicals and BP from botanicals, has been established to be a likely suitable extraction tool (Kadam et al., 2015). MAE and Sonication provide several advantages over traditional procedures, including a shorter extraction time, reduced solvent use, a better abstraction rate, and better results at a lesser cost. As a result, MAE provides a promising alternative to traditional extraction procedures, particularly for botanical extractions (Gupta et al., 2012).

MAE has been increasingly popular in the abstraction of bioactive chemicals, particularly from plant materials, in recent decades. Due to MW penetration into some materials to a certain penetration and interface with polar components. MAE can improve solvent infiltration into the model matrix and target component transfer from the matrix to the solvent (Azmir et al., 2013).

Protein secondary and tertiary structures can be altered by electromagnetic MW radiation. The primary process of MW-induced alteration is thought to be a rapid rise in temperature and the generation of reactive free radicals. Furthermore, the localization of charged groups in protein molecules can be affected by electromagnetic effects caused by MW radiation. MW radiation, in particular, can quickly permeate the core of complicated structures where cavitation is not possible. MW effects aid to accelerate fundamental changes in sonicated protein, while an established and constant cavitation process may help to decrease the constant heating phenomenon that can be a better way of modification than MW and ultrasound treatment separately. MW and ultrasound treatments worked well together in the production of biomaterials and the extraction of natural compounds, according to earlier research (Shamis et al., 2011).

16.4.5 Microwave-assisted extraction and applications

As the decline of extraction time and solvent usage and its high efficiency, MAE is currently extensively employed to take out natural chemicals from plant matrices. As a result, MAE is regarded as a green extraction process that is gaining popularity. To increase the flexibility of the approach and the improvement of green extraction methods, certain improved extraction methods using MWs as a heating source have been created in recent years; a portion of them will be described next. Some sensitive chemicals can be degraded by high temperatures, high irradiation power, and long irradiation times. MAE can be performed in an inert environment or under vacuum to help overcome this disadvantage, restricting the occurrence of oxygen in the reactor.

16.4.5.1 Nitrogen-protected microwave-assisted extraction

The use of a pressurized inert gas like nitrogen, in a locked system helps avoid oxidation of the active chemicals during the withdrawal process (Krise, 2011). The plant model is placed into the vessel along with a particular volume of solvent in nitrogen-protected microwave aided extraction. A vacuum pump is used to remove the air from the flask until it reaches a specified level of vacuum. The nitrogen from the gas cylinder is then pumped into the vessel. Ascorbic acid has been extracted from yellow pepper, guava, green pepper, and cayenne pepper using this method. The extraction was carried out for 10 minutes in a solution having 0.25 percent meta-phosphoric acid with 10 and a MW power of 400 W. In comparison to traditional MAE and solvent extraction procedures, the nitrogen-protected microwave aided extraction process greatly reduced or avoided ascorbic acid oxidation, resulting in a greater ascorbic acid extraction yield.

16.4.5.2 Vacuum microwave-assisted extraction

MAE in a vacuum system is known as vacuum microwave-assisted extraction (VMAE) (Gu et al., 2016). Under vacuum, the extraction solvent's boiling point is lower than at atmospheric pressure. As a result, the solvent may be reserved boiling and refluxing at a lesser temperature that is ideal for combining the model with the solvent and removing thermo-sensitive chemicals without degrading them. Furthermore, because most of the air in the extraction system is pushed out, oxygen-sensitive chemicals are avoided or reduced from oxidizing. This form of MAE improves mass transfer by boosting active component diffusion into the solvent through suction pressure. Certainly, the dissimilarity in pressure between the interior and external part of the cell wall improves the solute extraction (Xiao, 2012). As a result, VMA E is well suited to the removal of thermosensitive and oxidative chemicals at lesser temperatures and with small oxygen content.

16.4.5.3 Ultrasonic microwave-assisted extraction

The ultrasound/sonication assisted extraction (UAE) applied to maximize algal protein extraction has piqued curiosity. By utilizing acoustic cavitation events, this approach aids in cell wall collapse. Cavitation causes air bubbles to implode, resulting in micro turbulence that converts sound energy into mechanical energy, disrupting the cell wall (Bleakley et al., 2017; Dumay et al., 2016). In laboratory-scale experiments, two types of ultrasound equipment were used: an ultrasonic water bath (which is further often cheap) (Kadam et al., 2015). The key benefits of this technology are its quick processing time, the fact that it is a nonthermal procedure, and the fact that it does not require an organic solvent. However, using high ultrasonic power for a long time might generate heat, which can affect protein structure.

MAE is also known as ultrasonic microwave-assisted extraction can improve the mass transfer process in extraction (UMAE). Furthermore, the extraction is completed in less time and with less solvent use. (Chen et al., 2010). UMAE has been used to extract useful chemicals from a range of plants, including lycopene from tomatoes, vegetable oil (Cravotto et al., 2008), and polysaccharides (Lianfu et al., 2008). In a study comparing UMAE and traditional methods, researchers found that extracting polysaccharides under optimal UMAE circumstances amplified yield from 2.12 percent to 3.25 percent and purity from 64.03 percent to 73.16 percent, compared to 64.03 percent previously recorded by traditional hot water extraction.

16.5 Combination of novel extraction technologies

In addition to the mounting development toward environmentally welcoming operations, the food department is mainly worried in dropping assembly costs by either accelerating the procedure or enhancing yield. While there is no such thing as an ideal removal procedure based solely on technology, there are methods to attack an equilibrium between product quality, manufacturing costs, and solvent usage. New technologies including MW and ultrasound-aided extraction, are being used as potent tools to give greater productions and healthier products, as detailed in the preceding sections. Combinations of these technologies could be used to achieve a better balance.

16.6 Conclusion

Conventional and novel extraction technologies and their combinations are also highlighted of this chapter. Novel extraction technologies are used for getting maximum yield in less time, and improving quality and considered as eco-environmentally. Various researchers are now concentrating on the use of these innovative extraction techniques in combination. These processes might be an essential step toward the long-term production and use of BP from different sources. Moreover, combination of innovative extraction methods should be introduced at pilot scales for the extraction of BP.

References

Abu-Samra, A., Morris, J.S., Koirtyohann, S.R., 1975. Wet ashing of some biological samples in a microwave oven. Anal. Chem. 47 (8), 1475–1477.

Alarcon-Rojo, A.D., Carrillo-Lopez, L.M., Reyes-Villagrana, R., Huerta-Jimenez, M., Garcia-Galicia, I.A., 2019. Ultrasound and meat quality: A review. Ultrason. Sonochem. 55, 369–382.

Angiolillo, L., Del Nobile, M.A., Conte, A., 2015. The extraction of bioactive compounds from food residues using microwaves. Curr. Opin. Food Sci. 5, 93–98.

Al-Hilphy, A.R., Al-Temimi, A.B., Al Rubaiy, H.H.M., Anand, U., Delgado-Pando, G., Lakhssassi, N., 2020. Ultrasound applications in poultry meat processing: A systematic review. J. Food Sci. 85 (5), 1386–1396.

Aneja, K.R., Dhiman, R., Aggarwal, N.K., Aneja, A., 2014. Emerging preservation techniques for controlling spoilage and pathogenic microorganisms in fruit juices. Internat. J. Microbiol. 2014, 758942. https://doi.org/10.1155/2014/758942.

Apone, F., Barbulova, A., Colucci, M.G., 2019. Plant and microalgae derived peptides are advantageously employed as bioactive compounds in cosmetics. Front. Plant Sci. 10, 756.

Arrutia, F., Puente, Á., Riera, F.A., Menéndez, C., González, U.A., 2016a. Influence of heat pre-treatment on bsa tryptic hydrolysis and peptide release. Food Chem. 202, 40–48.

Arzeni, C., Martínez, K., Zema, P., Arias, A., Pérez, O.E., Pilosof, AMR., 2012. Comparative study of 67 high intensity ultrasound effects on food proteins functionality. J. Food Eng. 108 (3), 463–472. doi:10.1016/j.jfoodeng.2011.08.018.

Awad, T.S., Moharram, H.A., Shaltout, O.E., Asker, D.Y.M.M., Youssef, M.M., 2012. Applications of ultrasound in analysis, processing and quality control of food: A review. Food Res. Int. 48 (2), 410–427. doi:10.1016/j.foodres.2012.05.004.

Azmir, J., Zaidul, I.S.M., Rahman, M.M., Sharif, K.M., Mohamed, A., Sahena, F., …, Omar, A.K.M., 2013. Techniques for extraction of bioactive compounds from plant materials: A review. J. Food Eng. 117 (4), 426–436.

Barbero, G.F., Palma, M., Barroso, C.G., 2006. Determination of capsacinoids in peppers by microwave assisted extraction high performance liquid chromatography with fluorescence detection. Anal. Chim. Acta 578, 227–233.

Barnabas, I.J., Dean, J.R., Fowlis, I.A., Owen, S.P., 1995. Comparison of pressurized liquid extraction with soxhlet extraction for the analysis of polychlorinated dibenzo-p-dioxins and dibenzofurans from fly ash and environmental matrixes. J. Chromatogr. A. 815, 231–241.

Basilicata, M.G., Pepe, G., Sommella, E., Ostacolo, C., Manfra, M., Sosto, G., Pagano, G., Novellino, E., Campiglia, P., 2018. Peptidome profiles and bioactivity elucidation of buffalo-milk dairy products after gastrointestinal digestion. Food Res. Int. 105, 1003–1010.

Bean, S.R., Ioerger, B.P., Park, S.H., Singh, H., 2006. Interaction between sorghum protein extraction and precipitation conditions on yield, purity, and composition of purified protein fractions. Cereal Chem. 83 (1), 99–107.

Beazley, S.F. (2017). Effects of ultra-sonication process on digestibility of kafirin in sorghum.

Belwal, T., Ezzat, S.M., Rastrelli, L., Bhatt, I.D., Daglia, M., Baldi, A., Devkota, H.P., Orhan, I.E., Patra, J.K., Das, G., AtanasovDas, A.G., 2018. A critical analysis of extraction techniques used for botanicals: trends, priorities, industrial uses and optimization strategies. TRAC Trends Anal. Chem. 100, 82–102.

Betoret, E., Betoret, N., Vidal, D., Fito, P., 2011. Functional foods development: trends and technologies. Trends Food Sci. Technol. 22 (9), 498–508.

Bhat, Z.F., Kumar, S., Bhat, H.F., 2015a. Bioactive peptides from egg: A review. Nutr. Food Sci. 45, 190–212.

Bleakley, S., Hayes, M., 2017. Foods 6, 33.

Boateng, E.F., Nasiru, M.M., 2019. Applications of ultrasound in meat processing technology: A review. Food Sci. Technol. 7, 11–15.

Butnariu, M., Sarac, I., 2019. Functional food. Int. J. Nutr. 3 (3), 7–16.

Capriotti, A.L., Caruso, G., Cavaliere, C., Samperi, R., Ventura, S., Chiozzi, R.Z., Laganà, A., 2015. Identification of potential bioactive peptides generated by simulated gastrointestinal digestion of soybean seeds and soy milk proteins. J. Food Compos. Anal. 44, 205–213.

Capriotti, A.L., Cavaliere, C., Piovesana, S., Samperi, R., Aldo Laganà, A., 2016. Recent trends in the analysis of bioactive peptides in milk and dairy products. Anal. Bioanal.Chem. 408, 2677–2685.

Carrasco-Castilla, J., Hernández-Álvarez, A.J., Jiménez-Martínez, C., Gutiérrez- López, G.F., Dávila-Ortiz, G., 2012. Use of proteomics and peptidomics methods in food bioactive peptide science and engineering. Food Eng. Rev. 4, 224–243.

Chandrapala, J., Oliver, C., Kentish, S., Ashokkumar, M., 2012. Ultrasonics in food processing. Ultrason. Sonochem. 19 (5), 975–983.

Chemat, F., Chemat, N., Rombaut, A.G., Sicaire, A., Meullemiestre, A.S., Fabiano-Tixier, M., Abert-Vian, 2017. Ultrasound assisted extraction of food and natural products. Mechanisms, techniques, combinations, protocols and applications. A review. Ultrason. Sonochem. 34, 540–560. doi:10.1016/j.ultsonch.2016.06.035.

Chen, Y., Gu, X., Huang, S.Q., Li, J., Wang, X., Tang, J., 2010. Int. J. Biol. Macromol. 46, 429.

Cravotto, G., Boffa, L., Mantegna, S., Perego, P., Avogadro, M., Cintas, P., 2008. Ultrason. Sonochem. 15, 898.

Crozier, A., Jaganath, I.B., Clifford, M.N., 2006. Phenols, Polyphenols and tannins: An Overview. Blackwell Publishing Limited UK, Oxford, pp. 1–24 978-1-405-12509-3.

De-Mesa-Stonestreet, N.J., Alavi, S., Bean, SR., 2010. Sorghum proteins: The concentration, isolation, modification, and food applications of kafirins. J. Food Sci. 75 (5), 90–104.

Dogan, K., Akman, P.K., Törnük, F., 2019. Improvement of bioavailability of sage and mint by ultrasonic extraction. Int. J. Life Sci. Biotechnol. 2 (2), 122–135.

Dziki, D., Gawlik-Dziki, U., Biernacka, B., 2019. *Cistus incanus* L. as an innovative functional. Foods 8, 1–12. doi:10.3390/foods8080349.

Dziuba, B., Dziuba, M., 2014. Milk proteins-derived bioactive peptides in dairy products: Molecular, biological, and methodological aspects. Acta Scientiarum Polonorum, Technologia Alimentaria 13, 5–25.

Elias, R.J., Kellerby, S.S., Decker, E.A., 2008. Antioxidant activity of proteins and peptides antioxidant activity of proteins and peptides. Crit. Rev. Food Sci. Nutr. 48, 430–441.

Fan, S., Hu, Y., Li, C., Liu, Y., 2014. Optimization of preparation of antioxidative peptides from pumpkin seeds using response surface method. PLoS One 9, e92335. doi:10.1371/journal.pone.0092335.

Fang, G., Lau, H.F., Law, W.S., Li, S.F.Y., 2012. Systematic optimisation of coupled microwave-assisted extraction-solid phase extraction for the determination of pesticides in infant milk formula via LC–MS/MS. Food Chem. 134 (4), 2473–2480.

Fernandes, S.S., Coelho, M.S., de las Mercedes Salas-Mellado, M., 2019. Bioactive compounds as ingredients of functional foods: polyphenols, carotenoids, peptides from animal and plant sources new. In: Bioactive Compounds. Woodhead Publishing, Sawston, Cambridge, USA, pp. 129–142.

Fields, K., Falla, T.J., Rodan, K., Bush, L., 2009. Bioactive peptides: signaling the future. J. Cosmetic Dermatol. 8, 8–13.

Fitzgerald, R.J., Murray, B.A., Walsh, D.J., 2004. Hypotensive peptides from milk proteins. J. Nutr. 134 (4), S980–S988.

Fleurence, J., Levine, I., 2016. Seaweed in health and disease prevention. Academic Press, pp. 275–318.

Ganzler, K., Salgó, A., Valkó, K., 1986. Microwave extraction: A novel sample preparation method for chromatography. J. Chromatogr. A 371, 299–306.

Grosso, C., Valentão, P., Ferreres, F., Andrade, P.B., 2015. Alternative and efficient extraction methods for marine-derived compounds. Mar. Drugs. 13, 3182–3230. doi:10.3390/md13053182.

Gu, H., Chen, F., Zhang, Q., Zang, J., 2016. Application of ionic liquids in vacuum microwave-assisted extraction followed by macroporous resin isolation of three flavonoids rutin, hyperoside and hesperidin from Sorbus tianschanica leaves. J. Chromatogr. B 1014, 45–55.

Gupta, A., Naraniwal, M., Kothari, V., 2012. Modern extraction methods for preparation of bioactive plant extracts. Int. J. App. Natural Sci. 1 (1), 8–26.

Hema, G.S., Joshy, C.G., Shyni, K., Chatterjee, N.S., Ninan, G., Mathew, S., 2017. Optimization of process parameters for the production of collagen peptides from fish skin (*Epinephelus malabaricus*) using response surface methodology and its characterization. J. Food Sci. Technol. 54 (2), 488–496.

Hernández-Ledesma, B., Contreras, M.M., Recio, I., 2011. Antihypertensive peptides: Production, bioavailability and incorporation into foods. Adv. Colloid Interface Sci. 165 (1), 23–35.

Kadam, S.U., Álvarez, C., Tiwari, B.K., O'Donnell, C.P., Tiwari, B.K., Troy, D.J., 2015. Seaweed Sustainability. Academic Press, San Diego, pp. 243–269.

Kenakin, T., Miller, L.J., 2010. Seven transmembrane receptors as shapeshifting proteins: the impact of allosteric modulation and functional selectivity on new drug discovery. Pharmacol. Rev. 62 (2), 265–304.

Kitts, D.D., Weiler, K., 2003. Bioactive proteins and peptides from food sources. Applications of bioprocesses used in isolation and recovery. Curr. Pharm. Des. 9, 1309–1323.

Korhonen, H.J., 2009a. Bioactive milk proteins and peptides: from science to functional applications. Aust. J. Dairy Technol. 64 (1), 16–25.

Korhonen, H.J., 2009b. Milk-derived bioactive peptides: from science to applications. J. Funct. Foods 1 (2), 177–187.

Krise KM. (2011).The effects of microviscosity, bound water, and protein mobility on the radiolysis and sonolysis of hen egg white. Available at: https://etda.libraries.psu.edu/files/final_submissions/3497.

Lemes, A.C., Sala, L., Ores, J.D.C., Braga, A.R.C., Egea, M.B., Fernandes, K.F., 2016. A review of the latest advances in encrypted bioactive peptides from protein-rich waste. Int. J. Mol. Sci. 17, 950.

Lianfu, Z., Zelong, L., 2008. Ultrason. Sonochem. 15, 731.

López-Barrios, L., Gutiérrez-Uribe, J.A., Serna-Saldívar, S.O., 2014. Bioactive peptides and hydrolysates from pulses and their potential use as functional ingredients. J. Food Sci. 79 (3), R273–R283.

Ma, S., Yang, X., Zhao, C., Guo, M., 2018. Ultrasound-induced changes in structural and physicochemical properties of β-lactoglobulin. Food Sci. Nutr. 6 (4), 1053–1064.

Moller, N.P., Scholz-Ahrens, K.E., Roos, N., Schrezenmeir, J., 2008. Bioactive peptides and proteins from foods: Indication for health effects. Eur. J. Nutr. 47, 171–182.

Mora, L., Toldrá, F., 2012. Proteomic identification of small (<2000 Da) myoglobin peptides generated in dry-cured ham. Food Technol. Biotechnol. 50 (3), 343.

Mora, L., Reig, M., Toldrá, F., 2014. Bioactive peptides generated from meat industry by-products. Food Res. Int. 65 (Part C), 344–349.

Mustafa, A., Turner, C., 2011. Pressurized liquid extraction as a green approach in food and herbal plants extraction: A review. Anal. Chim. Acta 703, 8–18. doi:10.1016/j.aca.2011.07.018.

Nazari, B., Mohammadifar, M.A., Shojaee-Aliabadi, S., Feizollahi, E., Mirmoghtadaie, L., 2018. Effect of ultrasound treatments on functional properties and structure of millet protein concentrate. Ultrason. Sonochem. 41, 382–388.

O'Sullivan, J.J., Park, M., Beevers, J., Greenwood, R.W., Norton, I.T., 2016. Applications of ultrasound for the functional modification of proteins and nanoemulsion formation.

O'Sullivan, J.J., Murray, B., Flynn, C., Norton, I., 2016. The effect of ultrasound treatment on the structural, physical and emulsifying properties of animal and vegetable proteins. Food Hydrocoll. 53, 141–154. doi:10.1016/j.foodhyd.2015.02.009.

Ozuna, C., Paniagua-Martínez, I., Castaño-Tostado, E., Ozimek, L., Amaya-Llano, S.L., 2015. Innovative applications of high-intensity ultrasound in the development of functional food ingredients: Production of protein hydrolysates and bioactive peptides. Food Res. Int. 77, 685–696. doi:10.1016/j.foodres.2015.10.015.

Paré, J.J., Bélanger, J.M., Li, K., Stafford, S.S., 1995. Microwave-assisted process (MAP™): Application to the headspace analysis of VOCs in water. J. Microcolumn Sep. 7 (1), 37–40.

Popović, L., Peričin, D., Vaštag, Ž., Popović, S., Krimer, V., Torbica, A., 2013. Antioxidative and functional properties of pumpkin oil cake globulin hydrolysates. J. Am. Oil Chem. Soc. 90, 1157–1165.

Quansah, J.K., Udenigwe, C.C., Saalia, F.K., Yada, RY., 2013. The effect of thermal and ultrasonic treatment on amino acid composition, radical scavenging and reducing potential of hydrolysates obtained from simulated gastrointestinal digestion of cowpea proteins. Plant Foods Hum. Nutr. 68 (1), 31–38. doi:10.1007/s11130-013-0334-4.

Rodrigues, R., Trindade, F.G., Da Cunha, M., Gomes, V.M., 2015. Thionin-like peptide from *capsicum annuum* fruits: Mechanism of action and synergism with fluconazole against *candida* species. BMC Microbiol. 16, 12. doi:10.1186/s12866-016-0626-6.

Saeed, F., Hussain, M., Arshad, M.S., Afzaal, M., Munir, H., Imran, M., Tufail, T., Anjum, F.M., 2021. Functional and nutraceutical properties of maize bran cell wall non-starch polysaccharides. Int. J. Food Prop. 24 (1), 233–248.

Salleh-Mack, S.Z., Roberts, J.S., 2007. Ultrasound pasteurization: the effects of temperature, soluble solids, organic acids and pH on the inactivation of escherichia coli atcc 25922. Ultrason. Sonochem. 14, 323–329.

Shahidi, F., Zhong, Y., 2008. Bioactive peptides. J. AOAC Int. 91, 914–931.

Shamis, Y., Taube, A., Mitik-Dineva, N., Croft, R., Crawford, R.J., Ivanova, E.P., 2011. Specific electromagnetic effects of microwave radiation on Escherichia coli. Appl. Environ. Microbiol. 77 (9), 3017–3022.

Sharma, S., Singh, R., Rana, S., 2011. Bioactive peptides: A review. Int. J. Bioautomot. 15, 223–250.

Singh, B.P., Vij, S., Hati, S., 2014. Functional significance of bioactive peptides derived from soybean. Peptides 54, 171–179.

Solomon, A.T., Shimelis, A., 2020. Production and Processing of Antioxidant Bioactive peptides: A driving Force For the Functional Food market. Department of Food Science and Applied Nutrition. Heliyon Cell Press Elsevier Ltd, Cambridge, Massachusetts, U.S.

Sun, J., Mei, Z., Tang, Y., Ding, L., Jiang, G., Zhang, C., Bai, W., 2016. Stability, antioxidant capacity and degradation kinetics of pelargonidin-3-glucoside exposed to ultrasound power at low temperature. Molecules 21 (9), 1–12 2016.

Surati, M.A., Jauhari, S., Desai, K.R., 2012. A brief review: Microwave assisted organic reaction. Arch. Appl. Sci. Res. 4 (1), 645–661.

Szymanowska, U., Baraniak, B., 2019. Antioxidant and potentially anti-inflammatory activity of anthocyanin fractions from pomace obtained from enzymatically treated raspberries. Antioxidants 8 (8), 299. doi:10.3390/antiox8080299.

Tang, S., Hettiarachchy, N.S., Shellhammer, TH., 2002. Protein extraction from heat-stabilized defatted rice bran. 1. physical processing and enzyme treatments. J. Agric. Food Chem. 50 (25), 7444–7448. doi:10.1021/jf025771.

Téllez-Morales, J.A., Hernández-Santo, B., Rodríguez-Miranda, J., 2020. Effect of ultrasound on the techno-functional properties of food components/ingredients: A review. Ultrason. Sonochem. 61, 104787.

Tiwari, B.K., Álvarez, C., O'Donnell, C.P, 2015. Ultrasound applications for the extraction, identification and delivery of food proteins and bioactive peptides. Trends Food Sci. Technol. 46, 60–67.

Tiwari, B.K., C.P. O'Donnell, A. Patras, P.J. Cullen. 2008. Anthocyanin and ascorbic acid degradation in sonicated strawberry juice.

Udenigwe, C.C., Aluko, R.E., 2012. Food protein-derived bioactive peptides: Production, processing, and potential health benefits. J. Food Sci. 77 (1), R11–R24.

Walther, B., Sieber, R., 2011. Bioactive proteins and peptides in foods. Int. J. Vitam. Nutr. Res. 81 (2), 181.

Wang, N., Manabe, Y., Sugawara, T., Paul, N.A., Zhao, J., 2018. Identification and biological activities of carotenoids from the freshwater alga *oedogonium intermedium*. Food Chem. 242, 247–255. doi:10.1016/j.foodchem.2017.09.075.

Wei, Z., Kowalska, E., Verrett, J., Colbeau-Justin, C., Remita, H., Ohtani, B., 2015. Morphology-dependent photocatalytic activity of octahedral anatase particles prepared by ultrasonication–hydrothermal reaction of titanates. Nanoscale 7 (29), 12392–12404.

Xiao, X., W. Song, J. Wang and G. Li, Anal. Chim. Acta 2012, 712, 85.

Yu, L., Shearer, C., Shapter, J., 2016. Recent development of carbon nanotube transparent conductive films. Chem. Rev. 116 (22), 13413–13453.

Zaky, A.A., Chen, Z., Liu, Y., Li, S., Jia, Y., 2019. Preparation and assessment of bioactive extracts having antioxidant activity from rice bran protein hydrolysates. J. Food Measur. Character. 13, 2542–2548. doi:10.1007/s11694-019-00174-9.

CHAPTER SEVENTEEN

Recent advances in applications of sonication and microwave

Tejasvi Pandey[a], Anuradha Sandhu[a], Aditi Sharma[a] and Mohammad Javed Ansari[b]

[a]Department of Forensic science, School of Bioengineering and Biosciences, Lovely Professional University, Phagwara, India

[b]Department of Botany, Hindu College Moradabad (Mahatma Jyotiba Phule Rohilkhand University Bareilly), India

17.1 Introduction

Sonication and microwave technologies are gaining popularity in the preceding years due to their greener, efficient, and energy-efficient methodologies. These technologies have replaced the traditional approaches and are leading toward a sustainable future. Their utilization and applications in various fields have made them popular among all domains. Researchers, industries, and domestic households utilize these technologies and take advantage of them to obtain desired results. Sonochemistry revolves around chemical processes in a fluid or liquid medium that utilizes ultrasound irradiation which produces various physicochemical effects. These effects are thermal, chemical, and mechanical results from the cavitation phenomenon of sonication which happens by the formation and growth of microbubbles in a liquid state and then their sudden collapse. The microwave also utilizes the energy of 0.3–300 GHz frequency range of electromagnetic spectrum from microwave region by major phenomena of ionic conduction with dipolar polarization that exhibits dielectric mechanism of the microwave. The energy of microwave is absorbed by the material and then converted into heat which leads to bulk heating which is different from conventional conductive heating as here temperature also rises simultaneously for a whole (Chatel et al., 2019). This chapter focuses on the advances in applications of both sonication and microwave in various fields. Various applications of both ultrasonication and microwave are shown in Table 17.1.

17.2 Sonication

Due to increasing consumer demands, food quality, and nutritional value are a must demand in food technology. This reason is responsible for

Ultrasound and Microwave for Food Processing: Synergism for Preservation and Extraction.
DOI: https://doi.org/10.1016/B978-0-323-95991-9.00003-5

Table 17.1 Various functions of ultrasonication and microwave (Chatel et al, 2019).

Application	Ultrasonication	Microwave
Bulk metal use	Preferable	Not preferable
Reaction media	Organic and aqueous media	Microwave absorbing liquids and solvent-free procedures
Acceleration	Minutes to hours	Seconds to minutes
Scaling-up	Possible but still a challenge	Possible but still a challenge
Activation	Cavitation	Thermal effects
Chemical effects	Mechanistic switching, selective changes, waste reductions	Selective changes and waste reductions
Others	Cleaning, light emission, microstreaming, solubilization, etc.	Heating above boiling points and change in solvent properties etc.

the emerging nonthermal technologies rather than conventional thermal techniques. Ultrasound is one of these nonthermal techniques. Ultrasound is capable of preserving fluid foods and can be used solo or in a combination of various other preservation techniques. Sonication or ultrasonic energy is mostly used in analytical chemistry. Ultrasonic devices perform different functions and are thus different in operations. Sonication is a process of using ultrasonic frequency or sound energy (>20 kHz) for the extraction of particles, cells or fibers from a sample, and is thus also called ultrasonication (Dalimunthe et al., 2020). The process works by converting the electrical signal into physical vibration for disruptions for the purpose of dissolving, mixing, and removal of unwanted particles or gases from the liquids. Mostly, sonication is used for cleaning and separating different substances (Batista et al., 2009; Show et al., 2007). Production of active oxygen radicals and hydroxyl radicals occurs in the sonication process that oxidizes dibenzothiophene into sulfoxide and sulfone which diffuses the cell and transforms into 2-hydroxybiphenyl (Suryanto et al., 2018). The process of sonication uses ultrasonic sound waves. During the process, the ultrasound undergoes a series of compressions. Pressure is applied; due to which, thousands of vacuum bubbles are formed. When cavitation occurs, the bubble collapses. The collapsing of bubbles releases much energy for the disruption of molecular bonds. As the molecular interaction breaks, separation occurs, and mixing happens.

Sonication can be used in two ways: a sonication bath or a sonicator (horn or a probe). The sonicator consists of:

- A Generator: A generator converts electrical power into an electrical signal. This electrical signal is used to power the transducer.

- A Transducer: For converting an electrical signal into vibration, a transducer is used. After amplification of this vibration into longitudinal vibration that causes a cavity in the sample, the vibration is used in the probe tip. This cavitation makes it easier to break down the particles.
- A probe

The sonication process is governed by the principle that when bubbles collapse, they create cavitation which results in homogenization, dispersion, disintegration, extraction, emulsions, and sonochemical effects of liquids. When sonication or high-power ultrasound is introduced, it creates low-pressure regions called rarefactions and high-pressure regions called compressions. The process of ultrasound is used in combination with other techniques (Zinoviadou et al., 2015), if the technique is mild heat treatments (<100°C) it is called thermo-ultrasonication. If increased pressure is used with the combination of ultrasound, it is called manosonication, however, the use of ultrasound; increased pressure, and temperature is called manothermosonication (Knorr et al., 2004). This Mano sonication and manothermosonication are used for sterilization of liquid eggs and jams.

Sonication in food technology plays a very important role as it reduces contamination and preserves nutrition when used in a blend with mild heat (<50°C) and antimicrobials. For the preservation of food, low-frequency ultrasound (18–100 kHz, $\lambda = 145$ mm).

17.2.1 Types of sonicators

- Q700 sonicator: It is the most advanced model of sonicators. It provides a touch screen interface and a user-friendly interface. Sonicators are also reproducible, efficient, and reliable. It consists of a generator, convertor, 13 mm diameter probe, power cable, wrench set, and converter cable (Bioke et al., 2015).
- Q500 sonicator: This model is used for creating emulsions, homogenization, cell lysis, and nanoparticle dispersion. It consists of a generator, convertor, 13 mm diameter probe, converter cable, and wrench set.
- Q700/Q500 (direct horn options)
  - Replacement tips for standard probes: During use, these tips get eroded and with time can become less effective. Though are easily replaceable and can be easily removed.
  - Microtip probes: These probes are high-intensity thin probes that are used for small sample volume processing.
  - Coupler with stepped microtip: These are used when a small sample size and do not require high power.

- Extenders: These are so designed to fit into the vessels which are long-necked. Extenders are available in different lengths and have replaceable tips or are solid.
- Boosters are used for increasing the intensity of horns. Amplitude is increased by placing them between the converter and the horn.
- High gain horns: They are also called high-intensity horns. They increase the amplitude of horns to double and are attached directly to the converter.
- 4 tip horn: These 4 tip horn allows multiple sample processing at the same time and offers high intensity and is also effective for homogenization, mixing, and cell disruption.
- Sound enclosure: These sound enclosures are used to reduce loud noise from sonicators (about 20 dBa). They also consist of an internal support rod, and converter mounting system (Bioke et al., 2015).
- Flocells help in inline, continuous, large volume, batch sample processing. Dispersing and mixing are the main uses.
- Cup horn: These cup horns offer indirect sonications and perform functions like an ultrasonic water bath. These can process multiple samples. It also contains a chiller to keep the temperature cooler during the process.
- Recirculating chiller: These are used to maintain the temperature during the sonication process.

- Q700 sonicators: (Indirect horn options) microplate horns are larger than cup horns and are an indirect sonication device that can process 96 well microtiter plates at one time.
- Q125 sonicator: It is microprocessor-based and can be programmed. Its features include pulse mode and a digital display of both wattage and joules. It has other features like programmable operation, pulse mode, digital amplitude, elapsed time indicator, overload protection, etc.
- Q55 sonicator: It is compact and also costs lesser, it also occupies less space. Its features include the smallest unit available, thumb-switch or continuous operation, and simple and effective operation. Different features of various sonication models are discussed in Table 17.2.

17.2.2 Parameters influencing the sonication process

17.2.2.1 Frequency

The production of cavitation bubbles at low frequency is easier than their production at high frequency. To accomplish cavitation, as the sonic frequency increments, the power of the applied sound expands, guaranteeing

Table 17.2 Different features of different sonication models (Bioke, 2015).

Sonicators features	Q500	Q700	Q1375	Q125	Q55
Voltage	110v, 50/60 hz	110v, 50/60 hz	220 V, 50/60 Hz	110v, 50/60 hz	110v, 50/60 hz
Frequency	20 kHz	20 kHz	20 kHz	20 kHz	20 kHz
Programming	10 hours	72 hours	10 hours	10 hours	-
Dimensions	8*5.25*8.5 in.	8*5.25*8.5 in.	7 × 15 × 18.25 in.	8*13.75*5.75 in.	8*7.5*5.75 in
Power	500 watts	700 watts	1375 watts	125 watts	55 watts

that the firm powers of the fluid media survive, and voids are made (Rial-Otero et al., 2007). To initiate cavitation in water extra power is needed at 400 kHz than required at 10 kHz. The actual clarification for this is given by, at extremely high frequencies, the pattern of pressure, and decompression brought about by the ultrasonic waves turns out to be short to the point that the particles of the fluid cannot be isolated to shape a void and, subsequently, cavitation is not generally obtained (Pena-Farfal et al., 2004).

17.2.2.2 Intensity

The power of sonication is corresponding to the adequacy of ultrasonic source vibrations, and in that capacity, an addition in the abundance of vibration will prompt an expansion in the force of vibration and an increment in the sonochemical impacts (Louis-Marly et al., 2020). To accomplish the cavitation edge base force is required. This implies that higher amplitudes are not generally important to acquire the ideal outcomes. Furthermore, high amplitudes of sonication can prompt fast weakening of the ultrasonic transducer, bringing about fluid fomentation rather than cavitation, and in the helpless transmission of the ultrasound through the fluid media. Notwithstanding, the amplitude ought to be expanded when working with tests of high thickness, like blood (Capelo et al., 1998). This is because as the consistency of the example increments so does the obstruction of the example to the development of the ultrasonic gadget, for example, the tip of an ultrasonic test. Hence, focused energy (high amplitude) is expected to set the ultrasonic gadget to get the fundamental mechanical vibrations to advance cavitation in the sample. Several creators have concentrated on broadly the impact of sonication adequacy on the productivity of the strong fluid extraction of metals. There is understanding in the distributed information regarding this matter, and when in doubt, the higher the sufficiency the more analyte is extricated, for a given state of arrangement corrosiveness and sonication time (Lois-Marly et al., 2020; Capelo et al., 1998).

Along these lines, the analyte extraction increments ceaselessly with adequacy until an ideal is reached. This impact of abundance on the strong fluid extraction of metals can be stretched out to the next ultrasonic applications. In any case, harmony ought to be accomplished between the planned impacts and sonication plentifulness since, as clarified above, high amplitudes lead to high sonication forces, and high sonication powers can advance some undesired impacts, for example, analyte debasement. At long last, the utilization of high amplitudes doesn't generally prompt great outcomes. For instance, chromium is hard to extricate from natural examples utilizing ultrasonication, whatever the adequacy utilized for an ultrasonic test with a recurrence of 20 kHz and an intensity of 100 W.

17.2.2.3 Temperature

Dissolvable temperature assumes two parts in ultrasonication. One is, that the utilization of elevated temperatures assists with upsetting solid solute–lattice interactions, which include hydrogen holding, dipole attractions, and van der waals powers between the solute atoms furthermore dynamic locales on the grid (Cobley et al., 2008). Besides, quicker dissemination rates happen at higher temperatures. Then again, cavitation is improved at lower temperatures when the ultrasonic force of the generator is constant. It happens because the temperature of the dissolvable ascents so too does its fume pressure thus more dissolvable fume fills the cavitation bubbles, which then, at that point, will quite often implode less brutally, that is, the sonication impacts are less extreme than anticipated. Subsequently, a trade-off between cavitation and temperature should be accomplished. For instance, the extraction proportions of polycyclic sweet-smelling hydrocarbons from silt were expanded by somewhere in the range of 6% and 14% when ultrasonic extraction with a test was completed under noncooling conditions. Interestingly, when an attempt to speed up the enzymatic assimilation of proteins in acetonitrile utilizing an ultrasonic test at room temperature, the improvement in temperature brought about by ultrasonication prompted quick dissipation of the dissolvable, making test treatment inconceivable.

17.2.2.4 Solvent

The solvent used to perform test treatment with ultrasonication should be carefully picked. When in doubt, most applications are acted in water. Notwithstanding, other less polar fluids, like a few organics, can be additionally utilized, contingent upon the expected reason. Both dissolvable thickness and surface pressure are relied upon to repress cavitation. The higher the regular strong powers working inside a fluid (like high surface stain and high consistency) the additional troublesome it is to accomplish cavitation (Mason et al., 1992).

17.2.2.5 External pressure and bubbled gas

In case the outer strain is expanded, then, at that point, a more prominent ultrasonic energy is needed to initiate cavitation, that is, to break the dissolvable atomic powers (Mason et al., 1992; Lorimer et al., 1987). Moreover, there is an addition in the force of the cavitational bubble breakdown, and thus, an upgrade in sonochemical impacts is gotten. For a particular recurrence, there is a specific outer tension that will give an ideal sonochemical response. It should be focused on that most ultrasonic applications in scientific science are performed under climatic strain. Broken-up gas rises in a liquid can go about as cores for cavitation, leaning toward the ultrasonication interaction (Cobley et al., 2008). Be that as it may, ultrasonication can be utilized to degas a fluid. Hence, assuming that gas is utilized to build cavitation impacts it should be foamed ceaselessly into the dissolvable to keep up with the impact. Monoatomic gases like He, Ar, and Ne ought to be utilized.

17.3 Advances in sonication

17.3.1 Ultrasonic probe

The ultrasonic probe is immersed directly into the sample, unlike the ultrasonic bath in which the vessel container is immersed. Also, an ultrasonic probe produces 100x higher intensity than an ultrasonic bath. The ultrasonic probe consists of a generator, ultrasonic convertor, standard, and booster horns and probes. The function of generator is to convert the main voltage into electrical energy which is converted into mechanical vibrations by an ultrasonic converter. These mechanical vibrations have a fixed frequency. Horns may be standard or booster which is used to increase sonication amplitude. These detachable horns are used to transfer ultrasonic energy into the sample. These probes allow vibration from the horn to transmit and magnification of power. These probes are mainly made up of titanium alloy. They are thermo-resistant and noncorrosive and their treatment can be done in an autoclave. Although, these titanium probes contaminate the sample and thus can't be used in trace metal analysis. Probe shapes may vary uniform cylinder, exponential taper, linear taper, or cone and stepped.

17.3.2 Ultrasonic bath

Ultrasonic bath devices are incorporated by time control, temperature control, mode control, amplitude control, and frequency control. There are three types or classes of ultrasonic baths. The first type is the classic method. This common ultrasonic bath works with only one frequency that is, 40 kHz. The second operates at different frequencies that is, 25 kHz on the bottom and 40 kHz on the side. This ensures uniform distribution of power. The third system is rather advanced and consists of many features that include: dual-frequency of sonication, choice of frequencies, power regulation, three

Table 17.3 Comparison of sonoreactor with an ultrasonic probe and ultrasonic bath.

Parameters	Sonoreactor	Ultrasonic probe	Ultrasonic bath
Handling of sample	Low	High	Low
Degradation of organic samples	Medium	High	Low
Sonication time (second)	Up to 300	Up to 120	Up to several hours
Direct application	No	Yes	No
Solid-liquid extraction yield	Medium	High	Low
Amplitude (%)	20–100	10–100	10–100
Intensity of sonication at 1.5 ml vial (W)	0.5	15	0.01

operation modes: sweep; standard and degas, intensity control and heating, and timer. As an ultrasonic bath uses a vessel to be immersed into the sample, the shape of the vessel is also important for example a spherical base container reflects a huge amount of ultrasound waves. A comparison between a sonoreactor, ultrasonic probe, and an ultrasonic bath has been done in Table 17.3.

17.3.3 Other trends

In current times, cup horns and sonoreactor, and microplate horns are used as ultrasonic devices. These devices offer indirect sonication and are compared to high-intensity sonication water baths (Bender et al., 2004). In a cup horn, a titanium probe is fixed within an acrylic cup, which is filled with water, the cup contains the sample in which cavitation occurs. Cup horns allow multiple tubes processing at one time and are capable of sterile and pathogenic sample processing. The microplate horn is similar to a cup horn, though it allows high throughput applications.

The use of properties like piezoelectric and magnetostriction of material is currently used in ultrasound technologies and is captured and stored. An ultrasound generator with piezoelectric properties generates electric oscillations and converts them into mechanical oscillations. Magnetostrictive transducers are however, based on the exposure of materials with magnetic distortions to the magnetic field.

17.4 Applications

17.4.1 Application of direct and indirect ultrasonication

Ultrasonication can be applied in scientific science in two ways: straightforwardly to the sample or in a roundabout way through the dividers of the sample holder (Santos et al., 2007). In direct application, probes are immersed into the sample, performing ultrasonication straight over the arrangement with practically no boundary to be crossed by the ultrasonication wave other than the actual arrangement. This methodology has a few downsides. For example, test tainting with metals isolated from the test can be anticipated. Albeit current ultrasonic tests are produced using high immaculateness titanium, defilement by metals, for example, Cr or Al has been accounted for. Current ultrasonic probes produced using glass extraordinarily lessen this issue. Another weakness emerges from the way that most ultrasonic tests are utilized in open methodologies, that is, the example holder isn't fixed during test treatment. Thus, some unpredictable analytes can be lost (Patricio et al., 2006). For instance, when the substance of the 16 polycyclic fragrant hydrocarbons in the U.S. Environmental Protection Agency (USEPA) needs list was considered in silt, a huge part of the most unstable mixtures was lost because of the warming delivered by the sonication test utilized for the strong fluid extraction process. Aberrant application is performed, by and large, utilizing an ultrasonication shower, albeit present-day approaches exploit the incredible sonoreactor. In the two cases, the ultrasonic wave needs first to cross the fluid inside the ultrasonic gadget and afterward to cross the mass of the example holder. Subsequently, the ultrasonication force inside the example holder is lower than anticipated. As ultrasonic showers are not incredible gadgets, their applications are extraordinarily restricted by the absence of ultrasonic force. Indeed, numerous ultrasonic applications identified with showers can be connected to the warming delivered in the fluid that the shower contains–heat that is communicated to the example–rather than to genuine ultrasonic impacts, that is, cavitation (Patricio et al., 2006; Rial-Otero et al., 2007).

Notwithstanding, as referenced above, these days, a sonoreactor can be utilized rather than an ultrasonic shower in some normal applications. A sonoreactor works like an incredible, little ultrasonic bath it has been demonstrated to be a solid methodology for accelerating synthetic responses in present-day spaces of insightful science, like protein recognizable proof, and metal speciation.

17.4.2 Ultrasonic-assisted extraction

In sonication-assisted extraction, the formation of bubble and implosion results in high pressure and temperature which help in the release of solid matrix components into cell walls and mass transfer (Luna-Guevera et al., 2015). Ultrasonic extraction takes generally 30 min for extraction. The main

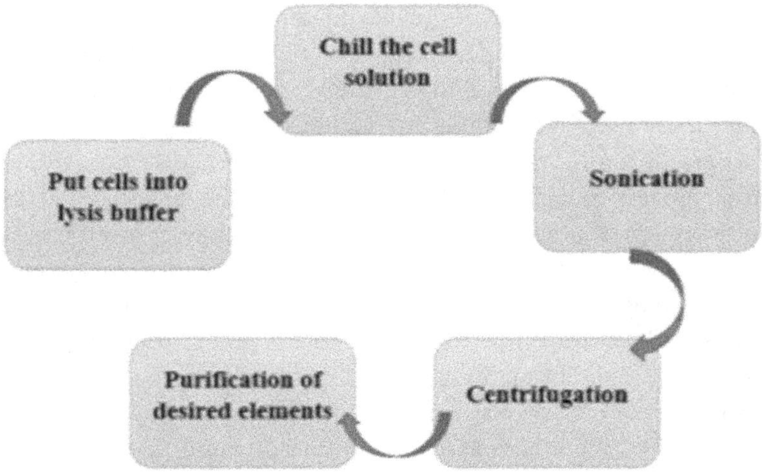

Figure 17.1 Ultrasonic assisted extraction procedure.

conditions in ultrasound-assisted extraction are the time of extraction, solvent amount and its polarity, and the amount and type of sample. Sonication cell lysis protocol is as follows (Fig. 17.1):

- Put the cell into a lysis buffer

    Centrifuge the cells into pellets and then keep these cells in lysis buffer which contains phenylmethylsulfonyl fluoride which is a serine protease inhibitor and helps in the nondegradation of the proteins exposed.
- Chill the cell solution

    The sample must be put on an ice bath due to heating during the procedure (Pakbaten et al., 2015).
- Sonication

    The sonication probe is immersed into the tube containing cells. The bottom of the tube must be so immersed that it doesn't touch the bottom of the tube. Also, the foaming must be prevented
- Centrifugation

    Centrifugation removes the cell debris which accumulates during sonication. This process at carried out at a temperature of 4 degrees.
- Purification of desired elements

    Desired elements are purified or obtained through this process.

17.4.3 Sonication in environmental science

Sonication is used to Degas or remove dissolved gas from liquid and break up microbial cells, and extract intracellular material. Domestic waste is treated through various biological processes like the activated sludge process, aerobic pond, and anaerobic respiration. However, the activated sludge process is more efficient. Due to problems like high operating costs and landfilling, new methods for reducing the volume and mass of excess sludge are growing. For this purpose, anaerobic respiration is used. However, anaerobic respiration is a slow process; therefore, a large volume of the digester and a long residence time in the digester is required. Anaerobic respiration takes place in four steps: hydrolysis, acidogenesis, acetogenesis, and methanogenesis. The process of biological sludge pretreatment for rupturing of cell walls and membrane and releasing of extracellular matter for further degradation is called sludge disintegration (Show et al., 2007), and this process is done using sonication. The sonication process reduces retention time by 25% and sludge is reduced to 30% as compared to conventional processes. The disintegration of organic compounds is done by strong alkali during the process of alkaline pretreatment. This process is used for the hydrolysis of lipids and amino acids, and is more efficient than thermal hydrolysis (Batista et al., 2009). Biological pretreatment is however, based on the enzymatic activity either in the system or externally. It is an enzyme-catalyzed reaction.

17.4.4 Sonication in biological science

Sonication is used in the lysis of virus cells, viral proteins, DNA, and RNA. Sonication is also used for lysis of cells, extraction of viruses, fragmentation of DNA and RNA (Crouse et al., 1993), and disperse virus clusters. Sonicators also have an important role in vaccine production for the creation of nanosized drug delivery platforms. Sonicators could be sized for research and development for lab-scale and industrial-scale production. Nanoemulsions are produced via ultrasonication through the process of emulsion droplet size reduction (Fig. 17.2) (Singh et al., 2017).

Other biological application of sonication includes (Chen et al., 1993):

- Lysis of bacterial or viral cells for the release of DNA/RNA, protein and antigen, etc. (Crouse et al., 1993).
- Bacterial and virus-cell destruction.
- Preparation of nanosized drug delivery system.
- Increment of bioavailability by the formation of nanoparticles, nanoemulsions, and liposomes.
- Ultrasonic dispersion and deagglomeration for creating homogeneous suspensions.
- Protein purification
- Degassing

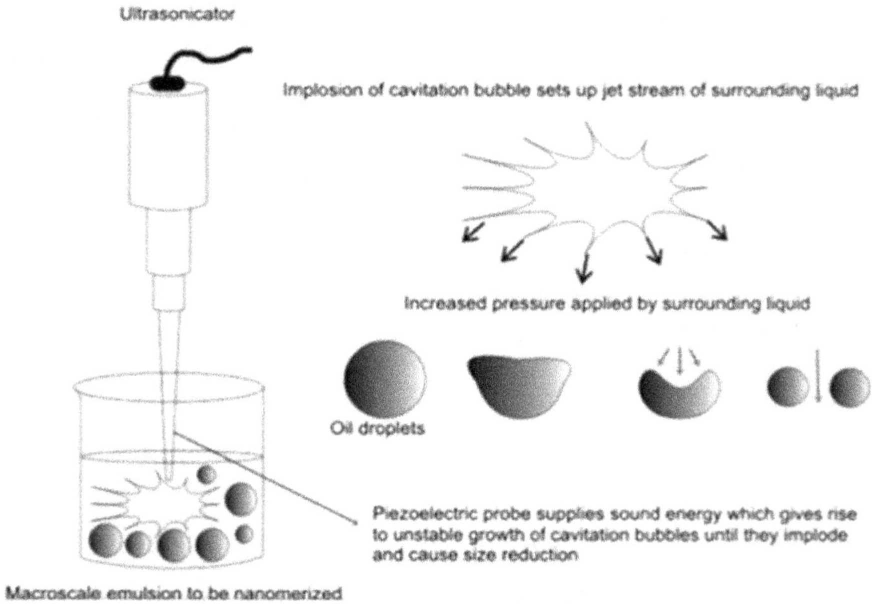

Figure 17.2 Emulsion droplet size reduction via ultrasonication used for producing nano emulsions (Singh et al., 2017).

- Sonochemistry
- Liposomes
- Crystallization
- Emulsification
- Extraction of oils and pigments
- Transesterification
- Solubilize compounds
- Virus isolation
- Vaccine manufacturing

17.4.5 Sonication in chemical analysis

Sonication is used in the chemical analysis by (Chen et al., 1993; Singh et al., 2017):

- Increase in accuracy of solid matrix dispersion technique.
- Accelerating enzymatic reactions.
- Accelerating solid-liquid extraction of organic species.
- Increasing the performance of extraction procedures like sold-liquid and microextraction.
- Lessen elemental speciation schemes.

- It also enhances liquid-liquid extraction techniques.
- Sonication also enhances elemental extraction like solid-liquid elemental extraction.

17.4.6. Sonication in laboratory applications (Moncada et al., 2012)

- Chemical stirring
- Product crushing
- Material mixing
- Cosmetics production
- Ultrasonic emulsification
- Logistics stirring
- Cell crushing

17.4.7 Sonication in food industry

Food preservation is an important aspect of food technology. Food preservation techniques include pasteurization, sterilization, and drying. These procedures result in problems such as vitamins and nutrients loss, larger processing time, fruit juice sedimentation, enzymatic browning, and microbial spoilage. Due to escalating demands of consumers for high-quality processed foods which are free from preservatives and additives, there is a need for nonthermal techniques for the preservation of food (Ravikumar et al., 2017). For this purpose, sonication can be used which is a nonthermal technique in which food nutritional quality is not destroyed. Sonication in the food industry helps in alcohol Chunhui, fine cosmetics particles, and nanoparticle preparation. Ultrasound can also be used to reduce the size of the globule of the fats. This increases the homogenization of food products.

The effect of ultrasound on fruits and vegetables is described as a few physiochemical changes as below:

- **Color**

  Polyphenols, carotenoids, chlorophylls, etc. are responsible for the color of the juice. Ultrasound is used to prevent color degradation caused due to processing and storing process. However, if the conditions are extreme ultrasound may also cause the degradation of food (Adekunte et al., 2010). So, thermosonication could be used preservation of juices.

- **Sensorial properties**

  Ultrasound might negatively affect milk quality, because of improvement of off-flavors and bothersome smells. Ultrasound treatment can likewise be applied for emulsification processes and, hence, the impact of US on the compound and sensorial attributes of various palatable oils has been analyzed.

During food emulsification and handling of sunflower oil, a metallic, and smelly scent has been distinguished for the sonicated tests (Chemat et al., 2004). In addition, huge deleterious changes in the synthesis of corn, olive oils, and soybean during food emulsification and handling have been accounted for and this impact was credited to the expanded number of revolutionaries created during US handling.

- **Viscosity**

The progressions in thickness can change contingent upon the ultrasound force. Food consistency may either increment or lessen and this impact may either be brief or super durable. As cavitation causes shear, an impermanent diminishing inconsistency is seen on account of thixotropic liquids. Assuming sufficient energy is applied, a super durable consistency is decreased because of changes in the atomic weight.

In food preservation, the process of ultrasound is a complex phenomenon, and consists of gas cavitation, due to which microbubbles are evolved in a liquid medium (Knorr et al., 2004). When high amplitude pressures are applied, cavitation takes place. There are one positive half and negative half of the cycle. In the negative half of the cycle, tensile force is applied to the liquid whereas in the positive half it undergoes compression which causes or results in the formation and increase in the size of microbubbles during these cycles. As sonication alone is not powerful for food spoilage agents, this is why it is used in combination with other preservation techniques. Ultrasound also affects physicochemical parameters, microbial inactivation, and nutritional parameters in liquid foods.

In many applications of ultrasound few are explained below:

- **Microbial inactivation:** Power ultrasound is used in multiple functions like reduction of spoilage and pathogenic micro-organisms and also removal of many harmful substances and thus it helps in preservation too. Micro-organisms are destroyed by various methods like mechanical effects, chemical effects, or physical effects. Mechanical effects are basically, the generation of turbulence, compression, shearing effect, and rarefaction. Chemical effects include cavitation and the formation of free radicals. Physical effects are those including temperature and pressure in and around food products (Moncada et al., 2012). Various micro-organisms are studied having different matrices under ultrasound as displayed in Table 17.4.

Table 17.4 Various micro-organisms studied in different matrices by ultrasound technique.

Matrices	Devices and operating conditions	Micro-organism studied	Reference
Apple juice	Ultrasonic cleaner, SB-500 at 25 kHz/ 70 μm, 20°C, 60 minutes	Total plate count and yeast and molds	(Abid et al., 2014)
Apple cider	Ultrasonic processor, 750 W at 20 kHz, 20°C, 20 minutes, 457 mW/ml	*Pichia membranifaciens, Wickerhamomyces anomalus, Zygosaccharomyces bailii*	(Baumann et al., 2005)
Milk	Ultrasonic processor S-4000, misoinix sonicators, at 20 kHz, 60 μm, amplitude, and 6 and 12 minutes	*E.coli, Staphylococcus aureus,*	(Herceg et al., 2012)
Liquid whole egg	Sonicator Virsonic Digital 475 at 40 kHz, 55°C, 5 minutes	*Salmonella enteritiuds*	(Huang et al., 2006)
Tomato juice	Vibra-cell, VC 1500, Sonics and materials Inc. at 20 kHz, 61 μm, amplitude, 40°C minutes	*Pichia fermentans*	(Adekunte et al., 2010)

- **Enhancing the shelf life and juice quality:** Due to many features of ultrasound like degassing, microbial inactivation, quality enhancement, prevention of juice sedimentation, and juice yield. The decrease of the microbial load is done by thermo sonication. The process of microbial inactivation can be affected by: the intensity of ultrasound waves, pH, temperature, water activity, processing time, and treatment temperature. The reactivity of ultrasound with micro-organisms varies. Mansonication (use of sonication in combination with pressure), and Manothermosonication (use of sonication with mild temperature and pressure) is used for increasing the shelf life of jams and liquid eggs.
- **Ultrasound-assisted drying:** Due to the limitations of traditional connective drying systems, the ultrasound-assisted drying technique is used. This drying is based on the composition of the product, time of exposure, environmental factors, and intensity of ultrasound.

- Another usage of ultrasound is in the pretreatment of vegetables before drying. This process helps in rehydration and less use of conventional techniques.

17.4.8 Sonochemistry

Sonochemistry applications are used in speeding up the reaction time. It can also be used to enhance the reactivity of metals like lithium, zinc, magnesium, and aluminum (Moore et al., 1987). They do so by removing the oxidized layer by ultrasonic exposure. Sonication also performs an important catalytic role in the reactions of powders. These powders are made finer by cavitation. Sonication could also be used in depolymerization.

17.4.9 Cell disruption

Ultrasound is transferred through various media like solid, liquid, or gases possessing elastic properties. This vibration causes the moment of the molecules which transmit to the other adjacent molecule. This creates a continuous moment. Due to increasing recombinant technologies and genetic engineering, the industrial-scale production of enzymes is successful. Although to isolate the intracellular microbial products, the used method should not only be capable of breaking the cell but also not inactivate the contents inside the cell. For the disruption of cells and releasing proteins from the cell, various methods like sonication, homogenization, bead milling, and nebulization are used (Schutte et al., 2001; Middelberg et al., 1995). Sonication is preferred as it has several advantages like low operating costs, ease of operation, and less training to handle. It is also easy to clean. The mechanical impacts of ultrasonication also bring about the inactivation of delivered items, for example, shear pressure created by whirlpools emerging out of shock waves. Various investigations on cell interruption have been conveyed utilizing the previously mentioned techniques, and scope of micro-organisms. Various models have been offered for cell crumbling just as for the quick arrival of intracellular protein, either recombinant or local (Melendres et al., 1993; sherker et al., 1989). The catalyst discharge process is for the most part answered to follow first-request energy. Be that as it may, the pace of delivery of a chemical relies upon its area inside the cell.

For instance, there is gradual delivery of alkaline phosphatase present in cell film while there is a momentary response during cell interruption to the alcohol dehydrogenase and L-lactic dehydrogenase situated in the cytoplasm. Subsequently, they may be profoundly impacted by the interruption technique. The seriousness of the interruption situations can have sway on the return of dynamic protein/compound recuperated simultaneously. Researchers investigated the impact of processes on maximal recuperation proficiency of dynamic protein and were researched utilizing one of them as of late favored mechanical interruption strategies, for example, ultrasonication (De Visscher et al.,1996; Ozbek et al., 2000).

17.5 Microwave

Microwaves are electromagnetic waves of 300 MHz to 300 GHz range frequency and 1 m to 1 mm range wavelength. Various communication applications like radar, cellular phone technology, television satellite, and noncommunication applications like heating and medicine are utilized from this range of electromagnetic spectra. The domestic microwave applications work at a frequency of 2.45 GHz. Federal Communications Commission has reserved various other frequencies like 5.8, 24.125, 61.25, 122.5, and 245.0 GHz for scientific, industrial, and medicinal applications (Zong et al., 2003). With advancements in the field of science and technology the demand for improved processes has increased that can be utilized by a vast variety of materials and can provide better results. The advanced methods need to be energy efficient and quick. Microwaves tend to be an advanced solution that can be utilized in various fields for different applications. Microwave processing is used for various types of materials and it is better than conventional processing. The surface is heated and then the heat is transferred into the materials by conduction, convection, and radiation phenomenon in conventional processing. Microwave processing deals with atomic level heating that generates volumetric heating in the processed materials. Microwave heating works on the principle of conversion of electromagnetic energy into heat within the material that moves away from the core of materials and towards the outer direction (Herceg et al., 2012; Zong et al., 2003). Applications of microwaves are utilized in several fields of chemistry, physics, biology, science, technology, and research. Various organic, inorganic, and polymer synthesis utilizes microwaves for better synthesis (Yuvaraj et al., 2015; Grewal et al., 2013). Microwave spectroscopy is also a widely used application of microwaves in physical research. Biological research also utilizes microwave irradiations in vitro by the production of heat in intra and extra cellular tissue fluids (Klinger et al., 1953).

Microwaves have been used in domestic households since the 1970s and 1980s for their various applications in food like heating, warming, baking, steaming, boiling, preservation, and pasteurization, etc (Ingole et al., 2015).

Superior heating levels, low cooking time, easy operation, uniform heating, minimal maintenance, and safer handling have made microwaves a popular device in various fields (Chandrasekraran et al., 2013).

17.5.1 Current advancements

17.5.1.1 A greener approach

Advancements in microwave technology have led to the achievement of a green technology tool in the microwave which benefits the environment and the researcher. Utilization of a set of principles that helps in the elimination or reduction of hazardous by-products during the manufacture, design, and application of any chemical substance is the main purpose of green chemistry. It helps to maintain a healthy and friendly environment that can result in sustainable development and growth. New technologies, processes, procedures, and methodologies are developing day-by-day to achieve desired results. It is the duty of a researcher to develop a process that follows the principles of green chemistry. Creating a nonhazardous, nonwasteful, efficient, and clean procedure is the ultimate goal. Energy consumption is also one of the major factors that are governed by green chemistry. Microwave reactors are the best options that use very short reaction times and selective reactions to give high-yields and outputs. Microwaves have proven to be energy efficient in terms of the synthesis of a product or during various chemical processes (Bassyouni et al., 2012).

Earlier it was believed that the compounds react only in a liquid or dissolved state so various solvents were also used for chemical synthesis which was damaging the environment. There were major problems relating to their disposal but this problem has been overcome by using the microwave irradiation technique where reactions can take place without the solvent. Clean chemical synthesis is achieved by coupling the microwave irradiation with the mineral-supported catalyst further results in high yields, more selectivity, more reaction rates, and easy manipulation of the reaction conditions. Microwave technologies have benefited various areas by providing a green technology beneficial for all (Ravichandran et al., 2011).

17.5.1.2 More than just a reaction

The outcomes of several researchers have accumulated that microwave irradiation is something other than a technique only for performing responses in a more limited time frame. It has been shown that various cycles and reactions cannot just be enhanced by this technique, however, can be

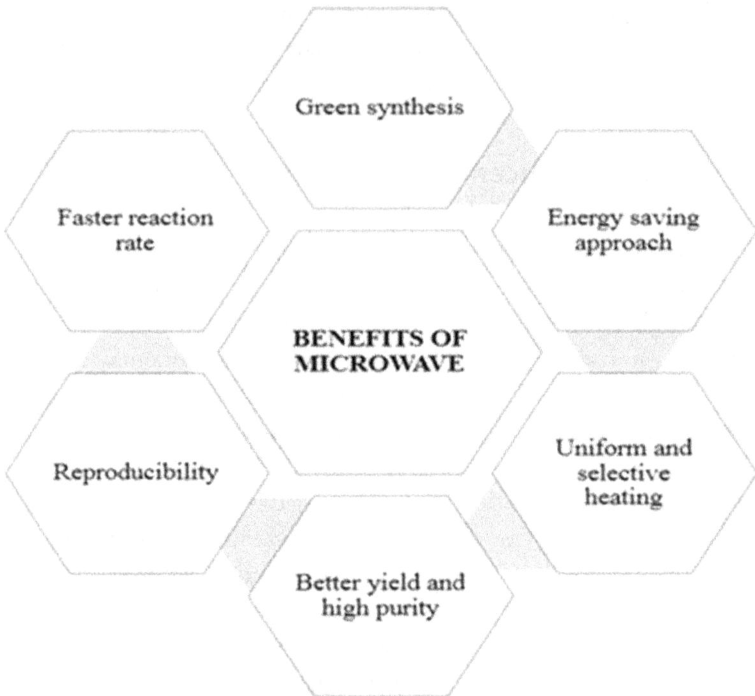

Figure 17.3 Various benefits of microwave.

controlled in the instance of serious responses, or as for chemo, regio or sound system selectivity might be changed or on the other hand, reversed (Langa et al., 1997).

17.5.1.3 Benefits of microwave

There are various benefits of microwaves (Fig. 17.3) like the yield is better, immaculateness is higher, response pace is fast, energy usage is lower, high reproducibility, uniformity in heating, and cleaner synthesis protocols are obtained.

17.6 Applications

17.6.1 Microwave-assisted extraction

One of the major utilization of microwaves is microwave-assisted extraction (MAE). In MAE the sample is heated in the presence of solvent where the subset of analytes is done from the sample into the solvent. The main advantages of this technique are its rapid heating system, low extraction time, increased throughput, and less solvent consumption. Closed vessels are preferred for extraction and elevated temperatures are used that transfers the targeted compounds into the matrix. Very few volumes of solvents generally 10–30 ml are utilized with a 15–30 min time duration for extraction which is ten times shorter than the conventional approaches for extraction. Various samples can also be extracted concurrently increasing the sample throughput using microwaves. Reproducibility and recovery of analytes are also great as compared to traditional methods (Eskilsson et al., 2000).

Various medicinal plants have been utilizing microwave-assisted extraction for their various analyte extraction such as furanocoumarins is extracted from pastinaca sativa fruits, coumarin and melilotic acid from flowering tops of melilotus officinalis L., phenolic compounds from hypericum perforatum and thymus vulgaris, and pigments from capsicum annum powders (Mandal et al., 2007). Various extracted products are steroids, essential oils, carotenoids, and flavonoids employing microwaves (Kaufmann et al., 2002).

17.6.2 Microwave-assisted organic synthesis

Organic synthesis has been one of the major functions of the microwave. The microwave offers a fast, clean, simple, economic, and efficient procedure for organic synthesis. Green chemistry principles utilization by microwave technology has been beneficial and utilized widely. A large number of organic reactions nowadays involve microwave heating which earlier involved traditional laboratory procedures that were tedious, time-consuming, and costly (Rana et al., 2014).

Many reactions namely heck reaction, hydrogenation of [beta]-lactams, epoxidation, diels-alder reaction, reductions, dehydration, condensations, ene reaction, cycloaddition reaction, hydrolysis, suzuki reaction, mannich reaction, esterification, protection, deprotection, and cyclisation reactions, etc are the organic reactions achieved by microwave (Grewal et al., 2013).

17.6.3 Microwave for radiopharmaceuticals

The applications of microwave-assisted organic synthesis can be utilized by the pharmaceutical industry when the synthesis happens at an elevated pressure to make radiopharmaceuticals. The radiopharmaceuticals are utilized as tracers during preclinical trials for generating a nuclear medical image. In these first trials multi-mode microwave oven has been utilized and the result was that the reaction rate was increased by using the microwave which makes the microwave an efficient tool for the generation of radiopharmaceuticals.

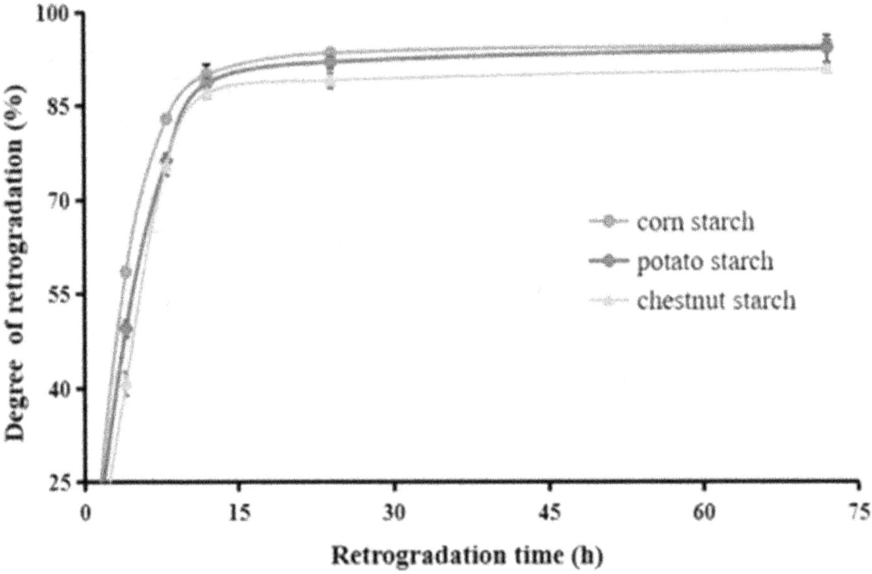

Figure 17.4 Impact of retrogradation time on the level of retrogradation after microwave irradiation method (Wang et al., 2019).

Microwave happens to be advantageous for such synthesis as it provides faster reaction and high yield owing to the short half-life of the reactants (Grewal et al., 2013).

17.6.4 Microwaves in food industry

Domestic and industrial (Figs. 17.4 and 17.5) usage of microwaves is increasing day by day due to their various advantages like high energy efficiency, very less starting time, controlled processes, selective heating mechanism, and space savings. The small rate of processing allows very few changes in nutritional changes in food properties making microwaves a friendly appliance to be used at home. Microwave energy finds its largest consumers in the food industry only. Applications like dehydration, processing, cooking, baking, melting, sterilization, pasteurization, and heating all utilize microwave energy (Joseph et al., 2017).

- **Microwave cooking**

Cooking being the major application of microwaves in the industry of food is highly relied on microwaves. Microwaves are also used in a combined state with a conventional oven for cooking purposes. Baking, heating, and reheating are popular methods (Joseph et al., 2017). Various researchers

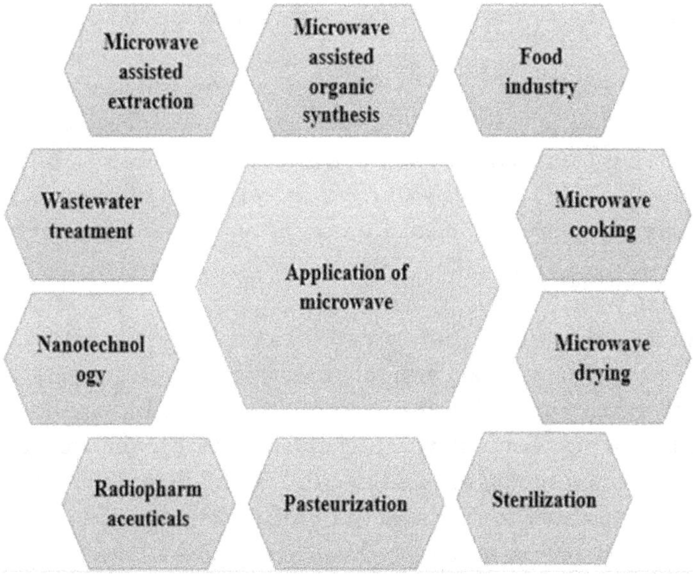

Figure 17.5 Applications of microwave technology.

have summarized the use and impacts of microwaves on numerous cooking factors like quality, color retention, and taste. Baking of bread and cooking of meat and rice are largely studied. Researchers studied the surface of bread and crust formation in the microwave. It was found that it is best to place susceptors at bottom of the sample in the microwave to get the best crust and browning of the food sample (Chandrasekaran et al., 2013).

The features of precooked bacon cooked in domestic and industrial microwave ovens were also studied by the researchers. It was found that parameters like power output and time of cooking had more effect on the superiority of food rather than factors such as type of bacon and its chemical composition. It was also concluded that the industrial oven provides more uniform cooking owing to uniformity in magnetron position as compared to the domestic oven (James et al., 2006).

Skipjack tuna was taken to study the chemical changes in its cholesterol content and omega-3 polyunsaturated fatty acids while cooking, canning, frying, and microwave heating. The contents of processed fish were contrasted to that of raw fish. The cholesterol content was found to be largely increased with canning, slightly increased with microwave heating, and decreased with frying. It was concluded that to retain the beneficial omega-3 fatty acid contents one must opt for microwave heating and boiling (Stephan et al., 2010).

Nowadays microwave energy effects on the digestibility of starch have also been investigated. After microwave irradiation, retrogradation occurs and its effect on the retrogradation degree of starch is also one the important factors. Many studies focus on the influence of retrogradation on the digestive and physicochemical properties of starch when treated with microwave radiation. Researchers have also evaluated starches of corn, potato, and chestnut with different types of crystallinity and their retrogradation process during the time of microwave irradiation retrogradation. Various outcomes were obtained like with microwave irradiation and retrogradation with less than the time of 8 h the in vitro digestibility of starches increased. The resistant starch content was higher after 24 h compared to that of naive starch which contributed to decreasing the digestibility of starch. It was observed that after the microwave irradiation retrogradation process the digestive performance of starch greatly varied making it low digestible which is better for diabetic patients and has good health benefits also (Wang et al., 2019).

- **Microwave drying**

To preserve the food microwave drying is done. Drying helps in the preservation of food items so that they can be saved for a longer period. Solar/air drying, convection drying, spray drying, smoking, drum drying, explosive puffing, freeze-drying, and osmotic drying are the different drying methods for dehydration purposes. Different drying procedures can cause different undesired effects such as nutrition loss, microbial growth, and oxidation of food also, so an appropriate method should be selected (Joseph et al., 2017).

Microwave drying is useful as it shows fast drying rates and has also shown improvement in the quality of food items. The controlled energy absorption levels can be utilized for careful heating that can heat the targeted or interior parts of the sample that contains moisture and is devoid causing an effect on the exterior or nontargeted parts. Microwave drying also finds its application through the falling rate period which is the rate-limiting period that can cause shrinkage and very less moisture content. Microwave drying comes with a volumetric heating mechanism that creates vapors inside and forces the water outside to prevent the shrinkage of food. To achieve high-quality food and better drying it can be said that microwave energy can be combined with various drying methods also (Zhang et al., 2006).

- **Sterilization and pasteurization**

Destroying and inactivation of the micro-organisms is the sole purpose of pasteurization and sterilization which add to food safety and the long shelf life of the product. To kill or inactivate a pathogen, the food is kept at a specific temperature for a certain period. In pasteurization, thermal treatment is done to destroy the vegetative form of micro-organisms like bacteria. Various fluid juices and milk sterilization has been done by using microwave pasteurization. Coconut water, apple juice, apple cider, grape water juice, sweet potato, and milk has been examined for the inactivation of enzymes and microbes by microwave pasteurization (Salazer-Gonzalez et al., 2012). There are various benefits of microwaves in the field of pasteurization and sterilization mainly microwaves being time-efficient but there are some shortcomings also. Nonuniform distribution of energy, unpredictability, and hard monitoring makes their use on the industrial scale very less (Joseph et al., 2017).

17.6.5 Microwave in nanotechnology

Nanoparticles, nanomaterials, and various nanocomposites utilize microwave-assisted methods that are greener and more efficient. Microwave-assisted methods use homogenous heating that has an upper hand that provides good nucleation and a great growth environment which is essential for nanomaterial synthesis. Nowadays, researchers are redesigning various strategies to synthesize smart nanomaterials that can be utilized in various fields of nanotechnology. Various 1D, 2D, and 3D nanomaterials have been synthesized using microwave-assisted synthesis deprived of any reducing agents. Nanomedicine, nanoelectronics, and biomaterials are the various fields that benefit from the same. Different adjustments in activation energy barriers cause nucleation and crystal growth to occur sooner in the microwave-assisted synthesis of nanoparticles. Also, the homogenous and rapid nucleation leads to increase viable development of ideal nuclear planes via dipole rotation under the microwave. It has also been utilized for the union of core-shell nanoparticles that produce high crystalline, homogenous, and in-center core-shell nanoparticles (Kumar et al., 2020).

Fe-cored carbon nanocapsules were prepared by researchers by microwave metal discharge method. The synthesis with this method provided more stability, high purity, and enhanced presentation. The influence of microwave power and irradiation time was observed where microwaves at different powers of 800, 1000, 1200, 1400, 1600, 1800, 2000, 2400, and 2600 W were used. Microwave metal discharge has become a potential tool to synthesize nanomaterials as this method gives various advantages, some of them being strong selective heating, plasma effects, and photocatalytic activity. This method also helps in the elimination of various toxic volatile organic emissions (Wang et al, 2018).

Carbon dots were blended in one pot employing microwave illumination utilizing green regular material kelp being the primary carbon source and ethylenediamine as nitrogen dopant. Great water solvency and astounding optical properties were shown by carbon dots prepared from the microwave. The CDs showed high affectability to pH as well as displayed solid selectivity and affectability for the detection of Co2+ (Zhao et al., 2019).

17.6.6 Microwave in treatment of waste water

Microwave heating can be used for the treatment of sludge and organic pollutants. Microwave utilizes oxidation reaction which happens to be a strong absorption agent leading to a catalytic reaction. Ferric oxide as a catalyst has been used by some researchers for the treatment of wastewater showing the removal of 50% COD and 80% color from water. Fe-Zr catalyst has also been used in combination with H2O2 oxidant treatment of n-butyric acid. Microwave has also been utilized in combination with UV to give microwave-induced UV treatment to wastewater. The researchers performed an experiment where they use oxidant sodium hypochlorite for the treatment and found that 95% TOC was removed from wastewater. In addition to this microwave technology has also been utilized for treating pollutants. Adsorption of activated carbon and other pollutants can be achieved by microwave as it makes pollutants fully degraded and also the adsorbent utilized is renewable. This technique also happens to be very cost-effective.

17.6.7 Microwave sintering of ceramics

Microwave sintering of ceramics is also one of the advanced applications of microwave technology. This has advantages over traditional processes as it produces an enhanced rate of reactions and diffusion kinetics that yields better properties. Alumina is the most popular ceramic used and full sintering of alumina is achieved by utilizing microwave technology. In the conventional heating method, the sintering of alumina requires at least 1600 uC, and a time of 2 h of soaking to get the same densification degree whereas in microwave full density can be achieved at 200 uC less than that of conventional. Electronic ceramic devices utilizing zinc oxide varistors have been processed under microwave heating. In microwave heating, it was observed that there was a reduction in the time of the cycle, and more enhanced electrical properties were obtained having greater densities, uniform microstructures, and finer grain size.

Zirconia ceramics have also been sintered using a multi-mode microwave. Transparent ceramics have also been developed by using the microwave as it provides fully dense ceramic in a shorter duration. Microwave processing has sintered hydroxyapatite into a transparent ceramic in 10 min at 1100 uC. Spinel and alumina were also sintered to obtain transparent ceramics (Aggarwal et al., 2010).

17.7 Summary

Technological advancements and the need for hours have made both sonication and microwave technology most desired among other traditional approaches. These technologies utilize the principles of green chemistry and have made the synthesis and protocols more efficient, and cleaner taking a step towards a sustainable future. Both sonication and microwave find their various applications in chemical extraction and synthesis. The chemical reactions carried out by them are safe, more time, and energy-efficient. Sonication and microwave energy have also shown their major importance in the food industry as both technologies are being utilized at the domestic level also. Food processing, heating, drying, baking, pasteurization, and sterilization are the important applications of these technologies.

References

Abid, M., Jabbar, S., Hu, B., Hashim, M.M., Wu, T., Wu, Z., Khan, M.A., Zeng, X., 2014. Synergistic impact of sonication and high hydrostatic pressure on microbial and enzymatic inactivation of apple juice. LWT-Food Sci. and Technol. 59 (1), 70–76.

Adekunte, A.O., Tiwari, B.K., Cullen, P.J., Scannell, A.G.M., O'donnell, C.P., 2010. Effect of sonication on color, ascorbic acid, and yeast inactivation in tomato juice. Food Chem. 122 (3), 500–507.

Agrawal, D., 2010. Latest global developments in microwave materials processing. Mater. Res. Innovations 14 (1), 3–8.

Bassyouni, F.A., Abu-Bakr, S.M., Rehim, M.A., 2012. Evolution of microwave irradiation and its application in green chemistry and biosciences. Res. Chem. Intermed. 38 (2), 283–322.

Batista, B.L., Rodrigues, J.L., de Oliveira Souza, V.C., Barbosa Jr,, F., 2009. A fast ultrasound-assisted extraction procedure for trace elements determination in hair samples by ICP-MS for forensic analysis. Forensic Sci. Int. 192 (1-3), 88–93.

Baumann, A.R., Martin, S.E., Feng, H., 2005. Power ultrasound treatment of Listeria monocytogenes in apple cider. J. Food Prot. 68 (11), 2333–2340.

Bender, K., Farfân, M.J., Schneider, P.M., 2004. Preparation of degraded human DNA under controlled conditions. Forensic Sci. Int. 139 (2-3), 135–140.

Bioke, 2015. Sonica sonicators product catalog.

Capelo, J.L., Lavilla, I., Bendicho, C., 1998. Ultrasound-assisted extraction of cadmium from slurried biological samples for electrothermal atomic absorption spectrometry. J. Anal. At. Spectrom. 13 (11), 1285–1290.

Chandrasekaran, S., Ramanathan, S., Basak, T., 2013. Microwave food processing—A review. Food Res. Int. 52 (1), 243–261.

Chatel, G., Varma, R.S., 2019. Ultrasound and microwave irradiation: Contributions of alternative physicochemical activation methods to Green Chemistry. Green Chem. 21 (22), 6043–6050.

Chemat, F., Grondin, I., Sing, A.S.C., Smadja, J., 2004. Deterioration of edible oils during food processing by ultrasound. Ultrason. Sonochem. 11 (1), 13–15.

Chen, X.H., Hommerson, A.L., Zweipfenning, P.G.M., Franke, J.P., Harmen-Boverhof, C.W., Ensing, K., De Zeeuw, R.A., 1993. Solid phase extraction of morphine from whole blood by means of bond elut certify columns. J. Forensic Sci. 38 (3), 668–676.

Cobley, A.J., Mason, T.J., Robinson, J., 2008. Sonochemical surface modification: A route to lean, green, and clean manufacturing? J. Applied Surface Finish. 3 (4), 190–196.

Crouse, C.A., Ban, J.D., d'Alessio, J.K., 1993. Extraction of DNA from forensic-type sexual assault specimens using simple, rapid sonication procedures. Bio. Techniq. 15 (4), 641–644.

Dalimunthe, N.A., Alfian, Z., Wirjosentono, B., Eddiyanto, 2020. Developing sonication-based method as a preliminary qualitative study to analyze methamphetamine in the hair of the user at Datuk Etham rehabilitation center. AIP Conference Proceedings, 2267. AIP Publishing LLC.

De Visscher, A., Van Eenoo, P., Drijvers, D., Van Langenhove, H., 1996. Kinetic model for the sonochemical degradation of monocyclic aromatic compounds in aqueous solution. J. Phys. Chem. 100 (28), 11636–11642.

Eskilsson, C.S., Björklund, E., 2000. Analytical-scale microwave-assisted extraction. J. Chromatogr. A 902 (1), 227–250.

Grewal, A.S., Kumar, K., Redhu, S., Bhardwaj, S., 2013. Microwave assisted synthesis: A green chemistry approach. Intern. Res. J. Pharmaceut. and App. Sci. 3 (5), 278–285.

Huang, E., Mittal, G.S., Griffiths, M.W., 2006. Inactivation of Salmonella enteritidis in liquid whole egg using combination treatments of pulsed electric field, high pressure and ultrasound. Biosystems Eng. 94 (3), 403–413.

Herceg, Z., Režek Jambrak, A., Lelas, V., Mededovic Thagard, S., 2012. The effect of high intensity ultrasound treatment on the amount of Staphylococcus aureus and Escherichia coli in milk. Food Technol. and Biotechnol. 50 (1), 46–52.

Ingole, M.A.D., 2015. Application potential for developments in microwave oven systems. SSRG Intern. J. Electrical and Electronics Engineer. 2 (3), 11–13.

James, C., Barlow, K.E., James, S.J., Swain, M.J., 2006. The influence of processing and product factors on the quality of microwave precooked bacon. J. Food Eng. 77 (4), 835–843.

Joseph, O.I., 2017. Microwave heating in food processing. BAOJ Nutr. 3 (1), 1–9.

Kaufmann, B., Christen, P., 2002. Recent extraction techniques for natural products: Microwave-assisted extraction and pressurized solvent extraction. Phytochem. Analysis: An Intern. J. Plant Chemical and Biochem. Techniq. 13 (2), 105–113.

Klinger, H.H., 1953. Applications of microwaves in scientific research. J. Franklin Inst. 256 (2), 129–146.

Knorr, D., Zenker, M., Heinz, V., Lee, D.U., 2004. Applications and potential of ultrasonics in food processing. Trends Food Sci. Technol. 15 (5), 261–266.

Kumar, A., Kuang, Y., Liang, Z., Sun, X., 2020. Microwave chemistry, recent advancements, and eco-friendly microwave-assisted synthesis of nanoarchitectures and their applications: A review. Materials Today Nano 11, 100076.

Langa, F., de la Cruz, P., de la Hoz, A., Díaz-Ortiz, A., Díez-Barra, E., 1997. Microwave irradiation: More than just a method for accelerating reactions. Contemp. Org. Synth. 4 (5), 373–386.

Lorimer, J.P., Mason, T.J., 1987. Sonochemistry. Part 1—the physical aspects. Chem. Soc. Rev. 16, 239–274.

Louis-Marly, K.N., 2020. Study on ultrasound technologies concerning with nitrogen for decomposition of organic compound and for soil washing application.

Luna-Guevara, M.L., Hernández-Carranza, P., Luna-Guevara, J.J., Flores-Sánchez, R. and Ochoa-Velasco, C.E., 2015. Effect of sonication and blanching postharvest treatments on physicochemical characteristics and antioxidant compounds of tomatoes: Fresh and refrigerated (No. RESEARCH).

Mandal, V., Mohan, Y., Hemalatha, S., 2007. Microwave assisted extraction—an innovative and promising extraction tool for medicinal plant research. Pharmacogn. Rev. 1 (1), 7–18.

Mason, T.J., 1992. Practical Sonochemistry: User's Guide to Applications. *Chemistry and Chemical Engineering. Ellis Howood Ltd., New York, NY, USA*, p. 150.

Melendres, A.V., Honda, H., Shiragami, N., Unno, H., 1993. Enzyme release kinetics in a cell disruption chamber of a bead mill. J. Chem. Eng. Jpn. 26 (2), 148–152.

Middelberg, A.P., 1995. Process-scale disruption of micro-organisms. Biotechnol. Adv. 13 (3), 491–551.

Moncada, M., Aryana, K.J., Xu, Z.M., Boeneke, C., 2012. Comparative study of microbial counts and some volatile compounds of whole milk processed by sonication and high tempera ture short time (HTST) pasteurization. Milchwissenschaft 67 (4), 432–434.

Moore, C.M., Tebbett, I.R., 1987. Rapid extraction of anti-inflammatory drugs in whole blood for HPLC analysis. Forensic Sci. Int. 34 (1-2), 155–158.

Özbek, B., Ülgen, K.Ö., 2000. The stability of enzymes after sonication. Process Biochem. 35 (9), 1037–1043.

Pakbaten, S., Karimi, M., Elhamirad, A., Sheikholeslami, Z., 2015. Effect of sonication and emulsifier (E471) on emulsification and improvement of cup cake. Iranian Food Sci. & Technol. Res. J. 11 (1), 31–40.

Patricio, A., Fernandez, C., Mota, A.M., Capelo, J.L., 2006. Dynamic versus static ultrasonic sample treatment for the solid–liquid pre-concentration of mercury from human urine. Talanta 69 (3), 769–775.

Peña-Farfal, C., Moreda-Piñeiro, A., Bermejo-Barrera, A., Bermejo-Barrera, P., Pinochet-Cancino, H., de Gregori-Henríquez, I., 2004. Ultrasound bath-assisted enzymatic hydrolysis procedures as sample pretreatment for the multielement determination in mussels by inductively coupled plasma atomic emission spectrometry. Anal. Chem. 76 (13), 3541–3547.

Rana, K.K., Rana, S., 2014. Microwave reactors: A brief review on its fundamental aspects and applications. Open Access Library J. 1 (6), 1–20.

Ravikumar, M., Suthar, H., Desai, C., Gowda, S.A., 2017. Ultrasonication: An advanced technology for food preservation. Int. J. Pure Appl. Biosci 5, 363–371.

Ravichandran, S., Karthikeyan, E., 2011. Microwave synthesis-a potential tool for green chemistry. Intern. J. Chem. Tech. Res. 3 (1), 466–470.

Rial-Otero, R., Carreira, R.J., Cordeiro, F.M., Moro, A.J., Fernandes, L., Moura, I., Capelo, J.L., 2007. Sonoreactor-based technology for fast high-throughput proteolytic digestion of proteins. J. Proteome Res. 6 (2), 909–912.

Rial-Otero, R., Gaspar, E.M., Moura, I., Capelo, J.L., 2007. Gas chromatography mass spectrometry determination of acaricides from honey after a new fast ultrasonic-based solid phase micro-extraction sample treatment. Talanta 71 (5), 1906–1914.

Salazar-González, C., Martín-González, S., Fernanda, M., López-Malo, A., Sosa-Morales, M.E., 2012. Recent studies related to microwave processing of fluid foods. Food Bioprocess Technol. 5 (1), 31–46.

Santos, H.M., Capelo, J.L., 2007. Trends in ultrasonic-based equipment for analytical sample treatment. Talanta 73 (5), 795–802.

Schütte, H., Kula, M.R., 2001. Cell disruption and isolation of non-secreted products. Biotechnol. Set. 505–526.

Show, K.Y., Mao, T., Lee, D.J., 2007. Optimization of sludge disruption by sonication. Water Res. 41 (20), 4741–4747.

Singh, Y., Meher, J.G., Raval, K., Khan, F.A., Chaurasia, M., Jain, N.K., Chourasia, M.K., 2017. Nanoemulsion: Concepts, development and applications in drug delivery. J. Controlled Release 252, 28–49.

Skerker, P.S., Clark, D.S., 1989. Thermostability of alcohol dehydrogenase: Evidence for distinct subunits with different deactivation properties. Biotechnol. Bioeng. 33 (1), 62–71.

Stephen, N.M., Jeya Shakila, R., Jeyasekaran, G., Sukumar, D., 2010. Effect of different types of heat processing on chemical changes in tuna. J. Food Sci. Technol. 47 (2), 174–181.

Suryanto, H., Jenuardy, S., Kustono, D., Puspitasari, P., Lubis, D.Z., 2018. Sonication assisted synthesis of biocomposite from starch/nanoclay and its properties. Intern. Review of Mechan. Engineer. 12 (3), 272–278.

Wang, M., Sun, M., Zhang, Y., Chen, Y., Wu, Y., Ouyang, J., 2019. Effect of microwave irradiation-retrogradation treatment on the digestive and physicochemical properties of starches with different crystallinity. Food Chem. 298, 125015.

Wang, Y., Wang, W., Sun, J., Sun, C., Feng, Y., Li, Z., 2018. Microwave-based preparation and characterization of Fe-cored carbon nanocapsules with novel stability and super electromagnetic wave absorption performance. Carbon 135, 1–11.

Yuvaraj, N., Pradeep Kumar, M., 2015. Multiresponse optimization of abrasive water jet cutting process parameters using TOPSIS approach. Mater. Manuf. Processes 30 (7), 882–889.

Zhao, C., Li, X., Cheng, C., Yang, Y., 2019. Green and microwave-assisted synthesis of carbon dots and application for visual detection of cobalt (II) ions and pH sensing. Microchem. J. 147, 183–190.

Zhang, M., Tang, J., Mujumdar, A.S., Wang, S., 2006. Trends in microwave-related drying of fruits and vegetables. Trends Food Sci. Technol. 17 (10), 524–534.

Zinoviadou, K.G., Galanakis, C.M., Brnčić, M., Grimi, N., Boussetta, N., Mota, M.J., Saraiva, J.A., Patras, A., Tiwari, B., Barba, F.J., 2015. Fruit juice sonication: Implications on food safety and physicochemical and nutritional properties. Food Res. Int. 77, 743–752.

Zong, L., Zhou, S., Sgriccia, N., Hawley, M.C., Kempel, L.C., 2003. A review of microwave-assist polymer chemistry (MAPC). J. Microwave Power Electromagn. Energy 38 (1), 49–74.

CHAPTER EIGHTEEN

Opportunities and challenges of sonication-microwave for food processing, preservation, and extraction: An industrial point of view

Muhammad Afzaal[a], Farhan Saeed[a], Umber Shehzadi[a], Aftab Ahmed[a], Yasir Abbas Shah[a], Huda Ateeq[a], Maryam Bibi[a], Aasma Asghar[a], Numra Waris[a], Muzammal Hussain[a], Muhammad Amir[b], Hafsa Hanif[a], Ali Ikram[a], Rehman Ali[a] and Muhammad Ahtisham Raza[a]

[a]Department of Food Science, Government College University Faisalabad, Faisalabad, Pakistan
[b]National Institute of Food Science and Technology, University of Agriculture, Faisalabad, Pakistan

18.1 Introduction

Consumers have recently developed a greater awareness of the importance of maintaining a healthy lifestyle as a result of the widespread availability of scientific evidence on several concerns relevant to food characteristics regarding its safety. In the face of potential challenges, the primary drivers are technological innovation, and research, as well as the widespread adoption of newly developed technology (Ranjha et al., 2021). The evolution of the food processing business is driven by the changing expectations of consumers as well as the imperative to manufacture nutritious and high-quality foods. Emerging technologies provide the impression of being the optimal solution to all of the problems related to the attributes of food. Currently, there is a rise in the demand among market for natural, plant-based food products that are beneficial to their health. One example of this trend is the demand for fresh fruit juices that are microbially secure, in addition to having enhanced quality and longer shelf life.

The use of ultrasound is gaining popularity as a novel approach that can regulate, improve, and speed up operations without compromising the quality of the food (Bhat et al., 2011). Ultrasonic sound is defined as having a

Ultrasound and Microwave for Food Processing: Synergism for Preservation and Extraction.
DOI: https://doi.org/10.1016/B978-0-323-95991-9.00002-3

frequency that is higher than what the human ear is capable of hearing. It is possible to utilize it in a different industrial application, such as the processing of food, due to the frequency spectrum of ultrasonic waves. Ultrasonic low-power and high-frequency systems are utilized for nondestructive assessment of the physicochemical qualities of the foods, while high-power, and low-frequency ultrasonic systems have been used to improve the quality of food products (Ranjha et al., 2021).

Microwave heating and other advanced food processing technologies have grown in favor as alternatives to traditional food processing processes for different uses in the food industry. Microwave-assisted food processing techniques add a new perspective by using the benefits of microwave energy to reduce the shortcomings of traditional and emerging food processing technologies. Microwaves are commonly used in food processing for drying, tempering, sterilizing, and baking, with benefits like reduced processing time and operational costs, food component retention, and energy savings due to volumetric heating (Ekezie et al., 2017; Kim et al., 2012).

Nonthermal processing procedures increase food quality and security while preventing nutrient losses. Ultrasound, microwave, and high-pressure technologies have been widely employed in the food business, resulting in benefits such as increased mass transfer, environmental protection, and increased shelf-life. Even if there are still restrictions on the industrial implementation of these revolutionary strategies, several effective examples of controlling equipment and improving strategies have been given in recent studies. It has been proven that, depending on processing conditions and circumstances, novel technologies affect the physicochemical, and functional qualities of food products. The synergetic role of sonification-microwave technologies for various processes such as food processing, preservation, and extraction in food industries can yield positive outcomes.

18.2 Applications of sonication in food industries

Ultrasound (US) is a category of nonthermal technology that improves quality of food by enhancing mass and energy transfer mechanisms (Astráin-Redín et al., 2021). It is used in the food sector to manage manufacturing processes, discover flaws, assess food attributes, discover flaws, evaluate the condition of preservation, optimize fruit ripening parameters, and enhance the quantity of food component extraction. Ultrasound can be used to enhance traditional food technology unit operations by lowering chemical requirements and energy, resulting in a more environmentally-friendly choice. The uses of ultrasound in food processing are grouped into two categories: helping old technologies and replacing existing technologies. In the latter situation, using the ultrasound to improve many food techniques such as freezing, extraction, brining, thawing, dehydration, filtering, and oxidation improves processing efficiency and mitigates the drawbacks of typical processing technologies (Gallo et al., 2018). As a result of its exceptional qualities, ultrasonic technology is highly regarded for processing and preservation. Pulsed electric fields, ultrasound, ozone, and ultraviolet light are only a few of the potential/novel alternative ways that have been investigated in the last decade, but high-intensity ultrasound has sparked a lot of interest in food storage. Numerous scientific publications now support the use of ultrasonication, either alone or in combination, to produce safe food for customers (Boateng and Nasiru, 2019).

18.2.1 Food processing

The technique used to convert raw foods into different food items is categorized under food processing. This technology utilized one or more combination of processes that includes filtration, rehydration, foaming, degassing/deaeration, and many more. Ultrasound can be utilized in the food industry to enhance unit activities or in food product processing by either direct exposure or utilizing a tool. Ultrasound has been used to enhance the technology, when employed with many unit operations in the food sector. Filtration became an important technology in the industry of food for the isolation of particles from their mother liquor and producing solids-free liquid. A key challenge in this procedure is concentration polarization or fouling induced by the deposition of the filter cake or filtrate on the surface of the membrane. Ultrasound energy, on the other hand, is beneficial in combating this problem. The retentate layers deposited on the surface of the membrane, developing concentration polarization, are disturbed by ultrasound during the filtering process, although the membrane's intrinsic permeability is unaltered. As a result, flux increases while flow resistance decreases (Bhargava et al., 2021).

Proteins have key functional qualities in food technology; their amphiphilic nature, and have ability to form interfacial films help to stabilize systems like foams and emulsion. (Higuera-Barraza et al., 2016). In general, protein solubility, hydrophobicity, and softness influence foaming ability, whereas the concentration of protein, molecular interaction, and hydration influence foaming stability. Researchers found that using high-intensity ultrasound to maximize the foaming characteristics of dietary protein was successful. Ultrasound having high power (20 kHz) probe has been shown to have a significant impact on functional qualities of whey protein such as foamability and solubility (Sheng et al., 2018).

The usage of high-intensity ultrasound for sonochemistry, homogenizing, degassing, disintegration, and cleaning operations are common in process engineering. Cavitation is utilized in ultrasonic deagglomeration and dispersing to produce strong shear forces in the liquid, breaking up agglomerates of particles into single dispersed particles.

One of the most promising dehydration approaches that could be used instead of typical drying treatments is ultrasound-related technology. Ultrasound cavitation of water molecules in the food, microstreaming, pressure changes, and alternate rarefaction and compression on food are the primary concepts of ultrasound to promote food drying. Ultrasound has been widely used to dehydrate fruits and vegetables (Tao et al., 2019). Ultrasound waves penetrate the food sample and release a series of expansion and compression that can improve the process of internal moisture transfer and developed cavitation, which is useful for removing liquid that is strongly attached without overheating the food, preserving heat-sensitive food constituents (Li et al., 2020).

18.2.2 Food preservation

The technique of preserving food to considerably decrease or prevent deterioration (lack of, nutritional value, edibility or quality) caused or accelerated by micro-organisms is known as food preservation. Preserving food usually entails stopping the growth of fungi, bacteria, fungi, and other microbes and also delays the oxidation of fats which can be the reason for rancidity. The most frequent method for preserving perishable food like fruits and vegetables is drying.

The use of ultrasound for the drying technique helps speed up the drying process for fruits, vegetables, meat, and fish. It helps to protect product quality by reducing drying time and increasing heat and mass transfer rates (Kowalski et al., 2017). Ultrasound can activate a variety of mechanisms, including heating, turbulence, agitation, friction, and so on. A synergy effect occurs when ultrasound vibrations interact, which can result in increased convective drying effectiveness. As a result, innovative equipment, such as hybrid dryers with airborne ultrasonic generators, can speed up the drying of heat-sensitive products that are heat sensitive by improving mass and heat transfer and therefore minimizing energy usage (Szadzińska et al., 2019).

In the food industry, freezing is frequently employed as a productive method of preservation of food. In general, traditional methods of freezing include air contact freezing, and blast freezing, immersion freezing with a low freezing rate that could lead to the development of big ice crystals. Different freezing technologies like ultrasound-assisted freezing, electrically and magnetically disturbed freezing, high-pressure freezing, and microwave-assisted freezing has been researched for improving the quality of frozen food by developing evenly and small dispersed ice crystals. Ultrasound-assisted freezing is one of them, and it has been intensively researched during the last decade (Zhang et al., 2018). The ultrasonic quick-freezing technique has been studied to increase the productivity of frozen foods due to its multiple benefits of frozen foods (Zhu et al., 2020).

Traditional fermented foods play a critical role in consumer socio-economic development, food security, nutrition, and health. Traditional fermentation procedures are used to preserve food in numerous nations, increasing shelf life and adding nutrients to food products (Obafemi et al., 2022). The feasibility of using ultrasonic-assisted curing to improve sodium diffusion rates in salt pickled processes, for this process, standing waves are generated by the ultrasonic system, resulting in mechanical rather than thermal impacts. Ultrasonic treatment speeds up sodium diffusion in the pickled processes. Furthermore, as the most important part of food techniques, it may be used not only to enhance the properties of food but also can produce new products having good functionality (Naphona and Naphonb, 2018).

Ultrasound has been shown in several studies to improve quality aspects of fermented milk gels, including syneresis, textural profile, and water holding capacity (Akdeniz and Akalın, 2022). Processing factors including power, frequency, and treatment time determine the success of ultrasound food processing; however, additional variables like sample temperature and food qualities can also influence the final result. Many authors agree that at a low temperature, ultrasound alone, is insufficient to effectively ensure food safety and inactivate micro-organisms; therefore, they recommend combining ultrasound with pressure or heat, which improves favoring microbial inactivation and cavitation effects and alters the physicochemical state of cell membranes (Guimarães et al., 2021). The common food preservation technologies that are used in food industries are demonstrated in Table 18.1. The ultrasound technology plays a very crucial role in preserving food assisting these common technologies.

Table 18.1 Food preservation techniques in food industries.

Techniques	Working principal	Food products	Benefits	References
Drying	Heating, agitation, turbulence, etc.	Fruits and vegetables	Protect product quality by reducing drying time. Increasing heat and mass transfer rates.	(Kowalski et al., 2017)
Freezing	Uniform heat transfer	Fruits and vegetables	Speeding up the nucleation process. Increasing heat transfer efficiency.	(Zhu et al., 2020)
Salt pickling	Standing waves are generated resulting in mechanical impacts	Fruits and vegetables	Speeds up sodium diffusion in the pickled processes. Produce new products with unique functionality.	(Naphona and Naphonb 2018)
Microbial inactivation	Combination of ultrasound with heat and pressure	Dairy Products	Improves cavitation effects. Favoring microbial inactivation.	(Guimarães et al., 2021)

18.2.3 Extraction applications

Technological alternatives with long-term and green values have also been used in the food processing, cosmetics, and medicinal industries. Ultrasound-Assisted Extraction (UAE) for example, is a novel and occasionally green extraction technology made possible by technological advancements. Ultrasonic-assisted extraction technique can be used to scale up and improve bioactive component extraction efficiency. The efficiency of ultrasonic extraction is mainly reliant on the setting parameters of the equipment and the extraction rate used (Zhang et al., 2018). As a nonthermal extraction method, UAE is well suited to preserving the bioactive chemicals' functioning (Kumar, Srivastav et al., 2021). Acoustic waves have been used in the UAE to prepare food as a novel green technology and in the pharmacological and nutraceutical industries. In the ultrasound process, waves form cavitation bubbles at certain frequencies and amplitudes, which collapse when they reach a nonstable point, releasing high temperature and pressure. This phenomenon can rip cell walls, enabling metabolites to flow out (Lefebvre et al., 2021).

Some of the most recent extraction techniques include supercritical fluid extraction (SFE), controlled pressure drops procedure, pulse electric field extraction, subcritical water extraction, and microwave extraction. The Ultrasonic-assisted extraction technique has a great potential to be used in the food processing, cosmetics, and medicinal fields.

18.3 Sonication with combined technologies

Many researchers have tried to combine ultrasound with other antimicrobial treatments to enhance the efficiency of ultrasound on microbial and enzyme inactivation. SFE, soxhlet extraction, microwave-assisted extraction, and thermosensation (heat and ultrasound), are all favorable combinations (Baysal and Demirdoven, 2012).

18.3.1 Supercritical fluid extraction and Sonication

SFE is a new extraction technology that takes utilization of fluids' increased solvation power (above their critical point). Extraction happens as a result of the combined role of gas-like qualities such as mass transfer and liquid-like properties such as solvation, with a diffusion rate larger than liquids. SFE has several advantages, including preconcentration effects, sterilization with the promise of enhanced yields, and a simple operating manner. By combining ultrasound pretreatment and supercritical CO_2 extraction, Liu et al. (2020) found encouraging results for obtaining *Iberis Amara* seed oil.

Cured ham is a classic cuisine that has widespread recognition among consumers in various European countries. One of the micro-organisms (bacteria) that can be present in dry-cured ham is *Escherichia coli*. Because it can cause serious infections in humans, it must be eradicated, in this sense, the use of SC-CO_2, a nonthermal and environmentally friendly inactivation technique is crucial, that has proven to be a beneficial way of inhibiting pathogenic and food spoiling bacteria while having low effect on nutritional characteristics of food (Castillo-Zamudio et al., 2021).

18.3.2 Soxhlet extraction technology and ultrasound

Extraction is a crucial step in recovering phenolic chemicals from plant matrices. For several years, traditional procedures like as, boiling, soaking, hydro distillation, and soxhlet have all been utilized. In comparison to other technologies like soaking or filtration, soxhlet extraction is the most frequently utilizing technique to extract phenolic components owing to its lower process cost, reliability or ease of operation, suitable for initial and bulk extraction, better for full recovery of extracts, require less time, and solute.

Soxhlet extraction, in particular, is a well-established approach with superior performance when it is compared to other traditional extraction methods, with the exception that is well suited to the extraction of heat-sensitive compounds (Alara et al., 2018). However, in high-pressure soxhlet extraction, the extractants do not reach critical levels, which reduces the amount of time and solute used. Furthermore, several alternatives combining the various procedures previously mentioned have been proposed to address the major shortcomings in soxhlet extraction. To take benefit from both approaches while falling the number of soxhlet cycles, a hybrid of soxhlet and ultrasound aided extraction has been developed (Armenta et al., 2015). This novel application of ultrasound in the soxhlet process reduces extraction time while maintaining the quality of the extracts (Djenni et al., 2013).

18.3.3 Ultrasound combined with microwave-assisted extraction

As an alternative to standard extraction procedures, ultrasonic, and microwave-assisted extraction technology, the combination has recently attained a lot of interest. The use of ultrasound and microwave radiation together can provide several benefits, including increased extraction rates, and shorter processing times. Microwave-assisted extraction is a useful technology for extracting essential chemicals from a variety of matrices. Ultrasound extraction has various advantages, including greater mass transfer, cell destruction, better diffusion, and capillary effects. Combining these technologies can improve the efficiency of extraction and provide various synergistic benefits, such as shorter extraction times, requirements of lower solvents, improvement in energy efficiency, and cheaper prices (Ojha et al., 2020). Due to the synergistic effect that is developed by the amplification of mass (by ultrasound), and heat (by microwave) transmission, the extraction process can be accelerated with a reduction in extraction time by using simultaneous microwave and ultrasound irradiation. This process is more environmentally friendly due to the short reaction time and the use of green solvents (Sillero et al., 2020).

18.4 Challenges and opportunities in the sonication processing

Ultrasound technology has many challenges which require extensive research to fully utilize this technology and avoid potential losses related to food safety and quality. Many advanced studies concentrate on the commercially available ultrasonic frequency (20 or 40 kHz). It is vital to investigate the usage of various ultrasonic frequencies as well as variables such as temperature, acoustic strength, and treatment period. The ultrasound method is used to assess emulsion droplet ability and size to concentrate, observe the level of aqueous and suspensions phase, monitor fat crystallization, and monitor creaming properties in suspensions and emulsion.

The presence of small air bubbles in a sample can sometimes prevent an ultrasonic wave from propagating through the sample. Ultrasonic properties and various information regarding its thermo-physical properties like compressibility, densities, thermal conductivity, and heat capacity are needed to make theoretical predictions of a material. As a result, theoretically data analysis from systems having multiple elements with unknown attributes is limited. The ability of high-intensity ultrasound to affect the chemical and physical properties of food has been recognized for several years. Manufacturers have just lately begun to apply laboratory-scale technology to large-scale processing processes. The rising usage of ultrasound of high-intensity is mostly due to the low-cost equipment availability that has been proved to offer significant advantages over alternative methods.

Ultrasound requires additional energy input, which causes industrialists to reconsider employing this approach on a commercial scale. Ultrasound causes physicochemical effects that can lead to the evolution of off-flavors, changes in physical qualities, and component degradation, all of which can affect food quality. As a result of critical pressure and temperature circumstances, ultrasonication causes the production of radicals, which are responsible for alterations in food components (Madhu et al., 2019).

On the other hand, many researchers have demonstrated the efficiency of ultrasonic in the food business or food sector as an alternative, enrichment, and improvement of several conventional processing methods. However, combining ultrasound with other techniques gives better results with product properties and could be the subject of future research. For the further development of ultrasound for industrial usage, measurement adjustments and studies to investigate the role of sonic treatment in large production of food are necessary. In larger research, the safety characteristics and negative impacts of ultrasonography on individuals should be evaluated. Furthermore, commercializing and industrializing ultrasound requires a great amount of energy, preventing its usage in the food industry applications. As a result, ultrasonic researchers should focus on commercial and industrial food applications (Bhargava et al., 2021).

This technique is integrated with other preservation techniques to meet the requirements and aims of reducing processing time and increasing efficacy, industry applications are projected to grow. Temperature and ultrasound (thermosonication) and other nonthermal technologies will undoubtedly become increasingly important in the near future. The ultrasound impact on the extraction and biosynthesis of functional compounds and nutritional supplements that could be employed for the development of new bioactive components are important for future applications (Welti-Chanes et al., 2017)

18.5 Application of microwaves in food industries

Thermal technology is utilized in practically every field, and there are a variety of procedures for achieving the desired result. Microwaves are a variety of electromagnetic spectrums, and their wavelengths fall in between infrared and radio waves. Microwaves help the food industry save time and money while maintaining excellent sensory quality and making little variations in the nutritional content of food as compared to utilizing only traditional techniques (Chandrasekaran et al., 2013). The various application of microwaves in processing, preservation, and extraction of food are discussed.

18.5.1 Food processing

Microwave heating seems to have a various use in the processing of food for decades. Heat transmission methods such as conduction and convection are used in traditional heating. During microwave heating, however, the heat transmission processes are convection and radiation, resulting in volumetric heating (Kappe, 2013). Microwave ovens are used in baking to speed up the process, especially when combined with conventional and infrared ovens. Microwave baking has been studied since 1950, and various studies on acceptable formulations using a variety of emulsifiers, lipids, gums, starches, and enzymes have been conducted. Thus, microwave baking resulted in uniform internal moisture profiles and higher drying rates (Ahmad et al., 2001). Microwave baking cuts the time duration in half when compared to traditional oven timing for baking. These microwave baking technologies are marketed as cost-effective ways to make regular crusted bread, as well as prebaked or completely baked white crust loaves, with increased production capacity and energy efficiency.

Blanching is a method of color retention and enzyme inactivation that involves immersing food components in hot water, steam, or boiling solutions (having acids or salts). Microwave blanching of herbs such as marjoram and rosemary was accomplished by dipping the herbs in a little amount of water and then exposing them to microwaves. Microwave blanched samples were shown to have a higher retention properties condition than microwave dried food samples (without blanching). In addition, waste-assisted microwave treatment of fresh coriander leaves was developed to have an impact on the pathogenic bacteria such as *Salmonella typhimurium*, resulting in a 3–5 log cycle reduction in the microbe population (De et al., 2012).

Pasteurization is a heat treatment that kills harmful microorganisms in their vegetative state, such as bacteria. Pasteurization also inactivates the unwanted enzymes that cause clouding in some fruit juices (Chandrasekaran et al., 2013). Pasteurization of fresh juices through microwave and sanitation of milk by microwave were studied by Salazar-Gonzalez et al. (2012). Different fluid food samples like grape water juice, apple juice, apple cider, milk, sweet potato puree, and coconut water, were tested for enzyme and microbial inactivation (Salazar-Gonzalez et al., 2012). Another research claims that microwave pasteurization was utilized to destroy *Salmonella typhimurium* in the yolks of shell eggs (Shenga et al., 2010). Microwave irradiation for 18 s decreased germs by 24%, whereas wet heat treatment for 18 min lowered the level of micro-organisms by 38% (Shenga et al., 2010).

18.5.2 Food preservation

The microwave thawing of raspberries took 0.55 min and resulted in 4.50% drip loss, which was the least of the five thawing techniques tested. Microwave treatment also allowed the taste components to be preserved, as well as those that were most suitable for customers (Liu et al., 2020). Microwave thawing was the best approach for keeping the greater overall nutritional content in frozen baby mustard as compared to air, refrigerator thawing, or water. The microwave treatment resulted in a reduction of about when compared to the refrigerator and air thawing, 350 and 70 times are achieved (Zhang et al. 2021).

When high temperatures are given to food samples for an extended period, it loses flavor, color, and volatile aromatics, deteriorates texture, and affect the overall features of the product (Wray and Ramaswamy, 2015). Microwave energy-based drying technologies that use volumetric heating can address these challenges related to conventional drying. To address these challenges, it is possible to combine several drying processes with a microwave system. Microwave-assisted convection drying, microwave-assisted freeze-drying, and microwave-vacuum drying equipment are available on a commercial scale. In the drying of pasta, crackers, or chips, the microwave hot air-drying method is utilized as the last drying step to ensure that the product is completely dry and that the moisture content is leveled (Osepchuk, 2002). The application of microwave in food processing and preservation are demonstrated in Table 18.2.

18.5.3 Microwave extraction applications

The microwave-assisted extraction abbreviated as (MAE) technique is considered effective for the extraction method of bioactive compounds. This technique is beneficial for the solvent and sample matrix by rotation of the molecular dipoles (Sun et al., 2007). The characteristics of the extract and microwave extraction rates are based on the temperature, dielectric potentials of the food sample, and solubility of the absorbed component present in the solvent. In the extraction process, the most important factor is temperature, the extraction and diffusion rate are higher at the higher temperature. But during extraction of heat-sensitive food components over a long extraction

Table 18.2 Application of microwave in food processing and preservation.

Food item	Technique	Application results	Reference
cooking of rice, meat, and baking of bread	Microwave heating	Comparison between microwave heating and conventional heating. Improve cooking parameters (quality, taste, and color retention).	(Kutlu et al., 2021)
Emulsifiers, fats, gums, starches, and enzymes.	Microwave baking	Microwave baking resulted in uniform internal moisture profiles and higher drying rates	(Ahmad et al., 2001)
In microwave packaging, glass, paper, and ceramics were used.	Microwave sterilization	Reduction in bacterial count such as *Pseudomonas fragi* and Escherichia coli.	(Guillard et al., 2010; Maktabi et al., 2011)
Raspberries potatoes In frozen baby mustard	Microwave thawing	Less drip loss, preserve the characteristic flavor in raspberries. Resulting in a loss of structure in potatoes. Preserving higher nutritional values in frozen mustard.	(Liu et al., 2020; Phinney et al., 2017; Zhang et al., 2021)
Fruits and vegetables pasta, crackers, or chips	Microwave drying	Short processing time, rapid energy dissipation, and efficient drying.	(Osepchuk, 2002)

time this may be undesirable. However, microwave extraction is an advantageous method with high extraction rates, decreased extraction duration, fast heating rate of the compound, less solvent absorbance, higher achievement of target compounds, and having the highest potential for processing (Wang et al., 2010).

In plants, the evaluation of phenolic compounds is considered essential by using different solvents in aromatic plants, conventional (reflux) extraction among microwave extraction of phenolic substances has been studied. In the conventional method, the level of phenolic components that are extracted was lowered with lowering polarity of the solvents, 50% methanol, 50% acetone, and ethyl acetate/water (40:20, v/v). The application of microwave extraction changes the previous order into 50% acetone, 50% methanol, water, and ethyl acetate/water. The results showed that the more effective technique is microwave-assisted extraction compared to the conventional method. The extracted level of phenolic compounds was higher, used less solvent, and extraction time was decreased (Proestos and Komaitis, 2008). Microwave hydro diffusion and gravity (MHG) technology is used for the extraction of essential oils and also for extremely effective technology in achieving phenolic compound-rich fruit extracts (Moraes et al., 2022).

Like other technologies, solvent-free microwave extraction technique depends on the idea of green technology. The extracted sample (without solvents or water) is placed in a microwave cavity. The water which was available in the food sample heated and degrade the cell wall of the sample, and helps the extraction within a short period (Chemat et al., 2015). Microwave extraction is utilized for the isolation of bioactive compounds including extracting phenolic antioxidants from peanut skin, phenolic compounds from blueberry leaves, pectin from apple pomace, and anthocyanin from red raspberries, grape polyphenols, and zearalenone extract from wheat and corn (Routray and Orsat 2014).

Microwave extraction has potential and is being explored as an alternate method for extracting bioactive chemicals from food and industrial leftovers. The most recent trend is the waste recovery of active industrial food processing leftovers as possible sources of bioactive compounds. As a byproduct of apple juice pressing, apple pomace is considered a source of dietary fiber and phenolic substances such as flavanols, dihydrochalcones, and hydroxycinnamates. Comparative research was conducted on four apple types utilizing several extraction methods such as maceration, pressured

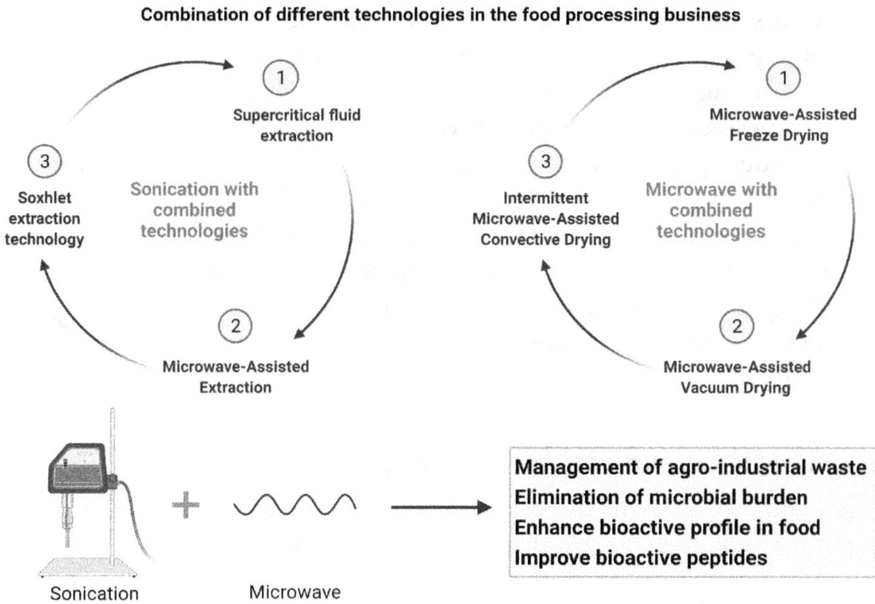

Figure 18.1 The combination of different technologies in the food processing business.

liquid extraction, and microwave aided extraction with various solvents, indicating that microwave-assisted extraction with ethanol was a good approach for antioxidant extraction (Angiolillo et al., 2015).

18.6 Microwave combined with other technologies

Microwave technology is characterized by homogeneous energy transfer, minimal energy usage, and quick heating, which helps food products retain much of their nutritious content. However, the combination of the microwave with other technologies like microwave-assisted freeze drying, microwave-assisted vacuum drying, and intermittent microwave-assisted convective drying (IMCD) are shown in Fig. 18.1.

18.6.1 Microwave-assisted freeze drying

The freeze-drying technique is used for the removal of moisture and supporting rehydration, but freeze-drying is a costly process with long time duration, lower productivity, and higher energy costs (Duan et al., 2016). The primary advantage of microwave-assisted freeze drying is important due to its short processing time duration, efficient drying, extreme energy savings, rapid energy dissipation, and volatile retention when compared to freeze-drying samples (Duan et al., 2016). A hybrid drying technology having freeze-drying and microwave vacuum drying was used for the drying of duck eggs (white proteins). The researchers demonstrated that the hybrid technology enhanced the drying potential and made protein powders (better color and low density). Correspondingly, the drying of Fuji apple by using this hybrid technology and the results revealed that the double-step technology may lessen the total freeze-drying time by 45% without effect on the nutritional value while comparing to entirely used freeze-drying (Zhou et al., 2014).

18.6.2 Microwave-assisted vacuum drying

The significance of microwave-assisted vacuum drying is enhanced by combining the technique and the benefit of microwave heat with vacuum increases because of its heat absorption and lesser pressure. It provides advantages such as quick evaporation of liquid that decreases changes in chemo physical properties in dried food samples (Ekezie et al., 2017). The microwave-assisted vacuum drying can be applied to a range of food samples. Recently, frozen blueberries were observed regarding their quality after drying by using hot air convective drying with microwave convective drying. When compared to other technologies, the final results showed that blueberries dried utilized hot air convective drying at 90°C with microwave-assisted vacuum drying displayed enhanced retention of anthocyanins and radical scavenging activity (Zielinska and Michalska, 2016).

18.6.3 Applications of intermittent microwave-assisted convective drying

Convective drying is a technique for dehydration of food, but it has drawbacks such as slower drying levels and decreased sample quality because of thermal degradation occurrence (Onwude et al., 2016). Due to certain resultant advantages, it has been planned to use microwave radiation with convective drying. Usually, the microwave can be integrated into convective drying significantly, and this hybrid technology is known as IMCD. This technology involves the processing of microwaves at pulsed rates accompanied by convective drying because the continuous processing of microwaves can damage the quality of the sample owing to overheating. Therefore, the

IMCD can be used by using the appropriate intermittency. Thus, enhancing product quality and energy effectiveness, and moisture profile by allowing redistribution of temperature within the food products (Kumar, 2015).

18.7 Challenges and opportunities in microwave processing

In processed materials, the microwave is responsible for thermal damage due to the nonuniform heating nature, which requires special tooling design. It is difficult to get an actual-time measurement of experimental material because microwave energy absorption in food materials changes with the increasing temperature. Microwave energy is a green technology that reduces processing time by combining energy efficiency. The application of microwaves in food is considered important due to the good nutritional profiles in a short time. But less time in the microwave for processing food may prevent the development of desirable sensory and textural characteristics. Consequently, the efficiency of microwave processing may depend on its combination with other techniques used for processing. The quality of microwave-processed products can be improved by using the latest technologies like infrared, radiofrequency, and jet impingement (Chhanwal et al., 2019).

Many studies carried out concerning microwave utilization and the challenges to predict the actual temperature distribution profile. Additionally, during microwave heating, the variation in the dielectric properties can affect the effectiveness of processing (Zhang et al., 2001), as well as undesired effects, occur on the textural characteristics of the food products like hardening of beef and softening of vegetables (Raaholt et al., 2017). Therefore, the degradation kinetics and appropriate processing of microwave usage should be made to ensure the preservation of nutritional components, inactivation of micro-organisms, and sensory and textural characteristics.

Furthermore, in the drying process, microwave technology may be demonstrated with accurate mitigation technologies like the use of additives. While microwave technique mostly results in increasing efficiency of the textural properties, particularly rice has indicated high stickiness, hardness, and chewiness values with microwave heating and an optimum parameter are also essential (Chin et al., 2020). Even though microwave roasting, it is difficult to achieve the fine-tuned level of all the nutritional characteristics because it requires improvement in the techno-functional pro and bio functional properties. After microwave processing, it also shows improved bioactive properties of alfalfa seeds, and these seeds presented the foaming properties hence it was not acceptable for the use in the food industry where foaming is important (Sahni et al., 2021).

18.8 Synergetic effect of sonication-microwave

Some food items absorb microwaves whole or partially, whereas others reflect and transfer them. In closed containers at high pressures, microwave increased extraction rate, and high dielectric liquids can be conducted. Because the MWs rapidly heat these solvents, the thermal gradient in the reaction medium will induce mass transfer between extractants and feedstock. This method can be used to make materials that are resistant to high temperatures (Gude, 2015). The microwave and ultrasonic technologies work together to lessen sample digestion and dissolve time for analytical purposes. Long-term sonication in plant molecule extraction can be harmful, because degradation potential, and unwanted side reactions, especially with sensitive compounds have been observed (Cintas et al., 2015).

Extraction through sonication-microwave makes some limitations, this hybrid technology has not been appropriated for heat-sensitive bioactive components of food or plants. Moreover, US-MW equipment is costly and not easy to use, only trained persons can be used. Both approaches are tough to scale up, and the challenge raises when used in combination with MW and US. The important challenge of this hybrid technology is that microwaves heat the reaction mixture, but ultrasound is recognized for its uneven effects, lowering the temperature and improving the results. It is concluded that heterogeneous processes are a great fit for this type of technology. Different applications of US-MW technologies for industrial purposes are discussed.

18.8.1 US-MW management of agro-industrial waste

The biological conversion of agricultural or industrial waste to reduced carbon energy biofuel has been measured as a potential way to make a supplementary product while also treating waste. In recent studies, researchers have observed the assembly of hydrogen (H_2) by using the dark fermentation technique and palm oil mill wastewater as a source of carbon. To boost fermentative H2 generation from high potency agricultural or industrial waste and semi-solid slurry, various pretreatment procedures (biological and physicochemical) have been used (Loh et al., 2019). Some researchers investigated in their study the effects of ultrasonic-microwave hybrid technology as a pretreatment of palm oil mill effluent (POME) slurry before fermentation on hydrogen (H2) generation and chemical oxygen demand extraction efficiency in fermentation. Through the use of this combined technology palm oil mill effluent before anaerobic digestion improves the conversion of complicated organic waste to hydrogen. Seed sludge pretreatment using sonication and microwave was demonstrated to have a significant influence on collective H2 generation or chemical oxygen demand diminution from POME wastewater. Sonication combined with microwave pretreatment of slurry inoculums had shown to be successful in neutralizing the organic content of POME wastewater (Mishra et al., 2021)

18.8.2 Elimination of microbial burden

Fruit juices provide a diverse range of nutrients as well as a complex nutritional combination that is beneficial to the health of consumers. For methods meant to manage the efficiency of disinfection treatments, *Cryptosporidium, Escherichia coli, Salmonella,* and *Listeria monocytogenes,* are considered significant harmful microbes depending on the kind of juice. Kernou et al. (2021) suggested that the high-temperature short-time technique was developed in the food production industry to eliminate harmful microorganisms in juicy foods while maintaining the value of food. Microwave and sonication processes are used to achieve a large reduction in microbial burden in a short amount of time and they improve pasteurization procedures for drinks, especially orange juice drinks, wherever acid and high-temperature resistant organisms like *E. coli* are present. The study estimated the combined effect of microwaves and ultrasounds treatment to inactivate the cells of harmful microbes especially *Escherichia coli* in the soft juices (Kernou et al., 2021).

18.8.3 Enhance bioactive profile in food

Environmental (processing, drying, extraction, and storage) and physiological (exposure of bioactive components with different food complexes) factors impact the successful delivery of bioactive components to target tissues in the human body. As a result, the cultivation, processing, packaging, usage, and storage of these substances should be closely managed and rechecked over time, until they reach the required portion in the human system (Urbani et al., 2016). As a result, extracting and purifying important substances such as bioactive chemicals naturally present in plants is entirely dependent on extraction and purification techniques (Garavand et al., 2017).

Garavand et al. (2019) worked on the extraction and nanoencapsulation of saffron bioactive ingredients by using sonication-microwave technologies. It can be utilized for fast extraction of bioactive components (polyphenols) and selecting the best possible level of solvents to attain greater effectiveness. This technique is used to provide good quality extracts that can take short time for extraction, produce a greater yield, easy to operate, and are economically feasible as compared to traditional extraction techniques. Sonication technique has benefits due to ease of operation, good extraction rate, appropriate for heat no sensitive bioactive compounds, less pollution cause, easily use for other extraction methods, and increase the bioavailability of bioactive components (Garavand et al., 2019).

Nadeem et al. (2021) have studied the higher time for sonication from 2 to 8 min. This study demonstrated that a synergistic approach of this hybrid technology had a positive effect on the nutritional quality of fruit juice, increasing total phenolic and flavonoids or increasing the power and antioxidant properties of fruit juice, and reducing microbial load. The hybrid technique of sonication and microwave can be used to treat phalsa beverages with higher nutritional retention without using chemical preservatives (Nadeem et al., 2021).

18.8.4 Improve bioactive peptides

Nowadays, the food processing industry is concentrating on the production of new food products that are nutritious, quick, safe, and efficient with good organoleptic and functional potential. High-intensity and low-frequency ultrasonic techniques have been effectively used to enhance a variety of unit operations and affect the functional characteristics of proteins like solubility, gelling, foaming, and emulsification. Traditional bioactive compound extraction strategies are time-consuming and inefficient.

Recent methods for extraction like microwave and ultrasonication combined assisted extraction can enhance extraction yield while improving bioactive ingredients. The combination of ultrasonic and microwave-aided extraction technology used to improve the extraction of food items demonstrated that a high yield of proteins and strong proteolytic activity of pineapple byproduct could be obtained in a short amount of time. Bioactive peptides produced from pineapple and its byproducts can be employed in food production (Mala et al., 2021).

18.9 Conclusion

Microwave energy is a green technology that reduces processing time by combining energy efficiency. The application of microwaves in food is considered important due to the good nutritional profiles in a short period. The advanced sonication microwave discussed in this chapter provides a great potential for processing and preservation of a wide variety of foodstuff as compared to traditionally processed food in terms of safety, sensory quality, and nutritional parameters. The use of these advanced food processing technologies extends the food shelf life without the addition of preservatives and minimally affects the color, flavor, aroma, and food functional qualities. However, still, a lot of research and further investigations are required in the field to face the recent challenges in the food industry.

References

Ahmad, S.S., Morgan, M.T., Okos, M.R., 2001. Effects of microwave on the drying, checking and mechanical strength of baked biscuits. J. Food Eng. 50 (2), 63–75.

Akdeniz, V., Akalın, A.S., 2022. Recent advances in dual effect of power ultrasound to microorganisms in dairy industry: activation or inactivation. Crit. Rev. Food Sci. Nutr. 62 (4), 889–904.

Alara, O.R., Abdurahman, N.H., Ukaegbu, C.I., 2018. Soxhlet extraction of phenolic compounds from Vernonia cinerea leaves and its antioxidant activity. J. Appl. Res. Med. Aroma. Plants 11, 12–17.

Angiolillo, L., Del Nobile, M.A., Conte, A., 2015. The extraction of bioactive compounds from food residues using microwaves. Curr. Opin. Food Sci. 5, 93–98.

Armenta, S., Garrigues, S., de la Guardia, M., 2015. The role of green extraction techniques in green analytical chemistry. TrAC Trends Anal. Chem. 71, 2–8.

Astráin-Redín, L., Alejandre, M., Raso, J., Cebrián, G., Álvarez, I., 2021. Direct contact ultrasound in food processing: impact on food quality. Front. Nutr. 8, 633070.

Ballard, T.S., Mallikarjunan, P., Zhou, K., O'Keefe, S.F, 2009. Optimizing the extraction of phenolic antioxidants from peanut skins using response surface methodology. J. Agric. Food Chem. 57 (8), 3064–3072.

Baysal, T., Demirdoven, A., 2012. Ultrasound in food technology. In: Chen, D., Sharma, SK, Mudhoo, A. (Eds.), Handbook on Applications of Ultrasound: Sonochemistry for Sustainability, CRC Press, pp. 163–182 ISBN 9781439842065.

Bhargava, N., Mor, R.S., Kumar, K., Sharanagat, V.S., 2021. Advances in application of ultrasound in food processing: A review. Ultrason. Sonochem. 70, 105293.

Bhat, R., Kamaruddin, N.S.B.C., Min-Tze, L., Karim, A.A., 2011. Sonication improves kasturi lime (*Citrus microcarpa*) juice quality. Ultrason. Sonochem. 18 (6), 1295–1300.

Boateng, E.F., Nasiru, M.M., 2019. Applications of ultrasound in meat processing technology: A review. Food Sci. Technol. 7 (2), 11–15.

Castillo-Zamudio, R.I., Paniagua-Martínez, I., Ortuño-Cases, C., García-Alvarado, M.A., Larrea, V., Benedito, J., 2021. Use of high-power ultrasound combined with supercritical fluids for microbial inactivation in dry-cured ham. Innov. Food Sci. Emerg. Technol. 67, 102557.

Chandrasekaran, S., Ramanathan, S., Basak, T., 2013. Microwave food processing—A review. Food Res. Int. 52 (1), 243–261.

Chemat, F., Fabiano-Tixier, A.S., Vian, M.A., Allaf, T., Vorobiev, E., 2015. Solvent-free extraction of food and natural products. TrAC Trends Anal. Chem. 71, 157–168.

Chhanwal, N., Bhushette, P.R., Anandharamakrishnan, C., 2019. Current perspectives on nonconventional heating ovens for baking process—A review. Food Bioprocess Technol. 12 (1), 1–15.

Chin, L., Therdthai, N., Ratphitagsanti, W., 2020. Effect of microwave cooking on quality of rice berry rice (*Oryza sativa L.*). J. Food Qual. 2020, 9 4350274.

Cintas, P., Tagliapietra, S., Caporaso, M., Tabasso, S., Cravotto, G., 2015. Enabling technologies built on a sonochemical platform: Challenges and opportunities. Ultrason. Sonochem. 25, 8–16.

De La Vega-Miranda, B., Santiesteban-López, N.A., López-Malo, A., Sosa-Morales, M.E., 2012. Inactivation of Salmonella Typhimurium in fresh vegetables using water-assisted microwave heating. Food Control 26 (1), 19–22.

Djenni, Z., Pingret, D., Mason, T.J., Chemat, F., 2013. Sono–Soxhlet: In situ ultrasound-assisted extraction of food products. Food Anal. Methods 6 (4), 1229–1233.

Duan, X., Liu, W.C., Ren, G.Y., Liu, L.L., Liu, Y.H., 2016. Browning behavior of button mushrooms during microwave freeze-drying. Drying Technol. 34 (11), 1373–1379.

Ekezie, F.G.C., Sun, D.W., Han, Z., Cheng, J.H., 2017. Microwave-assisted food processing technologies for enhancing product quality and process efficiency: A review of recent developments. Trends Food Sci. Technol. 67, 58–69.

Gallo, M., Ferrara, L., Naviglio, D., 2018. Application of ultrasound in food science and technology: A perspective. Foods 7 (10), 164.

Garavand, F., Madadlou, A., Moini, S., 2017. Determination of phenolic profile and antioxidant activity of pistachio hull using high-performance liquid chromatography–diode array detector–electro-spray ionization–mass spectrometry as affected by ultrasound and microwave. Int. J. Food Prop. 20 (1), 19–29.

Garavand, F., Rahaee, S., Vahedikia, N., Jafari, S.M., 2019. Different techniques for extraction and micro/nanoencapsulation of saffron bioactive ingredients. Trends Food Sci. Technol. 89, 26–44.

Gude, V.G., 2015. Synergism of microwaves and ultrasound for advanced biorefineries. Res. Efficient Technol. 1 (2), 116–125.

Guillard, V., Mauricio-Iglesias, M., Gontard, N., 2010. Effect of novel food processing methods on packaging: Structure, composition, and migration properties. Crit. Rev. Food Sci. Nutr. 50 (10), 969–988.

Guimarães, J.T., Scudino, H., Ramos, G.L.P.A., Oliveira, G.A.R., Margalho, L.P., Costa, L.E.O., Freitas, M.Q., Duarte, M.C.K.H., Sant'Ana, A.S., Cruz, A.G., 2021. Current applications of high-intensity ultrasound with microbial inactivation or stimulation purposes in dairy products. Curr. Opin. Food Sci. 42, 140–147.

Higuera-Barraza, O.A., Del Toro-Sanchez, C.L., Ruiz-Cruz, S., Márquez-Ríos, E., 2016. Effects of high-energy ultrasound on the functional properties of proteins. Ultrason. Sonochem. 31, 558–562.

Kappe, C.O., 2013. How to measure reaction temperature in microwave-heated transformations. Chem. Soc. Rev. 42 (12), 4977–4990.

Kernou, O.-N., Belbahi, A., Amir, A., Bedjaoui, K., Kerdouche, K., Dairi, S., Aoun, O., Madani, K., 2021. Effect of sonication on microwave inactivation of Escherichia coli in an orange juice beverage. J. Food Proc. Eng. 44 (5), e13664.

Kim, J., Mun, S.C., Ko, H.U., Kim, K.B., Khondoker, M.A.H., Zhai, L., 2012. Review of microwave assisted manufacturing technologies. Int. J. Precis. Eng. Manuf. 13 (12), 2263–2272.

Kowalski, S.J., Mierzwa, D., Stasiak, M., 2017. Ultrasound-assisted convective drying of apples at different process conditions. Drying Technol. 35 (8), 939–947.

Krishnamurthy, K., Khurana, H.K., Soojin, J., Irudayaraj, J., Demirci, A., 2008. Infrared heating in food processing: An overview. Compre. Rev. Food Sci. Food Saf. 7 (1), 2–13.

Kumar, C., 2015. Modelling intermittent microwave convective drying (IMCD) of food materials (Doctoral dissertation). Queensland University of Technology.

Kumar, C., Joardder, M.U.H., Farrell, T.W., Millar, G.J., Karim, M.A., 2016. Mathematical model for intermittent microwave convective drying of food materials. Drying Technol. 34 (8), 962–973.

Kumar, K., Srivastav, S., Sharanagat, V.S., 2021. Ultrasound assisted extraction (UAE) of bioactive compounds from fruit and vegetable processing by-products: A review. Ultrason. Sonochem. 70, 105325.

Kutlu, N., Isci, A., Sakiyan, O., Yilmaz, A.E., 2021. Extraction of phenolic compounds from cornelian cherry (Cornus mas L.) using microwave and ohmic heating assisted microwave methods. Food Bioprocess Technol. 14 (4), 650–664.

Lefebvre, T., Destandau, E., Lesellier, E., 2021. Selective extraction of bioactive compounds from plants using recent extraction techniques: A review. J. Chromatogr. A 1635, 461770.

Li, R., Huang, L., Zhang, M., Mujumdar, A.S., Wang, Y.C., 2014. Freeze drying of apple slices with and without application of microwaves. Drying Technol. 32 (15), 1769–1776.

Li, Y., Wang, X., Wu, Z., Wan, N., Yang, M., 2020. Dehydration of hawthorn fruit juices using ultrasound-assisted vacuum drying. Ultrason. Sonochem. 68, 105219.

Liu, L., Lv, C., Meng, X., Xin, G., Li, B., 2020. Effects of different thawing methods on flavor compounds and sensory characteristics of raspberry. Flav. Fragrance J. 35 (5), 478–491.

Loh, L.M., Yan, Y., Yap, P.W., Nadarajan, R.U.P.I.N.E.E., Ong, A.S.H., 2019. Palm oil mill effluent as alternate carbon source for ammonia removal in wastewater treatment. Sains Malaysiana 48 (4), 871–876.

Madhu, B., Srinivas, M.S., Srinivas, G., Jain, S.K., 2019. Ultrasonic technology and its applications in quality control, processing and preservation of food: A review. Curr. J. Appl. Sci. Technol. 32 (5), 1–11.

Maktabi, S., Watson, I., Parton, R., 2011. Synergistic effect of UV, laser and microwave radiation or conventional heating on E. coli and on some spoilage and pathogenic bacteria. Innov. Food Sci. Emerg. Technol. 12 (2), 129–134.

Mala, T., Sadiq, M.B., Anal, A.K., 2021. Comparative extraction of bromelain and bioactive peptides from pineapple byproducts by ultrasonic-and microwave-assisted extractions. J. Food Process Eng. 44 (6), e13709.

Matsuura, K., Hirotsune, M., Nunokawa, Y., Satoh, M., Honda, K., 1994. Acceleration of cell growth and ester formation by ultrasonic wave irradiation. J. Ferment. Bioeng. 77 (1), 36–40.

Mishra, P., Wahid, Z.A., Singh, L., Zaid, R.M., Tabassum, S., Sakinah, M., Jiang, X., 2021. Synergistic effect of ultrasonic and microwave pretreatment on improved biohydrogen generation from palm oil mill effluent. Biomass Convers. Biorefin. 12, 1–8.

Moraes, D.P., Farias, C.A., Barin, J.S., Ballus, C.A., Barcia, M.T., 2022. Application of Microwave Hydrodiffusion and Gravity for Phenolic Compounds Extraction from Fruits. Food Bioprocess Technol. 1–12.

Nadeem, M., Ghaffar, A., Hashim, M.M., Murtaza, M.A., Ranjha, M.M.A.N., Mehmood, A., Riaz, M.N., 2021. Sonication and microwave processing of phalsa drink: a synergistic approach. Int. J. Fruit Sci. 21 (1), 993–1007.

Naphona, N., Naphonb, P., 2018. Salt diffusion enhancement in the salt pickled lime processing by ultrasonic treatment technique. Int. J. Appl. Eng. Res. 13 (6), 4268–4272.

Nemes, S.M., Orsat, V., 2010. Screening the experimental domain for the microwave-assisted extraction of secoisolariciresinol diglucoside from flaxseed prior to optimization procedures. Food Bioprocess Technol. 3 (2), 300–307.

Obafemi, Y.D., Oranusi, S.U., Ajanaku, K.O., Akinduti, P.A., Leech, J., Cotter, P.D., 2022. African fermented foods: Overview, emerging benefits, and novel approaches to microbiome profiling. NPJ Sci. Food 6 (1), 1–9.

Ojha, K.S., Aznar, R., O'Donnell, C., Tiwari, B.K., 2020. Ultrasound technology for the extraction of biologically active molecules from plant, animal, and marine sources. TrAC Trends Anal. Chem. 122, 115663.

Onwude, D.I., Hashim, N., Chen, G., 2016. Recent advances of novel thermal combined hot air drying of agricultural crops. Trends Food Sci. Technol. 57, 132–145.

Osepchuk, J.M., 2002. Microwave power applications. IEEE Trans. Microwave Theory Techniq. 50 (3), 975–985.

Pallaroni, L., Von Holst, C., Eskilsson, C., Björklund, E., 2002. Microwave-assisted extraction of zearalenone from wheat and corn. Anal. Bioanal.Chem. 374 (1), 161–166.

Pan, X., Niu, G., Liu, H., 2003. Microwave-assisted extraction of tea polyphenols and tea caffeine from green tea leaves. Chem. Eng. Process. 42 (2), 129–133.

Peighambardoust, S.H., Karami, Z., Pateiro, M., Lorenzo, J.M., 2021. A review on health-promoting, biological, and functional aspects of bioactive peptides in food applications. Biomolecules 11 (5), 631.

Phinney, D.M., Frelka, J.C., Wickramasinghe, A., Heldman, D.R., 2017. Effect of freezing rate and microwave thawing on texture and microstructural properties of potato (*Solanum tuberosum*). J. Food Sci. 82 (4), 933–938.

Proestos, C., Komaitis, M., 2008. Application of microwave-assisted extraction to the fast extraction of plant phenolic compounds. LWT-Food Sci. Technol. 41 (4), 652–659.

Raaholt, B.W., Hamberg, L., Isaksson, S., 2017. Continuous tubular microwave heating of particulate foods at high temperatures. J. Microwave Power Electromag. Energy 51 (4), 259–285.

Ranjha, M.M.A.N., Irfan, S., Lorenzo, J.M., Shafique, B., Kanwal, R., Pateiro, M., Arshad, R.N., et al., 2021. Sonication, a potential technique for extraction of phytoconstituents: A systematic review. Process 9 (8), 1406.

Routray, W., Orsat, V., 2014. MAE of phenolic compounds from blueberry leaves and comparison with other extraction methods. Ind. Crops Prod. 58, 36–45.

Sahni, P., Sharma, S., Singh, B., 2021. Impact of different processing treatments on techno and biofunctional characteristics of dhaincha (Sesbania aculeate). Food Sci. Tech. Int. 27 (3), 251–263.

Salazar-González, C., Martín-González, S., Fernanda, M., López-Malo, A., Sosa-Morales, M.E., 2012. Recent studies related to microwave processing of fluid foods. Food Bioprocess Technol. 5 (1), 31–46.

Seyhun, N.A.Dİ.D.E., Ramaswamy, H., Sumnu, G., Sahin, S., Ahmed, J., 2009. Comparison and modeling of microwave tempering and infrared assisted microwave tempering of frozen potato puree. J. Food Eng. 92 (3), 339–344.

Sheng, L., Wang, Y., Chen, J., Zou, J., Wang, Q., Ma, M., 2018. Influence of high-intensity ultrasound on foaming and structural properties of egg white. Food Res. Int. 108, 604–610.

Shenga, E., Singh, R.P., Yadav, A.S., 2010. Effect of pasteurization of shell egg on its quality characteristics under ambient storage. J. Food Sci. Technol. 47 (4), 420–425.

Sillero, L., Prado, R., Labidi, J., 2020. Simultaneous microwave-ultrasound assisted extraction of bioactive compounds from bark. Chem. Eng. Process.-Process Intensif. 156, 108100.

Sun, Y., Liao, X., Wang, Z., Hu, X., Chen, F., 2007. Optimization of microwave-assisted extraction of anthocyanins in red raspberries and identification of anthocyanin of extracts using high-performance liquid chromatography–mass spectrometry. Eur. Food Res. Technol. 225 (3), 511–523.

Szadzińska, J., Łechtańska, J., Pashminehazar, R., Kharaghani, A., Tsotsas, E., 2019. Microwave- and ultrasound-assisted convective drying of raspberries: Drying kinetics and microstructural changes. Drying Technol. 37 (1), 1–12.

Tao, Y., Han, M., Gao, X., Han, Y., Show, P.-L., Liu, C., Ye, X., Xie, G., 2019. Applications of water blanching, surface contacting ultrasound-assisted air drying, and their combination for dehydration of white cabbage: Drying mechanism, bioactive profile, color and rehydration property. Ultrason. Sonochem. 53, 192–201.

Urbani, E., Blasi, F., Simonetti, M.S., Chiesi, C., Cossignani, L., 2016. Investigation on secondary metabolite content and antioxidant activity of commercial saffron powder. Eur. Food Res. Technol. 242 (6), 987–993.

Wang, Y.L., Xi, G.S., Zheng, Y.C., Miao, F.S., 2010. Microwave-assisted extraction of flavonoids from Chinese herb Radix puerariae (Ge Gen). J. Med. Plants Res. 4 (4), 304–308.

Welti-Chanes, J., Morales-de la Peña, M., Jacobo-Velázquez, D.A., Martín-Belloso, O., 2017. Opportunities and challenges of ultrasound for food processing: An industry point of view. Ultrasound: Advances for food processing and preservation 457–497.

Wen, C., Zhang, J., Zhang, H., Dzah, C.S., Zandile, M., Duan, Y., Ma, H., Luo, X., 2018. Advances in ultrasound assisted extraction of bioactive compounds from cash crops–A review. Ultrason. Sonochem. 48, 538–549.

Wray, D., Ramaswamy, H.S., 2015. Novel concepts in microwave drying of foods. Drying Technol. 33 (7), 769–783.

Zhang, F., Lin, P.-X., Xia, P.-X., Di, H.-M., Zhang, J.-Q., Wang, Z.-H., Li, Z.-Q., Huang, S.-Y., Li, H.-X., Sun, B., 2021. The effect of different thawing methods on the health-promoting compounds and antioxidant capacity in frozen baby mustard. RSC Adv. 11 (17), 9856–9864.

Zhang, H., Datta, A.K., Taub, I.A., Doona, C., 2001. Electromagnetics, heat transfer, and thermokinetics in microwave sterilization. AlChE J. 47 (9), 1957–1968.

Zhang, P., Zhu, Z., Sun, D.W., 2018. Using power ultrasound to accelerate food freezing processes: Effects on freezing efficiency and food microstructure. Crit. Rev. Food Sci. Nutr. 58 (16), 2842–2853.

Zhao, Y.Y., Yi, J.Y., Bi, J.F., Chen, Q.Q., Zhou, M., Zhang, B., 2019. Improving of texture and rehydration properties by ultrasound pretreatment for infrared-dried shiitake mushroom slices. Drying Technol. 37 (3), 352–362.

Zhou, B., Zhang, M., Fang, Z., Liu, Y., 2014. A combination of freeze drying and microwave vacuum drying of duck egg white protein powders. Drying Technol. 32 (15), 1840–1847.

Zhu, Z., Zhang, P., Sun, D.W., 2020. Effects of multi-frequency ultrasound on freezing rates and quality attributes of potatoes. Ultrason. Sonochem. 60, 104733.

Zielinska, M., Michalska, A., 2016. Microwave-assisted drying of blueberry (Vaccinium corymbosum L.) fruits: Drying kinetics, polyphenols, anthocyanins, antioxidant capacity, colour and texture. Food Chem. 212, 671–680.

CHAPTER NINETEEN

Effect of synergism of sonication and microwave on fermentation and emulsification processes

Muhammad Aamir[a], Muhammad Afzaal[b], Farhan Saeed[b], Atka Afzal[b], Yasir Abbas Shah[b], Ifra Tariq[b], Ifrah Usman[b], Muhammad Nouman[b] and Gulzar Ahmad Nayik[c]

[a]National Institute of Food science and Technology, University of Agriculture, Faisalabad, Pakistan
[b]Institute of Home and Food Sciences, Government College University, Faisalabad, Pakistan
[c]Department of Food Science and Technology, Government Degree College, Shopian, Jammu and Kashmir, India

19.1 Introduction

Food fermentation is a process in which complex compounds (organic compounds) are transferred into the simpler organic compounds through enzyme action and organic catalyst (produced by microorganisms such as molds, bacteria and yeast (Corma et al., 2007). Consumer's demand for safe and good quality food, together with industrial and for energy-efficient processes leads to the formation of novel technologies in food fermentation (Pereira and Vincente, 2010). Other than convention techniques, novel techniques like microwave and ultrasound are proved to process efficient and productive in the fermentation of food (Chemat and Khan, 2011; Chisti, 2003). Modern models of ultrasound provide more comprehensive and selective benefits (improve extraction time, easy to integrate in industries, productive and environment friendly) (Singla and Sit, 2021). Ultrasound low and high-frequency techniques in fermentation are studied for years. Research shows that to monitor the fermentation process, high-frequency ultrasound (nondestructive analytical method) is typically used (Novoa-Díaz et al., 2014). Whereas low-frequency ultrasound also helps to improve the fermentation rate (Nguyen et al., 2012), degassing or deaeration of alcoholic beverages (Chemat and Khan, 2011), wine maturation and aging (Tao et al., 2014), and pasteurization (Gracin et al., 2016). Microwave technology during the winemaking process shortens the fermentation time. Moreover, microwave treatment is known to be the best pretreatment tool in

Ultrasound and Microwave for Food Processing: Synergism for Preservation and Extraction.
DOI: https://doi.org/10.1016/B978-0-323-95991-9.00004-7

fermentation processes, enabling the production of valuable components (Ojha and Tiwari, 2016).

Just like fermentation, emulsification is also an important process that is used to modify and preserve food. Emulsification is a structure-forming technique practicable in food, cosmetics, as well as in pharmaceutical industries. Over the past decade, ultrasonication become a growing interest and together with emulsion production it is also investigated as an extensive process parameter (ultrasonic power, acoustic filed etc) (Kentish et al., 2008; Jafari et al., 2007). As an alternative of conventional homogenization, ultrasound minimizes the size of fat constitute and stabilize the emulsion and cream by the interaction of fat globules (Carrillio-Lopez et al., 2021). Many ultrasonic gadgets are used in the emulsification providing a number of advantages over other traditional methods (Li et al., 2018). Cavitation treatment during the emulsification process is the main reason behind its efficiency. However, microwave technology (thermal/nonthermal) during emulsification in some products add in the emulsifying properties but in other products (crude oil) promotes the demulsification.

Even though individual effects of microwave and ultrasound are quite effective but nowadays scientists are working on synergism, which is basically the combination of two or more technologies, processes or tools that put a larger effect together as compared to the sum of individual effects. The combination of microwave and ultrasound provides a cost-effective, energy-efficient, and reliable approach for a large number of synthetic variations. Synergism effects of us and mw have been investigated into the different processing treatments. This chapter highlights the isolated physical effect of ultrasound and microwave in fermentation and emulsification and how their combined effects can assist the process (fermentation and emulsification) and influence the product during the different stages of processing.

19.2 Ultrasound and its application in food fermentation

19.2.1 Fundamentals of ultrasound

Ultrasounds are sound waves that have frequencies higher than the human hearing limit of 16 kHz (Ojha and Tiwari, 2016). These US waves can be categorized depending on their frequencies as diagnostic ultrasound (>1 MHz) high frequency or extended range for sonochemistry (20 kHz–2 MHz) and power ultrasound (20–100 kHz). Ultrasonics can also be

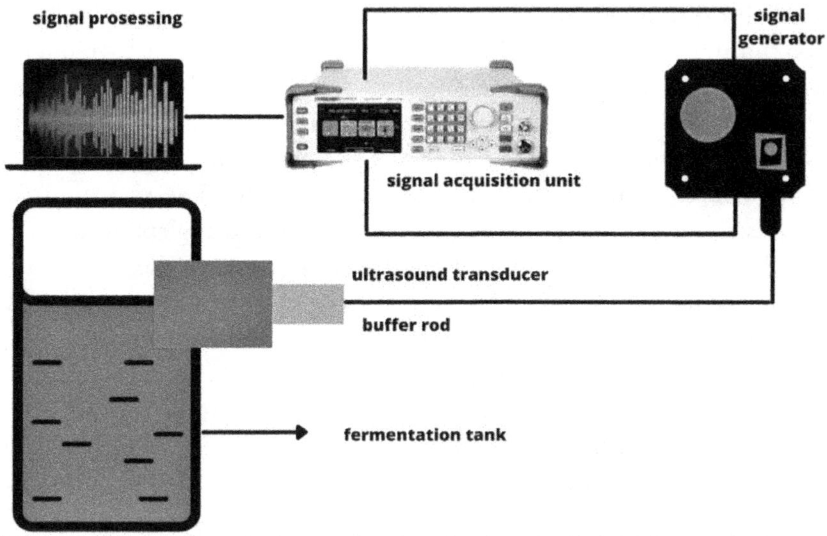

Figure 19.1 Ultrasound based fermentation process tank.

categorized as low intensity (less than 1 Wcm^{-2}) or high intensity (10–1000 Wcm^{-2}) sonication depending on the application. low power levels employed in high-frequency ultrasound induce no or mild chemical and physical changes in the transporting medium, making it suitable for quality control and food analysis techniques. Low-frequency ultrasound employs a significantly greater power to exert required physiochemical modifications for several bioprocesses, whereas no or minimum effects on food ingredients or reaction media are exerted by high-frequency ultrasound due to their much lower power level. Ultrasound categorized as Low frequency is a frequently used type employed in food fermentation. Ultrasonication has a number of advantages in food processing, including shorter processing time, improved precision and reproducibility, lower costs, improved product quality, and reduction in several downstream purification processes (Chemat et al., 2017). A transducer generates low-frequency ultrasound that is transferred via a probe-based device or ultrasonic bath operates at multiple frequencies (Fig. 19.1). The transducer's placement is critical in process intensification for uniformly distributing ultrasonic energy. In bath system arrangement the major advantage is that the transducers do not come in touch with the material, but the vessels and surroundings suffer large acoustic energy losses. Finding the optimal acoustic energy need to be applied in food fermentation is the fundamental aspect in working with power ultrasound.

The absorption, transmittance, and reflection of sound waves when carried through a medium are the fundamental principles of high-frequency ultrasonography used for food technical analysis and monitoring. Ultrasonic velocity, acoustic absorption, and acoustic impedance are all important ultrasound parameters that can be used to characterize and evaluate different physicochemical characteristics of fermentation media. The two most frequently used methods in ultrasound sensors for monitoring applications are pulse-echo and continuous wave ultrasound. In the pulse-echo mechanism, a transducer is affixed to a fermentation vessel pass out an ultrasound signal via the sample present in the vessel, then walls of the fermentation vessel reflect back the signal to the transducer. In the continuous wave mechanism, electric pulses delivered through a chamber are converted into ultrasonic waves by one transducer which are, then these ultrasonic signals are translated into electric signals by the other transducer and then an oscilloscope record them at the end.

19.2.2 Ultrasound physiochemical effects

Physicochemical effects associated with ultrasound of different frequencies are agitation, vibration, shock waves, pressure, microjets, acoustic streaming, shear forces and rarefaction, cavitation, and the free radicals' formation. Various aspects, influence the ultrasound efficiency on micro-organisms and enzymes utilized in fermentation, such as microbial ecology (e.g., composition and medium and microbes types), ultrasonic factors (e.g., ultrasound frequency and power), treatment time, temperature, and pH (Moncada et al., 2012). Studies have revealed that micro-organisms can be inactivated by ultrasound. Although, microbial inactivation takes place at low frequencies but these cells are very slightly affected and have no influence on their viability when exposed to a greater frequency range. Radel et al. (2000), for example, found no substantial alterations in the vitality of yeast cells but when treated at 2.2 MHz ultrasonic frequency, some cellular component shuffling was detected (Radel et al., 2000).

The physical and chemical impacts of ultrasound rays cause cell membrane thinning/disruption, resulting in microbial inactivation. Ultrasound rays have different effects depending on micro-organism type. Gram-positive bacteria, for example, are highly resistant to ultrasound than Gram-negative bacteria, probably because the former's cell wall is thicker and stronger owing to the presence of cross-linking between peptidoglycan and teichoic acid (Monsen et al., 2009). The enzyme activity and development of beneficial bacteria have been shown to be enhanced by ultrasound, leading to the formation of biologically active macromolecules. The adverse effect of sonication on micro-organisms is due to the pore formation on their cell membranes, thereby facilitating in and out movement of necessary nutrients and toxic compounds respectively across these porous membranes (Yeo and Liong, 2013). High power or exposure time of sonication leads to sudden cell death or inactivation of cells due to cellular content drainage.

Enzyme activity is affected by sonication is an enzyme-specific phenomenon; O'Donnell et al. studied the ultrasound effect on several enzymes (O'donnell et al., 2010). The effect of sonication on enzyme activity stimulation and/or retardation has not been well comprehended yet. However, controlled denaturation, alterations in protein conformation, or accelerated collision probability of the substrate and enzyme can all be influenced by ultrasound (Szabó and Csiszár, 2013)

19.2.3 Mechanism of action of ultrasound during fermentation process

Ultrasonication is one of the widely utilized technologies in fermentation for process intensification to time optimization of fermentation, increased output and product quality, and, lately, for on-site assessment of fermentation parameters throughout the process. The underlying principle of ultrasonication in fermentation for optimizing the process is centered on the cavitation phenomena. In ultrasonication treatment, movement, growth, nucleation, and implosive rapid collision of microscopic vapors or gas bubbles induced by the pressure difference generated owing to the effect of passage of ultrasonic waves through the medium is known as acoustic cavitation (Martinez-Guerra et al., 2014).

More entrance of materials while expansion of bubbles than the exit of materials during bubbles' compression is favored by larger surface area, and the thickness of the boundary layer transferring mass expand during bubble compression and vice versa during an expansion (Leong et al., 2016). This phenomenon has both physicochemical ramifications that usefully improve the reaction kinetics (Martinez-Guerra et al., 2014), and it may also demonstrate why power ultrasound is used in most fermentation procedures that are ultrasound-assisted. Micro convection caused by ultrasound cavitation induces structural alterations in the secondary structure of intracellular and extracellular enzymes, resulting in protein unfolding of enzymes, revealing the amino acid residues or inner hydrophobic groups, and ultimately reducing the substrates' availability for the binding site, enhancing enzymatic activity (Bundhoo and Mohee, 2018). During ultrasonicated fermentation, acoustic cavitation in the medium performs as a reaction catalyst, speeding up chemical reactions at molecular or even at the cellular levels by improving effective substrate mixing/homogenization, chemical bonds' dissolution, cell clusters' break down, and enhancing emulsification rate and dispersion of cellular material. Just because substrates and products move more quickly through the cell membrane, inhibitory substances do not concentrate in the medium for too long, enhancing the enzyme inhibitory tolerance participating in the metabolic pathway. Tizazu et al. found that by using *Candida tropicalis* during sugar cane bagasse fermentation, morphological alterations occur in the secondary composition of intracellular enzymes that participate in xylitol metabolism cause the unfolding of enzyme, thus expose its inner hydrophobic groups that consist of the substrate binding sites, resulting in rise in biomass by 17% and an utmost yield of xylitol (0.56 g/g) (Tizazu et al., 2018).

Huang et al. in their research on the morphology of *C. tropicalis* found that 1 h ultrasonication treatment on a 100 ml sample with 120 W/L power density and 28-kHz frequency resulted in an increase in the number of yeast cells (Huang et al., 2017). When compared to the nonsonicated sample, the hyphae length, as well as its number, was increased. The enhanced biomass of cells in sonicated sample was credited to improved cell porosity and increased dissolved oxygen, both of which are consequences of ultrasonic therapy on the cell morphology. A morphological comparison of *C. tropicalis* between ultrasound-treated cells and nonsonicated cells revealed that these waves provide the activation energy via acoustic cavitation, needed by the yeast cells for forming their nascent mycelia and consequently oval-shaped yeast cells. Ultrasonication of yeast cells disperses and separates their mycelia into single cells, allowing for a more efficient/effective use of oxygen and nutrients.

19.3 Application of ultrasounds in fermented foods

19.3.1 Dairy fermented products

Ultrasound was shown to enhance the favorable properties as compared to the traditional treatments (Yu et al., 2021). ultrasound technology has immense use in dairy products and the fermentation of milk. Many studies on milk and its derivatives' ultrasonicated fermentation have been conducted, and the findings have shown promising outcomes in terms of both process efficiency and product quality. Milk is mostly pasteurized before using in the fermentation of dairy products. Solo or combined effect of low-frequency ultrasonication with external pressure (manosonication), heat (thermosonication), or both (manothermosonication), has been shown to upgrade milk safety and meet the desired 5 log declining trend in pathogenic microbes such as *Escherichia coli* and *Listeria innocua*. (Lee et al., 2009). Low-frequency ultrasonication of milk has been shown to cause favorable physical and chemical variations in macromolecules, such as homogenization, pasteurization, decreased yogurt fermentation time (Wu et al., 2000) enzyme modification and enhanced rheological characteristics of yogurt (Vercet et al., 2002). Before inoculation of starting culture, sonication of milk enhances viscosity, water binding capacity, and reduces water loss (syneresis). However, milk sonication after its inoculation has been demonstrated to exert no positive effects on syneresis despite a 30 min reduction in the time of fermentation (Wu et al., 2000). Homogenization induced by ultrasound may lead to a shift in the water-binding capacity of milk proteins, resulting in enhanced water holding capacity.

Fermentation time is defined as the period from inoculation of culture of bacteria (lactic acid) into milk until acidification reaches at 4.7 pH (Puvanenthiran et al., 2002). Ultrasonication has been effectively employed in fermentation of milk technology to expedite the procedure and minimize the duration of fermentation. Dolatowski et al. also claimed that in yogurt making usage of short frequency ultrasound can reduce the manufacturing time up to 40% (Dolatowski et al., 2007). Nguyen et al. in their studies have shown the effects of stimulating milk fermentation under high-intensity ultrasound (Nguyen et al., 2009). It showed that the time needed for the bifidobacteria strain (*Bifidobacterium breve*) employed to reach acidification was lowered by 2–5 h at the ultrasonic treatment for 15 min, 100-W power, 20-kHz frequency. Applications of ultrasound in dairy fermentation can be observed in Table 19.1.

This stimulating impact was shown to be influenced by both the ultrasonication time and the bifidobacteria strain. Another study about the high-intensity ultrasonication effects on milk fermentation, Nguyen et al. discovered that it increased transgalactosylation through milk bifidobacteria and lactose hydrolysis, resulting in considerably high production of sugars for example, galactose, glucose, and organic acids for example, acetic and lactic acids throughout a 24 h of fermentation process (Nguyen et al., 2012). It was reported by Barukcic et al. when the culture of *Lactobacillus acidophilus* is treated with ultrasonication under 84 W of power and 20 kHz of frequency for 150 s there is a decrease in time of fermentation of

Table 19.1 Application of ultrasound in dairy fermentation.

Elements/products	Ultrasound treatments	Implementations	Sources
Probiotic fermented milk	20 kHz for (7–30 min) at 100 W	Improved fermentation rates	(Nguyen et al., 2012; Nguyen et al., 2009)
High-protein fermented milk	ultrasonic device (Sonopuls HD 2200; Bandelin, Germany), 30 W 22.5 W cm^{-2}	reduced pH Water Holding capacity was reduced effect on particle size reduction	(Körzendörfer, and Hinrichs 2019)
Yogurt (during fermentation milk sonicated)	Ultrasonic water bath (300 W)	Sonicated set gels exhibited syneresis increased particle numbers,	(Körzendörfer et al., 2017)
Protein-enriched Greek yogurt	Ultrasonic device (Sonopuls HD 2200) 20 kHz, 20 W	US on fermentation of yogurt shows a drop in G' value. Exhibiting weaker structure of gels Reduce gel firmness	(Körzendörfer et al., 2019)
Skim milk	US treatment (power levels 80, 100, 120, 140, 160 and 180 w/l) Time in hours (0.5, 1, 2, 3 and 4 h) Fermented milk-28 kHz Ultrasonic pulsed model (UPM) = on time (100 s) + off time = (10 s) + 100 W/L power + 30 min After fermentation time about 9 at temperature 37°C	Boost enzyme activities When ultrasound remove the fermentation effects also disappear	(Huang et al., 2019)
	Ultrasonic device 20 kHz 200 W ultrasonic output + 1 s pulse mode	reduced viscosity improved the visually perceived coarseness softening of the set gel	(Körzendörfer et al., 2019)

(*continued on next page*)

Skim dry milk	Ultrasound (22 ± 1.65 kHz) Intensity 60 W/L = 1 min 90 W/L = 3 min 120 W/L = 5 min	Modify the accumulation of biologically active compounds increased antioxidant activity	(Potoroko et al., 2018)
	Ultrasonic homogenizer probe transducer (30 kHz)	improved final product quality Boost fermentative process by 10%	(Shershenkov and Suchkova, 2015)
Milk (in fresh form)	Ultrasonic system + Fermentor (200 kHz and 17.2 kW/m^2) Time range = 2, 4, 6, 8 ,12 h Temperature range = 37 + 0.2°C	High degree of lactose hydrolysis	(Wangand Sakakibara, 1997)
	Ultrasound generator, power = 20, 50, and 100. 450 W and 225 W for minutes (1, 6 and 10)	High viable cell count Reduce fermentation time by 0.5 h almost Improved homogenization water-holding capacity also improved	(Niamah, 2019)
Nonfat dry milk	Sonicator 200 kHz	decreased the viable cell count 55% vs. 35.6% in control fermentation it hydrolyzed Lactose	(Sakakibara et al., 1994)
	Sonicator (200 kHz) + 17.2 kW/m^2	high cell viabilities High degrees of lactose hydrolysis	(Wang et al., 1996)
Feta-type cheese	Ultrasonic Homogenizer = 20, 40 and 60 kHz Time = 20 min Intensity = 80%	Highest level of acidity accelerated lipolysis and proteolysis	(Staszak and Wieszczycka, 2022)

reconstituted sweet whey by 30 min (Barukčić et al., 2015). Abesinghe et al. demonstrated that ultrasonic can assist to reduce milk fermentation time, lowering costs, and speeding up the process while also enhancing the texture and consistency of milk gels (Abesinghe et al., 2019). Peng, 2010 has reported that because of the short fermentation time rearrangement of the yogurt gel structure is reduced, However, the acidification rate is the primary predictor of the extent of rearrangement in milk gel (Peng, 2010). In milk fermentation, ultrasound increased cell membrane permeability, and microbe multiplication. When counts of *B. infantis* and *B. breve* were compared before and after ultrasonicated fermentation, Nguyen et al. discovered a substantial improvement in probiotic multiplication (Nguyen et al., 2009). Ultrasonic frequency of 30 kHz and power of 100-W at amplitudes (20%, 60%, and 100% amplitude) employed for 2–3 min in milk aided lactobacilli growth with colony counts surpassing 9 log cfu/ml along with increased activity of βglucosidase in cells, allowing bioconversion of isoflavone glucosides in soymilk to bioactive compounds aglycones (Ewe et al., 2012). Another research revealed, ultrasonicated *L. fermentum* BT 8633 grew 3.23–9.4% faster in soymilk supplemented with biotin when set against the control, with improved intra- and extracellular inversion in aglycones from isoflavones glucosides during fermentation when treated with ultrasound for 2 min under 100-W power, 30-kHz frequency (Ewe et al., 2012). Huang et al. to stimulate fermentation of skim milk, used low-intensity ultrasound, and applied ultrasonic conditions of100 W/L power, 28-kHz frequency for 30 min yielded 49.5 and 43.5% rise in peptide composition of the media and viable cells of inoculum, respectively(Huang et al., 2019). These accretions could be affiliated with activation of the extracellular enzymes during sonication of fermented skim milk medium as compared to the nonsonicated treated sample.

During ultrasonication of milk fermentation and its derivatives product quality has been reported to enhance significantly. High amplitude ultrasound improved the consistency and viscosity of yogurt, owing to the protein denaturation and homogenization of fat globules in milk during ultrasonic treatment (Nguyen and Anema, 2017). Ultrasound alters the texture of the end product thus, affecting the gelation process during fermentation when administered prior to the processing. Thus, according to Riener et al. ultrasonic pretreatment of milk can improve fat homogenization during fermentation while also aiding in manufacturing commercial set yogurt with reduced milk solids supplementation (Riener et al., 2009). Researchers studied, the proliferation of bacteria cells is the reason for physicochemical alterations during the fermentation, which results in an overall alteration in the intrinsic behavior of the medium, by increasing its thickness, nonlinear elevation in attenuation coefficient and modifying its ultrasound wave transmission (Resa et al., 2007). Ultrasound monitoring of the fermentation process provides a contactless and nondestructive control, thus reducing the potential contamination risk and presenting an alternative tool for healthy processing control.

19.3.2 Alcohol/alcoholic beverages

The manufacturing of alcoholic beverages necessitates the employment of certain micro-organisms, but there is always the spoilage risk due to the invading of undesired bacteria or yeasts. Traditional spoilage prevention measures include the use of preservatives (e.g., dimethyl dicarbonate, sulphur dioxide), elimination of "rogue" micro-organisms, thermal pasteurization, or filtration to remove spoilage microbes. Alternative treatments are typical of interest because such involvements are occasionally followed by the presence of off-flavors. *S. cerevisiae* can be ultrasonically treated to improve its fermentation performance by increasing its proliferation, permeability of cell membrane, and alcohol tolerance. The impact of low-intensity ultrasonication on *S. cerevisiae* was researched by Dai et al. who discovered that optimum ultrasonic parameters of 140 W/L power, 28 kHz frequency and treatment time of 1 h boosted *S. cerevisiae* biomass by 127.03% (Dai et al., 2017). A substantial upsurge in the growth of *S. cerevisiae* was reported by Zhang et al. under a constant frequency (23 kHz) ultrasonic treatment (Zhang et al., 2019). Low intensity ultrasound promoted the *S. cerevisiae* growth (Lanchun et al., 2003). Its development and reproduction could be hampered by high alcohol content as at higher alcohol percentage over 10% growth of *S. cerevisiae* is inhibited. During alcoholic fermentation, low-intensity ultrasonic therapy improves cell membrane permeability. It was also reported that extracellular nucleic acid, fructose-1,6-diphosphate and protein escalated by 15.05%, 220%, and 19.6% respectively, after a 3–4 h ultrasonic therapy at 28 kHz frequency (Dai et al., 2017). Some applications of ultrasound in Alcoholic fermented products have been stated in Table 19.2.

Alcoholic fermentation procedures supported by ultrasound treatment have yielded higher alcohol yields. Ultrasonic treatment at a frequency of 20 kHz resulted in a gain in ethanol concentration with an intensity of 11.8 W/cm^2 (Sulaiman et al., 2011). According to Nitayavardhana et al.

Table 19.2 Application of ultrasound in alcoholic fermented products.

Elements/ products	Ultrasound treatments	Implementations	Sources
Grape wine	Ultrasonication = 90 kHz Time = 1 h	Slight increase of visible absorption Tannin concentration increase	(Singleton and Draper, 1963)
Rice wine	Wine fermentation 20 kHz ultrasound 1-week duration	Ethyl acetate compound increase	(Chang and Chen, 2002)
		Reduction in Alcohol content + Acetalde-hyde content	
Red wine	Ultrasound = 26 kHz	Wine flavor modification microorganisms cell spoilage potential diminish	(Luo et al., 2012)
	Time = 20 min		
	Ultrasonic processor = 24 kHz Continuous flow treatment	lactic acid bacteria and *Brettanomyces* reduced	(Gracin et al., 2016)
	ultrasound 50 kHz 50 min/week	Massive Rise in the concentration of polysaccharides	(Kulkarni et al., 2015)
Model wine	40 kHz ultrasonic bath system	After 20 h ultrasonic treatment Yeast viability heavily affected	(Martín et al., 2013)

during fermentation, there is a direct correlation between increased ethanol yields and sonication time (Nitayavardhana et al., 2010). In comparison with non-sonicated samples, ultrasound-assisted simultaneous fermentation and saccharification at 37 kHz frequency and 200 W power, considerably enhanced the yield and percentage of bioethanol reported by Ofori-Boateng and Lee (Ofori-Boateng and Lee, 2014). Incomparable research, with sonication treatment there was an increase of 1.8 times in bioethanol output, as compared to a nonsonicated sample. Therefore, ultrasound technology offers a viable option for this procedure that is both rapid and cost-effective. Ultrasonic waves have the substantial capacity to help with wine aging and maturation. Chang and Chin, has also reported a rise in temperatures and pressure by ultrasonic waves, which consequently catalyze the ageing process and also assist the development of wine taste and flavor (Chang and Chen, 2002). Fermentation processes are being monitored using ultrasonic wave-based analytical techniques. Using ultrasound technique during fermentation, evaluated the concentrations of sugar and ethanol (Schöck and Becker, 2010).

19.4 Ultrasound assisted emulsification mechanism

Emulsifying properties can be raised by proper ultrasound treatments, which also improve the emulsion's zeta potential (Liu et al., 2013). Ultrasound emulsification equipment has now been extensively used in a variety of applications (Li et al., 2018), as they provide several benefits over traditional emulsification processes for example, high energy effectiveness, good emulsion strength, and ease of operation and cleaning are only a few of them. Acoustic cavitation involves the development and collision of air bubbles in a process caused by ultrasound equipment. The release of reactive radicals that promote chemical reactions can be facilitated by the collapsing of air bubbles (Leong et al., 2017). The contact surface collapse of a cavitation bubble involving two immiscible phases forms a more stable emulsion with reduced energy (Udagama, 2009). The average size and bubble density increase as the acoustic power rose. This might enhance the shear forces caused by bubbles collapsing, allowing oil droplets to be disrupted more easily. So, the size of the oil droplets is reduced as a result of these modifications, which improves emulsion stability (Cavalieri et al., 2018). According to Cucheval & Chow, The physical consequences of ultrasound caused by acoustic cavitation will promote oil droplet disruption, promoting the production of steady O/W emulsions with tiny droplet sizes (Cucheval and Chow, 2008) (Fig. 19.2). The emulsions are made in two phases using high-intensity ultrasound (HIU). First, large droplets are transformed from the dispersed phase to the continuous phase. The emulsions are made in two phases using HIU. Huge droplets of the dispersed phase develop first in the continuous. Second, the large droplets can gradually disperse due to the continual physical shearing and acoustical cavitation (Leong et al., 2018).

The emulsification process might be influenced by cavitation efficiency in ultrasonic treatment. Sharma & Mudhoo, reported some parameters (ultrasound frequency, temperature, power, and viscosity or surface tension of the medium) that can affect cavitation efficiency (Chen et al., 2011). If the applied frequency is raised, for example, the power must be increased

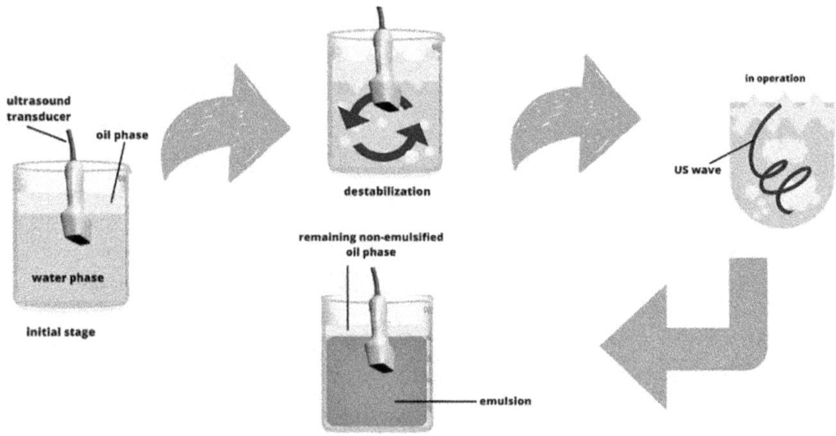

Figure 19.2 Size of lipid droplet during emulsification under ultrasound application.

to reach the same level of cavitation activity. Cavitation in a liquid is hard to achieve at higher frequencies (Cavalieri et al., 2018) because of the shorter compression and rarefaction cycles. much higher frequencies (in the MHz range) are employed to reverse the process of emulsification (de-emulsification) (Cavalieri et al., 2018; Pangu and Feke 2009). However, low-frequency ultrasound range, particularly 20 kHz, has been demonstrated to be effective in generating stable emulsions. Furthermore, Chen et al. reported the boosts in the sonochemical effects by enhancing the sonication power (Chen et al., 2011). The mean size and bubbles density grew when the acoustic power rose. This might enhance the shear forces caused by bubbles collapsing, allowing oil droplets to be disrupted more readily. These modifications minimize the size of oil droplets, improving emulsion stability (Cavalieri et al., 2018). Some researchers also found the variable parameters (e.g., treatment time, processing volume and amplitude to (high pressure homogenization) at pressure ranges from 20 MPa to 120 MPa) to compare the energy economy and ultrasonic emulsification efficiency (Li et al., 2018).

The generation of double emulsions might be difficult because they are thermodynamically unstable. In the W/O/W emulsion, excessive oil droplet reduction might accelerate the escape of the entrapped aqueous phase, limiting the system's shelf life (Leong et al., 2017). So, HIU can generate oil droplets of adequate size that can be controlled by controlling the sonication frequency, power or duration. Leong, Martin, and Ashok Kumar, by utilizing sunflower oil and skim milk manufacture W/O/W emulsions effectively with the application of ultrasound (Leong et al., 2017). During the process of secondary emulsification different ultrasonic power levels of 2, 6, 10, 18 and 26 W were applied for 5 s to adjust the droplets size of secondary emulsion (Leong et al., 2017). It was reported that at the power of 18 or 26 W, double emulsions were shown to be stable for 7 days. It's probable that increasing ultrasonic strength or duration denatures casein in skim milk to some extent.

19.4.1 Ultrasound effect on the formation of different emulsions

Various emulsions, like protein, stabilized emulsions, protein-polysaccharide, and polysaccharides emulsions have been prepared using HIU, as indicated in Table 19.3. Owing to their amphiphilic nature, a variety of protein-based emulsifiers are utilized to stabilize emulsions (McClements, 2004). During emulsification, protein molecules are disrupted and adsorbed on the oil droplets' surface. By generating a protein barrier this could minimize interfacial tension at the oil/water interfaces and facilitate steady protein emulsions development (Sui et al., 2017).

It has been discovered by a large number of scientists that emulsifying properties could be improved by HIU of food proteins by improving their adsorption and solubility efficacy at the interfaces (Higuera-Barraza et al., 2016; Zhou et al., 2016). The characteristics of several polysaccharides have been altered using ultrasound. For example, apple pectin has found a decrease in molecular weight and methylation level by using HIU (Zhang et al., 2013). Jambrak et al. researched physical characteristics of the corn starch under the impact of ultrasound and found that Ultrasound treatment lessened the gelatinization viscosity and molecules granular structure was also destroyed (Jambrak et al., 2010). Sonication was employed to create oil-water emulsions by using cellulose crystals as stabilizers because ultrasound facilitated the structural disruption of cellulose crystals. Furthermore, Meirelles et al. observed smaller droplet sizes of emulsions stabilized with cellulose crystals under ultrasound than those prepared with the Ultraturrax homogenizer (Meirelles et al., 2020). Regardless of the fact that by reducing the droplet size proteins could maintain emulsions, several of these emulsions were found to be vulnerable to environmental stressors including ionic strength and pH (Taha et al., 2019). While Shamsara et al. reported, polysaccharides, on the other hand, developed stable emulsions across a variety of environmental positions, including ionic strength, pH and temperature (Shamsara et al., 2017). As a result, the formation of emulsions by combining polysaccharides and proteins could be effective with tiny droplets and great environmental

Table 19.3 Application of ultrasound in different emulsions.

Emulsions	Ultrasound conditions	Emulsifiers	Oil phase	Source
Protein stabilized emulsions	24 kHz (20, 40, 60 Sec)	Whey protein	D-limonene (20%)	(Mahdi et al., 2006)
	20 kHz 176 W (8 min)	Dairy content	Flax seed oil (21%)	(Shanmugam and Ashokkumar 2014)
	750 W (6 min)	Hazelnut protein	Clove essential oil (3%)	(Taha et al., 2019)
	20 kHz (6 min)	Bovine gelatin	Rapeseed oil (10%)	(Gul et al., 2018)
	25 kHz (15 min)	Almond protein	Corn oil (30%)	(Zhu et al., 2018)
	65–195W (10-20 min)	Rice bran protein	Soybean oil (10%)	(Meirelles et al., 2020)
Polysaccharide stabilized emulsions	20 kHz, 525W, (4 min)	Cellulose crystals	Sunflower oil (10%)	(Sun et al., 2019)
	24 kHz, 175W, (25 min)	Jujube polysaccharides	Nettle essential oil (10%)	(Gharibzahedi, 2017)
	19 kHz, 320 W, (2 min)	Gum Arabic	Annatto seed oil	(Silva et al., 2015)
	20 kHz, (20 min)	Hydroxypropylmethyl cellulose HPMC	Sunflower oil	(Camino and Pilosof, 2011)
	30 kHz, (40s)	Pectin	n-hexadecanen-dodecane	(Alba et al., 2016)
Protein-polysaccharide complexes stabilized emulsions	160 W, (2 min)	Whey protein; gum	Lime essential oil	(Campelo et al., 2017)
	700W, 20 kHz, (15 min)	WPC-pectin	Chia oil	(Vélez-Erazo et al., 2018)
	150 W, (7 min)	Sodium caseinate-corn starch hydrolysates	Resveratrol and Palm oil	(Consoli et al., 2018)
	25 kHz, 5–15 min	potato starch	sunflower oil	(Silva and Sato, 2019)

resilience. Thus Rodriguez Patino & Pilosof, found the alterations in the adsorbed layers of interfacial characteristics and the dispersed layer stabilization by the synergistic impacts of protein and polysaccharides (Patino and Pilosof, 2011). The outcome of ultrasonication on the emulsifying and physicochemical properties of SPI-citrus pectin complexes were examined by the Ma et al. and they found that under the influence of acoustic cavitation. the structural modifications, enhanced emulsifying properties of SPI and CP and improved electrostatic interactions (Ma et al., 2019).

The graft reaction of glucomannan with peanut protein isolate (PPI) is also accelerated by ultrasonic treatment (Li et al., 2014). The ultrasound effect and concentration of gelatin on the characteristics of alginate stabilized emulsions or potato starch were examined by Silva and Sato (Silva and Sato, 2019). Silva and Sato, also examined the effects of ultrasound on biopolymer's properties (reduced interfacial tension, zeta potential values and viscosity) that resulted in tiny droplet emulsions with greater stability. This was most probably due to the fact that ultrasound cavitation might improve the emulsifying characteristics of polysaccharides and gelatin biopolymers by facilitating intermolecular interactions (Dickinson, 2008).

19.4.2 Dairy/dairy products

Ultrasound has been used in several ways to enhance the attributes and applications of milk and milk products concerned to the dairy industry, which includes safety, emulsification, homogenization, fermented products (yogurt), and production of fermented beverages. Ultrasound impacts the fermentation and deterioration of milk components, through microbes, functional, technological, and sensory features of dairy products have all received special attention. Ultrasonication's ability to reduce fat particle size is dependent on ultrasound power (Freitas et al., 2006). Emulsification is an important process of reducing the droplet size in dairy. Basically, the initial stage of processing is called reduction of size (Udagama, 2009). First, turbulence caused by acoustic vibration caused fat droplets to disperse into the liquid phase of milk. Later, sonicated cavitation aids in the disintegration of fat droplets (Leong et al., 2016). Owing to casein micelles interaction with the ruptured membrane, sonication has the ability to deteriorate the milk fat globule membrane, resulting in a reduction of fat globule size and modifying the fat globule's granular surface (Kaci et al., 2014). They believe that ultrasonication forces large existing droplets to break and fall into the aqueous phase. Where cavitation, the collapse of micro-bubbles in response to ultrasonic waves, reduces the size of those droplets. The liquid is surrounded by highly localized turbulence as a result of the collapse. Power ultrasound (with 20 kHz, 600 W) has been observed to increase the gelation reconstitution, emulsification, solubility, and ultrasonicated milk protein concentrate considerably (Yanjun et al., 2014). Prior to spray drying the changes were shown in the structural attribute of ultrasound treated micellar protein extract and in functional properties (Zhang et al., 2018). As the ultrasonication period was extended, there was a considerable rise in conductivity, emulsification, gelation, and solubility.

According to several authors, milk proteins (being natural emulsifiers) maintain emulsions created with ultrasonic aid (Amiri et al., 2018). According to Shanmugam and Ashok Kumar, whey proteins help to stabilize ultrasonicated-emulsified oil globules, resulting in gels with finner structural characteristics compare to those made without any sonicated treated pasteurized skim milk (Shanmugam and Ashok Kumar, 2014). Similar findings were represented by Amiri et al. according to which ultrasonicated whipped cream constitutes lowers fat globule size (Amiri et al., 2018). As a result, denatured proteins extend to surround the globules, restricting air bubbles and enhancing the density and symmetry of whipped cream. According to Chandrapala et al. the size of the fat globules in milk (fresh skim milk) in exposure to ultrasound for 60 min (1 h) was reduced to 10 nm (Chandrapala et al., 2012). When compared to typical homogenization at temperature 50°C, the creation of coagulated fat globules (ultrasonicated cream) at low temperatures could minimize the potential energy requirements (Chandarapala et al., 2016). In ultrasound-treated goat milk, Karlovic et al. found that the proteins in milk, particularly caseins, bind to the surface of the membrane of fat globules, act as intrinsic emulsifiers (Karlović et al., 2014). In this situation, milk proteins treated with ultrasonication have shown diverse emulsifying effects (O'Sullivan et al., 2014). After 28 days, milk protein solely develops the smaller-sized droplets containing emulsion with zero coalescences. Regardless of the reduction in protein size, emulsions made from sodium caseinate, and isolated whey protein had almost the same diameters of the droplet as handled without ultrasonography. Aslan and Dogan investigated the ultrasound impact on the creation of dairy emulsions (14% sucrose, 7, 10, and 15% olive oil, 11% skimmed milk powder, 0.3% mono-diglycerides and 0.2% xanthan gum as emulsifiers) (Aslan and Dogan, 2018). they analyze mechanical agitation and HIU application for emulsifying olive oil in milk-based media. They discovered that as compare to emulsions created by mechanical agitation, the HIU promotes the apparent viscosity of a drop. The size reduction of the oil droplets encourages a more effective distribution of the proteins in the system. HIU has been shown to boost whey protein solubility (Khatkar et al., 2018), improve emulsifying characteristics, and decrease rigidity (Shen et al., 2017). Mongenot et al. employed ultrasonication to investigate emulsify cheese odor and discovered that it produced an unstable smell in butyric acid flavor (Mongenot et al., 2000). Their analysis of the fragrance profile revealed that when estimating the efficacy of carriers and emulsification processes, each fatty acid must be taken into account. Ultrasound lowers fat globules, making it a viable alternative to standard homogenization. In creams and emulsions, the fat-protein connection promotes stability.

19.5 Microwave and its application in food fermentation

19.5.1 Fundamentals of microwave

Microwave heating is a new food processing technique that depends on the thermal reduction of bacteria populations in foods to preserve them. Heating occurs when at specific frequencies electromagnetic radiation interacts with dielectric materials (also known as volumetric heating) as, unlike the mechanisms of conduction or convection, microwave radiation directly absorbs the material, generates massive heat, resulting in high energy efficiency and shorter heating durations (Zhu et al., 2007). This technique has a lot of uses in the food sector since it has a lot of advantages over traditional heating, like faster processing, more energy efficiency, and better food quality. Microwave heating in foods is a complex process involving microwave propagation, which is directed by Maxwell's equations for electromagnetic waves, microwave and food interaction, led by heat dissipation and dielectric properties, followed by basic mass and heat transfer concepts (Tang et al., 2002). Heat generation within the material is one of the unique abilities of microwaves that cannot be achieved by any other method (Fu, 2004).

Microwave radiation is also categorized as nonionizing radiation to differentiate it from other types of radiation (gamma rays, X-rays), which are classified as ionizing radiation because they produce positive and negative charged atoms or molecules (Parker, 2003). A dipolar interaction emerges when microwaves penetrate the material; polar molecules and water molecules align themselves with the electrical current. Intermolecular friction is created when molecules attempt to vibrate at such frequency.

As a result, the microwave field produces a significant amount of energy and heating results. Heat is thus produced and circulated through the food components, and depending on the temperature causing the heat to spread on all sides (Hossan et al., 2010). Convection is also used to transmit heat in liquid foods. Conduction and convection heat transfer took place as volumetric heating, although the time scales are various.

19.5.2 Mechanism of action of microwave during the fermentation process in food products

Microwave heating (MWH) is a technology that uses electromagnetic radiation at frequencies between 300 to 300,000 MHz to heat objects. Many studies focus on the possibility of a variety of microwave uses in fermentation processes, including the biohydrogen generation, hydrolases production, the reduction in production of accumulative gas, and the safety of fermented products (Ramaswamy and Tang, 2008). In terms of ion conduction and dipole rotation, microwave heating has a direct influence on macromolecules (Ramaswamy and Tang, 2008).

Several possible microwave-assisted fermentation methods have been examined by various researchers. Microwave-assisted extraction (MAE) is also one of the potential applications among them (Chemat et al., 2012; Rombaut et al., 2014), preferably as pretreatment in the extraction of high added-value compounds such as lipids (Dai et al., 2014), pectins (Maran et al., 2014), anthraquinones (Hemwimon et al., 2007), essential oils (Farhat et al., 2010), and polyphenols (Aspéand Fernández 2011). Microwaves have been demonstrated to boost fermentation processes in a few small experiments. For example, during the fermentation of grapes, the effects of several treatments (yeast, microwave, conductive, microwave+ yeast, conductive+ yeast) were tested to see if the fermentation time could be decreased (Kapcsándi et al., 2013). The fermentation phase of samples was examined for changes in sugar, alcohol, and acidity for this purpose. As per scientists when the samples were processed with microwaves at the optimal circumstances, the fermentation time was shortened by a high percentage (up to 40%) compared to the control samples. There is a scarcity of knowledge on the commercial viability of these techniques at this point of development. The following are some of the most important microwave applications in fermentation processes (Fig. 19.3).

Furthermore, other analyzers have investigated the viability of using microwaves in various processes related to fermentation (e.g., reduction of cumulative gas production, hydrolytic enzymes creation, the generation

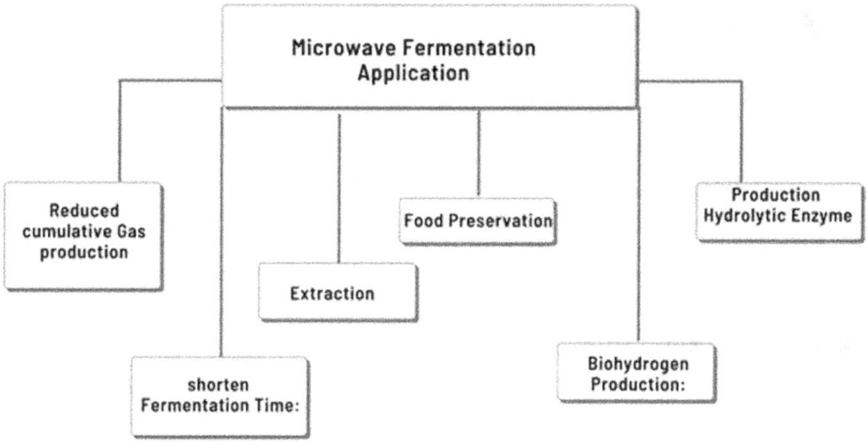

Figure 19.3 Microwave-assisted fermentation applications.

of biohydrogen, and fermented products preservation). MAE is based on the direct action of ionic conduction and dipole rotation on molecules (Pérez et al., 2014). When MAE is used, the sample absorbs the energy, particularly by polar molecules like water (aquas content), resulting in cell breakdown, which can aid in fermentation processes.

Other researchers examined the microwave pretreatment effects on synthesis of enzymes during Aspergillus oryzae solid-state fermentation of wheat (Melikoglu, 2012). Hydrolytic enzymes must be produced in large quantities for the production of various evolved products in order to drive proper cereal-based bioreactors, which are responsible for the conversion of macromolecules (such as proteins, phosphorus, and starch) into monomers (free amino acids, glucose, and free phosphorous) in cereals (Melikoglu, 2012). Microwaves, for example, can be a beneficial tool for improving hydrolytic enzyme synthesis, as the authors discovered a significant increase in the activities of protease and glucoamylase following microwave treatment for 30 sec and compared to untreated samples.

Using an in vitro gas generation approach, the effect of MW irradiation (1000 W) on the nutritional value and fermentation properties of tomato pomace for ruminants was investigated for a certain period of time (0–5 min) (Maheri-Sis et al., 2012). According to scientists, microwave irradiation decreased the high yield of cumulative gas along with fermentation during early incubation periods.

Microwaves as a pretreatment for enhancing fermentation processes and therefore enabling the generation of valuable molecules (e.g., biohydrogen

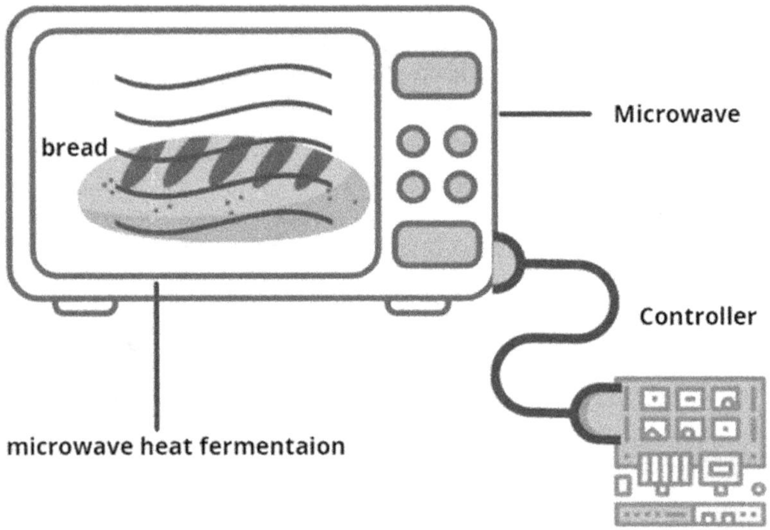

Figure 19.4 Microwave heating fermentation of bread.

production) has a significant renewable energy potential (Ethaib et al., 2015). It was studied that microwave-assisted lignocellulosic biomass pretreatment submerged in alkaline glycerol for the synthesis of fermentable sugar (Diaz et al., 2015). Microwave pretreatment of corn straw and rice husk submerged in water, alkaline glycerol, or aqueous glycerol, was used to improve enzyme hydrolysis. a significant increase in saccharification yields was discovered when biomass was microwave pretreated in alkaline glycerol rather than commercial cellulose cocktail *Celluclast* (for both maize straw and rice husk). They also discovered a huge alteration in the maize straw structure that had been microwave pretreated.

19.5.3 Wheat products

Temperature management is critical for the fermentation process, whether it's for yeast or sourdough. In fermented wheat products manufacturing, certain innovative thermal processes have proved helpful. The most promising heating technologies are microwave heating (MWH), ohmic heating, and radiofrequency heating. MWH changes the water molecules in the dough, which not only impacts the fermentation period, but also increases the overall fermentation quality of whole-wheat flour products. The energy absorption in the dough (particularly polar macromolecules like water) under the exposure of MWH improves the fermentation process Fig. 19.4.

As a result, MWH is most commonly utilized in the fermentation of liquid fermented products like yogurt, juice etc. (Gholamhosseinpour et al., 2020). Simultaneously, there is a scarcity of data on MWH's economic feasibility in the fermentation of flour products. Microwaves have been demonstrated to enhance fermentation processes in a few experiments. For example, some scientists investigated microwave heating effects on fermentation time and discovered that under the ideal circumstances, when the test sample was exposed to microwaves, fermentation time was decreased by 40% as compared to the control sample (Kapcsándi et al., 2013).

In another study after considering the microwave effects on enzymes in the course of solid-state fermentation; it has been investigated that after microwave pretreatment for about 30 seconds a significant rise in protease and glucose amylase activity and hydrolase yield as compare to untreated samples take place (Melikoglu, 2012).

Macromolecular hydrolase activity is essential for grain-based biological fermentation to work well, since in dough it facilitates the macromolecular compounds conversion, such as starch into fermentable sugars, protein into free amino acids and fat into fatty acids. MWH treatment has been shown to enhance parameters for fermented dough (rehydration ratio, bulk density, difference in color, and overall validity) (Patel et al., 2013). Microwave heating in wheat bran employs cellulose, lignin and other biomass to influence the generation of fermentable sugars (Wang and Lu 2013). MWH can also enhance the soluble dietary fiber's amount in the system (*E. garabinoxyla*) at the same time (Aguedo et al., 2013; Suwannasom and Ruangviriyachai 2011). This shows that MWH might help with whole-wheat fermented product development and quality management. It should be noted that the microwave heating in dough fermentation may be restricted under certain circumstances. First, microwaves have improver heat setting of high heat energy that can cause helpful bacteria in the fermentation process to become inactive (Gholamhosseinpour et al., 2020). Second, for the fermentation dough manufacturing at industrial scale the microwave heating equipment has to be modified and remanufactured. precisely, specialists are needed to operate and maintain them.

19.5.4 Effects of microwave on emulsification

Microwaves can influence the emulsification process in a variety of ways; by promoting the emulsification or by encouraging the demulsification under certain circumstances. It also depends on the type of microwave therapy, such as whether it is thermal or nonthermal. For example, sunflower protein after being extracted from industrially de-oiling of press cake was treated with a nonthermal microwave to observe how both structure and emulsifying properties were improved. When exposed to a nonthermal microwave, the polar amino acids concentration was reduced, and surface and solubility hydrophobicity were modified. In comparison to defrost mode treated and control samples, interfacial dilatational rheology studies demonstrated that microwaved treated samples generated a less stretchable, solid-like layer at the oil-water interface. It can be concluded that microwave treatment (nonthermal) might be regarded as a potentially simple, quick, "green" protein modification method (Subasi et al., 2022).

A study demonstrates that microwave heating increases the properties (both antioxidant and emulsifying) of ovalbumin glycated alongside glucose in solid-state (Tu et al., 2015). The goal of this finding was to determine the ovalbumin characteristics after it had been glycated with glucose and heated in the microwave. Microwaves (power levels of 480 and 640 W) were utilized to heat the solid-state ovalbumin-glucose system for the time range between 0, 5, 10, 15, 20, and 25 min, respectively. The absorbance of UV, emulsifying activity, intensity of browning, antioxidant activities, and emulsion stability of products (made of Maillard reaction) is directly proportional to the time and power (increased as increase in microwave treatment), whereas the pH of the system decreased. MW application reduces the reaction time significantly compared to old approaches. Implying that MW irradiation is unique as well as an effective way to induce the Maillard reaction in the dry state; increase both emulsification characteristics and antioxidation of proteins.

Egg albumen powder outperformed liquid eggs in terms of microbiological safety, convenience, and shelf life. Egg albumen proteins have the potential to produce stable foams and generate heat-induced through emulsifying capabilities and air incorporation, making them useful in a wide range of processed meals (Achouri et al., 2005). Microwave heating increases The EAI (Emulsifying activity index) of the OVA-glucose system. The EAI rose from 228 to 338 m^2/g after microwave treatment (480 W). Similarly, the EAI rose somewhat more with microwave treatment at 640 W than at 480 W. At both microwave power levels; the emulsion stability trend was nearly identical to the EAI (480 and 640 W). With increasing microwave heating time, an increase in the ESI (emulsion stability index) of samples was observed higher at 640 W than at 480 W. The findings were identical to Achouri et al. who discovered that glycation of the 11S-rich glycinin fraction improved its emulsifying ability, notably in the early and middle phases of the Millard reaction (Anchouri et al., 2005).

In the petroleum industry, the microwave irradiation method is utilized for demulsification. Klaila (Klaila, 1978) and Wolf (Wolf, 1986) were the first to introduce microwave heating in demulsification in their patent insights (Kuo and Lee, 2010). The inner phase of a water-oil emulsion (e.g., oily sludge) is water, which has a high level of dielectric loss than oil and hence absorbs more microwave (MW) radiation. Water stretched and the water-oil (W/O) interfacial film pushed to become thinner due to the microwave ability of energy absorption. The separation of water and oil can be achieved without the use of chemicals (Tan et al., 2007; Hu et al., 2013). Water oil (W/O) emulsion when heated with microwave, two main processes usually take place at the same time. A sudden rise in temperature is one of these, as it reduces the viscosity of emulsions and fractures the external coating of droplets. Molecular rotation, on the other hand, by rearranging the electrical charges around water molecules neutralizes the zeta potential, causing ions to migrate around droplets (Kuo and Lee 2010). Microwave potentials (power, duration, surfactant, pH, salt) together with other features of oily sludge, such as the water-oil ratio, can all prove that how well microwave heating works for oily sludge demulsification (Fortuny et al., 2007). Xia et al. found that in a relatively short period the demulsification productivity can reach 100% under microwave radiation (Xia et al., 2003). A researcher evaluated the demulsification properties of W/O emulsions when treated with a small amount of salt and discovered that it may increase the light transmittance of the separated water and effectively improve the demulsification efficiency (Xia et al., 2003). Microwave irradiation has been proven to be an efficient treatment of water-oil emulsions in multiple laboratory and field investigations. Hu et al. (2013) and Fang and Lai (1995) in a field test in order to demulsify 188 barrels of W/O emulsion in tanks, used a microwave heating method and the emulsion was separated into 146 barrels of oil and 42 barrels of water, it was concluded that water separation yield from emulsions was higher than when applying conventional heating (Hu et al., 2013, Fang and Lai 1995).

Fortuny et al. investigated a number of factors affecting crude-oil emulsions in the microwave demulsification process, including pH, salt, and water content (Fortuny et al., 2007). Microwave demulsification efficiency was observed to be better for emulsions with greater water content, excluding the fact that a high pH and salt content were present at the same time. Tan et al. used the microwave chemical approach to demulsify and separate crude-oil emulsions and discovered that the separation effect for crude-oil emulsions with high-water content was better than emulsions with lower-water-content (Tan et al., 2007). Under the condition of 50 ppm demulsifier, radiation duration of 10 s, and settling time of 1 min, the separation efficiency of this approach was similarly found to be around 95%. Kuo and Lee presented seawater-assisted microwave demulsification as a cost-effective origin of cations (Kuo and Lee 2010). The demulsifying efficiencies under optimum operating time conditions of 40 seconds of microwave irradiation at 700 W, settling time of 60 min, and additions of artificial seawater (12%, 32%, and 20%) of, an olive-oil emulsion, cutting-oil emulsion and a cutting-oil/olive-oil mix were reported to be 92.6%, 93.1%, and 93.2%, respectively. After adding inorganic acid, a fall in pH was observed to be resulted in a significant rise in demulsification efficiency, however, an increase in SDS, surfactant concentration, resulted in degradation efficiency. Microwave irradiation owing to its short heating duration attribute is high-energy-efficient and easy-to-control method for breaking emulsions. Aside from its excellent demulsification efficiency, the yield of light aromatic compounds can be increase as microwave generated direct heating inside the bulk medium and low temperature of the reactor wall might lead to a longer aromatization process.

19.6 Synergism of two or more technologies

The synergism phenomenon occurs as a result of the combination of tools, technologies, and reagents (Martinez-Guerra and Gude 2014). Combining two individual mechanisms may lead to (magnified impact) synergism (Haung et al., 2007). Following are some important criteria to address process intensification in the synergism mechanism (Van and Stankiewicz 2009).

- To maximize the effectiveness of intermolecular as well as intramolecular interaction.
- To ensure uniform heating and gradient less mixing
- To maximize specific surface area for mass and heat transfer
- To optimize driving forces

The most important issues in process intensification are energy, structure, synergy (in multi-functional developed tools the functional domain), time (Stankiewicz, 2006). To promote the process intensification as well as synergy among them; when combining the two conventional processes it is necessary to identify appropriate process configuration. It is not compulsory that all the process combinations lead to process intensification; although they provide

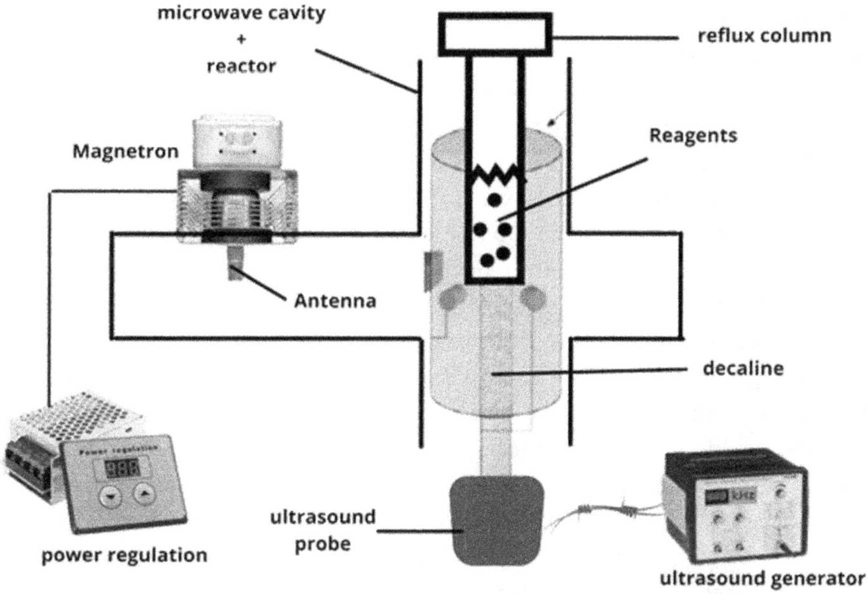

Figure 19.5 Systematic diagram of ultrasound-microwave combine system.

synergism effective. Alongside optimization and process control, various issues arise.

19.6.1 Synergism of ultrasound and microwave technologies and their applications

Many scientists are aware of microwave wave irradiation techniques; in addition to this many scientists are familiarized with ultrasound. Contrary to this, few scientists are familiar with both techniques. The nature and effects of both techniques are intrinsically varied. Between ultrasound and microwave, the conceptual diagram can be seen below Fig. 19.5 (Kappe, 2006; Mason and Lorimer 2002)

In recent years the focal point is ultrasound and microwave-mediated organic synthesis because of their remarkable effects in fewer reaction times and more product yields (Gude et al., 2011). In various engineering applications as well as chemistry these two process intensifications have been utilized. In a research, the palm oil mills effluent was treated individually with ultrasound and microwave and their combined effects. The result shows that ultrasound microwave combined treatment proves as much effective treatment comparatively; it provides much better results to neutralize the organic constitutes (Mishra et al., 2021). The most effective hybridized method (UMAE) is combining microwave-assisted extraction and ultrasound-assisted extraction which provide efficient and quick extraction (Sillero et al., 2020). Many researchers have focused on the combination of UMAE irradiation (Cravotto and Binello 2010). But the considerate ability of the hybrid process is yet not aptly utilized. In industrial as well as academic research we assume UMAE massive potential due to its significant efficiency and lesser extraction period. For the rapid sample preparation along with newer modern process intensification strategies, this is one of the most cost-efficient methods. In the extraction method to yield pectin from the sugar beet pulp (SBP), microwave-ultrasound assisted procedure was employed; comparison is done on the basis of the acid-base technique. From SBP the yield of pectin was observed higher. Moreover, better oxidative stability, emulsifying activity, molecular mass and viscosity were observed. Microwave ultrasound-assisted drying method is considered economical in case of energy, the shorter time required for drying, an internal pressure gradient which assists in the prevention of product shrinkage (Sumnu et al., 2005). Contrary to this drying can be done at low temperature as compared to conventional methods, hence minimizing the probability of degradation and oxidation. According to Szadzinska et al. stated that convective drying of the strawberries showed better results with combined ultrasound and microwave (Szadzińska et al., 2016). In addition to this hybrid process saved approx. 50% energy and 97% time. Many researchers found that the high cost of enzymes has urged them to find innovative technologies to elevate the industrial efficiency of the enzymes (Bashari, 2016). microwave-ultrasound assisted enzyme-based reactions is one of the emerging approaches. In an enzyme catalytic system microwave irradiation act by rearrangement of active sites and applicable substrate groups. The energy provided is utilized to reduce the distance between reactive species in substrate molecule as well as core active point of enzymes. Hence ensuring synchronization and during reactions enhances specificity and efficiency of enzymes (Bashari et al., 2016).

19.6.2 Ultrasound-microwave assisted technology in fermentation

Ultrasound together with microwave technology has a positive effect on fermentation. Even though very little research has been done yet on the combined effect of ultrasound and microwave on fermentation but as per studies, their effects result in a shortened time period and good quality products. For example, to preserve the nutritional values of some fermented products, according to research fermented black table olives were treated with ultrasound and microwave in order to preserve them. ultrasound pretreatment Effect on the parameters of microwave drying for example moisture ratio, drying rate, and effective diffusivity etc, in addition to this quality properties color parameters and total phenolic of black olives slices were noted (Bashari et al., 2016).

It was observed that pretreatment of ultrasound for 10 min, decreased drying time as compared to the control sample. Microwave drying had a diverse effect on the phenolic compounds of olives. In comparison to this ultrasound pretreatment proved to preserve color as well as phenolics of the samples. As a result of combination of drying with ultrasound more effective diffusivity value were achieved, because it accelerated drying rate (Gholamhosseinpour et al., 2020). In another observation it was seen that during fermentation individual and combined effects of microwave, ultrasound and the thermal treatment on milk were investigated and found that level of peptides increased during fermentation in all samples. No significant differences were observed between the samples which were treated by microwave + ultrasound and microwave (individually). But ultrasound + microwave treated samples contained more amounts of peptides Fig. 19.6(A). During fermentation, the peptide content of autoclaved milk was extremely less than other samples. Overall, during fermentation activity of proteolytic system of the lactic acid bacteria elevated the peptide content of all the samples.

It was observed that as fermentation time progressed antioxidant activity of all samples increased at a significant level (0.05). Sample processed with the ultrasound + microwave during all fermentation times; showed best antioxidant activity (0.05) as compared to other samples. In Fig. 19.6(B) the antioxidant activity of the samples increase as the fermentation time increased (Gholamhosseinpour et al., 2020). Hence antioxidants increased in all the samples to a significant level (0.05) and the best results were obtained with ultrasound + microwave. As a result, it was observed that combination of microwave with ultrasonication and autoclave treatments could elevate their effect on antioxidant properties of the samples. The antioxidative activity of ultrasonicated-microwave treatment are comparatively higher. Alpha-amylase inhibition ratio also increase under the USMW effects Fig. 19.6(C)

In the treated sample ultrasound-microwave had a synergistic effect on inhibition of the increasing amylase. At the end of the fermentation process

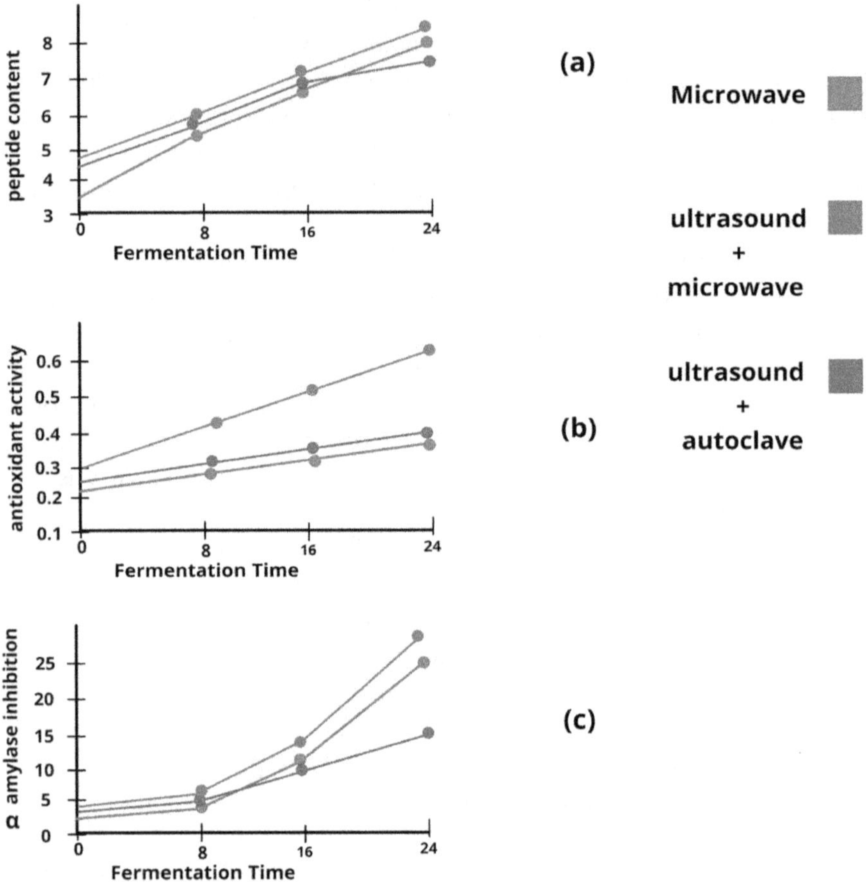

Figure 19.6 (A, B, C): Effects of different techniques (Microwave, ultrasound + microwave, ultrasound + autoclave) on the peptide contents/antioxidants and alpha amylase inhibition in fermentation time (h).

differences among other samples were also significant (0.05). In addition to this, autoclaved samples showed the lowest alpha-amylase inhibitory activity (Gholamhosseinpour et al., 2020).

19.6.3 Ultrasound-microwave assisted technology in emulsification

Microwave-ultrasonic synergistic irradiation can affect the emulsification process in multiple ways. Nowadays scientists are exploring the synergetic effect of novel technologies on different products and production methods. Microwave-ultrasonic assisted effects on emulsification are a new approach and need to explore more. According to research in raw seeds and nuts; study of the fatty acid (FA) profiles emulsification and thermal effects caused by ultrasonic irradiation and microwave led to elevated extractive properties of the n-hexane towards extraction of oils from sample matrix in suspension (Liu et al., 2013). Ultrasonic–microwave synergistic irradiation not only elevated extraction of the walnut kernel oil, moreover increased derivatization of fatty acids.

19.7 Conclusion

This chapter demonstrates a detailed summary of the latest research on the impact of microwave, ultrasound and their combined effects on fermentation and emulsification.

Conventional food fermentation and emulsification is time-consuming, and fermented food's maturation usually needs a huge space for long-term storage. conventional methods are now replaced with advanced technologies. Ultrasound tools are highly used in food industries to enhance the food efficiency and production of food with more refine quality. Apart from that ultrasound and microwave individually are also used in the formation of different emulsions.

These techniques also assist the stabilization of different food emulsions (protein-polysaccharide, polysaccharide and proteins). Microwaves also proved to reduce the fermentation time; however, microwave effects on emulsification vary to the subject. The combination of both microwave and ultrasound technologies emerged with more advantages such as time reduction, upgrade process efficiency and boost overall yield and extraction ratio. their assisted effects in emulsification and fermentation have been explained in the chapter. However, combine effects of ultrasound and microwave need to be explored more, as their synergism in the field of food science technology can prove revolutionary in the future. This processing can take the processing time from hours to minutes and minutes to a few seconds, together with energy and cost reduction. Effects of ultrasound and microwave on emulsifiers and fermented products should be studied tremendously in the future.

References

Abesinghe, A.M.N.L., Islam, N., Vidanarachchi, J.K., Prakash, S., Silva, K.F.S.T., Karim, M.A., 2019. Effects of ultrasound on the fermentation profile of fermented milk products incorporated with lactic acid bacteria. Int. Dairy J. 90, 1–14.

Achouri, A., Boye, J.I., Yaylayan, V.A., Yeboah, F.K., 2005. Functional properties of glycated soy 11S glycinin. J. Food Sci. 70 (4), C269–C274.
Aguedo, M., Vanderghem, C., Goffin, D., Richel, A., Paquot, M., 2013. Fast and high yield recovery of arabinose from destarched wheat bran. Ind. Crops Prod. 43, 318–325.
Alba, K., Sagis, L.M.C., Kontogiorgos, V., 2016. Engineering of acidic O/W emulsions with pectin. Colloids Surf. B 145, 301–308.
Amiri, A., Mousakhani-Ganjeh, A., Torbati, S., Ghaffarinejhad, G., Kenari, R.E., 2018. Impact of high-intensity ultrasound duration and intensity on the structural properties of whipped cream. Int. Dairy J. 78, 152–158.
Aslan, D., Dogan, M., 2018. The influence of ultrasound on the stability of dairy-based, emulsifier-free emulsions: rheological and morphological aspect. Eur. Food Res. Technol. 244 (3), 409–421.
Aspé, E., Fernández, K., 2011. The effect of different extraction techniques on extraction yield, total phenolic, and anti-radical capacity of extracts from Pinus radiata Bark. Ind. Crops Prod. 34 (1), 838–844.
Barukčić, I., Jakopović, K.L., Herceg, Z., Karlović, S., Božanić, R., 2015. Influence of high intensity ultrasound on microbial reduction, physico-chemical characteristics and fermentation of sweet whey. Innov. Food Sci. Emerg. Technol. 27, 94–101.
Bashari, M., Jin, Z., Wang, J., Zhan, X., 2016. A novel technique to improve the biodegradation efficiency of dextranase enzyme using the synergistic effects of ultrasound combined with microwave shock. Innov. Food Sci. Emerg. Technol. 35, 125–132.
Bundhoo, Z.M., Mohee, R., 2018. Ultrasound-assisted biological conversion of biomass and waste materials to biofuels: A review. Ultrason. Sonochem. 40, 298–313.
Camino, N.A., Pilosof, A.M., 2011. Hydroxypropylmethylcellulose at the oil–water interface. Part II. Submicron-emulsions as affected by pH. Food Hydrocoll. 25 (5), 1051–1062.
Campelo, P.H., Junqueira, L.A., Resende, J.V.D., Zacarias, R.D., Fernandes, R.V.D.B., Botrel, D.A., Borges, S.V., 2017. Stability of lime essential oil emulsion prepared using biopolymers and ultrasound treatment. Int. J. Food Prop. 20 (Sup1.), S564–S579.
Carrillo-Lopez, L.M., Garcia-Galicia, I.A., Tirado-Gallegos, J.M., Sanchez-Vega, R., Huerta-Jimenez, M., Ashokkumar, M., Alarcon-Rojo, A.D., 2021. Recent advances in the application of ultrasound in dairy products: Effect on functional, physical, chemical, microbiological and sensory properties. Ultrason. Sonochem. 73, 105467.
Cavalieri, F., Chemat, F., Okitsu, K., Sambandam, A., Yasui, K., Zisu, B., 2018. Handbook of Ultrasonics and Sonochemistry. Springer, Singapore.
Chandrapala, J., Martin, G.J.O., Zisu, B., Kentish, S.E., Ashokkumar, M., 2012. The effect of ultrasound on casein micelle integrity. J. Dairy Sci. 95 (12), 6882–6890.
Chandrapala, J., Ong, L., Zisu, B., Gras, S.L., Ashokkumar, M., Kentish, S.E., 2016. The effect of sonication and high pressurehomogenisation on the properties of pure cream. Innov. Food Sci. Emerg. Technol. 33, 298–307.
Chang, A.C., Chen, F.C., 2002. The application of 20 kHz ultrasonic waves to accelerate the aging of different wines. Food Chem. 79 (4), 501–506.
Chemat, F., Khan, M.K., 2011. Applications of ultrasound in food technology: processing, preservationand extraction. Ultrason. Sonochem. 18 (4), 813–835.
Chemat, F., Rombaut, N., Meullemiestre, A., Turk, M., Perino, S., Fabiano-Tixier, A.S., Abert-Vian, M., 2017. Review of green food processing techniques. Preservation, transformation, and extraction. Innov. Food Sci. Emerg. Technol. 41, 357–377.
Chemat, F., Vian, M.A., Cravotto, G., 2012. Green extraction of natural products: concept and principles. Int. J. Mol. Sci. 13 (7), 8615–8627.
Chen, D., Sharma, S.K., Mudhoo, A.eds., 2011. Handbook on applications of ultrasound: sonochemistry for sustainability. CRC press, Boca Raton.
Chisti, Y., 2003. Sonobioreactors: using ultrasound for enhanced microbial productivity. Trends Biotechnol. 21 (2), 89–93.

Consoli, L., Dias, R.A., Rabelo, R.S., Furtado, G.F., Sussulini, A., Cunha, R.L., Hubinger, M.D., 2018. Sodium caseinate-corn starch hydrolysates conjugates obtained through the Maillard reaction as stabilizing agents in resveratrol-loaded emulsions. Food Hydrocoll. 84, 458–472.

Corma, A., Iborra, S., Velty, A., 2007. Chemical routes for the transformation of biomass into chemicals. Chem. Rev. 107 (6), 2411–2502.

Cravotto, G. and Binello, A., 2010. Innovative techniques and equipments for flavors extraction-Application and effectiveness of ultrasound and microwaves.

Cucheval, A., Chow, R.C.Y., 2008. A study on the emulsification of oil by power ultrasound. Ultrason. Sonochem. 15 (5), 916–920.

Dai, C., Xiong, F., He, R., Zhang, W., Ma, H., 2017. Effects of low-intensity ultrasound on the growth, cell membrane permeability and ethanol tolerance of Saccharomyces cerevisiae. Ultrason. Sonochem. 36, 191–197.

Dai, Y.M., Chen, K.T., Chen, C.C., 2014. Study of the microwave lipid extraction from microalgae for biodiesel production. Chem. Eng. J. 250, 267–273.

Diaz, A.B., de Souza Moretti, M.M., Bezerra-Bussoli, C., Nunes, C.D.C.C., Blandino, A., da Silva, R., Gomes, E., 2015. Evaluation of microwave-assisted pretreatment of lignocellulosic biomass immersed in alkaline glycerol for fermentable sugars production. Bioresour. Technol. 185, 316–323.

Dickinson, E., 2008. Interfacial structure and stability of food emulsions as affected by protein–polysaccharide interactions. Soft Matter 4 (5), 932–942.

Dolatowski, Z.J., Stadnik, J., Stasiak, D., 2007. Applications of ultrasound in food technology. Acta Scientiarum Polonorum Technologia Alimentaria 6 (3), 88–99.

Ethaib, S., Omar, R., Kamal, S.M., Biak, D.A., 2015. Microwave-assisted pretreatment of lignocellulosic biomass: A review. J. Eng. Sci. Technol. 10, 97–109.

Ewe, J.A., Abdullah, W.N.W., Bhat, R., Karim, A.A., Liong, M.T., 2012. Enhanced growth of lactobacilli and bioconversion of isoflavones in biotin-supplemented soymilk upon ultrasound-treatment. Ultrason. Sonochem. 19 (1), 160–173.

Ewe, J.A., Wan-Abdullah, W.N., Alias, A.K., Liong, M.T., 2012. Effects of ultrasound on growth, bioconversion of isoflavones and probiotic properties of parent and subsequent passages of Lactobacillus fermentum BT 8633 in biotin-supplemented soymilk. Ultrason. Sonochem. 19 (4), 890–900.

Fang, C.S., Lai, P.M.C., 1995. Microwave heating and separation of water-in-oil emulsions. J. Microwave Power Electromag. Energy 30 (1), 46–57.

Farhat, A., Fabiano-Tixier, A.S., Visinoni, F., Romdhane, M., Chemat, F., 2010. A surprising method for green extraction of essential oil from dry spices: microwave dry-diffusion and gravity. J. Chromatogr. A 1217 (47), 7345–7350.

Fortuny, M., Oliveira, C.B., Melo, R.L., Nele, M., Coutinho, R.C., Santos, A.F., 2007. Effect of salinity, temperature, water content, and pH on the microwave demulsification of crude oil emulsions. Energy Fuels 21 (3), 1358–1364.

Freitas, S., Hielscher, G., Merkle, H.P., Gander, B., 2006. Continuous contact-and contamination-free ultrasonic emulsification—a useful tool for pharmaceutical development and production. Ultrason. Sonochem. 13 (1), 76–85.

Fu, Y.C., 2004. Fundamentals and industrial applications of microwave and radio frequency in food processing. Food Process.: Prin. Applic. 6, 79–100.

Gharibzahedi, S.M.T., 2017. Ultrasound-mediated nettle oil nanoemulsions stabilized by purified jujube polysaccharide: Process optimization, microbial evaluation and physicochemical storage stability. J. Mol. Liq. 234, 240–248.

Gholamhosseinpour, A., Hashemi, S.M.B., Jahromi, L.R., Sourki, A.H., 2020. Conventional heating, ultrasound and microwave treatments of milk: fermentation efficiency and biological activities. Int. Dairy J. 110, 104809.

Gracin, L., Jambrak, A.R., Juretić, H., Dobrović, S., Barukčić, I., Grozdanović, M., Smoljanić, G., 2016. Influence of high power ultrasound on Brettanomyces and lactic acid bacteria in wine in continuous flow treatment. Appl. Acoust. 103, 143–147.

Gude, V.G., Patil, P.D., Deng, S., Nirmalakhandan, N., 2011. Microwave enhanced methods for biodiesel production and other environmental applications. In: Green chemistry for environmental remediation. *Wiley Interscience, New York*, pp. 209–249.

Gul, O., Saricaoglu, F.T., Besir, A., Atalar, I., Yazici, F., 2018. Effect of ultrasound treatment on the properties of nano-emulsion films obtained from hazelnut meal protein and clove essential oil. Ultrason. Sonochem. 41, 466–474.

Hemwimon, S., Pavasant, P., Shotipruk, A., 2007. Microwave-assisted extraction of antioxidative anthraquinones from roots of Morindacitrifolia. Sep. Purif. Technol. 54 (1), 44–50.

Higuera-Barraza, O.A., Del Toro-Sanchez, C.L., Ruiz-Cruz, S., Márquez-Ríos, E., 2016. Effects of high-energy ultrasound on the functional properties of proteins. Ultrason. Sonochem. 31, 558–562.

Hossan, M.R., Byun, D., Dutta, P., 2010. Analysis of microwave heating for cylindrical shaped objects. Int. J. Heat Mass Transfer 53 (23-24), 5129–5138.

Hu, G., Li, J., Zeng, G., 2013. Recent development in the treatment of oily sludge from petroleum industry: a review. J. Hazard. Mater. 261, 470–490.

Huang, G., Chen, S., Tang, Y., Dai, C., Sun, L., Ma, H., He, R., 2019. Stimulation of low intensity ultrasound on fermentation of skim milk medium for yield of yoghurt peptides by Lactobacillus paracasei. Ultrason. Sonochem. 51, 315–324.

Huang, G., Tang, Y., Sun, L., Xing, H., Ma, H., He, R., 2017. Ultrasonic irradiation of low intensity with a mode of sweeping frequency enhances the membrane permeability and cell growth rate of Candida tropicalis. Ultrason. Sonochem. 37, 518–528.

Huang, K., Wang, S.J., Shan, L., Zhu, Q., Qian, J., 2007. Seeking synergistic effect—a key principle in process intensification. Sep. Purif. Technol. 57 (1), 111–120.

Jafari, S.M., He, Y., Bhandari, B., 2007. Production of sub-micron emulsions by ultrasound and microfluidization techniques. J. Food Eng. 82 (4), 478–488.

Jambrak, A.R., Herceg, Z., Šubarić, D., Babić, J., Brnčić, M., Brnčić, S.R., Bosiljkov, T., Čvek, D., Tripalo, B., Gelo, J., 2010. Ultrasound effect on physical properties of corn starch. Carbohydr. Polym. 79 (1), 91–100.

Kaci, M., Meziani, S., Arab-Tehrany, E., Gillet, G., Desjardins-Lavisse, I., Desobry, S., 2014. Emulsification by high frequency ultrasound using piezoelectric transducer: Formation and stability of emulsifier free emulsion. Ultrason. Sonochem. 21 (3), 1010–1017.

Kapcsándi, V., Neményi, M. and Lakatos, E., 2013. Effect of microwave treatment of the grape must fermentation process.

Kappe, C.O., 2006. The use of microwave irradiation in organic synthesis. From laboratory curiosity to standard practice in twenty years. CHIMIA Intern. J. Chem. 60 (6), 308–312.

Karlović, S., Bosiljkov, T., Brnčić, M., Semenski, D., Dujmić, F., Tripalo, B., Ježek, D., 2014. Reducing fat globules particle-size in goat milk: ultrasound and high hydrostatic pressures approach. Chem. Biochem. Eng. Q. 28 (4), 499–507.

Kentish, S., Wooster, T.J., Ashokkumar, M., Balachandran, S., Mawson, R., Simons, L., 2008. The use of ultrasonics for nanoemulsion preparation. Innov. Food Sci. Emerg. Technol. 9 (2), 170–175.

Khatkar, A.B., Kaur, A., Khatkar, S.K., Mehta, N., 2018. Characterization of heat-stable whey protein: Impact of ultrasound on rheological, thermal, structural and morphological properties. Ultrason. Sonochem. 49, 333–342.

Klaila, W.J., Frank T Sullivan Inc., 1978. *Method and apparatus for controlling fluency of high viscosity hydrocarbon fluids*. U.S. Patent 4,067,683.

Körzendörfer, A., Hinrichs, J., 2019. Manufacture of high-protein yogurt without generating acid whey–impact of the final pH and the application of power ultrasound on texture properties. Int. Dairy J. 99, 104541.

Körzendörfer, A., Nöbel, S., Hinrichs, J., 2017. Particle formation induced by sonication during yogurt fermentation–Impact of exopolysaccharide-producing starter cultures on physical properties. Food Res. Int. 97, 170–177.

Körzendörfer, A., Schäfer, J., Hinrichs, J., Nöbel, S., 2019. Power ultrasound as a tool to improve the processability of protein-enriched fermented milk gels for Greek yogurt manufacture. J. Dairy Sci. 102 (9), 7826–7837.

Kulkarni, P., Loira, I., Morata, A., Tesfaye, W., González, M.C., Suárez-Lepe, J.A., 2015. Use of non-Saccharomyces yeast strains coupled with ultrasound treatment as a novel technique to accelerate ageing on lees of red wines and its repercussion in sensorial parameters. LWT-Food Sci. Technol. 64 (2), 1255–1262.

Kuo, C.H., Lee, C.L., 2010. Treatment of oil/water emulsions using seawater-assisted microwave irradiation. Sep. Purif. Technol. 74 (3), 288–293.

Lanchun, S., Bochu, W., Zhiming, L., Chuanren, D., Chuanyun, D., Sakanishi, A., 2003. The research into the influence of low-intensity ultrasonic on the growth of S. cerevisiaes. Colloids Surf. B 30 (1-2), 43–49.

Lee, H., Zhou, B., Liang, W., Feng, H., Martin, S.E., 2009. Inactivation of Escherichia coli cells with sonication, manosonication, thermosonication, and manothermosonication: Microbial responses and kinetics modeling. J. Food Eng. 93 (3), 354–364.

Leong, T., Ashokkumar, M., Kentish, S., 2016. The growth of bubbles in an acoustic field by rectified diffusion. In: Handbook of ultrasonics and sonochemistry. Springer, Singapore, pp. 69–98.

Leong, T., Johansson, L., Mawson, R., McArthur, S.L., Manasseh, R., Juliano, P., 2016. Ultrasonically enhanced fractionation of milk fat in a litre-scale prototype vessel. Ultrason. Sonochem. 28, 118–129.

Leong, T.S., Martin, G.J., Ashokkumar, M., 2017. Ultrasonic encapsulation–a review. Ultrason. Sonochem. 35, 605–614.

Leong, T.S., Zhou, M., Kukan, N., Ashokkumar, M., Martin, G.J., 2017. Preparation of water-in-oil-in-water emulsions by low frequency ultrasound using skim milk and sunflower oil. Food Hydrocoll. 63, 685–695.

Leong, T.S.H., Manickam, S., Martin, G.J., Li, W., Ashokkumar, M., 2018. Ultrasonic production of nano-emulsions for bioactive delivery in drug and food applications. Springer, Berlin/Heidelberg, Germany.

Li, C., Huang, X., Peng, Q., Shan, Y., Xue, F., 2014. Physicochemical properties of peanut protein isolate–glucomannan conjugates prepared by ultrasonic treatment. Ultrason. Sonochem. 21 (5), 1722–1727.

Li, W., Leong, T.S., Ashokkumar, M., Martin, G.J., 2018. A study of the effectiveness and energy efficiency of ultrasonic emulsification. Phys. Chem. Chem. Phys. 20 (1), 86–96.

Liu, R.L., Song, S.H., Wu, M., He, T., Zhang, Z.Q., 2013. Rapid analysis of fatty acid profiles in raw nuts and seeds by microwave–ultrasonic synergistic in situ extraction–derivatisation and gas chromatography–mass spectrometry. Food Chem. 141 (4), 4269–4277.

Luo, H., Schmid, F., Grbin, P.R., Jiranek, V., 2012. Viability of common wine spoilage organisms after exposure to high power ultrasonics. Ultrason. Sonochem. 19 (3), 415–420.

Ma, X., Yan, T., Hou, F., Chen, W., Miao, S., Liu, D., 2019. Formation of soy protein isolate (SPI)-citrus pectin (CP) electrostatic complexes under a high-intensity ultrasonic field: Linking the enhanced emulsifying properties to physicochemical and structural properties. Ultrason. Sonochem. 59, 104748.

Mahdi Jafari, S., He, Y., Bhandari, B., 2006. Nano-emulsion production by sonication and microfluidization—a comparison. Int. J. Food Prop. 9 (3), 475–485.

Maheri-Sis, N., Eghbali-Vaighan, M., Mirzaaghazadeh, A., Ebrahimnezhad, Y., 2012. Effect of microwave irradiation on the fermentation characteristics and nutritive value of tomato pomace for ruminants using in vitro gas production technique. J. Bio. Sci. Biotechnol. 1 (3), 261–266.

Maran, J.P., Sivakumar, V., Thirugnanasambandham, K., Sridhar, R., 2014. Microwave assisted extraction of pectin from waste Citrullus lanatus fruit rinds. Carbohydr. Polym. 101, 786–791.

Martín, J.F.G., Guillemet, L., Feng, C., Sun, D.W., 2013. Cell viability and proteins release during ultrasound-assisted yeast lysis of light lees in model wine. Food Chem. 141 (2), 934–939.

Martinez-Guerra, E., Gude, V.G., 2014. Synergistic effect of simultaneous microwave and ultrasound irradiations on transesterification of waste vegetable oil. Fuel 137, 100–108.

Martinez-Guerra, E., Gude, V.G., Mondala, A., Holmes, W., Hernandez, R., 2014. Microwave and ultrasound enhanced extractive-transesterification of algal lipids. Appl. Energy 129, 354–363.

Mason, T.J., Lorimer, J.P., 2002. Applied sonochemistry: the uses of power ultrasound in chemistry and processing, 10. Wiley-Vch, Weinheim.

McClements, D.J., 2004. Protein-stabilized emulsions. Curr. Opin. Colloid Interface Sci. 9 (5), 305–313.

Meirelles, A.A.D., Costa, A.L.R., Cunha, R.L., 2020. The stabilizing effect of cellulose crystals in O/W emulsions obtained by ultrasound process. Food Res. Int. 128, 108746.

Melikoglu, M., 2012. Solid-state fermentation of wheat pieces by Aspergillus oryzae: effects of microwave pretreatment on enzyme production in a Biorefinery. Int. J. Green Energy 9 (6), 529–539.

Mishra, P., Wahid, Z.A., Singh, L., Zaid, R.M., Tabassum, S., Sakinah, M., Jiang, X., 2021. Synergistic effect of ultrasonic and microwave pretreatment on improved biohydrogen generation from palm oil mill effluent. Biomass Convers. Biorefin. 1–8.

Moncada, M., Aryana, K.J. and Boeneke, C., 2012. Effect of mild sonication conditions on the attributes of Lactobacillus delbrueckii ssp. bulgaricus LB-12.

Mongenot, N., Charrier, S., Chalier, P., 2000. Effect of ultrasound emulsification on cheese aroma encapsulation by carbohydrates. J. Agric. Food Chem. 48 (3), 861–867.

Monsen, T., Lövgren, E., Widerström, M., Wallinder, L., 2009. In vitro effect of ultrasound on bacteria and suggested protocol for sonication and diagnosis of prosthetic infections. J. Clin. Microbiol. 47 (8), 2496–2501.

Nguyen, N.H., Anema, S.G., 2017. Ultrasonication of reconstituted whole milk and its effect on acid gelation. Food Chem. 217, 593–601.

Nguyen, T.M.P., Lee, Y.K., Zhou, W., 2009. Stimulating fermentative activities of bifidobacteria in milk by highintensity ultrasound. Int. Dairy J. 19 (6-7), 410–416.

Nguyen, T.M.P., Lee, Y.K., Zhou, W., 2012. Effect of high intensity ultrasound on carbohydrate metabolism of bifidobacteria in milk fermentation. Food Chem. 130 (4), 866–874.

Niamah, A.K., 2019. Ultrasound treatment (low frequency) effects on probiotic bacteria growth in fermented milk.

Nitayavardhana, S., Shrestha, P., Rasmussen, M.L., Lamsal, B.P., van Leeuwen, J.H., Khanal, S.K., 2010. Ultrasound improved ethanol fermentation from cassava chips in cassava-based ethanol plants. Bioresour. Technol. 101 (8), 2741–2747.

Novoa-Díaz, D., Rodríguez-Nogales, J.M., Fernández-Fernández, E., Vila-Crespo, J., García-Álvarez, J., Amer, M.A., Chávez, J.A., Turó, A., García-Hernández, M.J., Salazar, J., 2014. Ultrasonic monitoring of malolactic fermentation in red wines. Ultrasonics 54 (6), 1575–1580.

O'donnell, C.P., Tiwari, B.K., Bourke, P., Cullen, P.J., 2010. Effect of ultrasonic processing on food enzymes of industrial importance. Trends Food Sci. Technol. 21 (7), 358–367.

Ofori-Boateng, C., Lee, K.T., 2014. Ultrasonic-assisted simultaneous saccharification and fermentation of pretreated oil palm fronds for sustainable bioethanol production. Fuel 119, 285–291.

Ojha, K.S., Tiwari, B.K., 2016. Novel food fermentation technologies. In: Novel Food Fermentation Technologies. Springer, Cham, pp. 1–5.

O'Sullivan, J., Arellano, M., Pichot, R., Norton, I., 2014. The effect of ultrasound treatment on the structural, physical and emulsifying properties of dairy proteins. Food Hydrocoll. 42, 386–396.

Pangu, G.D., Feke, D.L., 2009. Kinetics of ultrasonically induced coalescence within oil/water emulsions: modeling and experimental studies. Chem. Eng. Sci. 64 (7), 1445–1454.

Parker, R., 2003. Introduction to food science. Taylor & Francis, US.

Patel, D.N., Sutar, P.P., Sutar, N., 2013. Development of instant fermented cereal-legume mix using pulsed microwave vacuum drying. Drying Technol. 31 (3), 314–328.

Patino, J.M.R., Pilosof, A.M., 2011. Protein–polysaccharide interactions at fluid interfaces. Food Hydrocoll. 25 (8), 1925–1937.

Peng, Y., 2010. Impact of altering the gelation conditions on the physical properties of yogurt.

Pereira, R.N., Vicente, A.A., 2010. Environmental impact of novel thermal and nonthermal technologies in food processing. Food Res. Int. 43 (7), 1936–1943.

Pérez, L., Conde, E., Domínguez, H., 2014. Microwave hydrodiffusion and gravity processing of Sargassum muticum. Process Biochem. 49 (6), 981–988.

Potoroko, I., Kalinina, I., Botvinnikova, V., Krasulya, O., Fatkullin, R., Bagale, U., Sonawane, S.H., 2018. Ultrasound effects based on simulation of milk processing properties. Ultrason. Sonochem. 48, 463–472.

Puvanenthiran, A., Williams, R.P.W., Augustin, M.A., 2002. Structure and visco-elastic properties of set yoghurt with altered casein to whey protein ratios. Int. Dairy J. 12 (4), 383–391.

Radel, S., McLoughlin, A.J., Gherardini, L., Doblhoff-Dier, O., Benes, E., 2000. Viability of yeast cells in well controlled propagating and standing ultrasonic plane waves. Ultrasonics 38 (1-8), 633–637.

Ramaswamy, H., Tang, J., 2008. Microwave and radio frequency heating. Food Sci. Technol. Int. 14 (5), 423–427.

Resa, P., Bolumar, T., Elvira, L., Pérez, G., de Espinosa, F.M., 2007. Monitoring of lactic acid fermentation in culture broth using ultrasonic velocity. J. Food Eng. 78 (3), 1083–1091.

Riener, J., Noci, F., Cronin, D.A., Morgan, D.J., Lyng, J.G., 2009. The effect of thermosonication of milk on selected physicochemical and microstructural properties of yoghurt gels during fermentation. Food Chem. 114 (3), 905–911.

Rombaut, N., Tixier, A.S., Bily, A., Chemat, F., 2014. Green extraction processes of natural products as tools for biorefinery. Biofuels Bioprod. Biorefin. 8 (4), 530–544.

Sakakibara, M., Wang, D., Ikeda, K., Suzuki, K., 1994. Effect of ultrasonic irradiation on production of fermented milk with Lactobacillus delbrueckii. Ultrason. Sonochem. 1 (2), S107–S110.

Schöck, T., Becker, T., 2010. Sensor array for the combined analysis of water–sugar–ethanol mixtures in yeast fermentations by ultrasound. Food Control 21 (4), 362–369.

Shamsara, O., Jafari, S.M., Muhidinov, Z.K., 2017. Fabrication, characterization and stability of oil in water nano-emulsions produced by apricot gum-pectin complexes. Int. J. Biol. Macromol. 103, 1285–1293.

Shanmugam, A., Ashokkumar, M., 2014. Functional properties of ultrasonically generated flaxseed oil-dairy emulsions. Ultrason. Sonochem. 21 (5), 1649–1657.

Shen, X., Fang, T., Gao, F., Guo, M., 2017. Effects of ultrasound treatment on physicochemical and emulsifying properties of whey proteins pre-and post-thermal aggregation. Food Hydrocoll. 63, 668–676.

Shershenkov, B., Suchkova, E., 2015. Upgrading the technology of functional dairy products by means of fermentation process ultrasonic intensification. Agronomy Res. 13 (4), 1074–1085.

Sillero, L., Prado, R., Labidi, J., 2020. Simultaneous microwave-ultrasound assisted extraction of bioactive compounds from bark. Chem. Eng. Processing-Process Intensif. 156, 108100.

Silva, E.K., Gomes, M.T.M., Hubinger, M.D., Cunha, R.L., Meireles, M.A.A., 2015. Ultrasound-assisted formation of annatto seed oil emulsions stabilized by biopolymers. Food Hydrocoll. 47, 1–13.

Silva, K.C.G., Sato, A.C.K., 2019. Sonication technique to produce emulsions: The impact of ultrasonic power and gelatin concentration. Ultrason. Sonochem. 52, 286–293.

Singla, M., Sit, N., 2021. Application of ultrasound in combination with other technologies in food processing: A review. Ultrason. Sonochem. 73, 105506.

Singleton, V.L., Draper, D.E., 1963. Ultrasonic treatment with gas purging as a quick aging treatment for wine. Am. J. Enol. Viticulture 14 (1), 23–35.

Stankiewicz, A., 2006. Energy matters: alternative sources and forms of energy for intensification of chemical and biochemical processes. Chem. Eng. Res. Des. 84 (7), 511–521.

Staszak, K., Wieszczycka, K., 2022. Membrane applications in the food industry. Phys. Sci. Rev.

Subasi, B.G., Yildirim-Elikoğlu, S., Altay, İ., Jafarpour, A., Casanova, F., Mohammadifar, M.A., Capanoglu, E., 2022. Influence of non-thermal microwave radiation on emulsifying properties of sunflower protein. Food Chem. 372, 131275.

Sui, X., Bi, S., Qi, B., Wang, Z., Zhang, M., Li, Y., Jiang, L., 2017. Impact of ultrasonic treatment on an emulsion system stabilized with soybean protein isolate and lecithin: Its emulsifying property and emulsion stability. Food Hydrocoll. 63, 727–734.

Sulaiman, A.Z., Ajit, A., Yunus, R.M., Chisti, Y., 2011. Ultrasound-assisted fermentation enhances bioethanol productivity. Biochem. Eng. J. 54 (3), 141–150.

Sumnu, G., Sahin, S., Sevimli, M., 2005. Microwave, infrared and infrared-microwave combination baking of cakes. J. Food Eng. 71 (2), 150–155.

Sun, L.H., Lv, S.W., Chen, C.H., Wang, C., 2019. Preparation and characterization of rice bran protein-stabilized emulsion by using ultrasound homogenization. Cereal Chem. 96 (3), 478–486.

Suwannasom, P., Ruangviriyachai, C., 2011. Simultaneous quantification of Cd, Cu, Pb, and Zn in Thai fermented food by DPASV with a microwave digestion. Int. Food Res. J. 18 (2), 803–808.

Szabó, O.E., Csiszár, E., 2013. The effect of low-frequency ultrasound on the activity and efficiency of a commercial cellulase enzyme. Carbohydr. Polym. 98 (2), 1483–1489.

Szadzińska, J., Kowalski, S.J., Stasiak, M., 2016. Microwave and ultrasound enhancement of convective drying of strawberries: Experimental and modeling efficiency. Int. J. Heat Mass Transfer 103, 1065–1074.

Taha, A., Ahmed, E., Hu, T., Xu, X., Pan, S., Hu, H., 2019. Effects of different ionic strengths on the physicochemical properties of plant and animal proteins-stabilized emulsions fabricated using ultrasound emulsification. Ultrason. Sonochem. 58, 104627.

Tan, W., Yang, X.G., Tan, X.F., 2007. Study on demulsification of crude oil emulsions by microwave chemical method. Sep. Sci. Technol. 42 (6), 1367–1377.

Tang, J., Feng, H., Lau, M., 2002. Microwave heating in food processing. Adv. Bioprocess Eng. 1, 1–43.

Tao, Y., García, J.F., Sun, D.W., 2014. Advances in wine aging technologies for enhancing wine quality and accelerating wine aging process. Crit. Rev. Food Sci. Nutr. 54 (6), 817–835.

Tizazu, B.Z., Roy, K., Moholkar, V.S., 2018. Mechanistic investigations in ultrasound-assisted xylitol fermentation. Ultrason. Sonochem. 48, 321–328.

Tu, Z.C., Hu, Y.M., Wang, H., Huang, X.Q., Xia, S.Q., Niu, P.P., 2015. Microwave heating enhances antioxidant and emulsifying activities of ovalbumin glycated with glucose in solid-state. J. Food Sci. Technol. 52 (3), 1453–1461.

Udagama, R., 2009. Synthesis of polymer-polymer hybrids by miniemulsionpolymerisation and characterisation of hybrid latex. Université Claude Bernard-Lyon I, Lyon Doctoral dissertation.

Van Gerven, T., Stankiewicz, A., 2009. Structure, energy, synergy, time The fundamentals of process intensification. Ind. Eng. Chem. Res. 48 (5), 2465–2474.

Vélez-Erazo, E.M., Consoli, L., Hubinger, M.D., 2018. Mono and double-layer emulsions of chia oil produced with ultrasound mediation. Food Bioprod. Process. 112, 108–118.

Vercet, A., Oria, R., Marquina, P., Crelier, S., Lopez-Buesa, P., 2002. Rheological properties of yoghurt made with milk submitted to manothermosonication. J. Agric. Food Chem. 50 (21), 6165–6171.

Wang, D., Sakakibara, M., 1997. Lactose hydrolysis and β-galactosidase activity in sonicated fermentation with Lactobacillus strains. Ultrason. Sonochem. 4 (3), 255–261.

Wang, D., Sakakibara, M., Kondoh, N., Suzuki, K., 1996. Ultrasound-enhanced lactose hydrolysis in milk fermentation with Lactobacillus bulgaricus. J. Chem. Technol. Biotechnol.: Int. Res. Process Environ. Clean Technol. 65 (1), 86–92.

Wang, T.H., Lu, S., 2013. Production of xylooligosaccharide from wheat bran by microwave assisted enzymatic hydrolysis. Food Chem. 138 (2-3), 1531–1535.

Wolf, N.O., Conoco Inc, 1986. *Use of microwave radiation in separating emulsions and dispersions of hydrocarbons and water.* U.S. Patent 4,582,629.

Wu, H., Hulbert, G.J., Mount, J.R., 2000. Effects of ultrasound on milk homogenization and fermentation with yogurt starter. Innov. Food Sci. Emerg. Technol. 1 (3), 211–218.

Xia, L., Lu, S., Cao, G., 2003. Demulsification of emulsions exploited by enhanced oil recovery system. Sep. Sci. Technol. 38 (16), 4079–4094.

Yanjun, S., Jianhang, C., Shuwen, Z., Hongjuan, L., Jing, L., Lu, L., Uluko, H., Yanling, S., Wenming, C., Wupeng, G., Jiaping, L., 2014. Effect of power ultrasound pre-treatment on the physical and functional properties of reconstituted milk protein concentrate. J. Food Eng. 124, 11–18.

Yeo, S.K., Liong, M.T., 2013. Effect of ultrasound on bioconversion of isoflavones and probiotic properties of parent organisms and subsequent passages of Lactobacillus. LWT-Food Sci. Technol. 51 (1), 289–295.

Yu, Z., Su, Y., Zhang, Y., Zhu, P., Mei, Z., Zhou, X., Yu, H., 2021. Potential use of ultrasound to promote fermentation, maturation, and properties of fermented foods: A review. Food Chem. 357, 129805.

Zhang, L., Ye, X., Ding, T., Sun, X., Xu, Y., Liu, D., 2013. Ultrasound effects on the degradation kinetics, structure and rheological properties of apple pectin. Ultrason. Sonochem. 20 (1), 222–231.

Zhang, R., Pang, X., Lu, J., Liu, L., Zhang, S., Lv, J., 2018. Effect of high intensity ultrasound pretreatment on functional and structural properties of micellar casein concentrates. Ultrason. Sonochem. 47, 10–16.

Zhang, Z., Xiong, F., Wang, Y., Dai, C., Xing, Z., Dabbour, M., Mintah, B., He, R., Ma, H., 2019. Fermentation of Saccharomyces cerevisiae in a one liter flask coupled with an external circulation ultrasonic irradiation slot: Influence of ultrasonic mode and frequency on the bacterial growth and metabolism yield. Ultrason. Sonochem. 54, 39–47.

Zhou, M., Liu, J., Zhou, Y., Huang, X., Liu, F., Pan, S., Hu, H., 2016. Effect of high intensity ultrasound on physicochemical and functional properties of soybean glycinin at different ionic strengths. Innov. Food Sci. Emerg. Technol. 34, 205–213.

Zhu, J., Kuznetsov, A.V., Sandeep, K.P., 2007. Mathematical modeling of continuous flow microwave heating of liquids (effects of dielectric properties and design parameters). Int. J. Therm. Sci. 46 (4), 328–341.

Zhu, Z., Zhao, C., Yi, J., Cui, L., Liu, N., Cao, Y., Decker, E.A., 2018. Ultrasound improving the physical stability of oil-in-water emulsions stabilized by almond proteins. J. Sci. Food Agric. 98 (11), 4323–4330.

Index

Page numbers followed by "*f*" and "*t*" indicate, figures and tables respectively.

G

H

I

K

L

M